From Dietary Cholesterol to Blood Cholesterol

From Dietary Cholesterol to Blood Cholesterol

Editor

Frans Stellaard

MDPI

Basel • Beijing • Wuhan • Barcelona • Belgrade • Novi Sad • Cluj • Manchester

Editor
Frans Stellaard
University Hospital Bonn
Bonn, Germany

Editorial Office
MDPI
St. Alban-Anlage 66
4052 Basel, Switzerland

This is a reprint of articles from the Special Issue published online in the open access journal *Nutrients* (ISSN 2072-6643) (available at: https://www.mdpi.com/journal/nutrients/special_issues/From_Dietary_Cholesterol_to_Blood_Cholesterol).

For citation purposes, cite each article independently as indicated on the article page online and as indicated below:

Lastname, A.A.; Lastname, B.B. Article Title. *Journal Name* **Year**, *Volume Number*, Page Range.

ISBN 978-3-0365-8512-3 (Hbk)
ISBN 978-3-0365-8513-0 (PDF)
doi.org/10.3390/books978-3-0365-8513-0

Contents

About the Editor

Frans Stellaard

Dr. Frans Stellaard is an emeritus scientist who has been involved in the field of nutritional physiology and metabolism. He started his career in 1976 at the Argonne National Laboratories, Lemont, Ill, U.S.A (Prof. Peter Klein), where he got introduced in bile acid research, mass spectrometry and stable isotope technology. In 1979, dr. Stellaard continued bile acid research at the Ludwig Maximilians-University, Klinikum Grosshadern, Munich, Germany (Prof. Gustav Paumgartner). In 1986, he switched his focus to the development of mass spectrometry techniques for the prenatal and postnatal diagnosis and the study of newly discovered inherited metabolic disorders at the Free University Medical Center, Amsterdam, the Netherlands (Prof. Carel Jakobs). Dr. Stellaard finished his career from 1991 on at the University Medical Center Groningen, Groningen, the Netherlands. In this period, he participated in extensive research on various aspects of nutrient metabolism focusing on topics such as digestion, absorption and metabolism of carbohydrates, proteins, triglycerides, cholesterol, and bile acids. Furthermore, he introduced the ^{13}C-urea breath test for gastric Helicobacter Pylori detection in the Netherlands and developed a novel procedure to monitor the release of drugs from pharmaceutical devices intended to be delivered into the colon. His contributions were mainly based on the development of stable isotope investigations involving gas-chromatography/mass spectrometry and gas-chromatography/isotope ratio mass spectrometry analysis. After his retirement in 2011, he remained actively involved in research related to cholesterol, phytosterol and bile acid metabolism in collaboration with the Institute of Clinical Chemistry and Clinical Pharmacology, University Hospital Bonn, Bonn, Germany (Prof. Dieter Lütjohann) and the Department of Nutrition and Movement Sciences, Maastricht University Medical Center, Maastricht, the Netherlands (Prof. Jogchum Plat).

nutrients

MDPI

Editorial

From Dietary Cholesterol to Blood Cholesterol

Frans Stellaard [1,2]

[1] Department of Nutrition and Movement Sciences, NUTRIM (School of Nutrition and Translational Research in Metabolism), Maastricht University Medical Center, P.O. Box 5800 Maastricht, The Netherlands; f.stellaard@maastrichtuniversity.nl; Tel.: +31-652579392

[2] Institute of Clinical Chemistry and Clinical Pharmacology, University Hospital Bonn, Venusberg-Campus 1, 53127 Bonn, Germany

The *Nutrients'* Special Issue "From dietary cholesterol to blood cholesterol" aims to supply existing knowledge and novel new research data about human cholesterol (C) fluxes. The Special Issue contains eight review papers with updated knowledge concerning the complex endogenous and exogenous aspects of C metabolism as well as seven papers describing novel new information, all adding to a better understanding of C homeostasis.

To study the association between dietary C and blood C concentrations, one needs to understand the whole-body C metabolism. The general idea, that an enhanced intake of dietary C leads to enhanced C concentrations in blood, is far too simplified. Analyzing epidemiological studies and meta-analyses, Fernandez and Murillo [1] showed the lack in an association between dietary C and blood C. The explanation for this may be found in the regulatory mechanisms within C metabolism. The physiological lipid fluxes and C homeostasis were described by Stellaard et al. [2]. The intestinal flux of C does not solely consist of dietary C but is dominated by biliary C derived from hepatic secretion, temporary storage in the gallbladder and secretion from the gallbladder following neurohormonal stimuli. Furthermore, under physiological conditions, the intestinal fractional absorption rate (FAR) of C is highly variable, ranging from 20 to 80% [3,4]. Thus, only a high dietary C intake combined with a high biliary C secretion rate and a high C FAR may lead to a clear, high C input into the endogenous C. The endogenous C pool is not only determined by C absorption but also by endogenous C synthesis. Under physiological, dietary C intake conditions, C FAR and C synthesis are negatively associated, indicating a reduced C synthesis when C FAR is high and an induced C synthesis when C FAR is low [5]. Hepatic C metabolism determines C homeostasis, coordinating several C fluxes entering the liver as high density lipoprotein (HDL-C), low density lipoprotein (LDL-C), chylomicron remnant (CMR-C), hepatic C synthesis, and leaving the liver as very low-density lipoprotein (VLDL-C), biliary C secretion, and synthesis of bile acids (BAs). Thus, C synthesis and C FAR are the major determinants affecting the C input into the endogenous C pool when dietary C intake is not enhanced [2].

It is generally expected that enhanced C synthesis and C FAR are determinants of the serum C concentration and, predominantly, the LDL-C concentration. However, Stellaard et al. [6] demonstrated a lack in correlation between serum LDL-C and C synthesis or C FAR in healthy subjects, when measured directly or when approached by respective marker technologies. It is concluded that the effects of C synthesis and C FAR are diluted by the effects of the other mentioned hepatic C fluxes. Elevated serum LDL-C is considered the most predictive atherogenic criterium to express enhanced risk for CVD and LDL-C lowering therapy is generally started [7]. The LDL-C lowering target is determined by the additional risk factors such as smoking, obesity, and diabetes. Recently, it has become evident that cholesterol in LDL is not the only atherogenic factor. The C present in more triglyceride-rich lipoproteins (TRL) such as VLDL remnants, IDL remnants, and chylomicron remnants (CMR) exhibit high atherogenic potential [8]. Lütjohann et al. [9] described the potentials and limitations of determining the TRL-C with rapid, commercially available

Citation: Stellaard, F. From Dietary Cholesterol to Blood Cholesterol. *Nutrients* **2023**, *15*, 3086. https://doi.org/10.3390/nu15143086

Received: 3 July 2023
Accepted: 8 July 2023
Published: 10 July 2023

analytical techniques applicable in routine clinical diagnostic laboratories. The treatment of enhanced serum TRL-C requires the development of alternative treatment methods.

Mashnafi et al. [10] studied the effect of a low calorie diet on C synthesis and C FAR in obese male subjects using the marker technology. They concluded that diet induced weight reduction and led to reduced C synthesis and enhanced C absorption in conjunction with reduced total C and LDL-C in serum. Unfortunately, the authors did not measure the dietary C intake. A low calorie diet may include a low C intake and possibly an induced C FAR. Lytle et al. [11] studied the effects of low-fat ground beef and high-fat ground beef on the voluntary adaptation of the food composition, on the serum lipoprotein levels, and on lipoprotein composition. Interestingly, the overall fat intake in the low-fat intervention period was lower than at entry. The overall fat intake during the high-fat intervention period was only slightly higher than at entry. These effects are explained by the pan broiling of the beef patties, which led to significant fat loss. Both intervention diets contained less cholesterol compared with the intake values at entry. In both intervention periods, the serum LDL-C concentrations were lower than at entry. Unfortunately, no data were available for the daily flux of biliary C, nor for the C FAR. This study shows that alterations in dietary food composition are not automatically translated into expected, altered serum lipid profiles. Food preparation, food adaptation, and endogenous responses to the dietary changes affect the final results. Houttu et al. [12] observed important effects in three patients consuming extreme high-fat, low-carbohydrate diets. In two patients, the protein and fat intake expressed 37 and 61 E-% and carbohydrate intake less than 3 E-%. With 16, 9, and 12 mmol/L, the LDL-C concentrations in the three patients were at the level of homozygous familial hypercholesterolemic patients. Interestingly, dietary modifications towards reduction in animal fat intake clearly reduced their serum LDL-C values. Apparently, extreme high-animal fat diets do enhance serum LDL-C and the introduction of extreme reductions in animal fat intake do reduce LDL-C.

Due to required invasive methodologies, metabolic studies of lipid metabolism cannot be performed in humans. Experimental animals are used instead with a preference for mice. However, lipoprotein metabolism and lipid metabolism are different between humans and natural mice. Therefore, mice with humanized lipid metabolism have been developed. The latest model is described as APOE*3-Leiden.CETP mice, as described and applied by Paalvast et al. [13]. The authors analyzed the effect of a high-fat, high-C diet in the responding mice establishing an increased serum TG response and in the non-responding mice. Their main conclusions were that, in contrast to the responding animals, the non-responding animals developed a fat malabsorption possibly caused by a decreased production of BAs. The authors suggested that the variation in BA homeostasis may in part drive the phenotypic variation in the APOE*3-Leiden.CETP mice.

An important line of recent research on C metabolism is focused on the role of phytosterols. Phytosterols are present in plants and enter the body only via the diet. The majority are present as sitosterol and campesterol. These compounds have a similar chemical structure as C, differing in an additional methyl-group in the side chain. Phytosterols are transported similarly compared to C, with both transported in the intestine as in lipoproteins. However, the FARs of phytosterols are much lower than the C FAR, i.e., less than 10%. One research topic deals with the fact that a high phytosterol concentration in serum is expressed as a reflection of high C absorption. The ratio between plant sterol concentrations to total C in fasting serum is used as a surrogate marker for C absorption. Otherwise, phytosterol absorption has been shown to compete with C absorption at the level of uptake into micelles transporting sterols to the site of absorption. This has led to the development of food supplements containing phytosterols and/or stanols in order to reduce C absorption and lower serum LDL-C and therewith the enhanced risk for cardiovascular disease (CVD). However, the potential risk of elevated serum phytosterol concentrations for CVD development has been postulated in the literature. Windler et al. [14] discussed the pros and cons of phytosterol treatment and concluded that the regular dose of the supplement must be considered too low to introduce a significant risk for cardiovascular disease in the general population.

Non-C sterols represent a class of compounds combining C precursors in the C synthesis pathway (lanosterol, lathosterol, desmosterol, stigmasterol). The ratio between their concentrations and the total C concentration in fasting serum are considered markers for C synthesis. Van Brakel et al. [15] described an unexpected potential function of non-C sterols. In up to two year-old children, the authors were able to show that the odds of eczema were lower with higher non-C sterols in breast milk. The odds of allergic sensitization at age two were lower with higher concentrations of campesterol in breast milk.

BAs are hepatic metabolites of C and have many distinct functionalities. Synthesis of BAs is a major factor in the removal of endogenous C. Otherwise, they serve to produce mixed micelles and transport lipophilic compounds such as C and triglycerides through the intestine and C in bile. In recent years, BAs have been identified as signaling compounds in the regulation of many processes. At first, absorbing circulating BAs regulate the hepatic BA synthesis rate and control the rate of gallbladder relaxation. Secondly, they are involved in the regulation of many independent metabolic processes. These processes and mechanisms of action, involving mainly the farnesoid X receptor (FXR) and enterokine fibroblast growth factor 19 (FGF19), are described in detail by Di Ciaula et al. [16].

Disturbances in the gut microbiota have been repeatedly associated with CVD, including atherosclerosis and hypertension. Short-chain fatty acids, trimethylamine-N-oxide, and BAs have been identified as causal factors. The functionality of the BA pool is dependent on its composition. Different BAs have one to three hydroxyl groups in different positions. This determines the hydrophobicity of the individual BA pool. Per day, 30 to 50% of the primary BA pools (cholic acid and chenodeoxycholic acid) are malabsorbed and converted to secondary BAs. The bacterial products formed are determined by the gut microbial load and composition. The effects of microbial BA modulations on the etiology of cardiovascular diseases were discussed by Yntema et al. [17].

As mentioned, seven C fluxes are affecting the hepatic C homeostasis. The regulation of these fluxes is a critical factor in the establishment of the hepatic C pool and the flux of LDL-C being removed from the blood. The size of the hepatic C pool affects the regulation of hepatic C synthesis and the synthesis of the LDL receptor affecting the hepatic extraction of serum LDL. However, little is known about the regulation of the other C fluxes. Interestingly, in this Special Issue, two contributions point to the potential role of microRNAs (miRNAs) in hepatic regulation. miRNAs are small, single-stranded, non-coding RNA molecules containing 21 to 23 nucleotides. Konings et al. [18] performed a systematic review of the literature describing the involvement of miRNAs in the development of non-alcoholic fatty liver disease (NAFLD) associated with C metabolism due to accumulation of hepatic free C. They identified miR122, miR34a, miR21, and miR132 as potential candidates to play a role in the development of NAFLD via effects on cholesterol metabolism. Sidorkiewicz [19] contributed a review paper focusing on miR33 and described evidence for the involvement of miR33 in the regulation of reversed C transport via HDL production. A role for miR33 is also suggested for bile formation. Thus, microRNA research may clarify hepatic C homeostasis in the near future and offer new tools for treatment.

A field not covered so far is C metabolism in the brain. C cannot pass the blood–brain barrier. In the brain, C is synthesized mainly in the astrocytes and metabolized to 24S-C in the neurons. A disturbed brain C metabolism is associated with Alzheimer's disease (AD) and stimulation of cholesterol turnover in the brain was found to ameliorate the development of cognitive decline in AD mice. Martens et al. [20] described the upregulation of C efflux genes and downregulation of lipogenesis genes by incubation with lipophilic extracts of six European brown seaweed species. Their results confirm the described prevention of disease progression in AD mice by the Asian brown seaweed species *Sargassum fusiforme* [21]. Thus, supplementation with extracts of European brown seaweed species may become a strategy in the prevention and/or treatment of neurodegenerative diseases and possibly cardiometabolic and inflammatory diseases.

Funding: This research received no external funding.

Conflicts of Interest: The author declares no conflict of interest.

References

1. Fernandez, M.L.; Murillo, A.G. Is There a Correlation between Dietary and Blood Cholesterol? Evidence from Epidemiological Data and Clinical Interventions. *Nutrients* **2022**, *14*, 2168. [CrossRef]
2. Stellaard, F. From Dietary Cholesterol to Blood Cholesterol, Physiological Lipid Fluxes, and Cholesterol Homeostasis. *Nutrients* **2022**, *14*, 1643. [CrossRef]
3. Lütjohann, D.; Meese, C.O.; Crouse, J.R., 3rd; von Bergmann, K. Evaluation of deuterated cholesterol and deuterated sitostanol for measurement of cholesterol absorption in humans. *J. Lipid Res.* **1993**, *34*, 1039–1046. [CrossRef]
4. Bosner, M.S.; Ostlund, R.E., Jr.; Osofisan, O.; Grosklos, J.; Fritschle, C.; Lange, L.G. Assessment of percent cholesterol absorption in humans with stable isotopes. *J. Lipid Res.* **1993**, *34*, 1047–1053. [CrossRef]
5. Miettinen, T.A.; Kesäniemi, Y.A. Cholesterol absorption: Regulation of cholesterol synthesis and elimination and within-population variations of serum cholesterol levels. *Am. J. Clin. Nutr.* **1989**, *49*, 629–635. [CrossRef]
6. Stellaard, F.; Baumgartner, S.; Mensink, R.; Winkens, B.; Plat, J.; Lütjohann, D. Serum Low Density Lipoprotein Cholesterol Concentration Is Not Dependent on Cholesterol Synthesis and Absorption in Healthy Humans. *Nutrients* **2022**, *14*, 5370. [CrossRef]
7. Tertov, V.V.; Orekhov, A.N.; Martsenyuk, O.N.; Perova, N.V.; Smirnov, V.N. Low-density lipoproteins isolated from the blood of patients with coronary heart disease induce the accumulation of lipids in human aortic cells. *Exp. Mol. Pathol.* **1989**, *50*, 337–347. [CrossRef]
8. Koren, E.; Corder, C.; Mueller, G.; Centurion, H.; Hallum, G.; Fesmire, J.; McConathy, W.D.; Alaupovic, P. Triglyceride enriched lipoprotein particles correlate with the severity of coronary artery disease. *Atherosclerosis* **1996**, *122*, 105–115. [CrossRef]
9. Lütjohann, D.; Klör, H.-U.; Stellaard, F. Measurement of Serum Low Density Lipoprotein Cholesterol and Triglyceride-Rich Remnant Cholesterol as Independent Predictors of Atherosclerotic Cardiovascular Disease: Possibilities and Limitations. *Nutrients* **2023**, *15*, 2202. [CrossRef]
10. Mashnafi, S.; Plat, J.; Mensink, R.P.; Joris, P.J.; Kusters, Y.H.A.M.; Houben, A.J.H.M.; Stehouwer, C.D.A.; Schalkwijk, C.G.; Baumgartner, S. Effects of Diet-Induced Weight Loss on Plasma Markers for Cholesterol Absorption and Synthesis: Secondary Analysis of a Randomized Trial in Abdominally Obese Men. *Nutrients* **2022**, *14*, 1546. [CrossRef]
11. Lytle, J.R.; Price, T.; Crouse, S.F.; Smith, D.R.; Walzem, R.L.; Smith, S.B. Consuming High-Fat and Low-Fat Ground Beef Depresses High-Density and Low-Density Lipoprotein Cholesterol Concentrations, and Reduces Small, Dense Low-Density Lipoprotein Particle Abundance. *Nutrients* **2023**, *15*, 337. [CrossRef]
12. Houttu, V.; Grefhorst, A.; Cohn, D.M.; Levels, J.H.M.; Roeters van Lennep, J.; Stroes, E.S.G.; Groen, A.K.; Tromp, T.R. Severe Dyslipidemia Mimicking Familial Hypercholesterolemia Induced by High-Fat, Low-Carbohydrate Diets: A Critical Review. *Nutrients* **2023**, *15*, 962. [CrossRef]
13. Paalvast, Y.; Zhou, E.; Rozendaal, Y.J.W.; Wang, Y.; Gerding, A.; van Dijk, T.H.; de Boer, J.F.; Rensen, P.C.N.; van Dijk, K.W.; Kuivenhoven, J.A.; et al. A Systems Analysis of Phenotype Heterogeneity in APOE*3Leiden.CETP Mice Induced by Long-Term High-Fat High-Cholesterol Diet Feeding. *Nutrients* **2022**, *14*, 4936. [CrossRef]
14. Windler, E.; Beil, F.-U.; Berthold, H.K.; Gouni-Berthold, I.; Kassner, U.; Klose, G.; Lorkowski, S.; März, W.; Parhofer, K.G.; Plat, J.; et al. Phytosterols and Cardiovascular Risk Evaluated against the Background of Phytosterolemia Cases—A German Expert Panel Statement. *Nutrients* **2023**, *15*, 828. [CrossRef]
15. van Brakel, L.; Thijs, C.; Mensink, R.P.; Lütjohann, D.; Plat, J. Non-Cholesterol Sterols in Breast Milk and Risk of Allergic Outcomes in the First Two Years of Life. *Nutrients* **2022**, *14*, 766. [CrossRef]
16. Di Ciaula, A.; Bonfrate, L.; Baj, J.; Khalil, M.; Garruti, G.; Stellaard, F.; Wang, H.H.; Wang, D.Q.-H.; Portincasa, P. Recent Advances in the Digestive, Metabolic and Therapeutic Effects of Farnesoid X Receptor and Fibroblast Growth Factor 19: From Cholesterol to Bile Acid Signaling. *Nutrients* **2022**, *14*, 4950. [CrossRef]
17. Yntema, T.; Koonen, D.P.Y.; Kuipers, F. Emerging Roles of Gut Microbial Modulation of Bile Acid Composition in the Etiology of Cardiovascular Diseases. *Nutrients* **2023**, *15*, 1850. [CrossRef]
18. Konings, M.C.J.M.; Baumgartner, S.; Mensink, R.P.; Plat, J. Investigating microRNAs to Explain the Link between Cholesterol Metabolism and NAFLD in Humans: A Systematic Review. *Nutrients* **2022**, *14*, 4946. [CrossRef]
19. Sidorkiewicz, M. Is microRNA-33 an Appropriate Target in the Treatment of atherosclerosis? *Nutrients* **2023**, *15*, 902. [CrossRef]
20. Martens, N.; Zhan, N.; Voortman, G.; Leijten, F.P.J.; van Rheenen, C.; van Leerdam, S.; Geng, X.; Huybrechts, M.; Liu, H.; Jonker, J.W.; et al. Activation of Liver X Receptors and Peroxisome Proliferator-Activated Receptors by Lipid Extracts of Brown Seaweeds: A Potential Application in Alzheimer's Disease? *Nutrients* **2023**, *15*, 3004. [CrossRef]
21. Hu, P.; Li, Z.; Chen, M.; Sun, Z.; Ling, Y.; Jiang, J.; Huang, C. Structural elucidation and protective role of a polysaccharide from Sargassum fusiforme on ameliorating learning and memory deficiencies in mice. *Carbohydr. Polym.* **2016**, *139*, 150–158. [CrossRef]

nutrients MDPI

Review

From Dietary Cholesterol to Blood Cholesterol, Physiological Lipid Fluxes, and Cholesterol Homeostasis

Frans Stellaard [1,2]

[1] Department of Nutrition and Movement Sciences, NUTRIM (School of Nutrition and Translational Research in Metabolism), Maastricht University Medical Center, P.O. Box 5800 Maastricht, The Netherlands; f.stellaard@maastrichtuniversity.nl

[2] Institute of Clinical Chemistry and Clinical Pharmacology, University Hospital Bonn, Venusberg-Campus 1, 53127 Bonn, Germany

Abstract: Dietary cholesterol (C) is a major contributor to the endogenous C pool, and it affects the serum concentration of total C, particularly the low-density lipoprotein cholesterol (LDL-C). A high serum concentration of LDL-C is associated with an increased risk for atherosclerosis and cardiovascular diseases. This concentration is dependent on hepatic C metabolism creating a balance between C input (absorption and synthesis) and C elimination (conversion to bile acids and fecal excretion). The daily C absorption rate is determined by dietary C intake, biliary C secretion, direct trans-intestinal C excretion (TICE), and the fractional C absorption rate. Hepatic C metabolism coordinates C fluxes entering the liver via chylomicron remnants (CMR), LDL, high-density lipoproteins (HDL), hepatic C synthesis, and those leaving the liver via very low-density lipoproteins (VLDL), biliary secretion, and bile acid synthesis. The knowns and the unknowns of this C homeostasis are discussed.

Keywords: cholesterol; synthesis; absorption; bile acids; bile; hepatic; extrahepatic; lipoproteins; plant sterols; intestine

Citation: Stellaard, F. From Dietary Cholesterol to Blood Cholesterol, Physiological Lipid Fluxes, and Cholesterol Homeostasis. *Nutrients* **2022**, *14*, 1643. https://doi.org/10.3390/nu14081643

Academic Editor: Satoshi Nagaoka

Received: 14 March 2022
Accepted: 12 April 2022
Published: 14 April 2022

Publisher's Note: MDPI stays neutral with regard to jurisdictional claims in published maps and institutional affiliations.

1. Introduction

High dietary cholesterol (C) intake is considered unhealthy, since it may contribute to elevated concentrations of serum C and in particular of low-density lipoprotein C (LDL-C). This is associated with the development of atherosclerosis and cardiovascular events, such as heart attacks and strokes. In this review paper, attention will be paid to the potential role of dietary C in the control of serum C. Serum C is the balance between the input of C into the blood and the C efflux from blood via hepatic extraction and elimination. The main C fluxes are shown in Figure 1, and they will be discussed in detail hereafter. In this introduction, only a generally accepted short summary is described. Consumed animal fat contains C in both the free (FC) and esterified form (CE). Fat digestion is predominantly established by the activated release of pancreatic digestive enzymes after meal intake and their secretion into the small intestine. Triglycerides (TG) are cleaved to free fatty acids (FFA) and glycerol, and C-esters to free C and fatty acids. In parallel, bile is secreted due to gallbladder contraction, activated by cholecystokinin (CCK) release from the duodenal mucosa. Bile acids (BA) enable the establishment of mixed micelles in which the hydrophobic FFAs and C as well as xenosterols such as plant sterols can be transported to the active sites of absorption. Here, the compounds are released and taken up into the enterocytes where they are predominantly incorporated into chylomicrons (CM) after esterification. After release of CMs into the lymph system, conversion into chylomicron remnants (CMR) starts through interaction with Lipoprotein Lipase (LPL), which converts TG to FFA. The FFAs are released from CM and the CMRs are much smaller and denser as CMs and taken up into the liver and potentially also into extrahepatic tissues and macrophages. Besides oxidized LDL, CMRs are deposited in coronary artery plaques. The liver is the central organ to control C homeostasis. Interrelated processes occur in the liver. Extracted with the CMRs, TGs and a part of CEs are released and incorporated into very low-density lipoprotein (VLDL) particles, which are secreted into blood. Through LPL, VLDL is converted to LDL particles that are re-extracted by the liver or by extrahepatic tissues such as macrophages. The C mixes with the hepatic C pool from which molecules can be used for BA synthesis and for biliary C secretion. The liver also takes up high-density lipoprotein (HDL) particles or the C component from the HDL particles from blood that are formed in extrahepatic tissues. This source of C is mainly used for biliary C secretion [1,2]. In bile, C is incorporated into mixed micelles formed by BAs and phospholipids. Bile is largely stored in the gallbladder at night and between meals. After meal intake, biliary C mixes with dietary C. Thus, dietary C is only part of the total C flux moving through the intestine before absorption. The daily flux of biliary C through the intestine is 2–4 times larger than the flux of dietary C [3,4]. The input of C into the whole-body C pool originates from C absorption and whole-body C synthesis; C is synthesized by almost all cells, and it is eliminated from the body via BA synthesis and fecal excretion of C and its bacterial metabolites coprostanol and coprostanone. In health, input and elimination are in balance leading to a constant whole body C pool. Imbalance caused by excessive input or reduced elimination may lead to elevated serum C. Ideally, high C absorption is compensated by a reduced in vivo C synthesis and an enhanced BA synthesis. Vice versa, enhanced C synthesis compensates for extremely lowered C absorption.

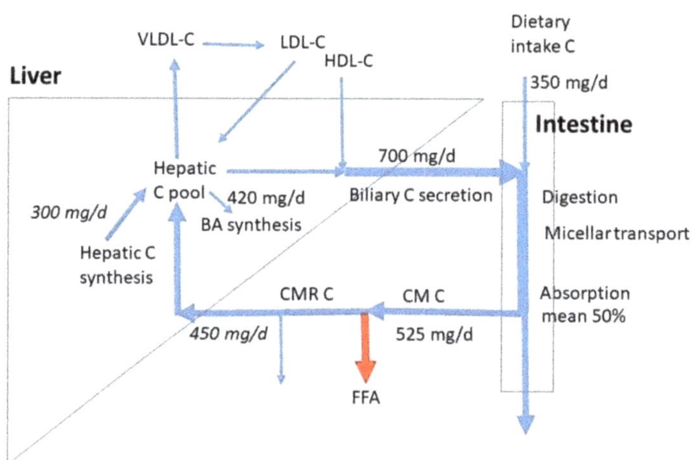

Figure 1. C fluxes in omnivore humans as published by Lütjohann et al. [4]. The numbers in italics are obtained by reasonable estimates based on the proportion of chylomicron remnants being extracted by the liver and the proportion of hepatic C synthesis to whole body synthesis. Chylomicron remnants not extracted by the liver are taken up by extrahepatic tissues including macrophages. CM = chylomicrons, CMR = chylomicron remnants, VLDL = very low density lipoprotein, LDL = low density lipoprotein, HDL = high density lipoprotein, FFA = free fatty acid, BA = bile acid. No numbers are available for the lipoprotein fluxes.

2. Dietary C Intake

In food, C is present in animal fat in the free form and predominantly in the esterified form. Animal fat is present in meat, fish, eggs, and in dairy products. In the western diet, daily C intake varies from approximately 100 to 400 mg/day, but excessive intakes up to 800 mg/day have been described [5]. C intake is much lower than the daily TGs intake of 60 to 150 g/day. In meat and fish, 100 g generally contains 40 to 120 mg of C. Restriction of C intake reduces the contribution to the daily intestinal C flux. Compared to omnivores, vegans have a 90% lower C intake, which results in a 13% lower serum LDL-C concentration [4]. This seemingly small effect on LDL-C is caused by a 35% increase in whole body C synthesis. This highlights the balance between absorption and synthesis.

Therefore, a reduction in dietary C intake alone is not sufficient to lower serum LDL-C to a larger extent as needed in some diseased conditions. In addition, the balance is important in situations in which C intake is high. In that case, reduction of C synthesis keeps LDL-C under control. A convincing example is described by Kern Jr [6]. The daily diet of an 88-year old single man consisted of 25 eggs cooked in the morning and consumed during the day. His serum C was normal due to a reduced fractional absorption rate (FAR) and synthesis rate as well as an increased BA synthesis. Apparently, this compensating control does not function effectively in all human beings. Egg consumption is an ongoing discussion. One egg contains 150 to 180 g of C in the yolk fraction, which is relatively high as compared to the average daily intake of 100 to 400 mg C. The increased risk of cardiovascular events associated with a higher egg intake may be linked to the accompanying food products. With breakfast, eggs may be consumed fried in butter with bacon and sausages or consumed cooked with bread. Chinese populations tend to eat eggs with diner together with vegetables and rice. Interestingly, the risk among the Chinese for all-cause mortality does not increase with increasing egg consumption up to 7 eggs/week [5]. This was in contrast with black and white Americans whose risk increases already from 3 eggs/week on. Moreover, egg intake has been shown to increase HDL-C in serum in parallel with LDL-C, which may partly compensate for the risk effect of a high LDL-C concentration. A recent meta-analysis by Kuang et al. [7] indicates that the unchanged HDL-C/LDL-C ratio may explain that egg consumption does not increase the risk for cardiovascular disease in the general population. Patients at risk for cardiovascular disease should limit egg consumption. In another recent review paper [8], the authors discuss the recommendations on egg consumption for patients at risk for cardiovascular disease. They conclude that the rate of C absorption is important. Indeed, the FAR varies individually from 20 to 80% [9,10]. A healthy subject consuming 300 mg C per day and absorbing 80%, absorbs 240 mg C per day. A healthy subject consuming 100 mg while absorbing 20%, absorbs only 20 mg. The difference in absorptive load of dietary C between the two subjects is 220 mg per day or 12-fold. The FAR affects not only dietary C but also biliary C. Thus, a high FAR may lead to elevated serum total C and LDL-C concentrations.

3. Known and Unknown C Fluxes in Humans

In Figure 1, average sizes of known C fluxes are indicated. The numbers are retrieved from the data for omnivores as published by Lütjohann et al. [4]. With a mean daily C intake of 350 mg/d, the subjects had a relatively high C consumption, when compared to the normal intake of 100 to 400 mg/d. The data show that the daily biliary C flux is twice the dietary C flux (700 mg/d). The total daily intestinal C flux is then 1050 mg/d. Based on the mean 50% FAR, the daily total flux of absorbed C is on average 525 mg/d. The majority will be taken up by the liver, the rest enters extrahepatic tissues including macrophages. The exact proportions are not known. We assumed 450 mg/d to be extracted by the liver. The measured whole-body C and bile acid synthesis rates were 840 and 420 mg/d. The exact contribution of hepatic C synthesis to whole-body synthesis in humans is not known. In the cynomolgus monkey, this value is around 20%, which may represent the value in humans [11]. Generally, in scholarly literature, a 50% contribution is assumed without references. Both options would lead to a hepatic synthesis of 165 and 420 mg/d. An in-between value of 300 mg/d was chosen as a compromise. Despite the clear high uncertainty, this value gives a rough estimate for hepatic synthesis. It must be realized that large inter-individual variations exist in dietary C intake, biliary C secretion, C FAR, and BA synthesis. The whole-body C synthesis compensates for the fecal loss of endogenous C and BA synthesis. The total influx of CMR-C into the liver and hepatic C synthesis together estimates 750 mg/d, much less than the sum of biliary C secretion rate of 700 mg/d and BA synthesis of 420 mg/d. The flux of C entering the liver in HDL particles may largely contribute to biliary C secretion. The hepatic LDL-C influx is smaller than the VLDL-C efflux. Daily flux values for the hepatic VLDL-C secretion, LDL-C uptake, and HDL-C uptake are not available. The VLDL secretion is mainly controlled by the triglyceride flux entering the liver with CMRs. The hepatic LDL uptake is mainly determined by VLDL secretion, fatty-acid release from VLDL, hepatic uptake of VLDL remnants, uptake of LDL into extrahepatic tissues, and hepatic LDL-receptor activity. Furthermore, biliary secretion is here defined as the daily flux of biliary C entering the intestine. This flux contains mostly C derived from bile stored in the gallbladder which is emptied several times a day. Intestinal biliary C is partly absorbed and recycled to the liver. Therefore, the amount of hepatic C newly secreted into bile is much lower than the flux of biliary C entering the intestine. The daily biliary C secretion rate into the intestine is dependent on the degree of gallbladder contraction, which may be 50 to 90% [12] and the number of gallbladder contractions being two to three per meal intake. The numbers depend on subject, meal size, and meal composition (fat and protein content). In case of extreme high fat meals, gastric emptying is delayed and the gallbladder may remain in a contracted state for many hours keeping the bile in continuous cycling. Thus, C homeostasis requires control over a complex system of fluxes.

Another largely unexplored aspect of C metabolism is the bacterial metabolism of C that escapes small intestinal absorption. Based on the fluxes presented in Figure 1, the daily flux of malabsorbed C averages 525 mg/d. The major metabolic metabolite is coprostanol [13]. The conversion rate, expressed as the coprostanol/cholesterol ratio in the feces of healthy subjects is high (high converters), but may be low in a minor group of subjects (low converters). Only a small number of bacterial strains has been identified that is able to produce coprostanol from cholesterol. This process has been shown to modify host C levels in serum [14]. The ratio of coprostanol/C in feces appears inversely related to the serum C concentration [15]. Diet factors may intervene with the colonic bacterial C metabolism as has been documented for milk polar lipids, reducing serum C concentration [16].

4. Fat Digestion and Intestinal Transport

The dominant lipids in the diet are TGs. In the western diet, the fat intake predominantly consisting of TG is between 60 and 150 g/day. The C intake comprises 100–400 mg/day. Considering an average TG intake of 100 g/d and a C intake of 200 mg/d, the TG/C ratio is 500. The digestion of TG and C require gastric emulsification and de-esterification with gastric but predominantly pancreatic lipases and cholesterol esterase and uptake in mixed micelles formed by BAs (Figure 2). The availability of BAs is determined by liver function and gallbladder motility. During gastric emulsification, fat is divided in droplets being slowly reduced to micro-droplets. These allow access to digestive enzymes and bile after their entrance into the duodenum. The appearance of fat in the duodenum leads to CCK release from the duodenal cells. The CCK stimulates secretion of pancreatic juice containing digestive enzymes and the activation of gallbladder contraction. Pancreatic juice and bile migrate to the duodenum. TGs are broken down to FFA, glycerol and monoacylglycerol, CE to C, and fatty acids. BAs initiate the formation of mixed micelles in which fatty acids and C can be packed for transportation through the small intestine. Conjugated plant sterols such as sitosterol and campesterol are also present in food and transported into the micelles in its free form. The daily intake of plant sterols is similar to the intake of C. It is generally assumed that in healthy subjects emulsification, de-esterification and micellar uptake are always highly efficient for TG and C. It is also assumed that the large load of TGs does not impair the processing of C. The micelles containing FFA and C move down the small intestine and FFA and C are delivered to the enterocytes, where they are taken up by specific transporter proteins.

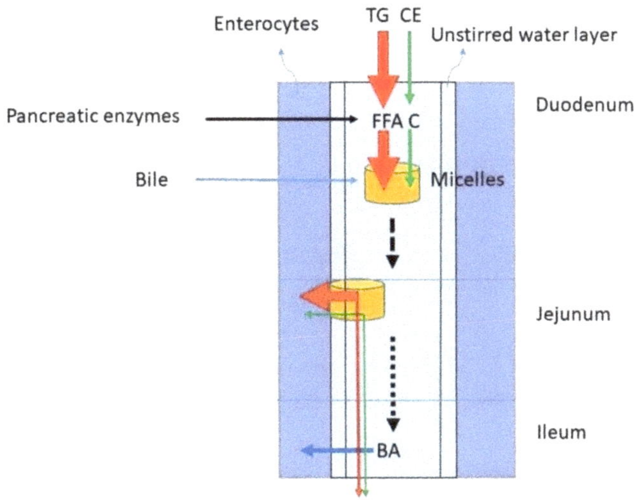

Figure 2. Fat digestion of TG and CE and micellar transport of FFAs and C to enable lipid uptake into the enterocytes. The red arrow represents TG metabolism, the green arrow C metabolism and the blue arrow BA absorption. TG = triglycerides, CE = cholesterol ester, FFA = free fatty acid, C = free cholesterol, BA = bile acid.

5. Absorption

The micelles allow lipids to be transported through the aqueous environment, to cross the unstirred water layer and to attach to the enterocyte in order to release free fatty acids and sterols into the cell (Figure 2); C is predominantly absorbed in the jejunum [17,18]. For sterols, Niemann-Pick C1 Like 1 (NPC1L1) is the selective uptake protein [19]. The enterocytes esterify free fatty acids and sterols back to triglycerides and sterol esters before incorporation into large lipid rich CMs carried by the ApoB-48 lipoprotein (Figure 3). After structuring and lipid loading of the chylomicron particles, they are secreted into the lymph. There are indications that C may be partly transported out of the enterocyte directly into blood via HDL [20,21]; CMs are transported through blood. A protein system consisting of ATP-binding cassette sub-family G member 5 and 8 (ABCG5/G8) has been shown to be effective in re-secreting part of the already taken up C and plant sterols back into the intestinal lumen [22,23]. The effect is larger on plant sterols. This may be explained by the smaller degree of esterification of plant sterols in the enterocyte. The overall FAR of TG in healthy subjects is high, i.e., >85%, mean values of >90% have been described [24,25]. In contrast, the overall FAR of C in healthy subjects is highly variable (20–80%), with a mean value of approximately 50% [9,10]. The large variation in the FAR of C is generally ascribed to the variable activities of NPC1L1 and ABCG5/G8. It has been shown that the whole length of the small intestine contains cells that enable absorption of the large FFA load. The smaller load of C may need a smaller area. The duodenum and jejunum are generally indicated as the regions for C absorption. However, whether FFA and C are dissociated in micelles is not known. On average, healthy subjects have an orocecal transit time of 4–6 h after consuming a solid meal. Part of this transit time is determined by gastric emptying, which is delayed when the food is enriched in fat. Reduced activity of intestinal ABCG5/G8 induced by gene mutations leads to enhanced absorption of C and plant sterols as shown in sitosterolemia patients [26,27]. The FARs for plant sterols are below 20% [28], which explains their low serum concentrations. Their low absorption rates are mainly due to the low intestinal re-esterification rate which makes the plant sterols more susceptible for re-secretion into the lumen by ABCG5/G8.

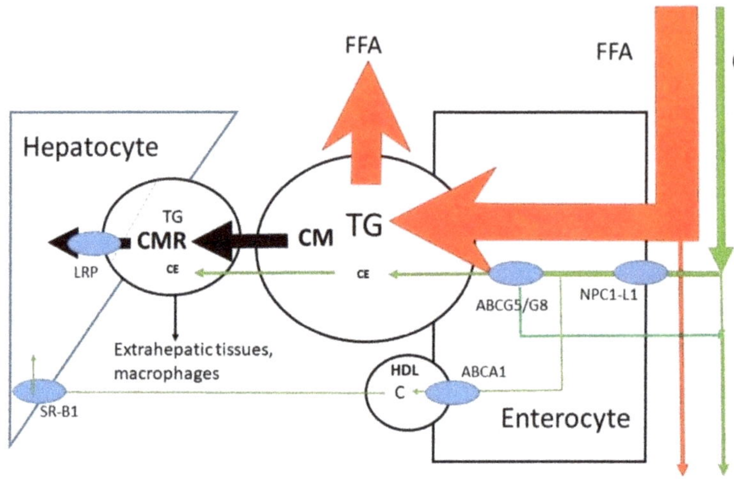

Figure 3. Uptake of FFA and C into enterocyte, CM secretion and conversion to CMR and uptake of CMR into the liver. Partial transport via HDL is also indicated. HDL= high density lipoprotein. TG = triglyceride, FFA = free fatty acid, C = free cholesterol, CE = cholesterol ester, CM = chylomicron, CMR = chylomicron remnant, ABCG= ATP-binding cassette sub-family G, NPC1L1 = Niemann–Pick C1 Like 1, ABCA1 = ATP-binding cassette transporter ABCA1, SR-B =Scavenger receptor class b type 1, LRP = LDL-receptor-related protein.

Both NPC1L1 and ABCG5/G8 are also involved in hepatic C secretion into bile [29,30]: NPC1L1 controls re-absorption of C from bile over the canalicular membrane back into the liver, and ABCG5/G8 regulates the C secretion into bile. General whole-body upregulation of ABCG5/G8 would enhance biliary C secretion, reduce absorption, and increase fecal C excretion, i.e., increased elimination. Upregulation of ABCG5/G8 will have a little effect on plant sterol absorption, but possibly a moderate effect on dietary C absorption. However, the combination with increased biliary C secretion may positively affect total C elimination. Plant stanols and sterols are being used as food supplements in order to lower C absorption [31,32]. This effect of plant sterols is generally ascribed to a competitive effect for the inclusion into mixed micelles. Plant sterols reduce the uptake of C into the micelles [33].

Chylomicron formation, loading and release appears to be a slow process. The TG produced at the level of the endoplasmic reticulum (ER) is either incorporated into pre-chylomicrons within the ER lumen or shunted to TG storage pools [34,35]. This protects the body against the large load of TG introduced by a meal. The absorption is spread out over the whole day including the fasting state. Isotope studies show that the appearance in blood of an orally administered bolus of labeled TG maximizes after approximately 8 to 12 h [24]. Oral administration of isotopically labeled C results in a maximum isotope enrichment in plasma after approximately 24 h [36]. The CMR formation kinetics is determined by the supply of CM and by LPL activity. In health, LPL activity is normally high, so that CMR formation is predominantly determined by the CM supply. It must be understood that CM supply and conversion of CM to CMR are determined by TG, not by C; C is just transported by CM and remains transported in CMR.

6. Hepatic C Metabolism

The CMRs that reach the liver are taken up by the LDL-receptor-related protein (LRP), and TG and CE are released (Figure 4). It may be predicted that 100 g TG and 200 mg C are consumed per day, that 400 mg/d biliary C is added, that 95% of TG and 50% of C are absorbed and that 10% of absorbed TG and 90% of absorbed C reach the liver. In that case 9.5 g TG and 270 mg C enter the liver per day with a TG/C ratio of approximately

35. To prevent fatty liver disease, TG is removed by incorporation into VLDL particles. In fasting serum VLDL, the average physiological TG/C ratio equals approximately 5 (w/w), which means that additional C is secreted and/or that part of TG in VLDL has already been de-esterified and released as free fatty acids at the moment of sample collection. Furthermore, the predictions assume that absorbed TG and C leave the enterocyte at the same time and with the same speed. This may not be true. The speed of incorporation of TG and C into chylomicrons may not be the same. The C in the liver, originating from CMR, may be secreted in VLDL, converted to bile acids or secreted into bile. The same may happen to *de novo* synthesized C in the liver and C that returned to the liver via LDL. The control over the distribution of these fluxes is largely unknown.

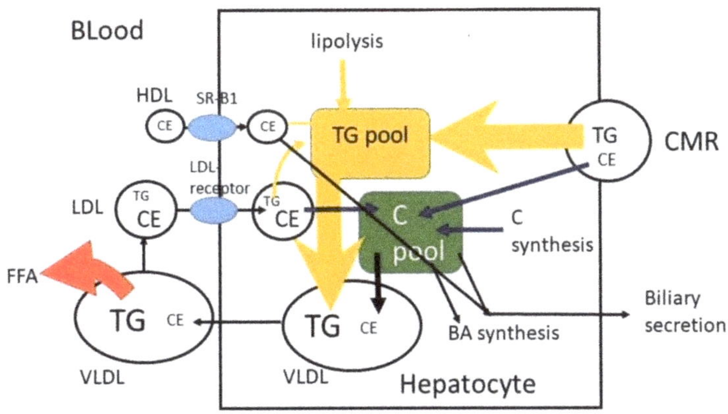

Figure 4. Hepatic C metabolism. TG = triglyceride, FFA = free fatty acid, C = free cholesterol, CE = cholesterol ester, CM = chylomicron, CMR = chylomicron remnant, BA = bile acid.

C is synthesized by a complex enzyme system. The rate limiting enzyme involved is 3-hydroxy-3-methylglutaryl coenzyme A reductase (HMG-CoA reductase). BAs are solely synthesized in the liver and circulate in the enterohepatic circulation with a very high FAR (~95%) in the terminal ileum. Two important pathways convert C to BAs [37]. The dominant one (neutral pathway, ~75%) involves 7α-hydroxylation of C as the first and rate limiting step. The second one (acidic pathway, ~25%) starts with 27-hydroxylation of C. BA synthesis has its own control. In the ileocyte the Farnesoid X receptor (FXR) is activated by BAs which leads to release of fibroblast growth factor 19 (FGF19) which inhibits the rate limiting enzyme in BA synthesis C-7α-hydroxylase. Decreased BA absorption induced by colesevelam and cholestyramine enhances BA synthesis and as a consequence hepatic C synthesis. Inhibited hepatic C synthesis induced by statins, does not reduce BA synthesis [38]. C synthesis and BA synthesis undergo diurnal variation [39–42]. C synthesis is predominantly stimulated at night, whereas BA synthesis is activated in the early evening.

7. Biliary C Secretion and Trans-Intestinal C Excretion (TICE)

In addition to BA synthesis, biliary C secretion is an approach to remove C from the body. It is unclear how biliary C secretion is controlled. Is C released from HDL its only substrate or is an increased hepatic C pool size an activator for induction? Gallbladder contraction creates an enhanced BA flux through the intestine and liver that is automatically forwarded toward the gallbladder. Also, biliary C is partly absorbed and recycled back to the liver. However, with an intestinal transit time of approximately 30 to 60 min, cycling of BAs is very rapid compared to the cycling of C [43]. Transport of absorbed C is strongly delayed by chylomicron metabolism. This may mean that during cycling after a meal, the bile becomes depleted in C and enriched in BAs. On the other hand, cycling BAs may have a stimulatory effect on hepatic C secretion into bile. In humans, measurement of biliary

C secretion requires very invasive techniques applying triple lumen intubation into the small intestine. The alternative technique is to determine dietary C intake, fecal excretion of C metabolites and BAs (sterol balance method) and determination of the C FAR. This allows calculation of total C loss and dietary C loss. The difference reflects biliary C loss. In mouse models, it was possible to estimate biliary BA and C secretion by puncturing the gallbladder after occlusion of the bile duct and collecting the bile secreted by the liver. It was observed that this flow was smaller than was calculated by the fecal balance method. An alternative route of C removal was defined as the trans-intestinal C excretion (TICE) to complete the balance [44,45]. In a later stage, a new method was developed to estimate biliary C secretion in humans since quantitative bile collection is not possible in humans; BA kinetics were measured with stable isotopes as well as the fractional turnover rate of BAs in plasma. These data were transferred to a biliary BA secretion rate. In bile, collected with CCK induced gallbladder contraction, biliary C secretion rate was calculated. Using this procedure also in humans a TICE flux could be demonstrated [36]. Intestinal ABCG5/G8 is most likely involved in removal of TICE C from the enterocyte. In parallel with biliary and dietary C, also C directly secreted into the intestinal lumen is subject to reabsorption [46]. Interestingly, this indicates incorporation of TICE C into micelles passing the small intestine, which makes the understanding of C homeostasis more complex. Otherwise, for balance studies, biliary C excretion and TICE do not necessarily need to be separated. In balance studies a total fecal C excretion rate is measured independent of the source. After correction for loss of dietary C, a value remains for loss of biliary C and TCE together.

8. Lipoprotein Metabolism

The hepatic metabolism of CMs and CMRs have been discussed in the previous sections. VLDL particles are secreted from the liver into blood. As for CMs, TGs in VLDL are de-esterified by LPL and FFAs are released and stored in adipocytes or muscles. The VLDL particles decrease in size and density and gradually turn over into VLDL-remnant particles of which LDL particles are the smallest. LDL and partly also VLDL remnants can be taken up by the liver, but also by extrahepatic tissues. The difference between VLDL-remnants and LDL is no longer strict. In addition to LDL-C, VLDL remnants C attribute to the risk for cardiovascular disease [47,48]. The terminology of non-HDL-C becomes popular and stands for all ApoB-100 lipoprotein bound C, i.e., LDL, VLDL and VLDL remnants [49]. In parallel, there is a trend to promote the measurement of ApoB-100 as a more valuable marker for atherosclerotic risk [50,51]. This is based on the knowledge that not only the C concentration inside lipoprotein fraction is important, but also the number of and the size of particles. VLDL and LDL consist of particles with different sizes. The more abundant and the smaller the LDL particles, the more atherogenic they are [52–54]. Independent of size, each particle contains one ApoB-100 unit. HDL transfers C from extrahepatic sites including macrophages and plaques to be delivered to the liver and secreted into bile for removal. However, HDL particles and LDL particles do not act independently. It has been demonstrated that in blood, CE can be transferred from HDL to VLDL and LDL by the cholesterol ester transfer protein (CETP) [55]. Thus, a high CETP activity lowers HDL-C and increases LDL-C. A low activity does the opposite, and it is protective against atherosclerosis. Medication to lower CETP is still under development, but it has not been successful so far [56,57].

9. C Flux Measurements in Humans

The first and principal C fluxes are synthesis and absorption. Whole body C synthesis is measured using the sterol balance method in which fecal C excretion, fecal BA excretion, and dietary C intake are measured [58]. Alternatively, C synthesis is measured via the mass isotopomer distribution analysis (MIDA) [59] applying stable isotopes. As C is a polymer of acetate units, infusion of ^{13}C-acetate leads to a continuously increasing isotope enrichment of the C pool. The increment of enrichment in serum C reflects the fractional C synthesis rate. In parallel, a methodology was developed employing deuterated water as the substrate [60].

Both the sterol balance method and isotope dilution techniques are not suitable for C synthesis measurement in large patient populations or patient care. For that purpose, surrogate marker technology was developed measuring intermediate metabolites in the C synthesis pathway. Squalene, lanosterol, desmosterol, and lathosterol have been studied, but the lathosterol concentration is the most generally accepted marker [61]. Stable isotope technology is used to measure C absorption; ^{13}C labeled or deuterated C is administered orally incorporated in a meal or a capsule during a meal. Two approaches have been introduced. One applies a seven-day continuous isotope feeding protocol combined with labeled sitostanol as minimal absorbable reference sterol and a three-day fecal collection period. From the ratio of enrichments of both labeled sterols in feces collected during the last three days, the C FAR can be calculated [9]. The second approach applies a single oral dosage of labeled C combined with a simultaneous intravenous dosage of a different labeled C. The ratio of both isotope enrichments in serum two to three days after administration allows the calculation of the C FAR [10]. Both approaches are unsuitable for studies in large patient populations or in patient care. Also, for C absorption surrogate markers have been evaluated. Cholestanol and various plant sterols have been studied of which the serum concentrations of campesterol and sitosterol are mostly used [62]. Plant sterols are much less efficient absorbed [63], but a high plant sterol absorption is associated with a high C absorption. Finally, BA synthesis is measured using the stable isotope dilution method administering ^{13}C or deuterated primary bile acids (cholic acid, chenodeoxycholic acid) in a single oral bolus and measuring the decay of isotope enrichment in serum during four days [64,65]. Alternatively, BA synthesis is measured as the rate of fecal excretion, since BAs escaping absorption are lost via feces as primary or secondary BAs. The concentration in serum of the intermediates in the BA synthesis pathway 7α-hydroxycholesterol and 7α-hydroxy-4-cholesten-3-one are used as markers for BA synthesis [66,67]. Generally, marker concentrations in serum are corrected for the total C concentration in order to correct for variable lipoprotein metabolism and LDL-receptor activity. It must be realized that a correlation between a marker/C ratio in a single fasting serum sample and the function measured over a four or a seven day period contains large variation. Thus, markers are valid for group comparisons but less valid for individual patients. Biliary C secretion cannot be measured directly in humans. Generally, it is calculated in the situation that fecal C excretion, dietary C intake, and the C FAR have been measured. The difference between total fecal C excretion and dietary C excretion is interpreted as the loss of biliary C. Via FAR, the biliary secretion may be calculated as discussed in Section 7. For C absorption, it may need discussion what flux is important. The measurement of C absorption presents a value for the C FAR, i.e., the percentage of the dosed C isotope that is absorbed. Important additional parameters are the dietary C intake and the biliary C secretion rate. Together with FAR they determine the daily fluxes of absorbed dietary and absorbed biliary C as well as the total daily C absorption rate. Dietary C intake is normally determined via a nutrition diary. Different research centers may use different information on C contents of food products and different protocols. Diaries also tend to underestimate food intake. The fluxes of C in the serum lipoproteins VLDL, LDL and HDL cannot be measured. The turnover of these lipoproteins has been measured by collecting the lipoproteins from blood, labeling these with isotopes, reinjecting these into the volunteer and measuring the disappearance from blood [68,69]. These techniques are only experimental, and they depend largely on extensive modeling. No alternative marker techniques are available. For patient care in the hospital so far only the classical parameters serum total C, LDL-C, HDL-C are available. In the near future the parameters serum non HDL-C and ApoB may be included in the selection of patients to be treated for prevention of cardiovascular disease. The surrogate markers for C absorption and C synthesis might be introduced in order to predict whether a patient is a high C absorber or high C synthesizer [70]. However, the validity of markers for this purpose may be limited.

10. Discussion

C homeostasis is the process that balances C input and C elimination. In detail, this includes all the steps in C metabolism as discussed above, involving transport of absorbed C in CMs and CMRs, endogenously synthesized C in hepatic and extrahepatic tissues, transport of C in VLDL, LDL and HDL particles, biliary C secretion and TICE as well as synthesis of BAs. Each step requires its own research. Many details are being unraveled at the cellular level discovering molecular mechanisms involved. Updates of information obtained in molecular mechanisms can be found in the recent review papers by Luo and Li [71,72]. Here we focused on the C fluxes in healthy conditions. The interactions between the different steps are so far largely unknown. Generally spoken, C absorption and C synthesis are balancing in that synthesis adjusts to changes in absorption and vice versa. A subject's absorption rate is determined by many factors as described above. Summarized they are:

1. C intake;
2. Digestion including gastric function, pancreatic function;
3. Hepatic BA synthesis and biliary secretion;
4. Biliary C secretion (NPC1-L1, ABCG5/G8);
5. Cholecystokinin (CCK) production and release;
6. Gallbladder function (contraction in response to CCK, relaxation);
7. Intestinal micelle formation, micellar uptake of C;
8. C uptake into enterocytes (NPC1-L1) and re-secretion into intestinal lumen (ABCG5/G8);
9. TICE;
10. CM formation, release and conversion to CMRs.

Every step involved has its own individual efficiency contributing to the daily flux of C entering the body C pool. The body C pool must sense excess or shortage of C being absorbed in order to upregulate or downregulate C synthesis. Extrahepatic tissues are able to synthesize C, take up C from blood as LDL particles, but they are largely unable to metabolize C. Their excess C must be removed via HDL transport. The liver is the intermediate in C homeostasis. Strongly reduced C absorption will reduce the hepatic C pool and stimulate hepatic C synthesis and increase the LDL-receptor activity stimulating C uptake. Excess C absorption may result into enhancement of the hepatic C pool, reduced hepatic C synthesis, enhanced BA synthesis and biliary C secretion but also enhanced VLDL secretion and LDL formation. In the ideal situation, serum C concentration does not increase. However, the hepatic responsiveness to strongly enhanced absorption may be impaired leading to increased serum LDL-C and total C concentrations. A reduction in the flux of absorbed C can be created by a reduction in dietary C intake and a reduction in C absorption. The latter option is achieved by pharmacological action of ezetimibe that reduces the intestinal NPC1L1 activity or by increased dietary intake of plant sterols or stanols as food supplements that compete with C uptake in micelles. However, plant sterols and stanols as food additives are under discussion [73,74]. A major drawback of reducing the C absorption rate remains compensation by increased synthesis. This means that the net effect is the difference between the reduced absorption and the increased compensatory synthesis. In order to highly efficiently lower serum C levels, reduction of absorption must be combined with a means to reduce C synthesis. In hypercholesterolemic patients, statin treatment is the first choice to lower serum C. In case of insufficient serum C reduction, ezetimibe treatment is added as a co-treatment. So far, only statins are available as the basic pharmacological tool to reduce C synthesis. For patients that do not tolerate statins, Bempedoic acid can now be prescribed, which blocks adenosine triphosphate-citrate lyase in the liver which is involved in C production [75]. When the combined treatment does not achieve sufficient LDL-C reduction, the next step is additional treatment with a proprotein convertase subtilisin/kexin type 9 (PCSK9) inhibitor [76]. The PCSK9 inhibitor binds to the LDL-receptor and inhibits receptor activity. This results in enhanced serum LDL-C levels. Two PCSK9 inhibitors are now on the market: Alirocumab and Evolocumab. Additionally,

Inclisiran has been developed, a small interfering RNA that inhibits translation of the PCSK9 protein [77].

11. Conclusions

The C homeostasis process is complex and highly regulated, including hepatic and extrahepatic C synthesis, uptake and release processes, intestinal absorption, various lipoprotein transport systems in blood, biliary secretion, and trans-intestinal excretion as well as BA synthesis. The major pathological focus is on serum C and in particular LDL-C in relationship to cardiovascular disease. Reduced dietary intake and dietary means to reduce C absorption have been shown to be effective tools to reduce serum C. However, their efficacies are limited due the contribution of biliary C and to the compensatory induction of C synthesis.

Funding: This research received no external funding.

Institutional Review Board Statement: Not applicable.

Informed Consent Statement: Not applicable.

Data Availability Statement: Not applicable.

Conflicts of Interest: The author declares no conflict of interest.

References

1. Nijstad, N.; Gautier, T.; Briand, F.; Rader, D.J.; Tietge, U.J. Biliary Sterol Secretion Is Required for Functional In Vivo Reverse Cholesterol Transport in Mice. *Gastroenterology* **2011**, *140*, 1043–1051. [CrossRef] [PubMed]
2. Lewis, G.F.; Rader, D.J. New Insights into the Regulation of HDL Metabolism and Reverse Cholesterol Transport. *Circ. Res.* **2005**, *24*, 1221–1232. [CrossRef] [PubMed]
3. Wilson, M.D.; Rudel, L.L. Review of cholesterol absorption with emphasis on dietary and biliary cholesterol. *J. Lipid Res.* **1994**, *35*, 943–955. [CrossRef]
4. Lütjohann, D.; Meyer, S.; Von Bergmann, K.; Stellaard, F. Cholesterol Absorption and Synthesis in Vegetarians and Omnivores. *Mol. Nutr. Food Res.* **2018**, *62*, e1700689. [CrossRef] [PubMed]
5. Pan, X.-F.; Yang, J.-J.; Lipworth, L.; Shu, X.-O.; Cai, H.; Steinwandel, M.; Blot, W.; Zheng, W.; Yu, D. Cholesterol and Egg Intakes with Cardiometabolic and All-Cause Mortality among Chinese and Low-Income Black and White Americans. *Nutrients* **2021**, *13*, 2094. [CrossRef] [PubMed]
6. Kern, F. Normal Plasma Cholesterol in an 88-Year-Old Man Who Eats 25 Eggs a Day. *N. Engl. J. Med.* **1991**, *324*, 896–899. [CrossRef]
7. Kuang, H.; Yang, F.; Zhang, Y.; Wang, T.; Chen, G. The Impact of Egg Nutrient Composition and Its Consumption on Cholesterol Homeostasis. *Cholesterol* **2018**, *2018*, 6303810. [CrossRef]
8. Kang, J.W.; Zivkovic, A.M. Are Eggs Good Again? A Precision Nutrition Perspective on the Effects of Eggs on Cardiovascular Risk, Taking into Account Plasma Lipid Profiles and TMAO. *J. Nutr. Biochem.* **2022**, *100*, 108906. [CrossRef]
9. Lütjohann, D.; Meese, C.O.; Crouse, J.R., 3rd; von Bergmann, K. Evaluation of deuterated cholesterol and deuterated sitostanol for measurement of cholesterol absorption in humans. *J. Lipid Res.* **1993**, *34*, 1039–1046. [CrossRef]
10. Bosner, M.S.; Ostlund, R.E., Jr.; Osofisan, O.; Grosklos, J.; Fritschle, C.; Lange, L.G. Assessment of percent cholesterol absorption in humans with stable isotopes. *J. Lipid Res.* **1993**, *34*, 1047–1053. [CrossRef]
11. Turley, S.D.; Spady, D.K.; Dietschy, J.M. Role of liver in the synthesis of cholesterol and the clearance of low density lipoproteins in the cynomolgus monkey. *J. Lipid Res.* **1995**, *36*, 67–79. [CrossRef]
12. Van Erpecum, K.J.; van Berge Henegouwen, G.P.; Stolk, M.F.; Hopman, W.P.; Jansen, J.B.; Lamers, C.B. Fasting gallbladder volume, postprandial emptying and cholecystokinin release in gallstone patients and normal subjects. *J. Hepatol.* **1992**, *14*, 194–202. [CrossRef]
13. Juste, C.; Gérard, P. Cholesterol-to-Coprostanol Conversion by the Gut Microbiota: What We Know, Suspect, and Ignore. *Microorganisms* **2021**, *9*, 1881. [CrossRef] [PubMed]
14. Villette, R.; Kc, P.; Beliard, S.; Tapia, M.F.S.; Rainteau, D.; Guerin, M.; Lesnik, P. Unraveling Host-Gut Microbiota Dialogue and Its Impact on Cholesterol Levels. *Front. Pharmacol.* **2020**, *11*, 278. [CrossRef]
15. Kenny, D.J.; Plichta, D.R.; Shungin, D.; Koppel, N.; Hall, A.B.; Fu, B.; Vasan, R.S.; Shaw, S.Y.; Vlamakis, H.; Balskus, E.P.; et al. Cholesterol Metabolism by Uncultured Human Gut Bacteria Influences Host Cholesterol Level. *Cell Host Microbe* **2020**, *28*, 245–257. [CrossRef]
16. Vors, C.; Joumard-Cubizolles, L.; Lecomte, M.; Combe, E.; Ouchchane, L.; Dra, J.; Raynal, K.; Joffre, F.; Meiller, L.; Le Barz, M.; et al. Milk polar lipids reduce lipid cardiovascular risk factors in overweight postmenopausal women: Towards a gut sphingomyelin-cholesterol interplay. *Gut* **2020**, *69*, 487–501. [CrossRef]

17. Sundaram, A.; Koutkia, P.; Apovian, C. Nutritional Management of Short Bowel Syndrome in Adults. *J. Clin. Gastroenterol.* **2002**, *34*, 207–220. [CrossRef]
18. Glaysher, M.A.; Ward, J.; Aldhwayan, M.; Ruban, A.; Prechtl, C.G.; Fisk, H.L.; Chhina, N.; Al-Najim, W.; Smith, C.; Klimowska-Nassar, N.; et al. The effect of a duo-denal-jejunal bypass liner on lipid profile and blood concentrations of long chain polyunsaturated fatty acids. *Clin. Nutr.* **2021**, *40*, 2343–2354. [CrossRef]
19. Altmann, S.W.; Davis, H.R., Jr.; Zhu, L.J.; Yao, X.; Hoos, L.M.; Tetzloff, G.; Iyer, S.P.; Maguire, M.; Golovko, A.; Zeng, M.; et al. Niemann-Pick C1 Like 1 protein is critical for intestinal cholesterol absorption. *Science* **2004**, *303*, 1201–1204. [CrossRef]
20. Iqbal, J.; Hussain, M.M. Evidence for multiple complementary pathways for efficient cholesterol absorption in mice. *J. Lipid Res.* **2005**, *46*, 1491–1501. [CrossRef]
21. Hussain, M.M. Intestinal lipid absorption and lipoprotein formation. *Curr. Opin. Lipidol.* **2014**, *25*, 200–206. [CrossRef] [PubMed]
22. Weggemans, R.M.; Zock, P.L.; Tai, E.S.; Ordovas, J.M.; Molhuizen, H.O.; Katan, M.B. ATP binding cassette G5 C1950G polymorphism may affect blood cholesterol concentrations in humans. *Clin. Genet.* **2002**, *62*, 226–229. [CrossRef] [PubMed]
23. Graf, G.A.; Yu, L.; Li, W.P.; Gerard, R.; Tuma, P.L.; Cohen, J.C.; Hobbs, H.H. ABCG5 and ABCG8 are obligate heterodimers for protein trafficking and biliary cholesterol excretion. *J. Biol. Chem.* **2003**, *278*, 48275–48282. [CrossRef] [PubMed]
24. Kalivianakis, M.; Minich, D.M.; Bijleveld, C.M.; van Aalderen, W.M.; Stellaard, F.; Laseur, M.; Vonk, R.J.; Verkade, H.J. Fat malabsorption in cystic fibrosis patients receiving enzyme replacement therapy is due to impaired intestinal uptake of long-chain fatty acids. *Am. J. Clin. Nutr.* **1999**, *69*, 127–134. [CrossRef]
25. Kalivianakis, M.; Elstrodt, J.; Havinga, R.; Kuipers, F.; Stellaard, F.; Sauer, P.J.; Vonk, R.J.; Verkade, H.J. Validation in an animal model of the carbon 13-labeled mixed triglyceride breath test for the detection of intestinal fat malabsorption. *J. Pediatr.* **1999**, *135*, 444–450. [CrossRef]
26. Lütjohann, D.; von Bergmann, K. Phytosterolaemia: Diagnosis, characterization and therapeutical approaches. *Ann. Med.* **1997**, *29*, 181–184. [CrossRef]
27. Berge, K.E.; von Bergmann, K.; Lutjohann, D.; Guerra, R.; Grundy, S.M.; Hobbs, H.H.; Cohen, J.C. Heritability of plasma noncholesterol sterols and relationship to DNA sequence polymorphism in ABCG5 and ABCG8. *J. Lipid Res.* **2002**, *43*, 486–494. [CrossRef]
28. Lütjohann, D.; Björkhem, I.; Beil, U.F.; von Bergmann, K. Sterol absorption and sterol balance in phytosterolemia evaluated by deuterium-labeled sterols: Effect of sitostanol treatment. *J. Lipid Res.* **1995**, *36*, 1763–1773. [CrossRef]
29. Zanlungo, S.; Nervi, F. Discovery of the hepatic canalicular and intestinal cholesterol transporters. New targets for treatment of hypercholesterolemia. *Eur. Rev. Med. Pharmacol. Sci.* **2003**, *7*, 33–39.
30. Yu, L.; Gupta, S.; Xu, F.; Liverman, A.D.; Moschetta, A.; Mangelsdorf, D.J.; Repa, J.J.; Hobbs, H.H.; Cohen, J.C. Expression of ABCG5 and ABCG8 is required for regulation of biliary cholesterol secretion. *J. Biol. Chem.* **2005**, *280*, 8742–8747. [CrossRef]
31. Gylling, H.; Miettinen, T.A. Serum cholesterol and cholesterol and lipoprotein metabolism in hypercholesterolaemic NIDDM patients before and during sitostanol ester-margarine treatment. *Diabetologia* **1994**, *37*, 773–780. [CrossRef] [PubMed]
32. Miettinen, T.A.; Gylling, H. Regulation of cholesterol metabolism by dietary plant sterols. *Curr. Opin. Lipidol.* **1999**, *10*, 9–14. [CrossRef] [PubMed]
33. Wong, N.C. The beneficial effects of plant sterols on serum cholesterol. *Can. J. Cardiol.* **2001**, *17*, 715–721. [PubMed]
34. Mansbach, C.M.; Siddiqi, S.A. The biogenesis of chylomicrons. *Annu. Rev. Physiol.* **2010**, *72*, 315–333. [CrossRef]
35. Auclair, N.; Melbouci, L.; St-Pierre, D.; Levy, E. Gastrointestinal factors regulating lipid droplet formation in the intestine. *Exp. Cell Res.* **2018**, *363*, 1–14. [CrossRef]
36. Jakulj, L.; van Dijk, T.H.; de Boer, J.F.; Kootte, R.S.; Schonewille, M.; Paalvast, Y.; Boer, T.; Bloks, V.W.; Boverhof, R.; Nieuwdorp, M.; et al. Transintestinal cholesterol transport is active in mice and humans and controls ezetimibe-induced fecal neutral sterol excretion. *Cell Metab.* **2016**, *24*, 783–794. [CrossRef]
37. Russell, D.W. The enzymes, regulation, and genetics of bile acid synthesis. *Annu. Rev. Biochem.* **2003**, *72*, 137–174. [CrossRef]
38. Sudhop, T.; Reber, M.; Tribble, D.; Sapre, A.; Taggart, W.; Gibbons, P.; Musliner, T.; von Bergmann, K.; Lütjohann, D. Changes in cholesterol absorption and cholesterol synthesis caused by ezetimibe and/or simvastatin in men. *J. Lipid Res.* **2009**, *50*, 2117–2123. [CrossRef]
39. Jones, P.J.; Schoeller, D.A. Evidence for diurnal periodicity in human cholesterol synthesis. *J. Lipid Res.* **1990**, *31*, 667–673. [CrossRef]
40. Parker, T.S.; McNamara, D.J.; Brown, C.; Garrigan, O.; Kolb, R.; Batwin, H.; Ahrens, E.H., Jr. Mevalonic acid in human plasma: Relationship of concentration and circadian rhythm to cholesterol synthesis rates in man. *Proc. Natl. Acad. Sci. USA* **1982**, *79*, 3037–3041. [CrossRef]
41. Gälman, C.; Angelin, B.; Rudling, M. Bile acid synthesis in humans has a rapid diurnal variation that is asynchronous with cholesterol synthesis. *Gastroenterology* **2005**, *129*, 1445–1453. [CrossRef] [PubMed]
42. Yoshida, T.; Honda, A.; Tanaka, N.; Matsuzaki, Y.; Shoda, J.; He, B.; Osuga, T.; Miyazaki, H. Determination of 7 al-pha-hydroxy-4-cholesten-3-one level in plasma using isotope-dilution mass spectrometry and monitoring its circadian rhythm in human as an index of bile acid biosynthesis. *J. Chromatogr. B Biomed. Appl.* **1994**, *655*, 179–187. [CrossRef]
43. Stellaard, F.; Lütjohann, D. Dynamics of the enterohepatic circulation of bile acids in healthy humans. *Am. J. Physiol. Gastrointest. Liver Physiol.* **2021**, *321*, G55–G66. [CrossRef] [PubMed]

44. van der Velde, A.E.; Vrins, C.L.; van den Oever, K.; Seemann, I.; Oude Elferink, R.P.; van Eck, M.; Kuipers, F.; Groen, A.K. Regulation of direct transintestinal cholesterol excretion in mice. *Am. J. Physiol. Gastrointest. Liver Physiol.* **2008**, *295*, G203–G208. [CrossRef]
45. Reeskamp, L.F.; Meessen, E.C.E.; Groen, A.K. Transintestinal cholesterol excretion in humans. *Curr. Opin. Lipidol.* **2018**, *29*, 10–17. [CrossRef]
46. Van de Peppel, I.P.; Bertolini, A.; van Dijk, T.H.; Groen, A.K.; Jonker, J.W.; Verkade, H.J. Efficient reabsorption of transintestinally excreted cholesterol is a strong determinant for cholesterol disposal in mice. *J. Lipid Res.* **2019**, *60*, 1562–1572. [CrossRef]
47. Heidemann, B.E.; Koopal, C.; Bots, M.L.; Asselbergs, F.W.; Westerink, J.; Visseren, F.L. The relation between VLDL-cholesterol and risk of cardiovascular events in patients with manifest cardiovascular disease. *J. Int. J. Cardiol.* **2021**, *322*, 251–257. [CrossRef]
48. Balling, M.; Afzal, S.; Varbo, A.; Langsted, A.; Davey Smith, G.; Nordestgaard, B.G. VLDL Cholesterol Accounts for One-Half of the Risk of Myocardial Infarction Associated With apoB-Containing Lipoproteins. *J. Am. Coll. Cardiol.* **2020**, *76*, 2725–2735. [CrossRef]
49. Sniderman, A.; McQueen, M.; Contois, J.; Williams, K.; Furberg, C.D. Why is non-high-density lipoprotein cholesterol a better marker of the risk of vascular disease than low-density lipoprotein cholesterol? *J. Clin. Lipidol.* **2010**, *4*, 152–155. [CrossRef]
50. Brown, W.V.; Myers, G.L.; Sniderman, A.D.; Stein, E. Should we use apoB for risk assessment and as a target for treatment? *J. Clin. Lipidol.* **2010**, *4*, 144–151. [CrossRef]
51. Sniderman, A.D.; Navar, A.M.; Thanassoulis, G. Apolipoprotein B vs. Low-Density Lipoprotein Cholesterol and Non-High-Density Lipoprotein Cholesterol as the Primary Measure of Apolipoprotein B Lipoprotein-Related Risk: The Debate Is Over. *JAMA Cardiol.* **2022**, *7*, 257–258. [CrossRef] [PubMed]
52. Ip, S.; Lichtenstein, A.H.; Chung, M.; Lau, J.; Balk, E.M. Systematic review: Association of low-density lipoprotein subfractions with cardiovascular outcomes. *Ann. Intern. Med.* **2009**, *150*, 474–484. [CrossRef] [PubMed]
53. Rajman, I.; Kendall, M.J.; Cramb, R.; Holder, R.L.; Salih, M.; Gammage, M.D. Investigation of low density lipoprotein subfractions as a coronary risk factor in normotriglyceridaemic men. *Atherosclerosis* **1996**, *125*, 231–242. [CrossRef]
54. Froyen, E. The effects of fat consumption on low-density lipoprotein particle size in healthy individuals: A narrative review. *Lipids Health Dis.* **2021**, *20*, 86. [CrossRef] [PubMed]
55. Oliveira, H.C.F.; Raposo, H.F. Cholesteryl Ester Transfer Protein and Lipid Metabolism and Cardiovascular Diseases. *Adv. Exp. Med. Biol.* **2020**, *1276*, 15–25. [PubMed]
56. Taheri, H.; Filion, K.B.; Windle, S.B.; Reynier, P.; Eisenberg, M.J. Cholesteryl Ester Transfer Protein Inhibitors and Cardiovascular Outcomes: A Systematic Review and Meta-Analysis of Randomized Controlled Trials. *Cardiology* **2020**, *145*, 236–250. [CrossRef]
57. Banerjee, S.; De, A. Pathophysiology and inhibition of cholesteryl ester transfer protein for prevention of cardiovascular diseases: An update. *Drug Discov. Today.* **2021**, *26*, 1759–1764. [CrossRef]
58. Grundy, S.M.; Ahrens, E.H., Jr. Measurements of cholesterol turnover, synthesis, and absorption in man, carried out by isotope kinetic and sterol balance methods. *J. Lipid Res.* **1969**, *10*, 91–107. [CrossRef]
59. Hellerstein, M.K.; Neese, R.A. Mass isotopomer distribution analysis: A technique for measuring biosynthesis and turnover of polymers. *Am. J. Physiol.* **1992**, *263*, E988–E1001. [CrossRef]
60. Jones, P.J.; Leitch, C.A.; Li, Z.C.; Connor, W.E. Human cholesterol synthesis measurement using deuterated water. Theoretical and procedural considerations. *Arterioscler. Thromb.* **1993**, *13*, 247–253. [CrossRef]
61. Kempen, H.J.; Glatz, J.F.; Gevers Leuven, J.A.; van der Voort, H.A.; Katan, M.B. Serum lathosterol concentration is an indicator of whole-body cholesterol synthesis in humans. *J. Lipid Res.* **1988**, *29*, 1149–1155. [CrossRef]
62. Tilvis, R.S.; Miettinen, T.A. Serum plant sterols and their relation to cholesterol absorption. *Am. J. Clin. Nutr.* **1986**, *43*, 92–97. [CrossRef] [PubMed]
63. Heinemann, T.; Axtmann, G.; von Bergmann, K. Comparison of intestinal absorption of cholesterol with different plant sterols in man. *Eur. J. Clin. Investig.* **1993**, *23*, 827–831. [CrossRef] [PubMed]
64. Stellaard, F.; Sackmann, M.; Sauerbruch, T.; Paumgartner, G. Simultaneous determination of cholic acid and chenodeoxycholic acid pool sizes and fractional turnover rates in human serum using 13C-labeled bile acids. *J. Lipid Res.* **1984**, *25*, 1313–1319. [CrossRef]
65. Hulzebos, C.V.; Renfurm, L.; Bandsma, R.H.; Verkade, H.J.; Boer, T.; Boverhof, R.; Tanaka, H.; Mierau, I.; Sauer, P.J.; Kuipers, F.; et al. Measurement of parameters of cholic acid kinetics in plasma using a microscale stable isotope dilution technique: Application to rodents and humans. *J. Lipid Res.* **2001**, *42*, 1923–1929. [CrossRef]
66. Hahn, C.; Reichel, C.; von Bergmann, K. Serum concentration of 7 alpha-hydroxycholesterol as an indicator of bile acid synthesis in humans. *J. Lipid Res.* **1995**, *36*, 2059–2066. [CrossRef]
67. Axelson, M.; Aly, A.; Sjövall, J. Levels of 7 alpha-hydroxy-4-cholesten-3-one in plasma reflect rates of bile acid synthesis in man. *FEBS Lett.* **1988**, *239*, 324–328. [CrossRef]
68. Magkos, F.; Sidossis, L.S. Measuring very low density lipoprotein-triglyceride kinetics in man in vivo: How different the various methods really are. *Curr. Opin. Clin. Nutr. Metab. Care.* **2004**, *7*, 547–555. [CrossRef]
69. Chan, D.C.; Watts, G.F. Recent advances in the investigation of lipoprotein metabolism using tracer methodology. *Clin. Lab.* **2006**, *52*, 353–361.
70. Lütjohann, D.; Stellaard, F.; Mulder, M.T.; Sijbrands, E.J.G.; Weingärtner, O. The emerging concept of "individualized choles-terol-lowering therapy": A change in paradigm. *Pharmacol. Ther.* **2019**, *199*, 111–116. [CrossRef]

71. Luo, J.; Yang, H.; Song, B.L. Mechanisms and regulation of cholesterol homeostasis. *Nat. Rev. Mol. Cell Biol.* **2020**, *21*, 225–245. [CrossRef] [PubMed]
72. Li, H.; Yu, X.H.; Ou, X.; Ouyang, X.P.; Tang, C.K. Hepatic cholesterol transport and its role in non-alcoholic fatty liver disease and atherosclerosis. *Prog. Lipid Res.* **2021**, *83*, 101109. [CrossRef] [PubMed]
73. Weingärtner, O.; Baber, R.; Teupser, D. Plant sterols in food: No consensus in guidelines. *Biochem. Biophys. Res. Commun.* **2014**, *446*, 811–813. [CrossRef] [PubMed]
74. Weingärtner, O.; Teupser, D.; Patel, S.B. The Atherogenicity of Plant Sterols: The Evidence from Genetics to Clinical Trials. *J. AOAC Int.* **2015**, *98*, 742–749. [CrossRef]
75. Pirillo, A.; Catapano, A.L. New insights into the role of bempedoic acid and ezetimibe in the treatment of hypercholesterolemia. *Curr. Opin. Endocrinol. Diabetes Obes.* **2022**, *29*, 161–166. [CrossRef]
76. Bagepally, B.S.; Sasidharan, A. Incremental net benefit of lipid-lowering therapy with PCSK9 inhibitors: A systematic review and meta-analysis of cost-utility studies. *Eur. J. Clin. Pharmacol.* **2022**, *78*, 351–363. [CrossRef]
77. Smith, K.W.; White, C.M. Inclisiran: A Novel Small Interfering RNA Drug for Low-Density Lipoprotein Reduction. *J. Clin. Pharmacol.* **2022**. [CrossRef]

nutrients

Review

Is There a Correlation between Dietary and Blood Cholesterol? Evidence from Epidemiological Data and Clinical Interventions

Maria Luz Fernandez [1,*] and Ana Gabriela Murillo [2]

[1] Department of Nutritional Sciences, University of Connecticut, Storrs, CT 06268, USA
[2] Department of Biochemistry, University of Costa Rica, San Jose 11501-2060, Costa Rica; anagabriela.murillo@ucr.ac.cr
* Correspondence: maria-luz.fernandez@uconn.edu

Abstract: Dietary cholesterol has been a topic of debate since the 1960s when the first dietary guidelines that limited cholesterol intake to no more than 300 mg/day were set. These recommendations were followed for several years, and it was not until the late 1990s when they were finally challenged by the newer information derived from epidemiological studies and meta-analysis, which confirmed the lack of correlation between dietary and blood cholesterol. Further, dietary interventions in which challenges of cholesterol intake were evaluated in diverse populations not only confirmed these findings but also reported beneficial effects on plasma lipoprotein subfractions and size as well as increases in HDL cholesterol and in the functionality of HDL. In this review, we evaluate the evidence from recent epidemiological analysis and meta-analysis as well as clinical trials to have a better understanding of the lack of correlation between dietary and blood cholesterol.

Keywords: dietary cholesterol; plasma cholesterol; lipoproteins; epidemiological studies; clinical interventions

Citation: Fernandez, M.L.; Murillo, A.G. Is There a Correlation between Dietary and Blood Cholesterol? Evidence from Epidemiological Data and Clinical Interventions. *Nutrients* **2022**, *14*, 2168. https://doi.org/10.3390/nu14102168

Academic Editor: Frans Stellaard

Received: 9 May 2022
Accepted: 20 May 2022
Published: 23 May 2022

1. Introduction

The relationship between dietary and blood cholesterol is very controversial and has been debated within the scientific community since the 1960s when the first guidelines for dietary cholesterol were published [1]. It was not until the 2015 dietary guidelines that the upper limits of dietary cholesterol were eliminated, based on more current information [2]. The guidelines from the 1960s were not based on epidemiological data, meta-analysis or clinical intervention; they were construed as a consensus based on the current information available at that time [1]. Animal studies that have been used to study effects of dietary cholesterol on atherosclerosis, oxidative stress, and inflammation are not reliable since the concentration of cholesterol used varied from 1.25 to 5% of the diet [3–5], which would be the equivalent of approximately 10,000 to 37,000 mg/day. These dietary challenges with these exorbitant amounts of dietary cholesterol cannot possibly have any clinical application. Since those early years, the information regarding dietary cholesterol in humans has been substantially increased by key findings from epidemiological data of large cohort studies including the Framingham study as early as the 1980s [6], the Nurses' study [7], National Health and Nutrition Examination Survey (NHANES) [8], and more recently others such as the Hellenic National Nutrition and Health Survey (HNNHS) [9]. In addition, clinical interventions in diverse populations including children [10], young adults [11,12], elderly people [13,14], obese individuals [15,16] metabolic syndrome populations [17,18], and diabetic patients [19,20] have demonstrated that the plasma biomarkers of coronary heart disease are not increased by dietary cholesterol (provided by eggs) but may result in the formation of less atherogenic lipoproteins [21,22].

There were two main objectives of this review: (1) To evaluate the most recent epidemiological evidence and meta-analysis that continue to support the lack of correlation between dietary and blood cholesterol and (2) to evaluate the effects of dietary cholesterol on plasma lipids and the atherogenicity of lipoproteins in recent clinical trials utilizing a dietary cholesterol challenge.

2. Epidemiolocal Evidence

Most studies that evaluate dietary cholesterol use eggs as a natural food source. The outdated dietary guidelines established a limit of 300 mg/day. Eggs contain approximately 180–200 mg of cholesterol in the yolk and have been identified as high cholesterol foods therefore they have been used to evaluate the effects of dietary cholesterol on blood cholesterol [8,23] However, observational and prospective studies have not found for the most part a direct relationship between egg consumption and blood cholesterol or cardiovascular disease (CVD) risk [24]. Although there are studies that have found correlations between dietary cholesterol, plasma cholesterol and heart disease risk [25,26].

Using data from the Prospective Urban Rural Epidemiology (PURE) study, Dehghan et al. evaluated egg consumption and CVD of individuals from 21 countries in a 9 year follow up [23]. The results showed that the higher egg intake (≥7 egg/week compared with <1 egg/week) was not significantly associated with blood lipids (including total cholesterol, HDL cholesterol, LDL cholesterol, total cholesterol/HDL cholesterol ratio, triglycerides, apolipoprotein (apo)A1, apo B, and apo B/apoA1 ratio, total mortality, or major CVD. Similar results were obtained with the data of other prospective studies: The ONTARGET (Ongoing Telmisartan alone and in Combination with Ramipril Global End Point Trial) and TRANSCEND (Telmisartan Randomized Assessment Study in ACEI Intolerant Subjects with Cardiovascular Disease). The three studies combined had more than 177,555 participants [23].

High blood pressure is considered a major risk factor for developing cardiovascular disease (CVD) [27]. Therefore, hypertensive patients must take further precautions to maintain a healthy diet, including, according to some researchers, reducing cholesterol intake. In a study conducted with data from the China Health and Nutrition Survey (CHNS) from 1991 to 2015, Wu et al. [28] evaluated the relationship between cholesterol intake from eggs and other sources and mortality among hypertensive patients. The study included 8095 participants who were followed-up for a mean of 11.4 years. Using regression analysis, the results indicated that people who consumed more than seven eggs per week had up to 29% lower mortality compared with patients who did not consume more than two eggs/week. Although egg consumption was inversely associated with mortality, cholesterol intake from other sources showed a positive association, suggesting that eggs may offer a protective effect that is not found in other cholesterol sources such as red meat, pork, cheese or butter, which also have a substantial amount of saturated fat, a nutrient positively linked to CVD [28,29]. Several authors have attributed this protective effect to eggs' anti-inflammatory properties [30,31] and vitamin, mineral, and antioxidant content [8,32,33].

A recent large-scale cross-sectional study examined the relationship between dietary cholesterol and dyslipidemia in 8358 Chinese adults [34]. A total of 2429 participants (29% of the sample) were diagnosed with dyslipidemia according to the 2016 Chinese adult dyslipidemia prevention guide. Among all examinees, the mean cholesterol intake was 213.7 mg/day. According to the linear regression analysis, higher cholesterol consumption was associated with lower plasma TG and higher HDL-cholesterol in women (before adjusting for nutrient components), and no associations were observed in men. When the analysis was divided by cholesterol sources, eggs but no other sources showed a significant inverse relationship with the risk of dyslipidemia. In general, the results of this study showed a null association between cholesterol intake and serum lipids concentrations, indicating that consumption of cholesterol rich foods do not increase the risk of altering blood lipids and therefore, developing CVD [34]. Another epidemiological study including individuals (n = 3558) from the Hellenic National and Nutrition Health Survey (HNNHS) concluded that the more frequent egg consumption decreased the odds of dyslipidemia compared to either no eggs or infrequent egg intake [9] supporting the lack of effect of dietary cholesterol on plasma lipids and lipoproteins.

Table 1 shows the main epidemiological studies proving the lack of correlation between dietary cholesterol and blood cholesterol.

3. Meta-Analysis

A recent meta-analysis conducted by Godos et al. [35] revised the data of 39 prospective cohort studies that evaluated the association between egg consumption and the risk of CVD, coronary heart disease (CHD), and stroke. The studies analyzed included patients from North America, Europe, Asia, and some multi-national cohorts. In relation with CVD incidence and/or mortality (14 studies), the analysis showed that the intake of up to six eggs (a vehicle of dietary cholesterol) per week had an inverse association with CVD events, when compared to no intake. A similar trend was observed for CHD incidence and mortality (16 studies), where the risk decreased when examinees had up to two eggs per week. No associations between egg intake and stroke were found in this study, however a positive and linear association was found with egg consumption and heart failure. Still, the authors concluded that there is no evidence that eggs play a role in the development of CVD [35].

Berger et al. [36] analyzed 40 studies published between 1979 and 2013 for their meta-analysis. The authors included cohorts with participants without CVD diagnosis, either healthy or with risk CVD factors such as high blood pressure, dyslipidemia, diabetes, or metabolic syndrome present at baseline. Clinical interventions that recruited healthy individuals (with no CVD risk factors and no lipid lowering drugs use) at the beginning of the trial were also included in the analysis. The results showed no association between dietary cholesterol and coronary artery disease (CAD), ischemic stroke, or hemorrhagic stroke. However, cholesterol intake did affect blood lipids by increasing both serum total cholesterol and LDL cholesterol, although changes in LDL-C were not statistically significant when the intervention intake was excessive (>900/day). It is important to mention that HDL cholesterol also was also significantly increased by dietary cholesterol, which means no net change in CVD risk [36]. Other blood lipids such as plasma triglycerides remained unaffected. According to the authors, one of the reasons why some studies have found a positive linear association between cholesterol consumption and CVD outcomes is that many interventions do not control for other nutrients that affect that risk, for example saturated fatty acids and calories from fat, which can increase with cholesterol source intake and are positively correlated with CVD risk; on the contrary, dietary fiber and vegetable protein show a negative association with CVD and in some dietary patterns, like the Western diet, these two nutrients are inversely correlated with cholesterol intake [37,38].

Drouin-Chartier et al. [39] reviewed three large cohort studies: the Nurses' Health Study (NHS) (1980–2012), NHS II (1991–2017) and The Health Professional Follow up Study (HPFS) (1986–2016) with the objective of establishing if there was an association between egg intake and the risk of developing CVD. In total, 14,806 subjects had diagnosed CVD (non-fatal myocardial infarction, fatal coronary heart disease or stroke) after the follow up. The analysis showed that most people consumed between one to five eggs per week and participants with a higher egg intake also had a higher BMI and were less likely to be under statin treatment. In this study, an increase of one egg per day was not associated with any CVD risk. In fact, in an updated meta-analysis that included multiple cohorts from the US, Europe, and Asia, moderate egg consumption was associated with no risk of developing CVD overall and lower risk in Asian populations [39].

Although is well stablished that dyslipidemia is a major CVD risk factor, altered blood lipids can also increase the risk for developing other chronic not transmittable diseases, such as type 2 diabetes mellitus (T2DM) [40,41]. In that matter, Drouin-Chartier et al. [42] also evaluated the association between egg intake and the risk of developing T2DM. Across the three large cohorts, higher egg intake was associated with lower prevalence of hypercholesterolemia, but 1 egg/day increase was associated with a 14% higher T2DM risk.

However, in a random-effects meta-analysis of 16 prospective cohort studies (6 American, 8 European, and 2 Asian), no significant association between egg consumption and T2DM risk was found. It is noteworthy that there was a significant geographical heterogenicity in the results. In fact, among US studies, for each egg per day, T2DM risk increased by 18% but this positive association was not found in the studies conducted in Europe or Asia. This is hypothesized to be due to the relationship between eggs and other foods consumed with them in different regions. For example, in the three large cohorts mentioned before, egg intake was positively associated with total calories consumed, red meat, bacon and processed red meats, refined grains, potatoes, full-fat milk, and coffee, which are reflective of the Western diet. As previously stated, there is a large body of evidence that links this dietary pattern to obesity and other chronic diseases, including CVD and T2DM [42,43], so that association may be due for egg consumption pattern and not eggs or cholesterol alone. This suggest that further studies that control for all possible dietary cofounders or that investigate the reasons for such geographical differences are needed. The main meta-analysis showing the lack of correlation between dietary cholesterol and blood cholesterol is presented in Table 1.

Table 1. Results from recent epidemiological studies and meta-analysis showing the lack of correlation between dietary cholesterol and blood cholesterol.

Population/Number of Studies	Association Assessed	Main Result	Reference (Year)
177,555 adults from PURE, TRASCEND and ONTARGET studies	Egg consumption with blood lipids and CVD	Higher egg intake is not associated with TC, LDL, TG, HDL, total mortality, or CVD.	[23] (2020)
8095 hypertense adults from the China Health and Nutrition Survey	Cholesterol intake from eggs and other sources and mortality	Cholesterol from eggs but not other sources is associated with lower mortality.	[28] (2020)
8358 Chinese adults	Dietary cholesterol and dyslipidemia	Cholesterol intake is associated with lower plasma TG and higher HDL-cholesterol in women, but not men. Cholesterol from eggs is associated with lower risk of dyslipidemia.	[34] (2022)
Three large cohort studies: NHS (1980–2012), NHS II (1991–2017) and HPFS (1986–2016). 16 prospective cohort studies (6 American, 8 European, and 2 Asian)	Egg intake and CVD risk	An increase of one egg per day is not associated with any CVD risk. Egg intake is associated with lower CVD risk in Asian populations.	[39] (2017)
39 prospective cohort studies from North America, Europe, and Asia	Egg consumption and the risk of CVD, CHD, and stroke	Consumption of six eggs per week has an inverse association with CVD events (but not stroke), when compared to no intake. No association is found for stroke.	[35] (2021)
40 studies with participants without diagnosed CVD		No association between dietary cholesterol and coronary artery disease (CAD), ischemic stroke, or hemorrhagic stroke. Dietary cholesterol increases total blood cholesterol, without affecting LDL/HDL ratio.	[36] (2015)
NHS (1980–2012), NHS II (1991–2017) and HPFS (1986–2016). 16 prospective cohort studies (6 American, 8 European, and 2 Asian)	Eegg intake and the risk of developing T2DM	Higher egg intake is associated with lower prevalence of hypercholesterolemia.	[42] (2020)

CVD: cardiovascular disease; CHD: coronary heart disease; TC: total cholesterol; LDL: low-density lipoprotein; HDL: high-density lipoprotein; TG: Triglyceride; NHS: the Nurses' Health Study; HPFS: the Health Professional Follow up Study.

4. Clinical Interventions. Effects of Dietary Cholesterol on Plasma Lipids and Lipoprotein Subfractions

Clinical interventions across the life cycle have been conducted in children, young adults, elderly individuals [10–14], and those with conditions that put them at risk for chronic disease including obesity [15,16] and metabolic syndrome [17,18] to evaluate the effects of cholesterol challenges on plasma lipid concentrations and lipoprotein metabolism.

Clinical interventions have also been conducted in people with type-2 diabetes [19,20]. The dietary cholesterol challenges have varied from adding between 200 mg/day of dietary cholesterol to 800 mg/day [10–20]. In all these interventions, eggs have been used as the vehicle for dietary cholesterol. In most of these studies, increases in HDL cholesterol have been observed [10–17,19]. In some studies, increases in LDL cholesterol have been reported following the cholesterol challenge [10,11,13,44]. However, the LDL/HDL ratio a very well-known biomarker for heart disease risk has either been maintained or has been decreased [10–20].

In a study following a protocol of 14 weeks in which young individuals consumed zero eggs for 2 weeks as the baseline, followed by one egg for 4 weeks, two eggs for 4 weeks and three eggs for the last 4 weeks, LDL cholesterol was lower or similar to baseline values during the whole intervention [12]. In contrast HDL cholesterol was higher than baseline all through the study from the intake of one to three eggs [12]. Other studies involving weight loss interventions [17], older people [14], overweight/obese individuals [15,16], and people with metabolic syndrome [17,18] that were challenged with two to three eggs per day (360–540 additional mg of dietary cholesterol) for extended periods of time, no increases in LDL cholesterol were observed, indicating that some individuals do have the ability to maintain LDL plasma cholesterol concentrations independent of the dietary cholesterol challenge.

Substituting high carbohydrate breakfast with eggs resulted in a similar lowering in plasma LDL cholesterol compared to baseline in 30 subjects following a randomized crossover study in which each breakfast was followed for 4 weeks [45]. However, lipoprotein subfractions were not different between dietary treatments [45].

Results from two randomized controlled studies conclude that the lack of effect of consuming 75 g or 150 g of eggs when compared to a no-egg diet on plasma LDL cholesterol was due to poor absorption of cholesterol from eggs [46]. Authors came to this conclusion by measuring the total cholesterol areas under the curve (AUC) between 0 and 10 hours [46]. Other studies have also shown no changes in LDL cholesterol and increases in HDL cholesterol after 56 individuals consumed one egg/day for 12 weeks [47] or in 12 sedentary young adults undergoing endurance exercise for 8 weeks [48]. In contrast, another study where subjects consumed either zero eggs (n = 34 per group) or two eggs per day five times a week for 14 weeks, no changes were observed in HDL cholesterol but also LDL cholesterol was not affected [49].

Regarding lipoprotein metabolism, dietary cholesterol leads to the formation of the large LDL particles that are known to be less atherogenic [32] and reduces the concentration of small LDL [10,16], which has been recognized as a highly atherogenic particle for its ability to become oxidized, penetrate the arterial wall and initiate the atherosclerosis process [50]. In the study conducted in Latino children who have higher concentrations of the small LDL, a shift toward larger LDL diameter and less of the small LDL subfractions was observed [10]. Increases in the large LDL after intake of 640 mg/d of cholesterol (3 eggs) resulted in higher concentrations of large LDL compared to 0 mg of additional cholesterol in young individuals [51,52], elderly subjects [53] and in obese individuals [54]. Increases in large LDL have also been noticed in young individual consuming up to 640 mg cholesterol per day [12]. Women and men who participated in a randomized clinical trial in which they consumed either three eggs or the equivalent amount of egg substitute, presented increases in the large buoyant less atherogenic LDL during the whole egg period [55].

Small HDL size has been correlated atherogenic dyslipidemia [56] highlighting the importance of increases in HDL size and in the number of large HDL particles following egg consumption. These increases in large HDL have been observed after a cholesterol challenge in the elderly [56]. Comparable results have been shown in young and healthy individuals [51,52] and in those with metabolic syndrome [57]. Interestingly, increases in the large HDL as well as compositional changes resulted in an HDL particle with increased cholesterol efflux capacity [22,58], which is also a better transporter for the carotenoids lutein and zeaxanthin in plasma [56,57], A summary of studies showing changes in lipoprotein subfractions and size following a cholesterol challenge is presented in Table 2.

Table 2. Beneficial Modifications in LDL and HDL size and subfractions due to dietary cholesterol.

	Dietary Cholesterol Intake and Population	Changes	Reference (Year)
LDL Diameter Compared to added 0 mg/d cholesterol	510 mg/day for 4 weeks in children	LDL diameter was larger	[10] (2005)
Large LDL compared to 0 added mg/cholesterol	640 mg/day for 4 weeks in elderly people	Higher concentrations of large LDL	[53] (2006)
Large LDL compared to 0 mg of added dietary cholesterol	210, 425, and 640 mg/day in young individuals for 4 weeks each	Higher concentrations of large LDL	[51] (2017)
Large LDL Compared to an oatmeal breakfast	640 mg/day for 4 weeks in young population	Higher concentrations of large LDL	[52] (2018)
Large LDL: Compared to 0 mg of added dietary cholesterol	640 mg/day for 4 weeks in an overweight/obese population	Higher concentrations of large LDL	[54] (2010)
Small LDL: Compared to 0 mg of added dietary cholesterol	210, 425, and 640 mg/day in young individuals for 4 weeks each	Lower concentrations of small LDL	[51] (2017)
Small LDL: Compared to 0 mg of dietary cholesterol	640 mg/day for 4 weeks in an overweight/obese population	Lower concentrations of small LDL	[54] (2010)
HDL Diameter: Compared to 0 mg of added dietary cholesterol	640 mg/day for 4 weeks in elderly people	Larger HDL diameter	[53] (2006)
Large HDL: Compared to 0 mg of added dietary cholesterol	210, 425, and 640 mg/day in young individuals for 4 weeks each	Higher concentrations of large HDL	[51] (2005)
Large HDL: Compared to an oatmeal breakfast	640 mg/day for 4 weeks in young population	Higher concentrations of large HDL	[52] (2018)
Large HDL: Compared to 0 mg of added dietary cholesterol	640 mg/day for 4 weeks in an overweight/obese population	Higher concentrations of large HDL	[54] (2010)

5. Mechanisms to Manage Dietary Cholesterol

The epidemiological data and the clinical interventions presented above clearly indicate the lack of correlation between dietary and blood cholesterol. These observations also suggest that the body has specific mechanisms to manage excesses of dietary cholesterol. The proposed mechanisms including decreased absorption or synthesis suppression were recognized early on [59] and they explain why dietary cholesterol cannot be extrapolated directly to plasma cholesterol. The absorption of dietary cholesterol varies according to each individual and it comprises cholesterol from food, biliary cholesterol, and to a certain extent intestinal epithelial sloughing [60]. The transport of cholesterol to the liver involves several steps including solubilization in micelles, transport to the enterocytes, incorporation into chylomicrons, and transport through lymph and blood vessels to the liver and other tissues. The absorption of cholesterol varies from 29 to 80% with an average of 60% [61]. Niemann-Pick C1-Like-1 (NPCL1), a receptor localized in the intestinal cells, has a major role in cholesterol absorption [61].

In terms of synthesis, intracellular cholesterol exerts a negative feedback regulation on HMG-CoA, the rate limiting enzyme of cholesterol synthesis. Excess cholesterol in cells leads to the suppression of HMG-CoA reductase activity [62]. HMG-CoA reductase can be regulated by changes in phosphorylation but most importantly by changes in transcription [63]. The transcriptional regulation involves the binding of sterol response element binding protein (SREBP) to sterol response elements located in the 5′ region of the CoA reductase gene [64]. This is also an important mechanism by which individuals maintain plasma cholesterol levels and can consume high amounts of dietary cholesterol and do not increase plasma concentrations. One clear case study that exemplifies the compensatory mechanisms to maintain cholesterol homeostasis is the case of the 88-year-old individual who consumed 25 eggs per day (about 4500 mg/day) and had no heart problems and normal plasma cholesterol levels [65]. The compensatory mechanisms were a marked reduction in cholesterol absorption, increased synthesis of bile acids, and reduced cholesterol synthesis [65].

Therefore, the handling of dietary cholesterol by the body can be explained by decreased absorption and down-regulation of synthesis. Figure 1 depicts the mechanisms by which the body handles dietary cholesterol and maintains plasma cholesterol homeostasis.

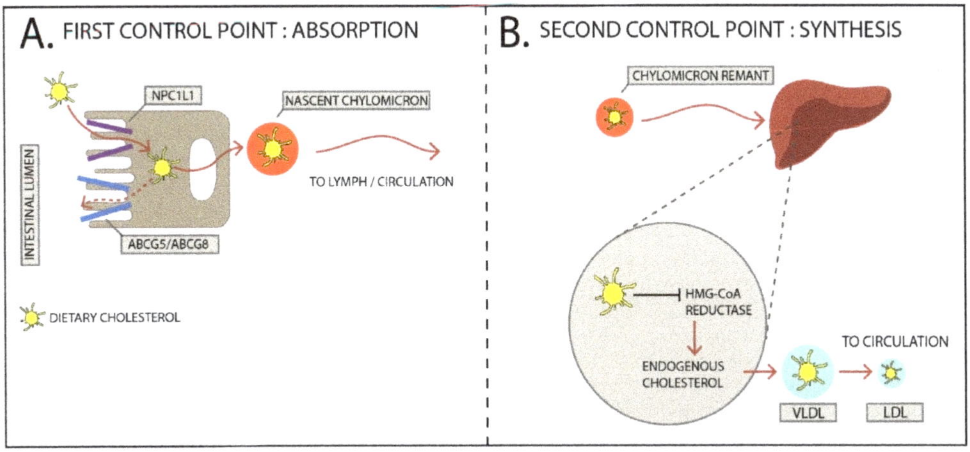

Figure 1. Mechanisms of how dietary cholesterol affects cholesterol metabolism. (**A**) Dietary cholesterol enters the enterocyte via NPC1L1 after being released from the micelle. However, some cholesterol is effluxed back to the intestinal lumen via ABCG5 and ABCG8 transporters explaining why only a percentage of the cholesterol consumed in the diet reaches the bloodstream. The remaining cholesterol gets packed into nascent chylomicrons, which enter the lymph and then the systemic circulation. (**B**) After losing most of its triglycerides, the cholesterol-loaded chylomicron remnant is removed by the liver. In the liver, free cholesterol inhibits HMG-CoA reductase, the rate limiting enzyme for endogenous cholesterol synthesis, Thus, if more cholesterol is consumed, less will be synthesized by the hepatocytes. VLDL: very-low-density lipoprotein; LDL: low-density lipoprotein; NPC1L1: Polytopic Niemann-Pick C1-like 1; ABCG5/ABCG8: ATP-binding cassette transporters G5/G8; HMG-CoA: 3-hydroxy-3-methyl glutaryl coenzyme A.

6. Conclusions

We confirm from the review of the literature on epidemiological data, meta-analysis, and clinical interventions where dietary cholesterol challenges were utilized that there is not a direct correlation between cholesterol intake and blood cholesterol. This lack of correlation is mainly due to the compensatory mechanisms exerted by the organism to manage excess dietary cholesterol, including decreases in cholesterol absorption and down-regulation of cholesterol synthesis. A great number of epidemiological studies and meta-analysis indicate that dietary cholesterol is not associated with CVD risk nor with elevated plasma cholesterol concentrations. Clinical interventions in the last 20 years demonstrate that challenges with dietary cholesterol do not increase the biomarkers associated with heart disease risk. Further, in the specific circumstances where eggs are the source of dietary cholesterol, an improvement in dyslipidemias is observed due to the formation of less atherogenic lipoproteins and changes in HDL associated with a more efficient reverse cholesterol transport. However, if the cholesterol sources are consumed with saturated and trans fats, as happens in the Western diet pattern, increases in plasma cholesterol may be observed. The most recent epidemiological data and clinical interventions for the most part continue to support the USDA 2015 dietary guidelines that removed the upper limit of dietary cholesterol.

Author Contributions: M.L.F. and A.G.M. equally contributed to the content of this review. Figure 1 was made by A.G.M. All authors have read and agreed to the published version of the manuscript.

Funding: This research received no external funding.

Institutional Review Board Statement: Not applicable.

Informed Consent Statement: Not applicable.

Data Availability Statement: Not applicable.

Conflicts of Interest: M.L.F. and A.G.M. declare no conflict of interest.

References

1. Committee on Nutrition. *American Heart Association. Diet and Heart Disease*; American Heart Association: Dallas, TX, USA, 1968.
2. U.S. Department of Health and Human Services and U.S. Department of Agriculture. *2015–2020 Dietary Guidelines for Americans*, 8th ed. 2015. Available online: https://health.gov/our-work/food-nutrition/previous-dietary-guidelines/2015 (accessed on 8 May 2022).
3. Vergnes, L.; Phan, J.; Strauss, M.; Tafuri, S.; Reue, K. Cholesterol and cholate components of an atherogenic diet induce distinct stages of hepatic inflammatory gene expression. *J. Biol. Chem.* **2003**, *278*, 42774–42784. [CrossRef] [PubMed]
4. Lee, Y.T.; Lin, H.Y.; Chan, Y.W.; Li, K.H.; To, O.T.; Yan, B.P.; Liu, T.; Li, G.; Wong, W.T.; Keung, W.; et al. Mouse models of atherosclerosis: A historical perspective and recent advances. *Lipids Health Dis.* **2017**, *16*, 12. [CrossRef] [PubMed]
5. Ishibashi, S.; Goldstein, J.L.; Brown, M.S.; Herz, J.; Burns, D.K. Massive xanthomatosis and atherosclerosis in cholesterol-fed low density lipoprotein receptor-negative mice. *J. Clin. Investig.* **1994**, *93*, 1885–1893. [CrossRef] [PubMed]
6. Dawber, T.R.; Nickerson, R.J.; Brand, F.N.; Pool, J. Eggs, serum cholesterol, and coronary heart disease. *Am. J. Clin. Nutr.* **1982**, *36*, 617–625. [CrossRef]
7. Hu, F.; Stampfer, M.J.; Rimm, E.; Manson, J.E.; Ascherio, A.; Golditz, G.A.; Rosner, B.A.; Spiegelman, D.; Speizer, F.E.; Sacks, F.M.; et al. A prospective study of egg consumption and risk of cardiovascular disease in men and women. *JAMA* **1999**, *281*, 1387–1894. [CrossRef] [PubMed]
8. Song, W.O.; Kerver, J.M. Nutritional Contribution of Eggs to American Diets. *J. Am. Coll. Nutr.* **2000**, *19*, 556S–562S. [CrossRef] [PubMed]
9. Magriplis, E.; Mitsopoulou, A.-V.; Karageorgou, D.; Bakogianni, I.; Dimakopoulos, I.; Micha, R.; Michas, G.; Chourdakis, M.; Chrousos, G.; Roma, E.; et al. Frequency and Quantity of Egg Intake Is Not Associated with Dyslipidemia: The Hellenic National Nutrition and Health Survey (HNNHS). *Nutrients* **2019**, *11*, 1105. [CrossRef]
10. Ballesteros, M.N.; Cabrera, R.M.; Saucedo, M.S.; Fernandez, M.L. Dietary cholesterol does not increase biomarkers for chronic disease in a pediatric population at risk from Northern Mexico. *Am. J. Clin. Nutr.* **2004**, *80*, 855–861. [CrossRef]
11. Missimer, A.; DiMarco, D.M.; Andersen, C.J.; Murillo, A.G.; Fernandez, M.L. Consuming 2 eggs per day, as compared to an oatmeal breakfast, decreases plasma ghrelin and maintains the LDL/HDL ratio. *Nutrients* **2017**, *9*, 89. [CrossRef]
12. DiMarco, D.M.; Missimer, A.; Murillo, A.G.; Lemos, B.; Malysheva, O.V.; Caudill, M.A.; Blesso, C.N.; Fernandez, M.L. Intake of up to 3 eggs per day increases HDL cholesterol and plasma choline while plasma trimethylamine-N-Oxide is unchanged in a healthy population. *Lipids* **2017**, *52*, 255–263. [CrossRef]
13. Greene, C.M.; Zern, T.; Wood, R.J.; Aggarwal, D.; Sharman, M.; Volek, J.S.; Fernandez, M.L. Maintenance of the LDL choles-terol: HDL cholesterol ratio in an elderly population given a dietary cholesterol challenge. *J. Nutr.* **2005**, *135*, 2793–2798. [CrossRef] [PubMed]
14. Vshwanathan, R.; Goodrwo-Kotyla, E.F.; Wooten, B.R.; Wilson, T.A.; Nicolosi, R.J. Consumption of 2 and 4 egg yolks/d for 5 wk increases macular pigment concentrations in older adults with low macular pigment taking cholesterol-lowering statins. *Am J. Clin. Nutr.* **2009**, *90*, 1272–1279. [CrossRef] [PubMed]
15. Wal, J.V.; Gupta, A.; Khosla, P.; Dhurandhar, N.V. Egg breakfast enhances weight loss. *Int. J. Obes.* **2008**, *32*, 1545–1551. [CrossRef] [PubMed]
16. Mutungi, G.; Ratliff, J.; Puglisi, M.; Torres-Gonzalez, M.; Vaishnav, U.; Leite, J.; Quann, E.; Volek, J.S.; Fernandez, M.L. Dietary cholesterol from eggs increas-es HDL cholesterol in overweight men consuming a carbohydrate restricted diet. *J. Nutr.* **2008**, *138*, 272–276. [CrossRef] [PubMed]
17. Blesso, C.N.; Andersen, C.J.; Barona, J.; Volek, J.; Fernandez, M.L. Whole egg consumption improves lipoprotein profiles and insulin sensitivity in individuals with metabolic syndrome. *Metabolism* **2013**, *62*, 400–410. [CrossRef]
18. Dibella, M.; Thomas, M.S.; Al-Yousef, H.; Millar, C.; Blesso, C.N.; Malisheva, O. Choline intake as supplement or as a com-ponent of eggs increases plasma choline and reduces interleukin-6 without modifying plasma cholesterol in participants with metabolic syndrome. *Nutrients* **2020**, *12*, 3120. [CrossRef]
19. Ballesteros, M.N.; Valenzuela, F.; Robles, A.E.; Artalejo, E.; Aguilar, D.; Andersen, C.J.; Valdez, H.; Fernandez, M.L. One Egg per Day Improves Inflammation when Compared to an Oatmeal-Based Breakfast without Increasing Other Cardiometabolic Risk Factors in Diabetic Patients. *Nutrients* **2015**, *7*, 3449–3463. [CrossRef] [PubMed]

20. Pearce, K.L.; Clifton, P.M.; Noakes, M. Egg consumption as part of an energy-restricted high-protein diet improves blood lipid and blood glucose profiles in individuals with type 2 diabetes. *Brit. J. Nutr.* **2011**, *105*, 584–592. [CrossRef]
21. Herron, K.L.; Vega-Lopez, S.; Conde, K.; Ramjiganesh, T.; Roy, S.; Shachter, N.S.; Fernandez, M.L. Pre-Menopausal Women, Classified as Hypo-or Hyper-Responders, do not Alter their LDL/HDL Ratio Following a High Dietary Cholesterol Challenge. *J. Am. Coll. Nutr.* **2002**, *21*, 250–258. [CrossRef]
22. Andersen, C.J.; Blesso, C.N.; Lee, J.Y.; Park, Y.K.; Fernandez, M.L. Egg consumption modulates HDL composition and in-creases the cholesterol accepting capacity of serum in metabolic syndrome. *Lipids* **2013**, *48*, 557–567. [CrossRef]
23. Dehghan, M.; Mente, A.; Rangarajan, S.; Mohan, V.; Lear, S.; Swaminathan, S.; Wielgosz, A.; Seron, P.; Avezum, A.; Lopez-Jaramillo, P.; et al. Association of egg intake with blood lipids, cardiovascular disease, and mortality in 177,000 people in 50 countries. *Am. J. Clin. Nutr.* **2020**, *111*, 795–803. [CrossRef] [PubMed]
24. Carson, J.A.S.; Lichtenstein, A.H.; Anderson, C.A.M.; Appel, L.J.; Kris-Etherton, P.M.; Meyer, K.A.; Van Horn, L. Dietary cholesterol and cardiovascular risk: A science advisory from the American heart association. *Circulation* **2020**, *141*, E39–E53. [CrossRef] [PubMed]
25. Zhuang, P.; Wu, F.; Mao, L.; Zhu, F.; Zhang, Y.; Chen, X.; Jiao, J.; Zhang, Y. Egg and cholesterol consumption and mortality from cardio-vascular and different causes in the United States: A population-based cohort study. *PLoS Med.* **2021**, *18*, e1003508. [CrossRef] [PubMed]
26. Zhong, V.W.; Van Horn, L.; Cornelis, M.C.; Wilkins, J.T.; Ning, H.; Carnethon, M.; Greenland, P.; Mentz, R.J.; Tucker, K.L.; Zhao, L.; et al. Associations of Dietary Cholesterol or Egg Consumption With Incident Cardiovascular Disease and Mortality. *JAMA* **2019**, *321*, 1081–1095. [CrossRef]
27. Fuchs, F.D.; Whelton, P.K. High Blood Pressure and Cardiovascular Disease. *Hypertension* **2020**, *75*, 285–292. [CrossRef]
28. Wu, F.; Zhuang, P.; Zhang, Y.; Zhan, C.; Zhang, Y.; Jiao, J. Egg and Dietary Cholesterol Consumption and Mortality Among Hypertensive Patients: Results From a Population-Based Nationwide Study. *Front. Nutr.* **2021**, *8*, 1–10. [CrossRef]
29. Sacks, F.M.; Lichtenstein, A.H.; Wu, J.H.Y.; Appel, L.J.; Creager, M.A.; Kris-Etherton, P.M.; Miller, M.; Rimm, E.B.; Rudel, L.L.; Robinson, J.G.; et al. Dietary fats and cardio-vascular disease: A presidential advisory from the American Heart Association. *Circulation* **2017**, *136*, e1–e23. [CrossRef]
30. Andersen, C.J. Bioactive Egg Components and Inflammation. *Nutrients* **2015**, *7*, 7889–7913. [CrossRef]
31. Hezaveh, Z.S.; Sikaroudi, M.K.; Vafa, M.; Clayton, Z.S.; Soltani, S. Effect of egg consumption on inflammatory markers: A systematic review and meta-analysis of randomized controlled clinical trials. *J. Sci. Food. Agric.* **2019**, *99*, 6663–6670. [CrossRef]
32. Blesso, C.N.; Andersen, C.J.; Bolling, B.W.; Fernandez, M.L. Egg intake improves carotenoid status by increasing plasma HDL cholesterol in adults with metabolic syndrome. *Food Funct.* **2012**, *4*, 213–221. [CrossRef]
33. Nimalaratne, C.; Wu, J. Hen Egg as an Antioxidant Food Commodity: A Review. *Nutrients* **2015**, *7*, 8274–8293. [CrossRef] [PubMed]
34. Pan, J.; Han, W.; Jiang, Y.; Wu, J.; Zhou, X. Association of dietary cholesterol and dyslipidemia in Chinese health examinees. *J. Heal. Popul. Nutr.* **2022**, *41*, 1–7. [CrossRef] [PubMed]
35. Godos, J.; Micek, A.; Brzostek, T.; Toledo, E.; Iacoviello, L.; Astrup, A.; Franco, O.H.; Galvano, F.; Martinez-Gonzalez, M.A.; Grosso, G. Egg consumption and cardiovascular risk: A dose–response meta-analysis of prospective cohort studies. *Eur. J. Nutr.* **2020**, *60*, 1833–1862. [CrossRef] [PubMed]
36. Hansel, B.; Giral, P. Cholestérol alimentaire et morbi/mortalité cardiovasculaire. *OCL* **2015**, *22*, D202. [CrossRef]
37. Millán, J.; Pintó, X.; Muñoz, A.; Zúñiga, M.; Rubiés-Prat, J.; Pallardo, L.F.; Masana, L.; Mangas, A.; Hernández-Mijares, A.; González-Santos, P.; et al. Lipoprotein ratios: Physiological significance and clinical usefulness in cardiovascular prevention. *Vasc. Heal. Risk Manag.* **2009**, *5*, 757. [CrossRef]
38. Torres, N.; Guevara-Cruz, M.; Velázquez-Villegas, L.A.; Tovar, A.R. Nutrition and Atherosclerosis. *Arch. Med. Res.* **2015**, *46*, 408–426. [CrossRef]
39. Drouin-Chartier, J.-P.; Chen, S.; Li, Y.; Schwab, A.L.; Stampfer, M.J.; Sacks, F.M.; Rosner, B.; Willett, W.C.; Hu, F.B.; Bhupathiraju, S.N. Egg consumption and risk of cardiovascular disease: Three large prospective US cohort studies, systematic review, and updated meta-analysis. *BMJ* **2020**, *368*, m513. [CrossRef]
40. Ahmmed, S.; Das Shuvo, S.; Paul, D.K.; Karim, M.R.; Kamruzzaman, M.; Mahmud, N.; Ferdaus, J.; Elahi, T. Prevalence of dyslipidemia and associated risk factors among newly diagnosed Type-2 Diabetes Mellitus (T2DM) patients in Kushtia, Bangladesh. *PLOS Glob. Public Heal.* **2021**, *1*, e0000003. [CrossRef]
41. Peng, J.; Zhao, F.; Yang, X.; Pan, X.; Xin, J.; Wu, M.; Peng, Y.G. Association between dyslipidemia and risk of type 2 diabetes mellitus in middle-aged and older Chinese adults: A secondary analysis of a nationwide cohort. *BMJ Open* **2021**, *11*, e042821. [CrossRef]
42. Drouin-Chartier, J.-P.; Schwab, A.L.; Chen, S.; Li, Y.; Sacks, F.M.; Rosner, B.; E Manson, J.; Willett, W.C.; Stampfer, M.J.; Hu, F.B.; et al. Egg consumption and risk of type 2 diabetes: Findings from 3 large US cohort studies of men and women and a systematic review and meta-analysis of prospective cohort studies. *Am. J. Clin. Nutr.* **2020**, *112*, 619–630. [CrossRef]
43. Kopp, W. How Western Diet And Lifestyle Drive The Pandemic Of Obesity And Civilization Diseases. *Diabetes Metab. Syndr. Obes.* **2019**, *12*, 2221–2236. [CrossRef] [PubMed]

44. Lemos, B.S.; Medina-Vera, I.; Blesso, C.N.; Fernandez, M.L. Intake of 3 Eggs per Day When Compared to a Choline Bitartrate Supplement, Downregulates Cholesterol Synthesis without Changing the LDL/HDL Ratio. *Nutrients* **2018**, *10*, 258. [CrossRef] [PubMed]
45. Maki, K.C.; Palacios, O.M.; Kramer, M.W.; Trivedi, R.; Dicklin, M.R.; Wilcox, M.L.; Maki, C.E. Effects of substituting eggs for high-carbohydrate breakfast foods on the cardiometabolic risk-factor profile in adults at risk for type 2 diabetes mellitus. *Eur. J. Clin. Nutr.* **2020**, *74*, 784–795. [CrossRef] [PubMed]
46. Kim, J.-E.; Campbell, W.W. Total cholesterol concentration in men and women: Results from 2 randomized controlled crossover studies. *Nutrients* **2018**, *10*, 1272. [CrossRef] [PubMed]
47. Mayurasakorn, K.; Srisura, W.; Sitphahul, P.; Hongto, P.-O. High-density lipoprotein cholesterol changes after continuous egg consumption in healthy adults. *J. Med. Assoc. Thail.* **2008**, *91*, 400.
48. Vislocky, L.M.; Pikosky, M.A.; Rubin, K.H.; Vega-López, S.; Gaine, P.C.; Martin, W.F.; Zern, T.L.; Lofgren, I.E.; Fernandez, M.L.; Rodriguez, N.R. Habitual consumption of eggs does not alter the beneficial effects of endurance training on plasma lipids and lipoprotein metabolism in untrained men and women. *J. Nutr. Biochem.* **2009**, *20*, 26–34. [CrossRef]
49. Rueda, J.M.; Khosla, P. Impact of Breakfasts (with or without Eggs) on Body Weight Regulation and blood Lipids in Uni-versity Students over a 14-Week Semester. *Nutrients* **2013**, *5*, 5097–5113. [CrossRef]
50. Khatana, C.; Saini, N.K.; Chakrabarti, S.; Saini, V.; Sharma, A.; Saini, R.V.; Saini, A.K. Mechanistic Insights into the Oxidized Low-Density Lipoprotein-Induced Atherosclerosis. *Oxid. Med. Cell. Longev.* **2020**, *2020*, 1–14. [CrossRef]
51. DiMarco, D.M.; Norris, G.H.; Millar, C.L.; Blesso, C.N.; Fernandez, M.L. Intake of up to 3 Eggs per Day Is Associated with Changes in HDL Function and Increased Plasma Antioxidants in Healthy, Young Adults. *J. Nutr.* **2017**, *147*, 323–329. [CrossRef]
52. Missimer, A.; Fernandez, M.L.; DiMarco, D.M.; Norris, G.H.; Blesso, C.N.; Murillo, A.G.; Vergara-Jimenez, M.; Lemos, B.S.; Medina-Vera, I.; Malysheva, O.V.; et al. Compared to an oatmeal breakfast, two eggs/day increased plasma carotenoids and choline without increasing trimethyl amine N-oxide concentrations. *J. Am. Coll Nutr.* **2018**, *37*, 140–148. [CrossRef]
53. Greene, C.M.; Waters, D.; Clark, R.M.; Contois, J.H.; Fernandez, M.L. Plasma LDL and HDL characteristics and carotenoid content are positively influenced by egg consumption in an elderly population1. *Nutr. Metab.* **2006**, *3*, 6. [CrossRef] [PubMed]
54. Mutungi, G.; Waters, D.; Ratliff, J.; Puglisi, M.; Clark, R.M.; Volek, J.S.; Fernandez, M.L. Eggs distinctly modulate plasma carotenoid and lipoprotein subclasses in adult men following a carbohydrate-restricted diet. *J. Nutr. Biochem.* **2010**, *21*, 261–267. [CrossRef] [PubMed]
55. Herron, K.L.; Lofgren, I.E.; Sharman, M.; Volek, J.S.; Fernandez, M.L. High intake of cholesterol results in less atherogenic low-density lipoprotein particles in men and women independent of response classification. *Metabolism* **2004**, *53*, 823–830. [CrossRef] [PubMed]
56. Pascot, A.; Lemieux, I.; Prud'homme, D.; Tremblay, A.; Nadeau, A.; Couillard, C.; Després, J.P. Reduced HDL particle size as an additional feature of the atherogenic dyslipidemia of abnormal obesity. *J. Lipid Res.* **2001**, *42*, 2007–2014. [CrossRef]
57. Blesso, C.N.; Andersen, C.J.; Bolling, B.; Fernandez, M.L. Egg intake improves carotenoid status by increasing HDL cholesterol in adults with metabolic syndrome. *Food Funct.* **2013**, *4*, 213–221.
58. Sawrey-Kubicek, L.; Zhu, C.; Bardagjy, A.S.; Rhodes, C.H.; Sacchi, R.; Randolph, J.M.; Steinberg, F.M.; Zivkovic, A.M. Whole egg consumption compared with yolk-free egg increases the cholesterol efflux capacity of high-density lipoproteins in overweight, postmenopausal women. *Am. J. Clin. Nutr.* **2019**, *110*, 617–627. [CrossRef]
59. McNamara, D.J.; Kolb, R.; Parker, T.S.; Batwin, H.; Samues, P.; Brown, C.D.; Ahrens, E.H., Jr. Heterogeneity of cholesterol homeostasis in man. Responses to changes in dietary fat quality and cholesterol quantity. *J. Clin. Investig.* **1987**, *79*, 1729–1739. [CrossRef]
60. Betters, J.L.; Yu, L. NPC1L1 and cholesterol transport. *Febs Lett.* **2010**, *584*, 2740–2747. [CrossRef]
61. Bosner, M.S.; Lange, L.G.; Stenson, W.F.; Ostlund, R.E., Jr. Percent cholesterol absorption in normal women and men quantified with dual stable isotopic tracers and negative ion mass spectrometry. *J. Lipid Res.* **1999**, *40*, 302–308. [CrossRef]
62. Altmann, S.W.; Davis, H.R., Jr.; Zhu, L.-J.; Yao, X.; Hoos, L.M.; Tetzloff, G.; Iyer, S.P.N.; Maguire, M.; Golovko, A.; Zeng, M.; et al. Niemann-Pick C1 Like 1 Protein Is Critical for Intestinal Cholesterol Absorption. *Science* **2004**, *303*, 1201–1204. [CrossRef]
63. Chambers, C.M.; Ness, G.C. Dietary cholesterol regulates hepatic 3-hydroxy-3-methylglutaryl gene expression in rats primarily at the level of translation. *Arch. Biochem. Biophys.* **1998**, *354*, 317–322. [CrossRef] [PubMed]
64. Brown, M.S.; Goldstein, J.L. The SREBP pathway: Regulation of cholesterol metabolism by proteolysis of a mem-brane-bound transcription factor. *Cell* **1997**, *89*, 331–340. [CrossRef]
65. Kern, F. Normal Plasma Cholesterol in an 88-Year-Old Man Who Eats 25 Eggs a Day. *N. Engl. J. Med.* **1991**, *324*, 896–899. [CrossRef] [PubMed]

nutrients

MDPI

Review

Investigating microRNAs to Explain the Link between Cholesterol Metabolism and NAFLD in Humans: A Systematic Review

Maurice C. J. M. Konings, Sabine Baumgartner, Ronald P. Mensink and Jogchum Plat *

Department of Nutrition and Movement Sciences, NUTRIM School of Nutrition and Translational Research in Metabolism, Maastricht University Medical Center, 6200 MD Maastricht, The Netherlands
* Correspondence: j.plat@maastrichtuniversity.nl; Tel.: +31-43-3881309

Abstract: Non-Alcoholic Fatty Liver Disease (NAFLD) is characterized by hepatic free cholesterol accumulation. In addition, microRNAs (miRNAs) might be involved in NAFLD development. Therefore, we systematically reviewed the literature to examine the link between miRNAs and cholesterol metabolism in NAFLD. Nineteen studies were retrieved by a systematic search in September 2022. From these papers, we evaluated associations between 13 miRNAs with NAFLD and cholesterol metabolism. Additionally, their diagnostic potential was examined. Four miRNAs (miR122, 34a, 132 and 21) were associated with cholesterol metabolism and markers for NAFLD. MiR122 was upregulated in serum of NAFLD patients, increased with disease severity and correlated with HDL-C, TAG, VLDL-C, AST, ALT, ALP, lobular inflammation, hepatocellular ballooning and NAFLD score. Serum and hepatic levels also correlated. Serum and hepatic miR34a levels were increased in NAFLD, and correlated with VLDL-C and TAG. Serum miR379 was also higher in NAFLD, especially in early stages, while miR21 gave ambiguous results. The diagnostic properties of these miRNAs were comparable to those of existing biomarkers. However, serum miR122 levels appeared to be elevated before increases in ALT and AST were evident. In conclusion, miR122, miR34a, miR21 and miR132 may play a role in the development of NAFLD via effects on cholesterol metabolism. Furthermore, it needs to be explored if miRNAs 122, 34a and 379 could be used as part of a panel in addition to established biomarkers in early detection of NAFLD.

Keywords: microRNA; cholesterol; human; NAFLD; NASH; lipoproteins

Citation: Konings, M.C.J.M.; Baumgartner, S.; Mensink, R.P.; Plat, J. Investigating microRNAs to Explain the Link between Cholesterol Metabolism and NAFLD in Humans: A Systematic Review. *Nutrients* **2022**, *14*, 4946. https://doi.org/10.3390/nu14234946

Academic Editor: Frans Stellaard

Received: 26 October 2022
Accepted: 18 November 2022
Published: 22 November 2022

1. Introduction

Due to unhealthy lifestyles, non-alcoholic fatty liver disease (NAFLD)—the hepatic consequence or even cause of metabolic syndrome (MetS)—has evolved into a serious health threat [1–3]. The disease progression of NAFLD is a process of impairment and deterioration, which can develop from hepatic steatosis to non-alcoholic steatohepatitis (NASH), then fibrosis, further to cirrhosis and, eventually, to hepatocellular carcinoma. The first steps of this process are reversible, but structural changes remain once fibrosis has developed [4]. Over the last decades, the prevalence of NAFLD has risen fast, replacing viral hepatitis as the most prevalent liver disease, affecting almost 25% of the adult population worldwide [5].

Since NAFLD can be considered as the hepatic component of MetS, it is strongly associated with hyperlipidemia [6,7]. To explore the molecular basis of hepatic alterations in NAFLD, animal models that mimic human conditions from a physiological and metabolic point of view have provided essential insights [8]. For example, in humanized apolipoprotein E2 knock-in (APOE2ki) mice and low-density-lipoprotein (LDL) receptor-deficient mice, NASH developed after only a few days of high-fat, high-cholesterol (HFC) feeding, in which the dietary cholesterol component was essential [9,10]. Not only in NASH, but also in Niemann–Pick type C1 disease—another disease characterized by hepatic inflammation—do lipids (e.g., free cholesterol) accumulate in the liver [11]. Therefore, exploring the role of hepatic free cholesterol accumulation as part of the molecular mechanisms responsible for hepatic inflammation warrants attention.

One potential link between cholesterol metabolism and NAFLD development are so-called microRNAs (miRNAs). MiRNAs are short, non-coding RNA molecules composed of 18 to 25 nucleotides that play an important role in regulating gene expression, and are therefore involved in many crucial biological processes [12,13]. Circulating miRNAs are extremely stable and protected from RNase degradation. They have therefore emerged as attractive candidate targets for interventions and as biomarkers for the early diagnosis of diseases and for monitoring of disease progression [14]. In recent reviews, the roles of miRNAs in NAFLD or cholesterol metabolism have been discussed [15–17]. However, no attention was paid to the link of miRNAs with cholesterol metabolism as it related to NAFLD in humans, which is the focus of the current systematic literature review. This knowledge may contribute to the understanding of whether (dietary) interventions targeting these miRNAs can modify hepatic cholesterol metabolism and, consequently, NASH development and/or progression. In addition, due to the invasive nature of a liver biopsy and the difficulty of using specific and sensitive ultrasound-based methods to differentiate between steatosis, inflammation and fibrosis at a large scale [18,19], there is an urgent need for novel non-invasive tools to diagnose and monitor the progress of the different stages in NAFLD. Therefore, the possibility to use specific miRNAs as an NAFLD biomarker will also briefly be discussed.

2. Methods

2.1. Search Strategy

A systematic literature search in three databases (Cochrane Central Register of Clinical Trials, Embase and Medline) was conducted in September 2022 to identify potentially relevant studies. Search terms consisted of keywords related to cholesterol metabolism, miRNAs and NAFLD. The following terms were used: Cholesterol (cholesterol or LDL-cholesterol or HDL-cholesterol or chylomicron-cholesterol) and NAFLD (NAFLD or NAFL or NASH or steatohepatitis or steatosis) and micro-RNA (miRNA or microRNA) and limited to humans.

2.2. Selection of Studies

Human studies that examined associations between cholesterol, miRNA and NAFLD were selected. The selection procedure consisted of 2 steps. In the first step, titles and abstracts of all retrieved papers were screened for the following aspects: (1) studies on humans with NAFLD, (2) plasma/serum and/or hepatic cholesterol measured, (3) miRNA measured either in plasma/serum and/or liver, (4) original research (e.g., no reviews, conference abstracts, proceedings or case reports), (5) written in English and (6) no duplicates. The selection was performed independently by two researchers (M.C.J.M.K. and S.B.). When inconclusive, eligibility was discussed by both researchers until agreement was reached. In the second step, the papers that were selected in step 1 were read as full papers to assess final eligibility and for data collection.

2.3. Data Collection

Data collection was performed using an a-priori-defined spreadsheet and included publication characteristics (reference number, first author, year of publication), study and patient characteristics (ethnicity, sample size, health status, NAFLD stage and age), analytical methods used and results.

3. Results

The systematic literature search resulted in 287 potentially relevant articles. Screening of titles and abstracts resulted in exclusion of 227 articles based on the predefined selection criteria. After reviewing the full texts of the remaining 40 articles, another 22 articles were excluded, since one or more parameters from the line of interest, i.e., miRNA, cholesterol and NAFLD, were missing. After the full selection procedure, a total of 19 studies were included. One article was retrieved from the reference list of one of the selected articles. A flowchart of the selection process is presented in Figure 1.

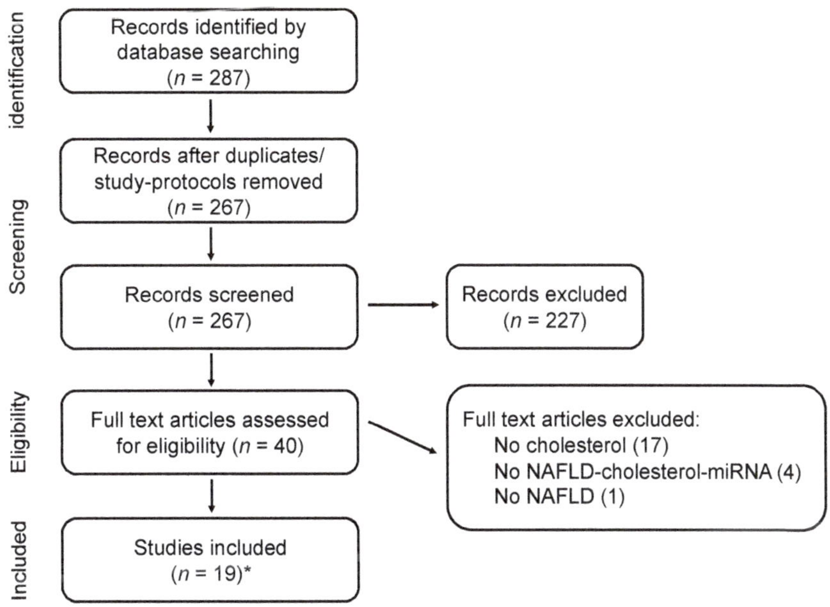

Figure 1. Flow chart of the study selection process. The literature search retrieved 287 potentially relevant papers; 228 were excluded after screening titles and abstracts, 40 articles were reviewed in full and ultimately 19 studies were included in the systematic review. * One study included from the reference list of the selected papers.

3.1. Study Characteristics and Selected miRNAs

In the 19 selected articles, 13 different miRNAs were evaluated. These 13 miRNAs (Table 1) have not been studied at the same level of detail. The most information was available for miR122 [20–27], which is one of the most abundant miRNAs in the liver and is possibly involved in hepatic disease progression; for miR34a [22,25,26,28,29], which is a critical tumor suppressor [30]; and for miR21 [22,25,26,31–33], which is a cancer-promoting miRNA targeting tumor-suppressor genes [34].

The remaining ten miRNAs, which were studied in less detail, were miR29a [24], miR379 [35], miR33a [21,31,36], miR33b and miR33b* [21,36], miR144 [36], miR451 [26], miR132 [37,38], miR129 [39] and miR486 [40]. MiRNA data were not uniformly expressed. For example, levels of miR122 were absolutely quantified against a calibration curve [22,24] or expressed relative to those of their controls [20,21,23,25–27]. MiR29a was only absolutely quantified [24], while miR34a and miR21 were absolutely quantified [22] or, like the other miRNAs, relatively expressed [20,21,23,25–29,31–33,35–37,39–41].

In the next paragraphs, the findings for the 13 different miRNAs have been summarized. For each miRNA, we have described their associations with cholesterol metabolism and NAFLD. When possible, additional information related to diagnostics, target genes and mechanisms has been provided. How these miRNAs may be involved in cholesterol metabolism in NAFLD have been summarized in Figure 2.

Table 1. miRNAs and their target genes.

miRNA	Target Genes
122 [24]	*CYP7A1, SRF, RAC1, RHOA, CCNG1, GTF2B, GYS1, NFATC2IP, ENTPD4, ANXA11, FOXP1, MECP2, NCAM1, TBX19, AACS, DUSP2, ATP1A2, MAPK11, AKT3, GALNT10, G6PC3, SLC7A1, FOXJ3, SLC7A11, TRIB1, DSTYK, PRKAB1, ACVR1C, PRKRA, PTP1B, P4HA1, ZNF395, SOCS1, HMOX1, CDK4*
34a [29]	*HNF4A, MTP, APOB, SREBP1C, ACC1, ACC2, HMGCR*
21 [32]	*PPARA*
21 [33]	*HMGCR*
379 [35]	Fibrosis and inflammation: *CAT, CTGF, IL10, PDGFA, PDGFRA, SMAD4, TGFBR1, THBS1* Energy management, including gluconeogenesis and lipogenesis: *CREB1, EIF4E, FOXO1, INSR, IGF1, IGF1R, ITPR2, PRKAA1, PRKAA2, RICTOR, SOCS1, TCF7L2* Cell survival and proliferation: *BCL2, CCNB1, HGF, PMAIP1, PTEN, YAP1* Signaling pathways: *HDAC2*
29a [24]	*DK6, RAN, BACE1, S100B, IMPDH1, GLUL, PPM1D, PIK3R1, LPL, CPEB3, CPEB4, ADAMTS9, TRIM63, MYCN, SERPINB9, DICER1, TNFAIP3, CDC42, PXDN, ITIH5, PTEN, ABL1*
144	Not reported
33a [36]	*CROT*
33b* [36]	*CROT*
33b	Not reported
451	Not reported
132 [37]	*ACHE, FOXO3, PTEN, SIRT1*
129	Not reported
486	Not reported

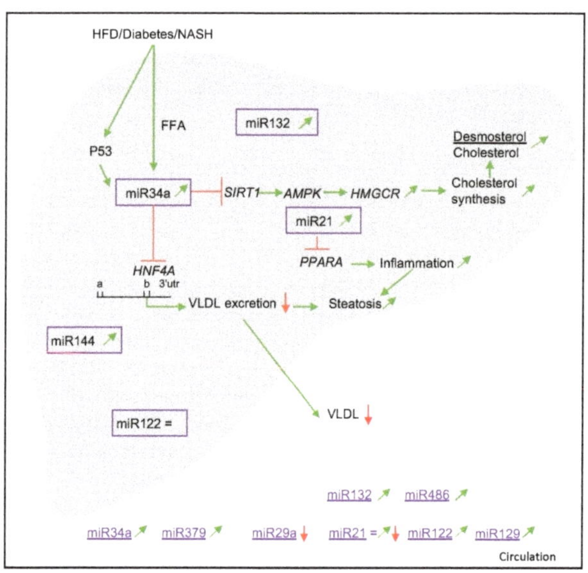

Figure 2. Schematic overview of potential mechanisms of miRNAs involved in cholesterol metabolism in humans. Red arrow is downregulated. Green arrow is upregulated, = is no change. HFD—high-fat diet; NASH—non-alcoholic steatohepatitis; FFA—free fatty acids; *SIRT1*—sirtuin1; *AMPK*—5′-adenosine-monophosphate-activated protein kinase; *HMGCR*—3-hydroxy-3-methylglutaryl-CoA reductase; *HNF4A*—hepatocyte nuclear factor 4 alfa; VLDL—very-low-density lipoprotein; *PPARA*—peroxisome proliferator-activated receptor alpha.

3.2. miR122

3.2.1. Associations with NAFLD and Cholesterol Metabolism

In all eight studies that measured miR122, serum/plasma miR122 concentrations were higher in NAFLD patients as compared to their controls [20–27]. In addition, concentrations increased with disease severity [21,22,24,26]. In one study, however, levels were increased in patients with mild and moderate steatotic/fibrotic stages as compared to those of patients with more severe steatotic/fibrotic stages, but still 4-5-fold higher than those of normal controls [23]. Moreover, serum miR122 correlated negatively with serum high-density lipoprotein cholesterol (HDL-C) concentrations [21,23] and positively with serum triacylglycerol (TAG) [20,23] and very-low-density lipoprotein cholesterol (VLDL-C) concentrations [20], total cholesterol (TC), low-density lipoprotein cholesterol (LDL-C) [22], alkaline phosphatase (ALP) [21], aspartate-aminotransferase (AST) [21–23], alanine-aminotransferase (ALT) [21–23] and with the NAFLD Activity Score (NAS) [24]. More results of these studies are presented in Table 2.

Table 2. Correlations between miRNA and NASH-related parameters.

MiR	With	Correlation	Where	Author
122	IL-1α	r = 0.250; p = 0.030; n = 75	serum	[20]
	TAG	r = 0.230; p = 0.048; n = 75	serum	[20]
	VLDL-C	r = 0.230; p = 0.048; n = 75	serum	[20]
	HDL-C	r = -0.305; p = 0.001; n = 65	serum	[21]
	ALP	r = 0.306; p = 0.021; n = 65	serum	[21]
	ALT	r = 0.351; p < 0.001; n = 65	serum	[21]
	AST	r = 0.367; p < 0.001; n = 65	serum	[21]
	Hepatocellular ballooning	r = 0.200; p = 0.035; n = 65	liver	[21]
	Lobular inflammation	r = 0.225; p = 0.017; n = 65	liver	[21]
	Liver	r = 0.253; p = 0.019; n = 65	serum	[21]
	ALT	r = 0.75; n = 34	serum	[22]
	AST	r = 0.55; n = 34	serum	[22]
	Fibrotic stage	r = 0.33; n = 34	serum	[22]
	Inflammation activation	r = 0.33; n = 34	serum	[22]
	LDL-C	r = 0.44; n = 34	serum	[22]
	TC	r = 0.36; n = 34	serum	[22]
	Fibrotic stage	r = 0.399; p < 0.002; n = 56	serum	[24]
	NAS	r = 0.306; p = 0.022; n = 56	serum	[24]
men	Severity of steatosis	normal vs. mild p < 0.001; n = 90 vs. n = 37	serum	[26]
		mild vs. severe p = 0.047; n = 37 vs. n = 11	serum	[26]
women	Severity of steatosis	normal vs. mild p = 0.002; n = 221 vs. n = 36	serum	[26]
		mild vs. severe p = 0.035; n = 36 vs. n = 8	serum	[26]
34a	Fibrotic stage	r = 0.41; n = 34	serum	[22]
	Inflammation activation	r = 0.43; n = 34	serum	[22]
	TAG	r = 0.43; n = 28	serum	[25]

Table 2. *Cont.*

MiR	With	Correlation	Where	Author
	VLDL-C	r = 0.44; n = 28	serum	[25]
21	Fibrosis	r = 0.461; p = 0.021; n = 19	liver	[41]
	Hepatic ballooning	r = 0.713; p < 0.001; n = 19	liver	[41]
	Lobular inflammation	r = 0.735; p < 0.001; n = 19	liver	[41]
	Steatosis	r = 0.539; p = 0.005; n = 19	liver	[41]
379	ALP	r = 0.278; p = 0.048; n = 53	serum	[35]
	TC (all participants)	r = 0.361; p = 0.039; n = 53	serum	[35]
	LDL-C	r = 0.285; p = 0.043; n = 53	serum	[35]
	Non-HDL-C	r = 0.286; p = 0.038; n = 53	serum	[35]
	TC (non-statin users)	r = 0.381; p = 0.045; n = 42	serum	[35]
29a	TAG	r = 0.144; p = 0.048; n = 46	serum	[24]
33b *	AST	r = 0.203; p = 0.046; n = 61	serum	[21]
	HDL-C	r = -0.276; p = 0.004; n = 61	serum	[21]
	Hepatic ballooning	r = 0.343; p = 0.001; n = 13	liver	[21]
	Lobular inflammation	r = 0.358; p < 0.001; n = 12	liver	[21]
	TAG	r = 0.279; p = 0.004; n = 61	serum	[21]
33a	HDL-C	r = -0.313; p = 0.004; n = 74	serum	[36]
144	HDL-C	r = -0.221; p = 0.043; n = 74	serum	[36]
129	TAG	r = 0.662; p < 0.001; n = 117	serum	[39]
	TC	r = 0.708; p < 0.001; n = 117	serum	[39]
132	ApoE	$\beta \pm$ SE = 0.038 $\pm$ 0.002; p = 0.017; n = 140	serum	[38]
	ALT	$\beta \pm$ SE = 0.005 $\pm$ 0.002; p = 0.018; n = 140	serum	[38]
	NAFLD	OR 3.08 (1.06, 8.99); p = 0.0392; n = 140	serum	[38]
	TAG	$\beta \pm$ SE = 0.072 $\pm$ 0.029; p = 0.015; n = 140	serum	[38]

Abbreviations: IL-1α—Interleukin 1 alfa; TAG—triacylglycerol; VLDL-C—very-low-density lipoprotein cholesterol; HDL-C—high-density lipoprotein cholesterol; ALP—alkaline phosphatase; ALT—alanine-aminotransferase; AST—aspartate-aminotransferase; LDL-C—low-density lipoprotein cholesterol; TC—total cholesterol; NAS—NAFLD Activity Score; Non-HDL-C—non-high-density lipoprotein cholesterol; apoE—apolipoprotein E; NAFLD—non-alcoholic fatty liver disease.

Auguet et al. [21] found a weak but significant positive correlation between hepatic and plasma miR122 levels. Additionally, a positive correlation between serum miR122 levels and the fibrotic stage has been reported [22,24]. Moreover, a positive correlation between plasma miR122 concentrations and lobular inflammation [21,22] and hepatocellular ballooning [21] was observed. This aligns with the observation of Abdel-Hamed et al. [20], who reported in NAFLD patients a positive correlation between serum miR122 and serum levels of IL-1α (Figure 3), an inflammatory cytokine. They also found that NAFLD patients were more frequent carriers of the high-risk allele DD (rs3783553), which is an insertion/deletion polymorphism in the 3′ UTR of the IL-1α allele. This polymorphism interrupts the binding of miR122, and Abdel-Hamed et al. [20] have reported in NAFLD patients an increased relative expression of miR122 in these DD-carriers compared to that in the low-risk genotype ID or II. Finally, Yamada et al. [26] have found a positive association between serum miR122 levels and severity of steatosis in both men and women.

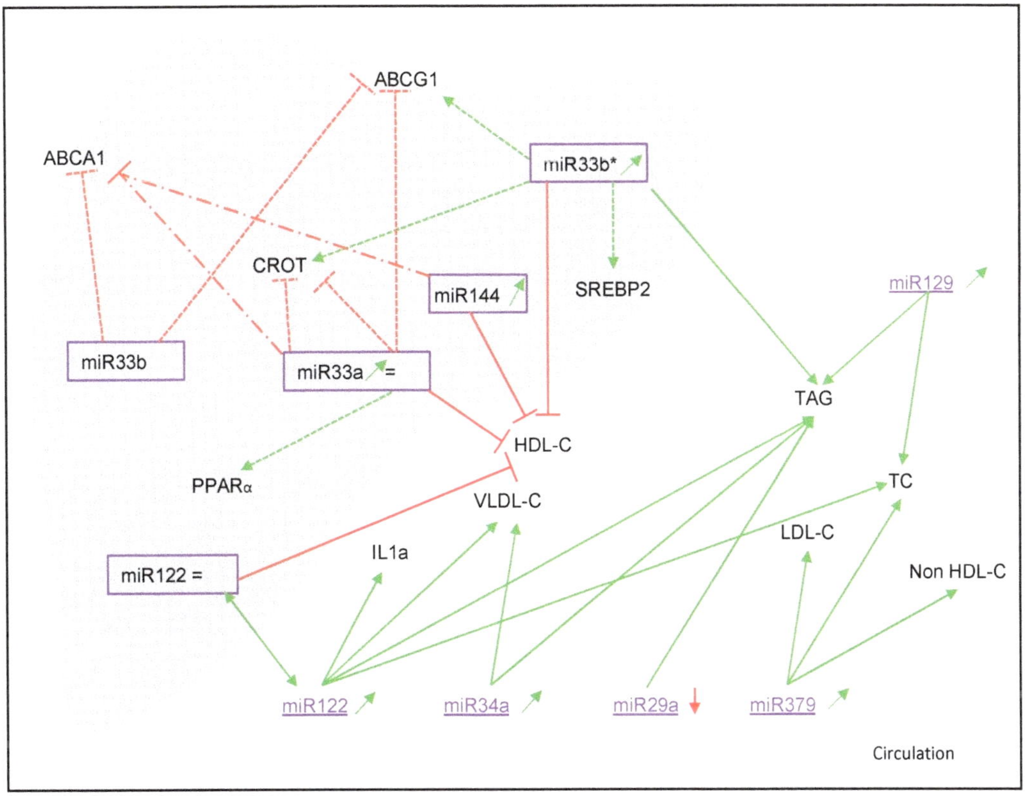

Figure 3. Relations between miRNAs, genes and cholesterol metabolism. Green: positive correlations; red: inverse correlations; small dotted lines: mRNA levels; broad–small dotted lines: protein levels; solid lines: other relations. Red arrow is downregulated. Green arrow is upregulated, = is no change; IL-1α—Interleukin 1 alfa; TAG—triacylglycerol; VLDL-C—very-low-density lipoprotein cholesterol; HDL-C—high-density lipoprotein cholesterol; LDL-C—low-density lipoprotein cholesterol; TC—total cholesterol; Non-HDL-C—non-high-density lipoprotein cholesterol; ABCA1—ATP-binding cassette transporter A1; ABGG1—ATP-binding cassette transporter G1; SREBP2—sterol regulatory element-binding protein 2; CROT—carnitine O-octanoyltransferase; PPARα—peroxisome proliferator-activated receptor alpha.

3.2.2. Diagnostics

Receiver Operation Characteristic (ROC) curves for serum miR122 showed an Area Under the Curve (AUC) of 0.83, which means that 83% of NAFLD patients were correctly discriminated from healthy controls [24]. Both Salvazo et al. [22] and Cermelli et al. [25] showed better, or at least similar, curve characteristics (Table 3) than for ALT, suggesting that serum miR122 levels can be used as a biomarker to discriminate NAFLD patients from healthy controls. Ye et al. [27] also showed an AUROC of 0.77 for miR122. Finally, Auguet et al. [21] concluded that the accuracy of miR122 in discriminating NAFLD from non-NAFLD patients resulted in an AUROC curve of 0.82. Moreover, for distinguishing advanced disease from a mild clinical form, the diagnostic value for miR122 was 0.76. Finally, using logistic regression analysis and after adjusting for BMI, HDL-C, TAG, AST and ALT, serum miR122 levels showed the highest odds ratio (OR: 2.19) to discriminate patients with and without hepatocellular ballooning.

Table 3. Receiver Operation Characteristics.

miR		AUC	Significance	Sensitivity	Specificity	PPV (%)	NPV (%)	Author
122	Hepatocellular ballooning [a,b]	0.76		74.4%	46.8%	46.8%	87.3%	[21]
	Lobular inflammation [a,c]	0.76		74.4%	46.8%	46.8%	87.3%	[21]
	NAFLD [a,d]	0.82		83.1%	69.8%	47.8%	92.5%	[21]
	ALT	0.91						[22]
	NAFLD-ss [e]	0.93						[22]
	NAFLD-ss [e] vs. NASH [f]	0.70						[22]
		0.83	$p < 0.001$	75.0%	82.4%			[24]
		0.86	$p = 0.001$, 95% CI = 0.77–0.95					[25]
34a		0.78	$p = 0.001$, 95% CI = 0.66–0.90					[25]
	ALT	0.83	$p = 0.001$, 95% CI = 0.73–0.94					[25]
	NAFLD-ss [e] vs. NASH [f]	0.76						[22]
379	NAFL	0.76						[35]
	NAFLD	0.72						[35]
	NASH	0.72						[35]
	Early stage NAFLD	0.74						[35]
	Advanced stage NAFLD	0.67						[35]
29a		0.68	$p = 0.007$	60.9%	82.4%			[24]
129	NAFLD	0.93		83.8%	92.7%			[39]

Abbreviations: NAFLD—non-alcoholic fatty liver disease; ALT—alanine-aminotransferase; NAFLD-ss—non-alcoholic fatty liver disease simple steatosis; NASH—non-alcoholic steatohepatitis. [a] optimum conditions selected. [b] discriminating hepato-cellular from non-hepato-cellular ballooning. [c] discriminating lobular inflammation from non-lobular inflammation. [d] discriminating NAFLD from control. [e] NAS score 1–4. [f] NAS score 5–7.

3.2.3. Target Genes and Mechanisms

Target genes of miR122 and the other miRNAs are shown in Table 1. For hepatic miR122 expression, no correlation was observed with genes directly related to lipid metabolism [21].

3.3. *miR34a*

3.3.1. Associations with NAFLD and Cholesterol Metabolism

Serum levels [22] and hepatic expression [25,26,28,29] of miR34a were significantly higher in NAFLD patients as compared to healthy controls. Moreover, serum miR34a levels in NAFLD patients were significantly increased with severity of liver steatosis [25], fibrosis and hepatic inflammation [22,25]. However, these findings were not supported by Yamada et al. [26]. In addition, a positive correlation has been found between serum miR34a and VLDL-C concentrations as well as between miR34a and TAG concentrations in NAFLD patients [25].

3.3.2. Diagnostics

Salvazo et al. and Cermelli et al. [22,25] reported that the diagnostic value of miR34a to discriminate NAFLD patients from healthy controls was comparable to that of ALT (AUC = 0.78/ 0.75 vs. 0.83).

3.3.3. Target Genes and Mechanisms

Hepatocyte nuclear factor 4 alfa (*HNF4A*) is a nuclear hormone receptor that controls the basal expression of many genes involved in bile acid, lipid, glucose and drug metabolism. Xu et al. [29] described that the miR34a-*HNF4A* pathway is highly active under conditions of metabolic stress such as diabetes, high-fat feeding and NASH. In addition, p53 is an oxidative-stress-inducible protein that upregulates miR34a expression in humans and also promotes liver steatosis in mice [29]. This protein was upregulated 6.7-fold in NASH patients ($p < 0.05$), which could at least partly explain the finding that serum miR34a levels were significantly higher in NAFLD patients as compared to healthy controls [25,28]. Moreover, Xu et al. [29] found that free fatty acids (FFA), cholesterol and p53 are all upstream activators for the miR34a-*HNF4A* pathway in diabetes, obesity and NASH.

In the study performed by Min et al. [28], 3-hydroxy-3-methylglutaryl-CoA reductase (*HMGCR*) phosphorylation was decreased in NAFLD and NASH patients, indicating higher activity of this enzyme [42]. Increased miR34a levels, as observed in NASH patients, suppress sirtuin 1 (*SIRT1*). This protein upregulates 5′ adenosine monophosphate-activated protein kinase (*AMPK*), a known positive HMGCR phosphorylation regulator. This suggests that miR34a activation translates into increased HMG-CoA reductase activity and higher endogenous cholesterol synthesis. The functional impact of the relatively dephosphorylated state of HMGCR in NASH patients was confirmed by a significant increase in circulating desmosterol/cholesterol ratios compared to controls ($p < 0.01$) [28].

3.4. *miR21*

3.4.1. Associations with NAFLD and Cholesterol Metabolism

Rodrigues et al. [32] showed significantly higher hepatic miR21 expression in NAFLD patients with increasing levels in the progression from steatosis to NASH ($p < 0.01$). Similar results were found in serum and skeletal muscle ($p < 0.05$). Moreover, the potential of serum miR21 concentrations as a biomarker for NAFLD stage was evaluated and a significant $\pm$3-fold ($p < 0.05$) increase in serum miR21 levels was found between patients with steatosis and NASH. Comparable results were reported by Loyer et al. [41], i.e., a 3-fold increase in liver tissue ($p < 0.01$), but only in NASH and not in simple steatotic livers. Compared to healthy controls, Yamada et al. [26] found significantly increased miR21 serum levels in male NAFLD patients ($p < 0.01$), but not in female patients. In contrast, Sun et al. [33] showed decreased serum miR21 levels in NAFLD patients compared to healthy controls ($p < 0.05$), while no differences between NAFLD patients and their controls were found [22,25]. Loyer et al. [41] found positive correlations between hepatic miR21 levels and hepatic ballooning ($p < 0.001$), lobular inflammation ($p < 0.001$), steatosis ($p < 0.01$) and fibrosis ($p < 0.05$). Lendvai et al. [31] found a positive correlation between serum miR21 with AST levels in the steatotic group, but not in hepatitis C, steatotic hepatitis C and normal liver groups.

3.4.2. Diagnostics

There were no diagnostic data provided related to miR21.

3.4.3. Target Genes and Mechanisms

In human hepatoma cells, *HMGCR* was a direct target of miR21 [33]. In line with these observations, HMGCR mRNA and protein levels in serum were significantly increased ($p < 0.01$ for mRNA, $p < 0.05$ for protein) in NAFLD patients as compared to healthy controls [33].

3.5. *Other miRNAs (miR379, miR29a, miR144, miR33a/b, miR33b*, miR451, miR132, miR129 and miR486)*

3.5.1. Associations with NAFLD and Cholesterol Metabolism

Okamoto et al. [35] described a 3.5-fold ($p < 0.05$) increase in serum miR379 levels in a Japanese population of NAFLD patients compared to those of control patients. When patients ($n = 79$) were divided into steatotic patients ($n = 9$) and NASH patients ($n = 70$), miR379 levels were significantly higher in the steatotic group than in controls, while levels were comparable in the NASH group. However, when patients were divided based on Brunt stages, e.g., patients with NAFLD early stage (non-alcoholic fatty liver (NAFL); Brunt stages 0–1, $n = 53$) and NAFLD advanced stage (Brunt stage 2–4, $n = 26$), they found in the early stage 3.65-fold ($p < 0.05$)—and in the NAFLD population, 4.87-fold ($p < 0.05$)—higher miR379 serum levels as compared to control patients ($n = 10$). MiR379 levels correlated positively with ALP, TC, LDL-C and non-HDL-C in early stage NAFLD patients ($n = 51$). Moreover, in the non-statin-treated patients of this group ($n = 42$), a significant positive correlation between serum TC and miR379 was found.

For miR29a, serum levels were significantly lower in NAFLD patients compared to control patients ($p < 0.01$) [24]. In addition, serum miR29a levels in patients with NAS scores < 4 were significantly lower than serum levels in healthy controls ($p < 0.05$). Unexpectedly, serum miR29a positively correlated with serum TAG concentrations within the NAFLD population.

Yamada et al. [26] found increased serum levels of miR451 in Japanese NAFLD patients compared to healthy controls, but no association with disease state. Hanin et al. [37] found a 13-fold upregulation of miR132 in postmortem hepatic tissues of NAFLD patients as compared to apparently healthy controls ($p < 0.01$). Moreover, a significant downregulation of its targets *FOXO3*, *PTEN* and *SIRT1* ($p < 0.05$ for all) and a significant upregulation of *P300* ($p < 0.05$) were observed.

Vega-Badillo et al. [36] demonstrated that hepatic miR144 had significantly higher expression in NASH patients than in control patients and patients with simple steatosis ($p < 0.05$ for both). Furthermore, miR33a was elevated in the simple steatosis group [31,36] and significantly higher in the NASH individuals ($p < 0.01$) compared to control patients [36]. Wang et al. [39] showed significantly higher miR129 levels in NAFLD patients compared to controls ($p < 0.001$). Serum levels of miR486 were also significantly increased in an overweight/obese group, which included children with NAFLD [40]. Hepatic miR33b* expression was higher in morbidly obese women with NAFLD as compared to morbidly obese women with a healthy liver and a comparable BMI ($p < 0.001$) [21]. This significantly higher expression was found in morbidly obese women with NASH ($p < 0.001$) and with simple steatosis ($p < 0.05$). Compared to women with a healthy liver, miR33b* expression was upregulated in moderately obese NAFLD women ($p < 0.05$) and the expression of miR33b* was higher in NASH patients ($p < 0.05$). Furthermore, in moderately obese women, hepatic miR33b* correlated positively with hepatic ballooning ($p < 0.005$) and lobular inflammation ($p < 0.001$). Circulating miR33b* serum levels were significantly higher in moderately and morbidly obese women as compared to normal weight women ($p < 0.001$); serum miR33b* correlated positively with concentrations of TAG and AST and negatively with that of HDL-C [36]. In contrast, Vega-Badillo et al. [36] observed an inverse correlation with plasma HDL-C and miR33a, and miR144, but not miR33b, levels. Lendvai et al. [31] showed a negative correlation between hepatic miR33a and serum AST and ALP levels. Wang et al. [39] found positive correlations between miR129 with TC and TAG but not with ALT. Logistic regression analysis suggested that miR129 can be independently influenced by TC. Al Azzouny et al. [40] showed in their multivariate analysis that miR486 expression was an independent predictor for NAFLD susceptibility.

3.5.2. Diagnostics

For miR379, Okamoto et al. [35] found AUROC values for diagnosing NAFLD (0.72), NAFL (0.76), NASH (0.72), early stage NAFLD (0.74) and advanced stage NAFLD (0.67). This data suggests that this miRNA would be slightly more specific in detecting early stages of NAFLD than later stages. Jampoka et al. [24] investigated miR29a as a possible biomarker to diagnose NAFLD and found an AUC of 0.68 ($p < 0.01$) with 60.9% sensitivity and 82.4% specificity. Wang et al. [39] found AUROC values for distinguishing NAFLD from healthy controls (0.93) with a sensitivity and specificity of 83.8% and 92.7%, respectively. There were no studies describing the diagnostic value of miR144, 33a, 33b and 33b*, 132 or 486.

3.5.3. Target Genes and Mechanisms

Okamoto et al. [35] identified 1423 potential target genes of miR379, which were classified according their function association or related biological processes according to Gene Ontology (GO) terms. Of all processes, biological regulation, metabolic and cellular processes covered over 70% of all pathways. After a more selective process, 27 genes were selected, which were grouped into four categories (Table 1). These categories were: fibrosis and inflammation (8 genes), energy management including lipogenesis (12 genes), cell survival and proliferation (6 genes), and cell signaling (1 gene).

Jampoka et al. [24] predicted the target genes of miR29a, as presented in Table 1. Vega-Badillo et al. [36] showed a significant inverse correlation between miR33a and hepatic ABCA1 mRNA ($p = 0.050$) and hepatic ABCA1 protein levels ($p < 0.01$). Moreover, miR33b inversely correlated with both hepatic ABCA1 mRNA ($p < 0.001$) and ABCG1 mRNA ($p < 0.005$). For miR144, an inverse correlation with hepatic ABCA1 protein ($p < 0.05$) was found. Auguet et al. [21] found positive correlations between miR33b* and *SREBP2* ($p < 0.005$) and *ABCG1* ($p < 0.01$), genes related to lipoprotein secretion. They also found positive associations for hepatic miR33a with *PPARA* ($p < 0.05$) expression and negative associations between hepatic miR33a and *ABCG1* ($p < 0.01$) expression.

3.6. Animal Data

Other than the human data, we also examined the animal data that were presented in the 19 selected articles (Figure 4). This figure shows what is mechanistically known from the animal data present in the selected papers and is useful to compare the overlap with the human condition. The animal data will be further used in the discussion.

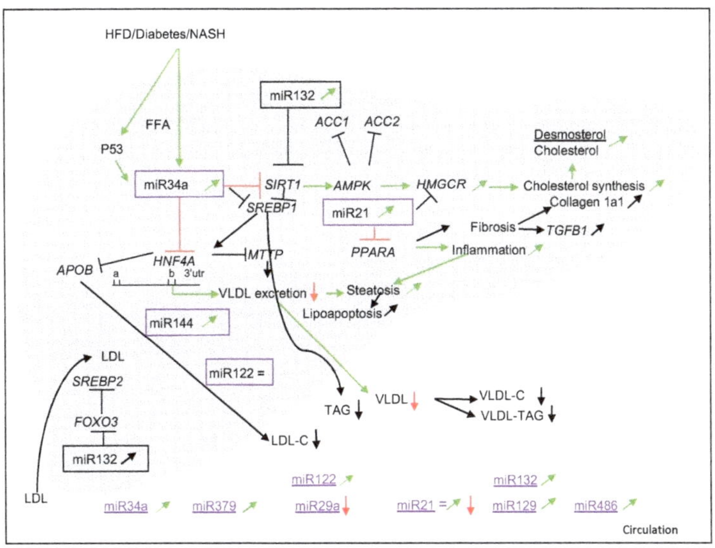

Figure 4. Schematic overview of potential mechanisms in animals (black) of miRNAs involved in cholesterol metabolism. Red arrow is downregulated. Green arrow is upregulated, = is no change. Data are derived as published in the papers selected for this review. HFD—high-fat diet; NASH—non-alcoholic steatohepatitis; FFA—free fatty acids; *SIRT1*—sirtuin1; *AMPK*—5′-adenosine-monophosphate-activated protein kinase; *HMGCR*—3-hydroxy-3-methylglutaryl-CoA-reductase, *HNF4A*—hepatocyte nuclear factor 4 alfa; VLDL-C—very-low-density lipoprotein cholesterol; LDL-C—low-density lipoprotein cholesterol; *PPARA*—peroxisome proliferator-activated receptor alpha; *ACC1/2*—Acetyl-CoA carboxylase1/2; *TGFB1*—transforming growth factor beta 1; *APOB*—apolipoprotein B; *MTTP*—microsomal triglyceride transfer protein; *SREBP1/2*—Sterol regulatory element-binding protein 1/2; *FOXO3*—forkhead-box protein O3.

4. Discussion

In this systematic review, we identified 13 miRNAs that may be associated with both NASH and cholesterol metabolism. It was suggested that four of these miRNAs (miR122, miR34a, miR21 and miR132) play a role in the development of NASH through effects on cholesterol metabolism. Therefore, it might be interesting to explore whether (dietary) interventions targeting these miRNAs can modify cholesterol metabolism, and hence, NASH development and/or progression. For the other nine miRNAs, the data was less conclusive. Furthermore, miR122, miR34a and miR379 may, in addition to traditional markers, be used in a panel for the early detection of NAFLD and/or monitoring of NAFLD development.

The most frequently studied miRNA that appeared in our literature search was miR122. In serum, miR122 concentrations were higher in patients with liver steatosis as compared to healthy controls, and concentrations were even higher in NASH patients [20,21,24–26,43]. Interestingly, Cheung et al. [44] showed lower hepatic miR122 expression in simple steatosis and NASH patients as compared to healthy controls, which resulted in an inverse correlation between liver and plasma levels of miR122 in NASH patients. This contrasts the findings of Auguet et al. [21], who demonstrated higher miR122 levels in morbidly obese patients with NASH compared to morbidly obese patients with simple steatosis and morbidly obese patients with normal livers. Studies in primary mouse hepatocytes and mouse models showed that phosphomevalonate-kinase (*PMVK*)—which catalyzes phosphorylation of mevalonate [45]—together with *HMGCS1*, *HMGCR* and *DHCR7* were target genes of miR122. This suggest that miR122 is involved in the regulation of cholesterol and fatty acid metabolism [46].

Another miRNA, which according to our search may be linked to both cholesterol metabolism and NAFLD, is miR34a. This miRNA is found in high levels in plasma and the liver, especially in patients with steatosis and NASH [47,48]. Furthermore, Xu et al. [29] found that upregulation of miR34a inhibits hepatic VLDL secretion, thereby promoting steatosis in which a role for *HNF4A* was postulated. Moreover, in rodents, the p53/miR34a/*SIRT1* pathway—or so-called proapoptotic pathway—is prevented by treatment with ursodeoxycholic acid (UDCA), which downregulates miR34a, in a cell-type-specific mechanism [47]. This suggests that inhibition of miR34a could prevent disease progression. Moreover, Ding et al. [49] showed in human cell lines and mice models not only that *PPARA*, its downstream genes and *SIRT1* are targets of miR34a but also that inhibition of miR34a decreased steatosis.

MiR21 was the third miRNA that appeared from our search. Rodrigues et al. [32] reported that miR21-knock-out mice—as compared to wild-type mice—who were fed a Western-type diet supplemented with obeticholic acid showed minimal steatosis, inflammation and lipo-apoptosis, which was most likely caused by an upregulation of *PPARA* and Farnesoid-X-Receptor (*FXR*) activation. In line with this, Loyer et al. [41] found that either knocking out miR21 or using an miR21 antagonist decreased liver injury, inflammation and fibrosis in mice. Moreover, they also reported increased hepatic expression of miR21 in NASH patients but not in steatotic patients, predominantly in inflammatory and biliary cells. Zhao et al. [50] showed that overexpression of miR21 stimulated extracellular-signal-related kinase1 (*ERK1*) signaling and epithelial–mesenchymal transition (EMT) of hepatocytes by targeting sprouty2 (*SPRY2*), which is a negative feedback regulator of multiple receptor kinases and *HNF4A* [50]. Changing this negative feedback mechanism may affect hepatic fibrosis development in NASH [50]. Although hepatic levels of miR21 were consistently higher in NASH and cirrhotic livers, differences in miR21 plasma or serum values were less consistent.

The fourth and final miRNA that we described was miR132. Although human data is limited, there are indications that hepatic miR132 expression is upregulated in liver biopsies from NAFLD patients compared to apparently healthy controls [37]. In mice, Hanin et al. [37] investigated the role of the direct target genes of miR132, which are involved in several steps of lipid and cholesterol metabolism. They concluded that the development of hepatic steatosis may be the consequence of affecting the various miR132 targets simultaneously when miR132 is elevated. Moreover, Zong et al. [38] found in the non-T2DM subgroup positive associations between serum miR132 and ALT, TAG, apoE and NAFLD.

Liver biopsies are the gold standard to diagnose the presence of NAFLD—more specifically, of NASH—and to monitor disease progression over time [51]. However, taking liver biopsies is not without risk, and there is a strong interest for other approaches to diagnose and monitor NASH development. Although different non-invasive imaging techniques and plasma biomarkers such as ALT and AST have been used, most patients with NAFLD remain asymptomatic and have non-elevated ALT levels [52]. Therefore, additional measurements such as plasma glucose, TAG, TC and lipoprotein cholesterol levels, in combination with BMI, fat distribution and family history [53,54] are frequently included to diagnose NAFLD. We here also explored whether the identified miRNAs could have an added value for diagnosing and monitoring NAFLD or its sub-stages. Our data suggests that three of the thirteen identified miRNAs could qualify for this purpose. Yamada et al. [26] found a positive correlation between steatosis severity and serum miR122 levels, suggesting that serum miR122 levels could be used to differentiate between healthy and diseased conditions. Jampoka et al. [24] also showed that serum miR122 levels were lower in patients with a NAFLD Activity Score (NAS) < 4 as compared to patients with a score > 4. Later, Ye et al. [27] confirmed the results of Auguet et al. [21] and Jampoka et al. [24] that miR122 levels are a more sensitive predictor than traditional serological markers for NAFLD. The question is whether we should rely on elevated miR122 levels only. It might be more attractive to include miR122 levels as part of a panel of different biomarkers to improve the accuracy of a non-invasive diagnosis of NAFLD [21]. Indeed, Ye et al. [27] developed a statistical model in which the traditional parameters such as TG, LDL-C, ALT and AST showed no significant change from baseline values until 4 weeks after onset of the disease. When miR122 was included in the model, it was possible to detect changes from week 1 after onset. Besides miR122, miR379 is another possible biomarker for diagnosing NAFLD. This miRNA had good AUROC values for all stages with slightly better performance for the earlier stages. However, a direct comparison of miR379 with more traditional biomarkers, or the added value when miR379 is included to a panel of traditional markers, is lacking. Finally, a third possible candidate miRNA that appeared from our search was miR34a. As for the other two miRNAs, plasma levels are increased with disease severity [47]. Cermelli et al. [22] already suggested that miR34a, as well as miR122, may represent novel, noninvasive biomarkers for diagnosis and histologically confirmed disease severity in patients with NAFLD. Pillai et al. [55] showed AUROCS of >0.80 between the plasma levels of five miRNAs (122, 34a, 375, 21 and 16) and three blood markers, and a positive correlation with different stages of metabolic dysregulation. Harrison et al. [56] even concluded that their NIS4 panel which consisted of miR34a and three other blood biomarkers has potential to reduce unnecessary liver biopsies in patients with lower risk of disease progression. Based on these findings, we suggest not to focus on one particular miRNA, but rather, to explore whether one or more of the identified miRNAs 122, 379 and 34a could be added to the traditional biomarkers to diagnose the presence and severity of NAFLD. Of course, this suggestion should be validated in future studies.

Compared with the relatively limited number of human studies using human samples, many more experiments have been carried out in animals. Data from animal studies that were part of the papers selected in our systematic literature search were therefore used to further explore possible underlying mechanisms to identify potential leads for future human research (Figure 4). For example, Xu et al. [29] showed that *HNF4A* was downregulated by miR34a in several mouse models and in human HepG2 cells. This consistency between species could relate to the fact that 3′-UTR binding sites in the *HNF4A* gene in mice and humans are highly conserved. Mutation experiments showed that miR34a binds to the second binding site (Figures 2 and 4) in 3′-UTR. Furthermore, overexpression of the human P53 protein increased miR34a expression in human HepG2 cells, thereby downregulating *HNF4A*. In addition, 13 lipid-related genes were significantly downregulated in NASH patients compared to healthy controls. Min et al. [28] showed an increased level of hepatic HMGCR mRNA and protein in NAFLD patients, which suggest that effects on *HMGCR* are transcriptionally regulated. This may be due to increased activation of *SREBP2*, the principal activator for *HMGCR*, which seems specific for NAFLD patients as it was not shown in weight-matched obese or hepatitis C patients as controls [28]. Potential mechanisms for increased activation of *SREBP2* could be the free cellular cholesterol content—which causes cellular lipotoxicity and consequently an inflammatory response, or may activate an unfolded protein response [57]—but these mechanisms remain to be elucidated. Loyer et al. [41] showed in mouse models that *PPARA* is targeted by miR21 and that downregulation of miR21 prevents NASH development. Min et al. [28] showed that miR34a can modulate HMGCR phosphorylation and may play a role in maintaining this enzyme in its active form. Ding et al. [49] showed in FFA-laden LO2 cells (normal human hepatocyte cell line) that miR34a downregulates *PPARA* expression. Therefore, it would be interesting to explore if there is an interplay between miR21/34a and the *PPARA* target *AMPK*. Though overlap exists between mice and human data, the abundance of data from in vitro and animal studies needs to be confirmed in humans to better understand the mechanistic link between miRNAs, cholesterol metabolism and NAFLD.

To conclude, we here show that miR122, miR34a, miR21 and miR132—all highly expressed in NASH—could relate to some of the cholesterol-related pathophysiological characteristics of NAFLD. Moreover, it is of interest to explore in more detail whether adding levels of plasma miRNAs (122, 34a and 379) to current noninvasive biomarkers improves the diagnosis of NAFLD and/or its different disease stages. Finally, we also identified several mechanistic leads from animal studies that need to be confirmed in human studies.

Author Contributions: S.B., J.P. and M.C.J.M.K. designed the study; M.C.J.M.K. and S.B. performed the literature search; M.C.J.M.K. tabulated all the data and wrote the draft manuscript; S.B., R.P.M. and J.P. reviewed and edited the manuscript. All authors have read and agreed to the published version of the manuscript.

Funding: This research received no external funding.

Conflicts of Interest: The authors declare no conflict of interest.

References

1. Ballestri, S.; Zona, S.; Targher, G.; Romagnoli, D.; Baldelli, E.; Nascimbeni, F.; Roverato, A.; Guaraldi, G.; Lonardo, A. Nonalcoholic fatty liver disease is associated with an almost twofold increased risk of incident type 2 diabetes and metabolic syndrome. Evidence from a systematic review and meta-analysis. *J. Gastroenterol. Hepatol.* **2016**, *31*, 936–944. [CrossRef] [PubMed]
2. Fan, J.G.; Farrell, G.C. Epidemiology of non-alcoholic fatty liver disease in China. *J. Hepatol.* **2009**, *50*, 204–210. [CrossRef] [PubMed]
3. Yki-Jarvinen, H. Non-alcoholic fatty liver disease as a cause and a consequence of metabolic syndrome. *Lancet Diabetes Endocrinol.* **2014**, *2*, 901–910. [CrossRef]
4. Starley, B.Q.; Calcagno, C.J.; Harrison, S.A. Nonalcoholic fatty liver disease and hepatocellular carcinoma: A weighty connection. *Hepatology* **2010**, *51*, 1820–1832. [CrossRef] [PubMed]
5. Wong, V.W.; Adams, L.A.; de Ledinghen, V.; Wong, G.L.; Sookoian, S. Noninvasive biomarkers in NAFLD and NASH—current progress and future promise. *Nat. Rev. Gastroenterol. Hepatol.* **2018**, *15*, 461–478. [CrossRef]
6. Kim, C.H.; Younossi, Z.M. Nonalcoholic fatty liver disease: A manifestation of the metabolic syndrome. *Clevel. Clin. J. Med.* **2008**, *75*, 721–728. [CrossRef] [PubMed]
7. Ludwig, J.; McGill, D.B.; Lindor, K.D. Review: Nonalcoholic steatohepatitis. *J. Gastroenterol. Hepatol.* **1997**, *12*, 398–403. [CrossRef]
8. Anstee, Q.M.; Goldin, R.D. Mouse models in non-alcoholic fatty liver disease and steatohepatitis research. *Int. J. Exp. Pathol.* **2006**, *87*, 1–16. [CrossRef] [PubMed]
9. Shiri-Sverdlov, R.; Wouters, K.; van Gorp, P.J.; Gijbels, M.J.; Noel, B.; Buffat, L.; Staels, B.; Maeda, N.; van Bilsen, M.; Hofker, M.H. Early diet-induced non-alcoholic steatohepatitis in APOE2 knock-in mice and its prevention by fibrates. *J. Hepatol.* **2006**, *44*, 732–741. [CrossRef] [PubMed]
10. Wouters, K.; van Gorp, P.J.; Bieghs, V.; Gijbels, M.J.; Duimel, H.; Lutjohann, D.; Kerksiek, A.; van Kruchten, R.; Maeda, N.; Staels, B.; et al. Dietary cholesterol, rather than liver steatosis, leads to hepatic inflammation in hyperlipidemic mouse models of nonalcoholic steatohepatitis. *Hepatology* **2008**, *48*, 474–486. [CrossRef]

11. Hendrikx, T.; Walenbergh, S.M.; Hofker, M.H.; Shiri-Sverdlov, R. Lysosomal cholesterol accumulation: Driver on the road to inflammation during atherosclerosis and non-alcoholic steatohepatitis. *Obes. Rev.* **2014**, *15*, 424–433. [CrossRef] [PubMed]
12. Rottiers, V.; Naar, A.M. MicroRNAs in metabolism and metabolic disorders. *Nat. Rev. Mol. Cell Biol.* **2012**, *13*, 239–250. [CrossRef] [PubMed]
13. Williams, M.D.; Mitchell, G.M. MicroRNAs in insulin resistance and obesity. *Exp. Diabetes Res.* **2012**, *2012*, 484696. [CrossRef] [PubMed]
14. Calvopina, D.A.; Coleman, M.A.; Lewindon, P.J.; Ramm, G.A. Function and Regulation of MicroRNAs and Their Potential as Biomarkers in Paediatric Liver Disease. *Int. J. Mol. Sci.* **2016**, *17*, 1795. [CrossRef] [PubMed]
15. Baffy, G. MicroRNAs in Nonalcoholic Fatty Liver Disease. *J. Clin. Med.* **2015**, *4*, 1977–1988. [CrossRef] [PubMed]
16. Dongiovanni, P.; Meroni, M.; Longo, M.; Fargion, S.; Fracanzani, A.L. miRNA Signature in NAFLD: A Turning Point for a Non-Invasive Diagnosis. *Int. J. Mol. Sci.* **2018**, *19*, 3966. [CrossRef] [PubMed]
17. Iravani, F.; Hosseini, N.; Mojarrad, M. Role of MicroRNAs in Pathophysiology of Non-alcoholic Fatty Liver Disease and Non-alcoholic Steatohepatitis. *Middle East J. Dig. Dis.* **2018**, *10*, 213–219. [CrossRef] [PubMed]
18. Dasarathy, S.; Dasarathy, J.; Khiyami, A.; Joseph, R.; Lopez, R.; McCullough, A.J. Validity of real time ultrasound in the diagnosis of hepatic steatosis: A prospective study. *J. Hepatol.* **2009**, *51*, 1061–1067. [CrossRef]
19. Sevastianova, K.; Hakkarainen, A.; Kotronen, A.; Corner, A.; Arkkila, P.; Arola, J.; Westerbacka, J.; Bergholm, R.; Lundbom, J.; Lundbom, N.; et al. Nonalcoholic fatty liver disease: Detection of elevated nicotinamide adenine dinucleotide phosphate with in vivo 3.0-T 31P MR spectroscopy with proton decoupling. *Radiology* **2010**, *256*, 466–473. [CrossRef]
20. Abdel-Hamed, A.R.; Mesbah, N.M.; Ghattas, M.H.; Abo-elmatty, D.M.; Saleh, S.M. Serum miRNA-122 expression in non-alcoholic fatty liver disease among Egyptian patients and its correlation with interleukin-1A gene polymorphism. *Meta Gene* **2017**, *14*, 19–23. [CrossRef]
21. Auguet, T.; Aragones, G.; Berlanga, A.; Guiu-Jurado, E.; Marti, A.; Martinez, S.; Sabench, F.; Hernandez, M.; Aguilar, C.; Sirvent, J.J.; et al. miR33a/miR33b* and miR122 as possible contributors to hepatic lipid metabolism in obese women with nonalcoholic fatty liver disease. *Int. J. Mol. Sci.* **2016**, *17*, 1620. [CrossRef] [PubMed]
22. Cermelli, S.; Ruggieri, A.; Marrero, J.A.; Ioannou, G.N.; Beretta, L. Circulating microRNAs in patients with chronic hepatitis C and non-alcoholic fatty liver disease. *PLoS ONE* **2011**, *6*, e23937. [CrossRef]
23. Hegazy, M.A.; Algwad, I.A.; Fadl, S.A.; Hassan, M.S.; Rashed, L.A.; Hussein, M.A. Serum micro-rna-122 level as a simple noninvasive marker of mafld severity. *Diabet. Metab. Synd. Obes.* **2021**, *14*, 2247–2254. [CrossRef] [PubMed]
24. Jampoka, K.; Muangpaisarn, P.; Khongnomnan, K.; Treeprasertsuk, S.; Tangkijvanich, P.; Payungporn, S. Serum miR-29a and miR-122 as potential biomarkers for non-alcoholic fatty liver disease (NAFLD). *MicroRNA* **2018**, *7*, 215–222. [CrossRef] [PubMed]
25. Salvoza, N.C.; Klinzing, D.C.; Gopez-Cervantes, J.; Baclig, M.O. Association of circulating serum MIR-34a and MIR-122 with dyslipidemia among patients with non-alcoholic fatty liver disease. *PLoS ONE* **2016**, *11*, e0153497. [CrossRef] [PubMed]
26. Yamada, H.; Suzuki, K.; Ichino, N.; Ando, Y.; Sawada, A.; Osakabe, K.; Sugimoto, K.; Ohashi, K.; Teradaira, R.; Inoue, T.; et al. Associations between circulating microRNAs (miR-21, miR-34a, miR-122 and miR-451) and non-alcoholic fatty liver. *Clin. Chim. Acta* **2013**, *424*, 99–103. [CrossRef]
27. Ye, D.; Zhang, T.; Lou, G.; Xu, W.; Dong, F.; Chen, G.; Liu, Y. Plasma miR-17, miR-20a, miR-20b and miR-122 as potential biomarkers for diagnosis of NAFLD in type 2 diabetes mellitus patients. *Life Sci.* **2018**, *208*, 201–207. [CrossRef] [PubMed]
28. Min, H.K.; Kapoor, A.; Fuchs, M.; Mirshahi, F.; Zhou, H.; Maher, J.; Kellum, J.; Warnick, R.; Contos, M.J.; Sanyal, A.J. Increased hepatic synthesis and dysregulation of cholesterol metabolism is associated with the severity of nonalcoholic fatty liver disease. *Cell Metab.* **2012**, *15*, 665–674. [CrossRef] [PubMed]
29. Xu, Y.; Zalzala, M.; Xu, J.; Li, Y.; Yin, L.; Zhang, Y. A metabolic stress-inducible miR-34a-HNF4alpha pathway regulates lipid and lipoprotein metabolism. *Nat. Commun.* **2015**, *6*, 7466. [CrossRef]
30. Slabakova, E.; Culig, Z.; Remsik, J.; Soucek, K. Alternative mechanisms of miR-34a regulation in cancer. *Cell Death Dis.* **2017**, *8*, e3100. [CrossRef]
31. Lendvai, G.; Jarmay, K.; Karacsony, G.; Halasz, T.; Kovalszky, I.; Baghy, K.; Wittmann, T.; Schaff, Z.; Kiss, A. Elevated miR-33a and miR-224 in steatotic chronic hepatitis C liver biopsies. *World J. Gastroenterol.* **2014**, *20*, 15343–15350. [CrossRef] [PubMed]
32. Rodrigues, P.M.; Afonso, M.B.; Simao, A.L.; Carvalho, C.C.; Trindade, A.; Duarte, A.; Borralho, P.M.; MacHado, M.V.; Cortez-Pinto, H.; Rodrigues, C.M.; et al. MiR-21 ablation and obeticholic acid ameliorate nonalcoholic steatohepatitis in mice. *Cell Death Dis.* **2017**, *8*, e2748. [CrossRef] [PubMed]
33. Sun, C.; Huang, F.; Liu, X.; Xiao, X.; Yang, M.; Hu, G.; Liu, H.; Liao, L. miR-21 regulates triglyceride and cholesterol metabolism in non-alcoholic fatty liver disease by targeting HMGCR. *Int. J. Mol. Med.* **2015**, *35*, 847–853. [CrossRef]
34. Feng, Y.H.; Tsao, C.J. Emerging role of microRNA-21 in cancer. *Biomed. Rep.* **2016**, *5*, 395–402. [CrossRef] [PubMed]
35. Okamoto, K.; Koda, M.; Okamoto, T.; Onoyama, T.; Miyoshi, K.; Kishina, M.; Matono, T.; Kato, J.; Tokunaga, S.; Sugihara, T.; et al. Serum miR-379 expression is related to the development and progression of hypercholesterolemia in non-alcoholic fatty liver disease. *PLoS ONE* **2020**, *15*, e0219412. [CrossRef] [PubMed]
36. Vega-Badillo, J.; Gutierrez-Vidal, R.; Hernandez-Perez, H.A.; Villamil-Ramirez, H.; Leon-Mimila, P.; Sanchez-Munoz, F.; Moran-Ramos, S.; Larrieta-Carrasco, E.; Fernandez-Silva, I.; Mendez-Sanchez, N.; et al. Hepatic miR-33a/miR-144 and their target gene ABCA1 are associated with steatohepatitis in morbidly obese subjects. *Liver Int.* **2016**, *36*, 1383–1391. [CrossRef]

37. Hanin, G.; Yayon, N.; Tzur, Y.; Haviv, R.; Bennett, E.R.; Udi, S.; Krishnamoorthy, Y.R.; Kotsiliti, E.; Zangen, R.; Efron, B.; et al. miRNA-132 induces hepatic steatosis and hyperlipidaemia by synergistic multitarget suppression. *Gut* **2018**, *67*, 1124–1134. [CrossRef] [PubMed]
38. Zong, Y.; Yan, J.; Jin, L.; Xu, B.; He, Z.; Zhang, R.; Hu, C.; Jia, W. Relationship between circulating miR-132 and non-alcoholic fatty liver disease in a Chinese population. *Hereditas* **2020**, *157*, 22. [CrossRef] [PubMed]
39. Wang, Y.; Feng, Y.; Zhang, H.; Niu, Q.; Liang, K.; Bian, C.; Li, H. Clinical Value and Role of miR-129-5p in Non-Alcoholic Fatty Liver Disease. *Horm. Metab. Res.* **2021**, *53*, 692–698. [CrossRef]
40. Al Azzouny, M.A.; Behiry, E.G.; Behairy, O.G.; Abd Ellraouf, H.A.; Elfallah, A.A. Serum microRNA-486-5p expression in obese Egyptian children and its possible association with fatty liver. *Diabetes Metab. Syndr.* **2021**, *15*, 102258. [CrossRef] [PubMed]
41. Loyer, X.; Paradis, V.; Henique, C.; Vion, A.C.; Colnot, N.; Guerin, C.L.; Devue, C.; On, S.; Scetbun, J.; Romain, M.; et al. Liver microRNA-21 is overexpressed in non-alcoholic steatohepatitis and contributes to the disease in experimental models by inhibiting PPARalpha expression. *Gut* **2016**, *65*, 1882–1894. [CrossRef]
42. Burg, J.S.; Espenshade, P.J. Regulation of HMG-CoA reductase in mammals and yeast. *Prog. Lipid Res.* **2011**, *50*, 403–410. [CrossRef] [PubMed]
43. Pirola, C.J.; Fernandez Gianotti, T.; Castano, G.O.; Mallardi, P.; San Martino, J.; Mora Gonzalez Lopez Ledesma, M.; Flichman, D.; Mirshahi, F.; Sanyal, A.J.; Sookoian, S. Circulating microRNA signature in non-alcoholic fatty liver disease: From serum non-coding RNAs to liver histology and disease pathogenesis. *Gut* **2015**, *64*, 800–812. [CrossRef]
44. Cheung, O.; Puri, P.; Eicken, C.; Contos, M.J.; Mirshahi, F.; Maher, J.W.; Kellum, J.M.; Min, H.; Luketic, V.A.; Sanyal, A.J. Nonalcoholic steatohepatitis is associated with altered hepatic MicroRNA expression. *Hepatology* **2008**, *48*, 1810–1820. [CrossRef] [PubMed]
45. Vincent, R.; Sanyal, A. Recent Advances in Understanding of NASH: MicroRNAs as Both Biochemical Markers and Players. *Curr. Pathobiol. Rep.* **2014**, *2*, 109–115. [CrossRef]
46. Esau, C.; Davis, S.; Murray, S.F.; Yu, X.X.; Pandey, S.K.; Pear, M.; Watts, L.; Booten, S.L.; Graham, M.; McKay, R.; et al. miR-122 regulation of lipid metabolism revealed by in vivo antisense targeting. *Cell Metab.* **2006**, *3*, 87–98. [CrossRef] [PubMed]
47. Castro, R.E.; Ferreira, D.M.; Afonso, M.B.; Borralho, P.M.; Machado, M.V.; Cortez-Pinto, H.; Rodrigues, C.M. miR-34a/SIRT1/p53 is suppressed by ursodeoxycholic acid in the rat liver and activated by disease severity in human non-alcoholic fatty liver disease. *J. Hepatol.* **2013**, *58*, 119–125. [CrossRef]
48. Zarfeshani, A.; Ngo, S.; Sheppard, A.M. MicroRNA Expression Relating to Dietary-Induced Liver Steatosis and NASH. *J. Clin. Med.* **2015**, *4*, 1938–1950. [CrossRef]
49. Ding, J.; Li, M.; Wan, X.; Jin, X.; Chen, S.; Yu, C.; Li, Y. Effect of miR-34a in regulating steatosis by targeting PPARalpha expression in nonalcoholic fatty liver disease. *Sci. Rep.* **2015**, *5*, 13729. [CrossRef]
50. Zhao, J.; Tang, N.; Wu, K.; Dai, W.; Ye, C.; Shi, J.; Zhang, J.; Ning, B.; Zeng, X.; Lin, Y. MiR-21 simultaneously regulates ERK1 signaling in HSC activation and hepatocyte EMT in hepatic fibrosis. *PLoS ONE* **2014**, *9*, e108005. [CrossRef]
51. Becker, P.P.; Rau, M.; Schmitt, J.; Malsch, C.; Hammer, C.; Bantel, H.; Mullhaupt, B.; Geier, A. Performance of Serum microRNAs -122, -192 and -21 as Biomarkers in Patients with Non-Alcoholic Steatohepatitis. *PLoS ONE* **2015**, *10*, e0142661. [CrossRef] [PubMed]
52. Browning, J.D.; Szczepaniak, L.S.; Dobbins, R.; Nuremberg, P.; Horton, J.D.; Cohen, J.C.; Grundy, S.M.; Hobbs, H.H. Prevalence of hepatic steatosis in an urban population in the United States: Impact of ethnicity. *Hepatology* **2004**, *40*, 1387–1395. [CrossRef] [PubMed]
53. Angulo, P. GI epidemiology: Nonalcoholic fatty liver disease. *Aliment. Pharmacol. Ther.* **2007**, *25*, 883–889. [CrossRef] [PubMed]
54. Mofrad, P.; Contos, M.J.; Haque, M.; Sargeant, C.; Fisher, R.A.; Luketic, V.A.; Sterling, R.K.; Shiffman, M.L.; Stravitz, R.T.; Sanyal, A.J. Clinical and histologic spectrum of nonalcoholic fatty liver disease associated with normal ALT values. *Hepatology* **2003**, *37*, 1286–1292. [CrossRef] [PubMed]
55. Pillai, S.S.; Lakhani, H.V.; Zehra, M.; Wang, J.; Dilip, A.; Puri, N.; O'Hanlon, K.; Sodhi, K. Predicting nonalcoholic fatty liver disease through a panel of plasma biomarkers and micrornas in female West Virginia population. *Int. J. Mol. Sci.* **2020**, *21*, 6698. [CrossRef] [PubMed]
56. Harrison, S.A.; Ratziu, V.; Boursier, J.; Francque, S.; Bedossa, P.; Majd, Z.; Cordonnier, G.; Sudrik, F.B.; Darteil, R.; Liebe, R.; et al. A blood-based biomarker panel (NIS4) for non-invasive diagnosis of non-alcoholic steatohepatitis and liver fibrosis: A prospective derivation and global validation study. *Lancet Gastroenterol. Hepatol.* **2020**, *5*, 970–985. [CrossRef]
57. Puri, P.; Mirshahi, F.; Cheung, O.; Natarajan, R.; Maher, J.W.; Kellum, J.M.; Sanyal, A.J. Activation and dysregulation of the unfolded protein response in nonalcoholic fatty liver disease. *Gastroenterology* **2008**, *134*, 568–576. [CrossRef] [PubMed]

nutrients

MDPI

Review

Recent Advances in the Digestive, Metabolic and Therapeutic Effects of Farnesoid X Receptor and Fibroblast Growth Factor 19: From Cholesterol to Bile Acid Signaling

Agostino Di Ciaula [1,†], Leonilde Bonfrate [1,†], Jacek Baj [2], Mohamad Khalil [1], Gabriella Garruti [3,†], Frans Stellaard [4], Helen H. Wang [5], David Q.-H. Wang [5] and Piero Portincasa [1,*]

1 Clinica Medica "A. Murri", Department of Biomedical Sciences & Human Oncology, University of Bari "Aldo Moro" Medical School, 70124 Bari, Italy
2 Department of Anatomy, Medical University of Lublin, 20-059 Lublin, Poland
3 Section of Endocrinology, Department of Emergency and Organ Transplantations, University of Bari "Aldo Moro" Medical School, 70124 Bari, Italy
4 Institute of Clinical Chemistry and Clinical Pharmacology, Venusberg-Campus 1, University Hospital Bonn, 53127 Bonn, Germany
5 Department of Medicine and Genetics, Division of Gastroenterology and Liver Diseases, Marion Bessin Liver Research Center, Einstein-Mount Sinai Diabetes Research Center, Albert Einstein College of Medicine, Bronx, NY 10461, USA
* Correspondence: piero.portincasa@uniba.it; Tel.: +39-328-4687215
† These authors contributed equally to this work.

Abstract: Bile acids (BA) are amphiphilic molecules synthesized in the liver (primary BA) starting from cholesterol. In the small intestine, BA act as strong detergents for emulsification, solubilization and absorption of dietary fat, cholesterol, and lipid-soluble vitamins. Primary BA escaping the active ileal re-absorption undergo the microbiota-dependent biotransformation to secondary BA in the colon, and passive diffusion into the portal vein towards the liver. BA also act as signaling molecules able to play a systemic role in a variety of metabolic functions, mainly through the activation of nuclear and membrane-associated receptors in the intestine, gallbladder, and liver. BA homeostasis is tightly controlled by a complex interplay with the nuclear receptor farnesoid X receptor (FXR), the enterokine hormone fibroblast growth factor 15 (FGF15) or the human ortholog FGF19 (FGF19). Circulating FGF19 to the FGFR4/β-Klotho receptor causes smooth muscle relaxation and refilling of the gallbladder. In the liver the binding activates the FXR-small heterodimer partner (SHP) pathway. This step suppresses the unnecessary BA synthesis and promotes the continuous enterohepatic circulation of BAs. Besides BA homeostasis, the BA-FXR-FGF19 axis governs several metabolic processes, hepatic protein, and glycogen synthesis, without inducing lipogenesis. These pathways can be disrupted in cholestasis, nonalcoholic fatty liver disease, and hepatocellular carcinoma. Thus, targeting FXR activity can represent a novel therapeutic approach for the prevention and the treatment of liver and metabolic diseases.

Keywords: bile acids; nonalcoholic fatty liver disease; FGF15/19; FXR; agonist; nuclear receptors

Citation: Di Ciaula, A.; Bonfrate, L.; Baj, J.; Khalil, M.; Garruti, G.; Stellaard, F.; Wang, H.H.; Wang, D.Q.-H.; Portincasa, P. Recent Advances in the Digestive, Metabolic and Therapeutic Effects of Farnesoid X Receptor and Fibroblast Growth Factor 19: From Cholesterol to Bile Acid Signaling. *Nutrients* **2022**, *14*, 4950. https://doi.org/10.3390/nu14234950

Academic Editor: Arrigo Cicero

Received: 28 October 2022
Accepted: 20 November 2022
Published: 22 November 2022

1. Introduction

Bile acids (BA) are components of human bile, a brownish or olive-green liquid made of water, organic solutes, and inorganic salt, at a pH of 7.6–8.6. Bile is the major excretion route of cholesterol which appears as solubilized cholesterol and as BA, i.e., degradation products. The three main lipids in bile are bile salts, phospholipids, and nonesterified cholesterol. Free cholesterol accounts for 97% of all sterols in bile, the rest are cholesterol precursors and dietary phytosterols. BA are amphipathic molecules synthesized as "primary" BA from cholesterol in the pericentral hepatocytes. Following their synthesis, primary BA are conjugated with the compound taurine (2-aminoethanesulfonic acid) and the amino acid glycine; this step increases BA solubility in aqueous solutions, such as bile. Primary-conjugated BA are actively secreted from the hepatocyte into bile and are mainly and actively absorbed in the terminal ileum. A small amount will travel to the colon and will undergo the microbiota-dependent biotransformation into "secondary" and "tertiary" BA, which will be passively absorbed. By this dynamic mechanism, BA undergo continuous enterohepatic circulation with 4–12 cycles/day. The fecal loss is minimal at every cycle (5%) and must be compensated by the de novo synthesis of primary BA in the liver (~200–600 mg/daily). The synthesis of BA is modulated by negative feedback mechanisms controlled by the farnesoid X receptor (FXR) in the liver and in the intestine [1,2]. This pathway is strictly linked with the enterohepatic circulation of BA [3], with the signaling role of BA [4], and with the composition and abundance of gut microbiota [5].

Because of this scenario, the role of BA is not simply limited in the emulsification and absorption of dietary fat and fat-soluble vitamins. BA also regulate the proliferation of epithelial cells [6,7], gene expression [8,9], epigenetic mechanisms [10,11], fibrogenesis [12], lipid [13] and glucose metabolism [14]. These effects derive from the role of BA as endogenous ligands and from their ability to activate specific receptors. Besides FXR, involved receptors also include the membrane-associated G-protein-coupled bile acid receptor-1 (GPBAR-1, also known as TGR5 or M-BAR) [6,15], and sphingosine-1-phosphate receptor 2 (S1PR2) [16,17] in the intestine, in the liver, in the muscle and in the brown adipose tissue [18,19]. In the latter decades, the recognition of BA as signaling molecules has shed new light on the complex pathophysiological mechanisms and potential therapeutic implications. Consequences of disrupted BA homeostasis include cholestasis [20], hepatic steatosis, liver fibrosis, and liver tumor [21,22]. Notably, therapeutic manipulation of the BA-FXR axis is paving the way to innovative and potent therapeutic tools [3,23,24], potentially able to act at multiple levels (i.e., BA composition, mitochondrial function, modulation of gut microbiota, glucose and lipid homeostasis, liver inflammation) [25–33]. The present review depicts the complex relationship linking BA and FXR, and how its modulation can become a valid therapeutic target for several liver diseases.

2. The Enterohepatic Circulation and Kinetics of BA and the FXR–FGF19 Dynamics

The cycling of BA between the liver and the intestine is defined enterohepatic circulation, while the total burden of BA in the enterohepatic circulation represents the circulating BA pool [34]. The complex process operating in the enterohepatic circulation of BA has been extensively described in previous papers [19,21]. Briefly, the enterohepatic circulation implies that BA reach the terminal ileum and the colon as well. In the ileum, BA act as signaling molecules and agonists to the nuclear receptor FXR. FXR has antimicrobial effects and protects the intestinal barrier. FXR activation is linked, in humans, with the transcription of the enterokine fibroblast growth factor 19 (FGF19) or the ortholog FGF15 in mice. FGF19 enters the portal flow and has effects on the gallbladder and liver [21]. BA activate FXR with the following rank order: chenodeoxycholic acid (CDCA) > lithocholic acid (LCA) = deoxycholic acid (DCA) > cholic acid (CA) in the conjugated and unconjugated forms [35].

FGF19 acts as agonist of the hepatic FGF receptor 4 (FGFR4)/β-Klotho and activates c-Jun *N*-terminal kinase/extracellular signal-regulated kinase (JNK/ERK). This pathway inhibits the expression of CYP7A1 and CYP8B1 and further hepatic BA synthesis, in concert with the FXR- small heterodimer partner (SHP) inhibitory pathway [36]. BA travelling from the intestine into the portal tract enter the liver via the sodium taurocholate cotransporting polypeptide (NTCP) and organic anion transporting polypeptide (OATP) transporters. FXR works with other nuclear receptors involved in BA homeostasis, namely retinoid X receptor (RXR), SHP, liver receptor homologous-1 (LRH-1), and liver X receptor (LXR) [18]. FXR is the main sensor of BA and regulates synthesis, secretion, and metabolism of BA in the liver, ileum and colon [36–38]. In the liver, FXR regulates target gene transcription by binding to RXRs as a heterodimer [39]. This binding leads to increased transcription of the SHP expression. SHP, in turn, inhibits CYP7A1 expression by blocking transactivation of the hepatic activators LRH-1 and hepatic nuclear factor 4 (HNF-4), at the promoter [39]. This pathway ultimately prevents the activation of target genes involved in the synthesis of fatty acids and BA. By contrast, at low concentrations of BA, LRH-1 operates together with LXR and stimulates the synthesis of BA [40–42]. FXR transcriptionally activates the enzymes involved in BA conjugation to glycine or taurine (bile acid-CoA synthetase [BACS] and bile acid CoA: amino acid *N*-acetyltransferase [BAT]) [43], regulates hepatic BA secretion by bile salt export pump (BSEP), and hepatic phospholipid secretion by ABCB4. BA re-entering the liver also interact with GPBAR-1 on the macrophagic Kupffer cells, together with the pathway stimulated by the FGFR4/β-Klotho dimer. In addition, the activation of FXR induces expression of BA detoxification enzymes (i.e., cytosolic sulfotransferase 2A1 (SULT2A1 [44]), aldol-keto reductase 1 B7 (AKR1B7 [45]), cytochrome P450 3A4/3a11 (CYP3A4/Cyp3a11), and UDP-glycosyltransferase 2B4 (UTG2B4)) [46]. In the kidney (proximal tubule), circulating BA undergo uptake by the apical sodium/dependent bile acid transporter (ASBT). Glomerular filtration of BA are modulated by MRP 2, 3, 4 transporters [47].

Experimental models in animals show that the pathogenesis of NASH is likely linked with a markedly altered composition of BA in the enterohepatic- rather than in the systemic circulation, with specific reduction of secondary BAs (mainly DCA), and scarce presence of FXR and TGR5 ligands in the portal blood. In this case, the major role of enterohepatic BA in the pathogenesis of NASH is confirmed by the protection from NASH secondary to dietary correction (i.e., DCA supplementation) of the BA profile [48].

3. Regulation of BA Homeostasis: The Role of Gut Microbiota

The gut represents a dynamic interface between the internal and the external body environment [49], with several stimuli continuously interacting with the intestinal barrier [50]. The gut microbiota is sensitive to dietary habits and nutrients such as dietary fiber, proteins, fat, carbohydrates, [51]. They are also sensitive to external toxic agents, which include smoking [52], ethanol consumption [53], and environmental pollutants such as heavy metals and pesticides [54–57]. All these factors can affect the diversity and relative abundance of the gut microbiota [58] during the process of enterohepatic circulation, when the primary BA synthetized and secreted by the liver are transformed into secondary BA in the colon [58,59]. BA and gut microbes interact within a continuous bidirectional crosstalk in health but also in disease [50,60,61]. Gut dysbiosis can disrupt BA homeostasis and can change the composition of the BA pool, while the increased production of deoxycholic acid (DCA), a cytotoxic secondary BA, can damage the composition of gut microbiome [9]. BA bind to FXR and this step produces antimicrobial peptides (AMPs) (i.e., angiogenin 1 and RNase family member 4). These AMPs play an active role in the inhibition of gut microbial overgrowth and intestinal barrier dysfunction [62]. BA can also modulate the gut microbiome by stimulating the growth of BA-metabolizing bacteria or inhibiting other bile-sensitive bacteria. Gut *Eubacterium lentum, Ruminococcus gnavus* and *Clostridium perfringens* decrease the antimicrobial effect of BA via the iso-BA pathway by transforming DCA and LCA into iso-DCA and iso-LCA (3b-OH epimers). The secondary DCA has hydrophobic and cytotoxic profiles with detergent effects on the bacterial cell membranes and antimicrobial properties [63].

Changes of the microbiota and therefore BA pool can interfere with activation of signaling pathways [64] involved in intestinal, metabolic homeostasis and tumorigenesis. Dietary changes associated with low short-chain fatty acids have been linked with an high risk for cancer development [65,66]. Colorectal cancer represents another example since, in humans, dietary habits can vary BA conjugation. For example, diets enriched in animal protein favor taurine conjugation while vegetarian diets favor glycine conjugation. In this scenario, TCA can to stimulate gut microbes able to convert taurine and cholic acid to hydrogen sulfide and DCA, which act as genotoxin and tumor-promoter, respectively [65].

Dysbiosis and variations of the BA pool composition and size can play a role in alcohol associated liver disease (ALD) [67], in NAFLD, obesity, and type 2 diabetes [58,68–70].

Compared to healthy controls, higher levels of total faecal BA, primary CA and CDCA, and higher BA synthesis have been reported in patients with NASH, who also showed a higher ratio of primary to secondary BA, but a similar ratio of conjugated to unconjugated BA. Patients with NASH were also characterized by a decreased count of *Bacteroidetes* and *Clostridium leptum*. The count of *C. leptum* increased with fecal unconjugated LCA and decreased with unconjugated CA and CDCA. Taken together, these findings point to a link between NAFLD, dysbiosis and altered BA homeostasis, which puts patients at risk of further hepatic injury [71].

Dysbiosis can be also implicated in tumorigenesis via BA-induced changes. DCA increases on a western diet and becomes a predisposing factor to intestinal carcinogenesis. In DCA-treated APC (min/+) mice, a disrupted intestinal microbiota was associated with altered gut barrier, low-grade gut inflammation and tumor progression. Fecal microbiota transplantation from DCA-treated mice to Apc (min/+) mice increased tumor multiplicity, caused inflammation, recruited the M2 phenotype tumor-associated macrophages, and activated the tumor-associated Wnt/beta-catenin signaling pathway. Notably, the antibiotic-induced depletion of the microbiota blocked DCA-induced intestinal carcinogenesis [72]. Another important study found that a 2-week food exchanges profoundly affected the microbiota, metabolome profile, mucosal biomarkers of cancer risk when African Americans were fed a high-fibre, low-fat African-style diet and rural Africans were fed a high-fat, low-fibre western-style diet. In the African Americans the study found a protective profile by saccharolytic fermentation and butyrogenesis and suppressed secondary BA synthesis [66].

4. The Microbiota–BA–FXR Axis

The BA-FXR axis protects the liver for BA overload and potential harmful effects of BA upon accumulation in the hepato-biliary-intestinal tract [73]. Additional aspects of such interaction include systemic metabolic effects deriving from FXR activation and FGF19 secretion [19]. Thus, changes in the profile of the BA pool and primary/secondary BA ratio can produce multiple consequences, including altered metabolic pathways and liver damage [71,74,75]. In this context, a major role is played by microbiota-induced deconjugation of BA. This is the case when the murine primary BA Tauro-beta-muricholic Acid (TβMCA), a natural occurring FXR antagonist, will be deconjugated by the microbiota. This step will alleviate FXR signaling in ileum leading, in the liver, to inhibition of BA synthesis [60,76]. In normal conditions with a functioning enterohepatic circulation of a physiological BA pool, the FXR-induced gene expression including *Ang1*, *iNos* and *Il18* in ileum, has antimicrobial action, enteral protection and inhibition of bacteria damage to the intestinal mucosa. In rodents, the biliary obstruction that causes small intestinal bacterial overgrowth (SIBO) can be reversed by administration of BA [77,78]. A study in the cholestatic model of mice also reported protection by a potent synthetic agonist of FXR [79]. Metabolic studies in mice show that the gut microbiota promotes diet-induced obesity via FXR signaling [62,75,76]. In an animal model of NAFLD induced by high-fat diet, the antibiotic treatment decreased BSH-encoding *Lactobacillus*, increased the synthesis of the FXR antagonist TβMCA, and improved insulin resistance and liver steatosis [80]. In humans with newly diagnosed type 2 diabetes, naïve treatment with metformin modified gut microbiota, decreasing *Bacteroides fragilis* and increasing the BA glycoursodeoxycholic acid (GUDCA) in the gut. These changes were paralleled by an inhibition of FXR signaling, pointing to GUDCA as an intestinal FXR antagonist able to improve metabolic dysfunction [81]. The inhibition of intestinal FXR is a key factor for the progression of NAFLD mediated by gut microbiome [75]. Mice fed a high milk- fat diet showed a shift in BA composition with increased TCA and expansion of *Bilophila wadsworthia*. This proteobacterium is recognized as a "bile-loving" microorganism associated to inflammatory bowel diseases [82,83].

In NAFLD rats, the administration of probiotics is able to significantly increase the expression of FXR, FGF15 mRNA, and protein in the liver, upregulating the diversity of gut microbiota, downregulating the abundance of pathogenic bacteria and finally alleviating NAFLD [26].

One note, changes in gut microbiota composition are possible following FXR/FGF19 modulation. In humans with biopsy-confirmed NASH, administration of the FGF19 analog aldafermin induced in the microbiota a dose-dependent enrichment in *Veillonella*, a rare commensal microbe which correlated with changes in serum bile acid profile, mainly in terms of decreased toxic BA [84].

Dietary lipids likely impact the gut microbial composition directly as a substrate or by shifting in BA composition [85]. The microbiota FXR signaling modulation was also assessed during the sub-ministration of the antioxidant Tempol, which resulted in a decrease of *Lactobacillus* and *Clostridium* (clusters IV and XIVa), accompanied by decreased BSH, accumulation of TβMCA and suppressed FXR signals [76]. Other nuclear receptors can also be involved during the interaction of microbiota with intestinal BA, such as the pregnane X receptor (PXR), vitamin D receptor (VDR), GPBAR-1 and Sphingosine-1-Phosphate Receptor 2 (S1PR2) [86–88].

5. Fibroblast Growth Factors FGF15 and Human Ortholog FGF19

Fibroblast growth factors FGF15 (rodent) and the human ortholog FGF19 are enterokines which play a major role in BA homeostasis and key metabolic functions, in concert with FXR [89]. The transcriptional regulation of *FGF19* involves the FXR during the enterohepatic circulation of BA [90,91]. Additional players include vitamin D receptor (VDR), pregnane X receptor (PXR), drugs, vitamins and cholesterol [92]. The sterol regulatory element-binding protein 2 (SREBP2) is a negative transcriptional regulator of *FGF19* [93]. FGFs belong to a family of at least 22 proteins with an effect on growth, development, and differentiation [90,94,95]. FGFs, currently grouped into six subfamilies, play an autocrine and paracrine role via activation of specific tyrosine kinase FGF receptors undergoing dimerization and activation of the cascade of intracellular signaling pathways [96]. The trio consisting of FGF15 (and the human ortholog FGF19), FGF21, and FGF23 act as circulating hormones [97]. FGF19 is the human protein encoded by the *FGF19* gene and has endocrine hormonal functions at a systemic level [90,98]. FGF19 was originally identified in the fetal brain and, with FGF15 only shares ~50% amino acid identity [99]. FGF15/19 are predominantly expressed in the small intestine, gallbladder, kidney, skin, cartilage, and brain [36,100,101].

Previous studies confirmed that BA or FXR agonists induced FXR responsive element (FXRE) activation and FGF19-dependent repression of CYP7A1 (and therefore BA synthesis) in human hepatocytes [102]. A further proof was that *Fgf15*-knockout mice and intestine-specific *Fxr*-knockout mice stimulated by FXR agonists could not repress CYP7A1 [103].

FGF19 displays two daily peaks (about 3pm and 9pm). The serum peak of FGF15/19 follows the increase of BA in the small intestine and increases about 1.5–2.0 h after the rise of postprandial serum levels of BA [104,105]. FGF19 is a small 25 kDA molecule with fasting levels varying from 49 to 590 pg per mL. Following CDCA activation of the intestinal FXR, serum levels of FGF19 increase by +250%. FGF19 half-life is 30 min short, likely dependent on renal elimination [106]. FGF19 circulating levels can either decrease or increase in extrahepatic cholestasis, inflammatory bowel disease, kidney disease, BA malabsorption, obesity, and liver steatosis. Therapies involving FGF19 seem promising. However, the translational value of these attractive therapeutic tools should be carefully verified in terms of possible hepatic tumorigenesis, as a consequence of their mitogenic potential [107,108]. The effect likely involves the FGF19-FGFR4 interaction, since blocking the receptor can prevent, in rodents, the development of hepatocellular carcinoma [109].

The function of secreted FGFs requires the interaction with the transmembrane Klotho proteins found at the fibroblast growth factor receptors (FGFRs) [106]. FGF19 binds to the receptor complex composed of the FGF receptor 4 (FGFR4) and a co-receptor β-Klotho, both highly expressed in liver. Whereas the carboxy-terminal domain of FGF19 is the segment specifically recognized by β-Klotho, the amino-terminal region is specific for the FGF19–FGFR interaction [110].

Affinity of FGF15/19 is greater for the FGFR4, which is primarily expressed in the liver, than for FGFR1, primarily expressed in the white adipose tissue (WAT) [111,112]. FGFR4 is also expressed in other cell types as macrophages, HSCs and some central neurons [113]. The human FGF19 is not effective on mouse FGFR4 [114]. Binding of FGF19 to the FGFR4–β-Klotho complex activates a signaling pathway which includes the small guanosine triphosphatase Ras, the extracellular signal–regulated protein kinase 1, 2 (ERK1, ERK2), JUN *N*-terminal kinase (JNK), fibroblast growth factor receptor substrate 2α (FRS2α) [115], the growth factor receptor-bound protein 2 (GRB2), cAMP-response element-binding protein (CREB) which is de-phosphorylated and inactivated [89,106,116].

FGF19 is also able to modulate gallbladder volume, which contributes to the pulsatile and dynamic function of the enterohepatic circulation in the fasting and postprandial period [117,118]. During the postprandial period, the fat-stimulated CCK release from the upper gut enterocytes promotes smooth-muscle-mediated gallbladder contraction and ejection of concentrated bile into the duodenum. This phenomenon is well visible during the ultrasonographic functional study of time-dependent changes of gallbladder volume in response to a meal, by following the rhythmic alternation of emptying-refilling episodes [117,119–121]. Upon gallbladder contraction, BA in the duodenum inhibit further CCK production and, in turn, gallbladder contraction [122]. BA arriving in the distal ileum, moreover, will stimulate the release of FGF19 which, by activation of the gallbladder FGFR4-β-Klotho, promotes gallbladder relaxation ready for the next filling with dilute hepatic bile. FGF-19 interacts with the liver FGFR4-β-Klotho leading to suppression of the synthesis of hepatic BA [21].

6. FXR–FGF19 and BA Homeostasis

The hepatic and intestinal expression of FXR contributes to the regulation of BA synthesis and homeostasis [103], via a negative gut-liver feed-back sustained by the intestinal FGF15/19. In the liver, mild activation of FXR and SHP suppresses the CYP7A1/*Cyp7a1* gene [40,123]. In parallel, the *Cyp8b1* gene repression via FXR depends equally on both intestinal and liver FXR [103]. The key role of FXR in BA homeostasis is testified by the regulatory capacity of inducing the expression of key transporters active in the enterohepatic circulation of BA, i.e., BSEP, IBABP and OSTα/β, while suppressing NTCP and ASBT [124–126]. The BA-FXR-FGF15/19-BA sequence has a fundamental role in governing BA homeostasis. *FGFR4* overexpression downregulates CYP7A1 and decreases the BA pool size [127].

By contrast, Knockout mice *Fgf15-*, *Fgfr4-* and *β-klotho (Klb)-* display impaired BA metabolism. Exogenous FGF19 fails to rescue this condition and fails to repress the function in CYP7A1 *Fgfr4-* and *Klb*-knockout animals [128,129]. The interaction between FGF15/19 and β-klotho-FGFR4 in the liver requires additional nuclear receptors to activate CYP7A1 repression. These nuclear receptors include the liver receptor homologue 1 (LRH-1) and hepatocyte nuclear factor 4α (HNF4α) [130] and SHP [131]. In addition to this complex picture, BA bind to ileal membrane-associated receptor GPBAR1. This activation is also associated with several extra-intestinal tissues to mediate host energy expenditure [132,133], glucose homeostasis [134], anti-inflammatory and immune responses [135,136].

7. FXR–FGF15/19 Pathway and Metabolic Effects

The axis FXR–FGF15/19 is also implicated in a number of metabolic functions which involve fat, glucose, glycogen, protein homeostasis, and energy expenditure [106].

7.1. Lipid Homeostasis

The pathways connecting cholesterol to BA synthesis are essential and contribute to the prevention of pathological amounts of cholesterol in the body [137], and also to metabolic homeostasis. Hepatic BA synthesis starts from cholesterol, and this step represents the main catabolic pathway of cholesterol metabolism in humans [21]. These pathways act through the "classic" neutral pathway (cholesterol 7α-hydroxylase, CYP7A1), contributing to about 75% of the total BA pool, and the "alternative" acidic pathway (sterol 27-hydroxylase, CYP27A1) contributing to about 25% of the total BA pool [21].

In this context, FXR has a relevant role in atherosclerotic risk factors, and strongly modulates the homeostasis of cholesterol due to the ability to inhibit CYP7A1.

Of note, increased serum levels of primary and secondary BA have been reported in patients with NAFLD, as compared with healthy controls [9]. This finding should suggest a role for the elevated BA production in the pathogenesis of NAFLD. This link, however, seems secondary to the increased proportion, in the BA pool, of the FXR antagonistic DCA, and to decreased levels of the FXR agonistic CDCA [9]. On the other hand, the cholesterol catabolism through BA synthesis has beneficial effects on metabolic homeostasis, and hepatic accumulation of cholesterol is a critical culprit for the development of NAFLD/NASH [138]. Transgenic mice overexpressing CYP7A1 in the liver are resistant to high-fat diet-induced obesity, fatty liver and insulin resistance, mainly through increased hepatic cholesterol catabolism and increased BA pool [139]. Further studies in FXR-deficient and CYP7A1-deficient mice confirmed the beneficial effects on hepatic inflammation of the increased CYP7A1 expression and BA synthesis secondary to activation of FXR [27].

FXR-deficient mice display increased content of hepatic lipids and increased plasma cholesterol and triglycerides [140]. By contrast the activation of FXR by the synthetic agonist GW4064 or by BA decrease liver steatosis and serum triglycerides [141,142]. Mechanisms include suppression of de novo fatty acid synthesis [141]. Primary hepatocytes incubated with the recombinant FGF19 protein display suppression of insulin-dependent stimulation of fatty acid synthesis with or without insulin [143]. FXR activation induces the SHP-mediated suppression of sterol regulatory element-binding protein 1c (SREBP1c), a step decreasing triglyceride load in hepatocytes [141]. This pathway also occurs with FGF19 via increased expression of SHP, increased signal transducer and activator of transcription 3 (STAT3) and decreased peroxisome proliferator-activated receptor γ coactivator-1β (PGC-1 β) [143]. An additional mechanism is the FXR-mediated activation of ApoCII gene transcription in the liver. This step, in turn, activates the lipoprotein lipase and decreases ApoCIII, a lipoprotein lipase inhibitor. The net effect is increased lipolysis of triglycerides in the vasculature [30,144]. Additional effects of FXR activation include increased triglyceride hydrolysis and clearance, increased fatty acid oxidation [28–30], and decreased hepatic export of very low-density lipoprotein (VLDL) [145].

Negative-correlations have been recently shown between stimulated intestinal FXR-FGF15 secretion and pathways linking hepatic cAMP regulatory element-binding protein (CREB) and peroxisome proliferator-activated receptor gamma, coactivator 1α (PGC1A). This effect indicates a possible downregulation of hepatic PGC1α by inactivation of CREB, finally resulting in suppression of fatty acid oxidation [146].

FXR contributes to the control of cholesterol homeostasis. Human cells lack enzymes able to degrade the ring structure of cholesterol and excess cholesterol must be excreted to avoid a harmful body accumulation. The excretion of cholesterol is mainly achieved in bile via the physiological carriers BA and phospholipids assembled with cholesterol as micelles and vesicles [147]. In addition, the concept of reverse cholesterol transport (RCT) encompasses the pathway transporting excess cholesterol accumulating within peripheral tissues back to the liver for biliary excretion into the feces [148]. FXR stimulates the RCT, as well as the trans-intestinal cholesterol excretion (TICE), the latter representing a significant alternative route to the biliary pathway of RCT. Mechanisms linking FXR to RCT include the changes in the BA pool hydrophobicity and their micellar function with cholesterol [148,149].

A critical role in the relationships between FXR, glucose, fatty acid metabolism and homeostasis seems to be played by the key enzyme pyruvate dehydrogenase kinase 4 (PDK4). In fact, an increased expression of this enzyme has been reported in FXR-null mice, and the inhibition of PDK4 expression alleviated lipid accumulation in hepatocytes both in vivo and in vitro [150].

7.2. Glucose Homeostasis and Gluconeogenesis

In both fed and fasted states, the liver is essential to maintain physiological blood glucose levels. Postprandially, exogenous glucose is used in liver to synthesize glycogen and triglycerides. During fasting, glucose is produced after gluconeogenesis and glycogenolysis. FGF15/19 release mediates the effects of FXR on glucose and lipid regulation. FGF15-deficient mice fail to maintain blood concentrations of glucose and normal postprandial amounts of liver glycogen [151]. Activation of the FXR-FGF15/19 axis is involved in glucose metabolism resulting in reduced hepatic gluconeogenesis and glycolysis, and paralleled by increased glycogen synthesis [152]. The mechanisms involve the downstream signaling via SHP-mediated suppression of transcription factors critically involved in gluconeogenesis [153]. Mechanisms might differ between species [154,155] or experimental models, and the role of FXR in glucose metabolism, although evident, is somewhat difficult to interpret. FXR-deficient mice show increased serum glucose levels and impaired glucose and insulin tolerance. By contrast, hepatic gluconeogenesis decreases, and insulin sensitivity improves if FXR is stimulated by feeding the primary BA cholate, treating with the agonist GW4064 or if FXR is overexpressed [142,156]. These results must be interpreted in the context of diet-induced obesity. In aged mice, hepatic loss of FXR and the FXR target SHP improves lipid and glucose homeostasis [157], likely due to an increase in autophagic gene expression (normally repressed by FXR postprandially) [157,158].

Additional sets of results originate from the manipulation of intestinal FXR. Mice with selective genetic deletion of intestinal FXR displayed protection against diet-induced diabetes and obesity [75,76].

Here, additional complex pathways might play a role. FXR might inhibit the secretion of the intestinal incretin Glucagon-like peptide 1 (GLP-1), as shown by administering the FXR agonist GW4064 [159]. The lack of intestinal FXR would therefore promote the intestinal L-cell release of GLP-1, which improves glycemic control and promotes weight loss.

In type 2 diabetic mice, the FXR antagonist Mebhydrolin improved glucose homeostasis by suppressing hepatic gluconeogenesis via FXR/miR-22-3p/PI3K/AKT/FoxO1 pathway. An additional effect was the promotion of glycogen synthesis through FXR/miR-22-3p/PI3K/AKT/GSK3β pathway [31].

Fexaramine is the intestine-specific FXR agonist that increases FGF15 signaling, leading to altered BA pool and increasing the level of the secondary tauro-conjugated BA taurolitocholic acid (TLCA) [160]. TLCA is a strong agonist of the membrane-associated receptor GPBAR-1 in the intestinal (ileum, colon) L-cell, and this step induces the secretion of Glucagon-like peptide 1 (GLP-1). This step appears to explain the improved glucose tolerance and insulin resistance, as well as the stimulation of browning of white adipocytes and energy expenditure i.e., produced heat through uncoupled electron transport in the mitochondria [161,162] by fexaramine in mice [163]. Additional pathways are likely active, since fexaramine treatment in GPBAR-1-deficient mice was also associated with increased energy expenditure [162].

The enterokine FGF15/19 also plays a role in glucose metabolism. FGF19 acts independently from the activity of insulin or the protein kinase Akt. FGF19 acts through a mitogen-activated protein kinase signaling pathway that activates components of the protein translation machinery and stimulates glycogen synthase activity. The processes are tightly regulated to maintain glucose homeostasis. As insulin, FGF15/19 inhibits gluconeogenic gene expression [105]. However, insulin acts through Akt-dependent phosphorylation and subsequent degradation of FOXO1, a transcription factor involved in fasting-mediated induction of gluconeogenic gene expression. Conversely, FGF19 operates promoting dephosphorylation and inactivation of the transcription factor cAMP regulatory element-binding protein (CREB). This effect blunts the expression of peroxisome proliferator-activated receptor gamma coactivator-1alpha (PGC-1α) and other genes involved in hepatic metabolism, such as glucose-6-phosphatase (G6pase). In human and rat hepatocytes and in mouse livers, activation of FXR increased glucose levels via expression of phosphoenolpyruvate carboxykinase (PEPCK) [105,164]. In turn, the overexpression of PGC-1α blocks the inhibitory effect of FGF15/19 on the expression of gluconeogenic gene. In support of this function, mice lacking FGF15 are not able to maintain proper blood concentrations of glucose [151].

Notably, mice treated with FGF19 or overexpressing FGF19 have a lower body weight despite elevated food intake. Mice had repression of acetyl-CoA carboxylase 2 (ACC2) and stearoyl-CoA desaturase 1 (SCD1), enhanced energy expenditure and were protected against diet-induced obesity. Decreased ACC2, in turn, decreases mitochondrial Malonyl-CoA levels, and is associated with the upregulation of Carnitine palmitoyltransferase I (CPT1), more availability of fatty acids for oxidation [165].

In addition, FGF19 inhibits the synthesis of hepatic fatty acid after suppressing SREBP1c activity [143]. The translational meaning of the above-mentioned studies requires extensive validation [166]. A role for FGF19 at the level of the central nervous system in improving glycemia and peripheral insulin signaling is also possible [167], and points to novel therapeutic targets with antidiabetic drugs. FGF15/19 works in tune with insulin (serum peak of 15 min), as a postprandial regulator of hepatic carbohydrate homeostasis.

7.3. FGF19 and Role in Glycogen Synthesis

Hepatic glycogen synthesis is also under the regulation of FGF15/19. In line with this function and compared to control wild-type mice, *Fgf15* knockout mice become glucose intolerant and store half as much glycogen in the liver. The defect is rescued by administration of FGF19 [151]. Diabetic mice lacking insulin have defective glycogen storage but FGF19 treatment rescues hepatic glycogen concentrations to normal levels. These models suggest that FGF19 activates insulin-independent pathways in regulating glycogen metabolism [151]. To promote the synthesis of glycogen and proteins, FGF19 activates the Ras/ERK pathway, whereas insulin activates the PI3K/Akt pathway [151].

Glycogen synthesis in the liver is regulated negatively by glycogen synthase kinase (GSK) 3α and GSK3β. The mechanism is modulated by the effects of BA on phosphorylation and inhibition of glycogen synthase (GS). Phosphorylation also inactivates GSK3 kinases, thus preventing the inhibition of GS and increasing glycogen synthesis. FGF19 acts directly on the liver and induces the phosphorylation of both GSK3α (Ser^{21}) and GSK3β (Ser^{9}), in parallel with a reduced phosphorylation of Ser^{641} and Ser^{645}. In this scenario, GS activity increases, liver glycogen content increases by about 30% without effect on liver cholesterol or triglycerides, as well as insulin or glucagon level. In accord with a direct function of FGF19 on the hepatocyte and glycogen synthesis, it is apparent that fed *Fgf15−/−* mice had >50% less hepatic glycogen and impaired glucose uptake from the circulation, as compared with wild-type animals. FGF19 administration rescued this phenotype completely [151].

Notably, maximal serum insulin levels occur within 1 h of a meal, serum FGF19 levels peak about 3 h after a meal [104] and, in human subjects, liver glycogen levels peak ~4 h after a meal [168]. The conclusive picture which arises from this experimental scenario is that insulin and FGF19 work in a coordinated temporal fashion, and that this synergy facilitates the postprandial storage of nutrients. FGF19, as compared to other anabolic enterokines, namely incretins, GLP-1 and GIP, appears to mimic insulin action independently and without stimulating its release.

7.4. FGF19 and Role in Protein Synthesis

In mice, the acute administration of FGF19 increases total protein synthesis by 18% and synthesis of albumin by 40%, while prolonged FGF19 administration increases levels of plasma albumin by 10% [151]. FGF19 stimulates hepatic protein synthesis with the phosphorylation of the ERK1/2, phosphorylation of eIF4B, eIF4E and the ribosomal protein S6 kinase (S6K1), an mTOR- dependent master regulator of muscle cell growth and therefore muscle weight [169]. In primary hepatocytes, the activation of FXR by the agonist INT-747 stimulates amino acids catabolism. In vivo, FXR activation increased ammonium clearance through induction of ureagenesis and glutamine synthesis [170]. In humans, the links between FGF19 and protein synthesis might play a critical role in muscle homeostasis. Lower FGF-19 levels and higher FGF-21 levels have been reported in elderly patients with- as compared to those without sarcopenia, impacting muscle strength [171]. This finding is paralleled by results deriving from animal models, in which antibiotic therapy induced skeletal muscle atrophy secondary to microbial dysbiosis, aberrant BA metabolism and inhibition of the FXR-FGF15 signaling. Of note, in this model, skeletal muscle loss was partly reversed by administration of FGF19 [172].

7.5. FGF19 and Role in Energy Expenditure

FGF19 has profound effects on overall metabolism and energy expenditure. The key role of FGF19 in maintaining the physiological homeostasis is testified at various levels and by different models. In the animal models using transgenic mice which express human FGF19 and have decreased fat mass. Mice did not become obese or diabetic when fed a high fat diet. This outcome appears to be the consequence of increased energy expenditure. Likely, FGF19 may increase energy expenditure via increase in brown adipose tissue (BAT) mass. In addition, reduced liver triglyceride levels derive from decreased liver expression of acetyl coenzyme A carboxylase 2 (ACC2) and, in turn, decreased levels of mitochondrially associated malonyl CoA levels and increased activity of carnitine palmitoyl transferase 1 (CPT1), with increased availability of fatty acids for β oxidation [173]. FGF19 increased metabolic rate in mice fed a high fat diet, and this effect was paralleled by reduced body weight and diabetes in leptin-deficient mice [165]. In the central nervous system FGF19 plays an additional role since it improves insulin sensitivity by reducing the activity of hypothalamic agouti-related peptide (AGRP)/neuropeptide Y (NPY) [174]. Thus, FGF15 and FGF19 appear to have several metabolic effects which can be summarized as weight loss, increased insulin sensitivity and thermogenesis. Serum concentrations of cholesterol and triglyceride also decrease with stimulation of FGF15/19. The translational value of such findings, however, requires additional evidence in humans. Postprandially, FGF19 and insulin promote protein and glycogen synthesis and suppresses hepatic gluconeogenesis [89]. At variance with insulin, however, lipogenesis is not stimulated by FGF19, due to different cellular signaling pathways. Taken together, this evidence suggests that FGF15/19 brings beneficial effects on metabolic syndrome and treatment of NASH. Modified FGF19 is beneficial in mouse models of NASH and cholestasis [175]. To harmonize somewhat conflicting results concerning FXR regulation of lipid/glucose metabolism, and possibly metabolic syndrome, it should be noted that FXR knockout mice develop fatty liver, increased serum concentrations of free fatty acids (FFAs) and glucose, and display insulin resistance [156]. In addition, overexpression or activation of hepatic FXR in diabetic db/db and wild-type mice was associated with decreased serum concentrations of FFAs and glucose, and with increased insulin sensitivity [142].

8. The Role of FXR in Liver Disease

8.1. Inflammation and Fibrosis

Experimental evidence show that FXR can have an anti-inflammatory function in the liver by reducing cholestasis and levels and accumulation of toxic BA [32,176]. The migration and infiltration of monocytes/macrophages is regulated by the chemokine monocyte chemoattractant protein-1 (MCP-1/CCL2) [177]. The synthetic FXR agonist WAY-362450 reduced MCP-1 expression and inflammatory cell infiltration in the liver of mice put on methionine-choline deficient (MCD) diet, which induces NASH. The reduction of hepatic fibrosis by WAY-362450 treatment was paralleled by a reduction in hepatic gene expression of fibrosis markers and was specifically linked to the presence of FXR [32]. The expression of various pro-inflammatory genes is induced by the transcription factor Nuclear factor kappa-light chain enhancer of activated B cell (NF-κB) [178]. Treatment with lipopolysaccharide

(LPS) induced in FXR KO mice a strong hepatic inflammation, with massive liver necrosis and marked increase in hepatic cytokine signaling molecules inducible nitric oxide synthase (iNOS), cyclooxygenase-2 (COX-2), and interferon-γ (IFN-γ) [176]. The use of FXR agonists in HepG2 cells and mouse primary hepatocytes suppressed NF-κB mediated inflammation in a FXR-dependent manner [176]. Notably, FXR activation mitigates the development of liver fibrosis due to an increased anti-fibrotic gene expression in hepatic stellate cells (HSCs). The pathway is based on the activation of FXR, induction of SHP, increased expression of peroxisomal proliferator activated receptor γ (PPARγ) with inactivation of HSCs [179,180].

In an animal model of cholestasis induced by parenteral nutrition, the administration of the FXR agonist GW4064 prevented hepatic injury and cholestasis. Treated animals showed a normalization of serum BA levels which were associated with increased expression of canalicular bile, of sterol and phospholipid transporters, and with suppression of macrophage recruitment and activation. These effects were secondary to the restoration of hepatic FXR signaling [181].

FGF15/19 can also play a role in inflammation. FGF15 deficiency is associated with the loss of FGF15-mediated suppression of BA synthesis, increased FXR activation, and reduced hepatic fibrosis [182]. In a human HSC cell line, LX2, FGF19 suppresses inflammation through modulating inhibitor of nuclear factor kappa B (IκB) activity without suppression of fibrogenic gene expression [182].

8.2. Cholestasis

The BA pool is made of amphipathic BA with either protective or toxic effects depending on the tight maintenance of the hydrophilic-hydrophobic balance [183]. Notably, BA structure is responsible for their double signaling function as protective molecules (i.e., proliferation in hepatocytes) or toxic molecules [184]. The process of cholestasis starts with decreased or abolished bile flow. Several conditions may be responsible for cholestasis, but the ultimate step leading to cholestatic liver injury is the intrahepatic accumulation of BA, a situation also defined as BA overload. Chronic BA overload will inevitably cause a progressive BA-induced damage, with the aid of additional toxic components [20,185]. Excess BA retention generates hepatocyte damage, steatosis, fibrosis and even liver tumorigenesis [21,22].

BA overload can develop at the hepatic and/or systemic level [186] when trans-hepatocyte BA flow is scarce because of decreased sinusoidal and/or canalicular BA transport [184,187], or in the presence of bile duct obstruction. Both extended (>70%) partial hepatectomy and massive hepatocyte loss [184,188–191] are a predisposing condition to BA overload [191,192]. Therefore, BA spillover will be evident into the systemic circulation, as confirmed in both animal and human models [19,187,193–196].

In the model where mice are put on a LCA-enriched diet [197] or to develop impaired pathways of SIRT1 [44], FGF receptors [198,199], SHP [200], FGF15 [201] studies show the expansion of a hydrophobic BA pool which interferes with the liver repairment capacity. Another example is that Cyp2c70-/- mice develop a more human-like hydrophobic BA pool with liver inflammation [202] and altered FXR signaling [203]. This is due to the presence of a more hydrophilic bile in mice, as compared to humans, since the enzyme CYP2c70 (missing in humans) converts CDCA to the more hydrophilic muricholic acid [204].

In line with such evidence, BSEP/abcb11-/- mice develop a mild non-progressive cholestasis [197], likely due to an enrichment of the BA pool with hyper-hydroxylated, less hydrophobic, and less cytotoxic BA [205]. In humans, progressive familial intrahepatic cholestasis (PFIC) is an autosomal recessive disease causing 15% of cases of neonatal cholestasis. The PFIC2 form has a mutation in the *ABCB 11* gene encoding BSEP. PFIC patients develop a disrupted secretion of BA from the hepatocytes, progressive hepatic fibrosis, liver cirrhosis and end stage liver disease requiring liver transplantation [206]. Thus, beneficial effects during cholestatic liver injury might derive from an increased hydrophilic profile of the BA pool, either in the animal model [207] and in PFIC children (i.e., by increasing the content of tetrahydroxy BA) [208]. BA overload will cause damage [209] via deranged mitochondrial function [210,211]. This step involves the release of cytochrome c and release of excessive reactive oxygen species (ROS) [212], plasma membrane damage, necrosis, apoptosis and cell death [213]. The indirect damage of excess BA retention involves inflammatory changes associated with cytokine release, neutrophils recruitment and macrophages activation [209,214].

Up to a certain limit, FXR-dependent adaptive responses will try to counteract the cholestatic liver damage. A first preventive mechanism includes the intrahepatic activation of the physiological BA sensor FXR by BA. Both basolateral and canalicular BA transporters and enzymes governing BA synthesis and conjugation are involved [193,215] and contribute to counteracting BA overload in the hepatocytes [216]. FXR-dependent mechanisms can modulate hepatocyte cell cycle progression [192], with a possible regulation of BA homeostasis [114], alcohol-related liver injury [217] and liver regeneration after partial hepatectomy [218,219]. FXR pathways can involve cholangiocyte cell cycle progression [219,220] during BA synthesis suppression [221,222]. The pro-inflammatory effect of BA is mediated by the intracellular assembly of the inflammasome, but FXR exhibits anti-inflammatory effects, because of the interaction with the NLRP3 protein machinery [223]. The modulatory effects of FXR on BA homeostasis and hepatocyte/cholangiocyte cell cycle progression are important to decrease liver BA uptake (inhibition of BA transporters), BA synthesis (suppression of CYP7A1/*Cyp7a1* gene [36,123] and *Cyp8b1* gene [103]), while stimulating BA excretion (activation of BA transporters [124,125]).

FXR is also expressed by mast cells infiltrating the liver during cholestasis, and promoting hepatic fibrosis. As shown by an animal model, FXR expressed by mast cells has a critical role in hepatic damage and ductular reaction, acting on BA homeostasis through disrupted intestinal and biliary FXR/FGF15 signaling [224].

9. FXR as a Therapeutic Target?

The tight interaction between BA-FXR-FGF19 and reflections on metabolism and inflammation in health and disease, has boosted the interest of research on the role of potential therapeutic approaches focusing on FGF19 and FXR stimulation.

9.1. FGF19 and FGF21 Variants

Scarce studies have focused on the role of FGF19 variants devoid of stimulatory effect on FGFR4 (because of potential tumorigenic activity). Mimetic molecules include chimeric molecules FGF19-4, 5 and 6 (mutagenesis in the *N*-terminus and in the heparin binding domains in amino acids 38–42), FGF19v (Conjugation of amino acids 1–20 of FGF21 with amino acids 25–194 of FGF19), M70 (3 amino acid substitutions and 5 amino acid deletions in the *N*-terminus). Such molecules are still able to modulate glucose metabolism but studies are restricted to animal models [225–227] with scarce studies in humans [226,228]. The translational value of FGF19 variants is still under evaluation and further evidence is awaited in this field.

FGF21 belong to the FGF19 subfamily of endocrine FGFs and, as FGF19, require the co-receptor β Klotho for binding and signaling through the FGF receptors [229]. As shown in experimental models, FGF21 is a negative regulator of BA synthesis, being able to decrease BA levels in the liver and in the small intestine, with a significant reduction of the BA pool size. These findings are paralleled by decreased colonic and fecal BA, with a concomitant increase in fecal cholesterol and fatty acid excretions [230]. The modulatory effect of FGF21 on BA synthesis seems independent of the FXR/FGF15 pathway [231]. Due to beneficial metabolic effects, FGF21 variants are emerging as promising therapeutic tools in metabolic diseases. The FGF21 variant LY2405319 has been tested in patients with obesity and type 2 diabetes, with beneficial effects on lipid metabolism, body weight, fasting insulin and adiponectin levels, but no significant reduction in fasting glucose levels [228]. Another long-acting FGF21 variant, PF-05231023, induced a body weight loss, an improvement in plasma lipoprotein profile and in adiponectin levels in overweight/obese subjects with type 2 diabetes, without effects on blood glucose. In the treated cohort, however, possible effects on bone formation and resorption were noticed [232]. In obese patients with hypertriglyceridaemia on atorvastatin, with or without type 2 diabetes, the same molecule reduced triglycerides in the absence of weight loss. In the treated group, however, serious adverse effects were noticed, causing the discontinuation of therapy in some participants [233]. Pegbelfermin (BMS-986036), a PEGylated FGF21 analog has been used in obese patients with type 2 diabetes predisposed to fatty liver, resulting in an improvement of the lipid profile, of fibrosis biomarkers and adiponectin levels, in the presence of mild adverse events [234]. Pegbelfermin has been also used in a phase 2a study in obese/overweight subjects and in NASH patients. In this trial, treatment significantly reduced hepatic fat fraction, in the presence of mild side effects (mainly diarrhea and nausea) [235]. Of note, Pegbelfermin promotes a significant reduction from baseline in serum concentrations of DCA and conjugates in patients with NASH and in overweight/obese adults, with a possible modulatory role on the synthesis of secondary BA, also acting on gut microbiome [236]. The extent of total BA decrease recorded in the cited study (about 20–30%) [236] is comparable to that obtained following treatment with FXR agonists [237], with the advantage to be selective for secondary BA [236].

9.2. FXR Agonists

Clinical trials are on the way using FXR modulators in chronic liver diseases such as primary biliary cholangitis, in cholestasis, nonalcoholic steatohepatitis (NASH), obesity, metabolic syndrome, hypertriglyceridemia, lipodystrophy. Additional trials include bile acid diarrhea, hepatitis B or association with reactivation of latent pro-virus (clinical trials.gov). Table 1 lists the main FXR agonists/modulators explored, at the moment, in clinical trials or experimental studies.

Table 1. FXR agonists/modulators mainly evaluated in clinical or experimental studies.

Obeticholic acid (approved for the treatment of primary biliary cholangitis)	[238–241]
Tropifexor (LJN452)	[242–249]
Cilofexor (GS-9674)	[250,251]
Vonafexor	[252]
Nidufexor	[253]
GW4064	[254]
MET409	[255]
TC-100	[256]
BMS-986339	[257]
HEC96719	[258]
WAY-450	[259]
WAY-362450	[260]
Px-102	[261]
Px-104	[262]
TERN-101	[263]
EDP-305	[264]
INT-767 (FXR-TGR5 dual agonist)	[25,265,266]

Most solid studies are reporting results with the steroidal molecule obeticholic acid (OCA), the 6α-Ethyl-Chenodeoxycholic Acid (6-ECDCA) and the non-steroidal Tropifexor (LJN452). OCA is modified from CDCA and is about 100 times more potent [238].

In male Wistar rats with cholestasis receiving i.v. infusion of LCA to impair bile flow, OCA alone did not induce cholestasis and during co-infusion with LCA, reversed the impairment of bile flow and protected hepatocytes from necrosis [238]. OCA is approved for the treatment of primary biliary cholangitis (PBC) especially in the subgroup of patients who fail to respond to UDCA [239–241].

Zucker (fa/fa) rats have a loss of function mutation in the hunger hormone leptin receptor [267], and suffer from hyperphagia and hyperleptinaemia resulting in obesity, insulin resistance, diabetes and fatty liver resembling NAFLD [267]. OCA treatment (10 mg/kg/day) over 7 weeks–meaning FXR activation -reversed insulin resistance, prevented body weight gain and fat deposition in the liver, reduced serum levels of triglycerides and aminotransferases, and improved liver histology [267]. Evidence suggests that the activation of FXR by OCA improves hyperglycemia through enhanced insulin secretion and glucose uptake by the liver. OCA increases insulin secretion in mouse β-TC6 cells and human pancreatic islets, while in β-TC6 cells OCA induces AKT (Protein Kinase B)-dependent translocation of glucose transporter 2 (GLUT2). This step increases glucose uptake by these cells [268]. The anti-inflammatory and anti-fibrotic properties of OCA became evident while investigating the NF-κB signaling pathway. In HepG2 cells stimulated with LPS or tumor necrosis factor alpha (TNFα), pretreatment with OCA at 3 uM inhibited the expression of cytokine-inducible enzymes COX-2 and iNOS [176]. OCA inhibited iNOS in LPS-treated primary mouse hepatocytes [176].

The agonist activity of OCA on FXR operates also in humans and trials are ongoing [33]. In NASH, OCA showed promising results, with an improvement of liver blood tests, a decreased extent of liver fibrosis and with no worsening of NASH [241]. The efficacy and safety of OCA have been evaluated in patients with type 2 diabetes and NAFLD by a phase IIa study using placebo (n = 23), 25 mg OCA (n = 20), or 50 mg OCA (n = 21) for 6 weeks (ClinicalTrials.gov, Number: NCT00501592). Treatment was well tolerated, and beneficial effects of OCA were observed in both groups with reduced γ-glutamyltransferase (GGT) and alanine aminotransferase (ALT) levels, dose-related weight loss, improved insulin sensitivity, elevated FGF19 serum levels, decreased BA precursor 7a-hydroxy-4-cholesten-3-one (C4) and endogenous BA. Results confirmed the OCA-dependent activation of FXR in human [269].

In the phase IIb trial Farnesoid X Receptor Ligand OCA in NASH Treatment (FLINT), OCA was tested in a multicenter, double-blind, randomized fashion. Non-cirrhotic NASH received 25 mg OCA (n = 141) daily or placebo (n = 142) for 72 weeks. OCA, compared with placebo, improved biochemical and histological features of NASH. NAFLD activity score improved by two points or greater (without worsening of fibrosis) in 45% of OCA patients vs. 21% in the placebo group. Side effects with OCA included pruritus and dyslipidemia [237].

The use of OCA in NASH patients awaits further safety and efficacy data [270]. The phase III trial REGENERATE (by Intercept) was designed to assess the effects of OCA on liver histology and clinical outcomes in 2065 biopsy- confirmed NASH patients. Groups included OCA 10 mg, 25 mg, or placebo for a total duration of six years. The interim analysis was performed by liver biopsy at 18 months in February 2019, and OCA achieved the primary endpoint of improving liver fibrosis without worsening of NASH [271].

Results from a REGENERATE 18-Month Interim Analysis in patients with NASH showed improvement of NASH and in health-related quality of life, despite the occurrence of mild pruritus early after the start of OCA therapy [272].

Tropifexor (LJN452), a non-steroidal FXR agonists reduced oxidative stress, steatosis, inflammation and fibrosis in the mouse models of NASH [242]. Tropifexor was safe and well-tolerated following single oral doses ranging from 10 μg to 3 mg. Circulating levels of FGF19 protein increased transiently in a dose-dependent fashion, pointing to a potent on-target FXR agonist activity [273]. Tropifexor evidenced ed a favourable tolerability and pharmacokinetic profile and induced a dose-dependent increase of FGF19 level, with no change in serum lipids in healthy volunteers [243]. In patients with primary bile acid diarrhoea, treatment with tropifexor 60 μg once daily showed an acceptable safety and tolerability profile. Tropifexor treatment decreased 7α-hydroxy-4-cholesten-3-one and bile acid concentration while increased FGF19 level [244]. Ongoing listed studies are focusing on efficacy and tolerability of tropifexor in patients with mild, moderate, severe hepatic impairment or NASH and fibrosis, and primary biliary cholangitis [274]. Recently, in patients with primary biliary cholangitis and inadequate response to UDCA, tropifexor induced a significant improvement in cholestatic markers, in the presence of mild to moderate adverse effects (mainly pruritus) [249].

Thus, further studies are needed with this agonist [245–248], also with respect to side effects, as compared with OCA. Furthermore, the safety, efficacy, tolerability, pharmacokinetics, pharmacodynamics of novel non-bile acid FXR agonist were intensively examined [275].

Cilofexor (GS-9674), another non-steroidal FXR agonist, evidenced safety and efficacy in non-cirrhotic NASH patients [250]. In detail, a double-blind, placebo-controlled, phase 2 trial was conducted on 140 NASH patients receiving orally cilofexor 100 mg, 30 mg, or placebo once daily for 24 weeks (ClinicalTrials.gov No. NCT02854605.). The study showed that cilofexor was safe and associated with significant attenuation of hepatic steatosis (lipid accumulation), liver enzymes (AST, ALT, and GGT), and bile acids synthesis without any significant alteration of fasting plasma levels of FGF19. In addition, cilofexor treatment was able to reduce different markers of fibrosis and liver stiffness which indicate that cilofexor may have potential antifibrotic effects. In noncirrhotic subjects with large-duct primary sclerosing cholangitis who underwent a 96-week open-label extension of a phase II trial, cilofexor treatment was safe and led to a significant improvement of liver biochemistry and biomarkers of cholestasis and cellular injury [251].

Despite promising results, the side effects and major obstacles of FXR agonist (mainly atherogenic risk, pruritogenic potency) and FGF19 variants (mainly increased appetite, diarrhea and nausea [232,235,276], altered bone homeostasis [232], increased blood pressure and heart rate [233]) still represent a significant matter of concern. Research needs to develop FXR agonists or modulators with beneficial anti-inflammatory effects and minimal metabolic actions. From this point of view, according to available results, the risk-benefit profile of FXR agonists can be modulated by structural optimization of FXR agonist. Harrison et al. [255] evaluated the effect of structurally optimized FXR agonist MET409 in NASH patients. At 12-week post treatment, MET409 reduced fat liver content and bile acid with no significant increase in FGF19 level. The low dose (50 mg) treatment of MET409 showed a 16% pruritus rate and a 9% LDL-C increase but with favorable reduction of fat liver content. The study suggested that the improvements in efficacy/tolerability of FXR agonist treatment could be achieved.

At the molecular level, one oral dose of Px-102, a nonsteroidal FXR agonist, decreased the synthesis of BA in healthy volunteers independently of increases in FGF19 [261] suggesting that that activation of hepatic FXR suppress BA synthesis, independently of FGF19.

The safety and efficacy of FXR agonist Vonafexor for the treatment of chronic hepatits B were also evaluated by a double-blind, placebo-controlled trial. Vonafexor was well tolerated overall with moderate gastrointestinal adverse effects (pruritus occurred in ~60% of subjects with twice-daily treatment compared with 16% in subjects with once-daily treatment). Vonafexor alone or combined with interferon-α2a showed an anti-viral effect by reducing HBV markers [252]. In non-diabetic NAFLD patients, a 4-weeks of treatment with the non-steroidal FXR agonist PX-104 improved liver enzymes and insulin sensitivity with no serious adverse events. Interestingly, PX-104 influenced gut microbiota by reducing *Coriobacteriaceae* abundance and total fecal BAs [262]. TERN-101 (FXR agonist) treatment in healthy volunteer was well-tolerated and showed a reduction in bile acid synthesis (decreases in serum 7α-hydroxy-4-cholesten-3-one) [263].

In an animal model (mice with obstructed BA flow), the novel FXR agonist TC-100 reduced BA pool size in serum and bile, with a shift to a more hydrophilic composition. The changes in BA pool were paralleled by a prevention of intestinal mucosal damage and by a progressive increase in Firmicutes:Bacteroidotes ratio, with increased *Akkermansia muciniphila* abundance [256].

In a recent preclinical study, the BMS-986339, a novel non-bile acid FXR agonist showed potent in vitro and in vivo activation of FXR, with anti-fibrotic efficacy and tissue-selective effects in vivo. The safety of this molecule, however, still requires to be adequately tested in humans [257].

HEC96719, another novel tricyclic FXR agonist, exhibits a potency of FXR activation superior to obeticholic acid, higher FXR selectivity and more favorable tissue distribution in liver and intestine. Although HEC96719 seems a promising tool in NASH treatment, its efficacy and safety profiles need further confirmations [258].

Interestingly, in a rat model of NAFLD/NASH, the administration of dapagliflozin, a sodium-glucose cotransporter-2 inhibitor, alleviated NASH also through a reduced de novo lipogenesis mediated by an upregulation of FXR/SHP and a downregulation of LXRα/SREBP-1c in the liver [277]. These findings parallel a recent observation in patients with T2D and NAFLD, in whom treatment with SGLT2-inhibitors was linked with improvement of liver steatosis and fibrosis markers and circulating pro-inflammatory and redox status [278].

The oral FXR agonist EDP-305 has been proposed for the treatment of NASH. Recent results from a double-blind phase II study in patients with fibrotic NASH showed a significant decrease in ALT levels and liver fat content in 12 weeks, with adverse effects (mainly pruritus) leading to drug discontinuation in about 20% of patients [264].

Interesting perspectives derive from experimental studies using FXR-TGR5 dual agonists [25,265,266]. Recently, in a mouse model of NASH, treatment with the FXR-TGR5 dual agonist INT-767 was able to prevent the progression of disease, with beneficial effects on liver mitochondrial function, lipid homeostasis, BA composition (i.e., decreased hydrophobicity index), liver inflammation and gut dysbiosis [25].

10. Conclusions and Future Perspectives

Cholesterol is eliminated from human body via biliary secretion and synthesis of primary BA, which are essential for digestion of fat and signaling functions on important receptors. BA homeostasis is a complex scenario which requires several pathways and dynamic events to maintain BA qualitative/quantitative profiles and function in health (Figure 1).

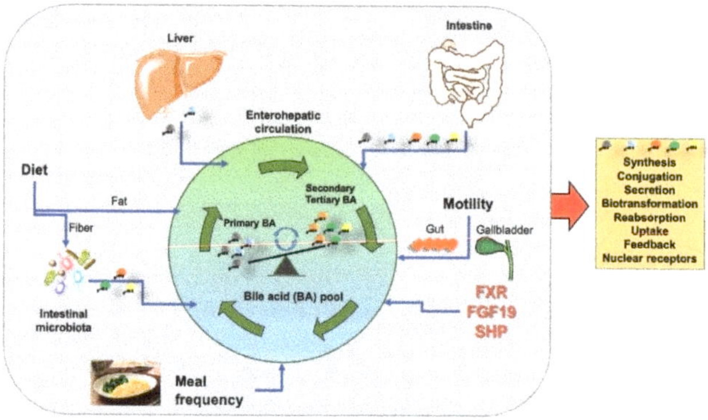

Figure 1. Summary of key factors involved in the composition of the bile acid pool. Several dynamic events are active daily either in the fasting and postprandial period (grey area) and contribute to shape the qualitative/quantitative profile of the bile acid pool, its maintenance and expansion. Such events work in concert with bile acid synthesis and related steps (orange box). Abbreviations: FGF-19, fibroblast growth factor 19; FXR, farnesoid X receptor; SHP, small heterodimer partner; primary bile acids are cholic and chenodeoxycholic acid (grey); secondary bile acids are deoxycholic acid (orange) and lithocholic acid (green); tertiary bile acid is ursodeoxycholic acid (yellow).

The effect of BA works in concert with the activation of the nuclear receptor FXR. This step, in turn, controls BA synthesis, excretion, and reabsorption. These pathways minimize over-accumulation of potentially toxic BA in the liver. Activation of FXR also brings several metabolic effects by regulating lipid metabolism, reducing hepatic gluconeogenesis, glycolysis, and increasing glycogen synthesis. In parallel, FXR activation has anti- inflammatory properties during liver injury. The FXR-FGF15/19 axis increases energy expenditure and glycogen synthesis and decreases gluconeogenesis and fatty acid synthesis (Figure 2). FXR and FGF19 have become promising targets for the treatment of NASH. Further studies are required in this important field.

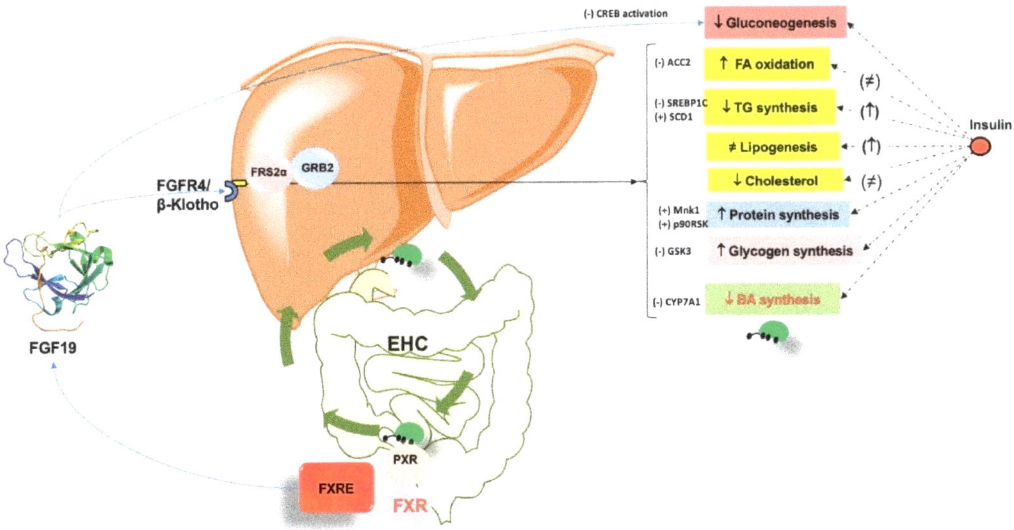

Figure 2. Summary of the metabolic effects of Fibroblast growth factor 19 (FGF19) in the liver. The role of insulin is shown for comparison. Bile acids during the enterohepatic circulation (EHC) activate intestinal FXR promoting FGF19 secretion. Circulating FGF19 signaling requires the presence of β-Klotho [279] which is the fibroblast growth factor receptor 4 (FGFR4) co-receptor required for liver-specific FGF19 actions [112]. The tissue-specific expression pattern of β-klotho and FGFR isoforms determines FGF19 metabolic activity [116]. In the hepatocyte, the signaling events allow the recruitment of cytosolic adaptors such as fibroblast growth factor receptor substrate 2α (FRS2α) and growth factor receptor-bound protein 2 (GRB2). The ultimate metabolic effects of FGF19 are depicted on the right within the boxes, with major regulatory elements listed. Legend; ACC2, acetyl-CoA carboxylase 2; CYP7A1, cholesterol-7α-hydroxylase; CREB cAMP-response element-binding protein; FXR, farnesoid X receptor; FXRE, FXR responsive element; GSK3, glycogen synthase kinase 3; Mnk1, protein kinase; p90 ribosomal S6 kinase; PXR, pregnane X receptor; PXRE, PXR response element; SCD1; stearoyl-CoA desaturase 1; SREBP1C, sterol regulatory element-binding protein 1C. (+) activation; (-) inhibition; ≠, unchanged.

Author Contributions: Conceptualization, A.D.C. and P.P.; literature review, A.D.C., L.B., J.B., M.K., G.G., F.S.; writing—original draft preparation, A.D.C., J.B., G.G.; writing—review and editing; supervision, H.H.W., D.Q.-H.W., F.S., P.P.; funding acquisition, P.P.; F.S. has particularly participated to the paragraph on bile acid kinetics. All authors have read and agreed to the published version of the manuscript.

Funding: This paper has been partly supported by funding to P.P. from the European Union's Horizon 2020 Research and Innovation program under the Marie Skłodowska-Curie Grant Agreement No. 722619 (FOIE GRAS), Grant Agreement No. 734719 (mtFOIE GRAS), Grant Regione Puglia, CUP H99C20000340002 (Fever Apulia), and Grant EUROSEEDS Uniba-S56-By-products Sustainable

Recovery 4 Health (BSR-4H): University of Bari Aldo Moro, 2022. The project SYSTEMIC "an integrated approach to the challenge of sustainable food systems: adaptive and mitigatory strategies to address climate change and malnutrition", Knowledge hub on Nutrition and Food Security, has received funding from national research funding parties in Belgium (FWO), France (INRA), Germany (BLE), Italy (MIPAAF), Latvia (IZM), Norway (RCN), Portugal (FCT), and Spain (AEI) in a joint action of JPI HDHL, JPI-OCEANS and FACCE-JPI launched in 2019 under the ERA-NET ERA-HDHL (n° 696295).

Institutional Review Board Statement: Not applicable.

Informed Consent Statement: Not applicable.

Acknowledgments: The authors are indebted to Gerard van Berge-Henegouwen, Dan Dumitrascu, Gustav Paumgartner, Thomas LaMont, Giuseppe Palasciano, Karel van Erpecum for continuous and restless scientific discussion. The authors thank Rosa De Venuto, Paola De Benedictis, Domenica Di Palo, Ilaria Farella, Elisa Lanza for their longstanding and skilful professional technical assistance.

Conflicts of Interest: The authors declare no conflict of interest.

Abbreviations

3β-HSD	3β-hydroxy-D5-C27-steroid dehydroxylase
6-ECDCA	6a-ethyl-chenodeoxycholic acid
ACC2	Acetyl CoA carboxylase
AGRP	Agouti-related peptide
AKR1C4	3a-hydroxysteroid dehydrogenase
AKR1D1	D4-3-oxosteroid 5b-reductase
AKT	Protein kinase B
ALT	Alanine aminotransferase
AMPs	Antimicrobial peptides
ASBT	Apical sodium-dependent bile salt transporter
BA	Bile acids
BAAT	Bile acid-CoA:amino acid *N*-acyltransferase
BACS	Bile acid CoA synthase
ABCB11	Bile acid export pump
BAT	Brown adipose tissue
BSEP	Bile salt export pump
BSH	Bile salt hydrolase
C4	7a-hydroxy-4-cholesten-3-one
CA	Cholic acid
CCK	Cholecystokinin
CCl4	Carbon tetrachloride
CDCA	Chenodeoxycholic acid
CF	Cycling frequency
COX-2	Cylcooxygenase-2
CREB	cAMP regulatory element binding protein
CYP7A1	Cholesterol-7α-hydroxylase
CYP27A1	Cytochrome P450 27A1
CYP2A12	Cytochrome P450 2A12
CYP2C70	Cytochrome P450 2C70
CYP7A1	Cytochrome P450 7A1
CYP7B1	Cytochrome P450 7B1
CYP8B1	Cytochrome P450 8B1
DBD	DNA binding domain
DCA	Deoxycholic acid
eIF4B	Eukaryotic initiation factor 4B
eIF4E	Eukaryotic initiation factor 4E

ER2	Everted hexanucleotide repeat separated by 2 nucleotides
ERK	Extracellular-signal-regulated protein kinase
FAR	Fractional absorption rate
FFA	Free fatty acid
FGF	Fibroblast growth factor
FGF15	Fibroblast growth factor 15
FGF19	Fibroblast growth factor 19
FGFR	Fibroblast growth factor receptor
FGFR1	Fibroblast growth factor receptor 1
FGFR4	Fibroblast growth factor receptor 4
FLINT	Farnesoid X receptor ligand obeticholic acid in NASH treatment
FXR	Farnesoid X receptor
FXRE	Farnesoid X receptor response element
FTR	Fractional turnover rate
GGT	Gamma-glutamyltransferase
GLP-1	Glucagon-like peptide 1
GLUT2	Glucose transporter 2
GPBAR-1	G-protein-coupled bile acid receptor-1
GS	Glycogen synthase
GSK3a	Glycogen synthase kinase 3a
HREs	Hormone response elements
HSC	Hepatic stellate cell
IkB	Inhibitor of nuclear factor kappa B
IBABP	Intestinal bile acid binding protein
IFN-g	Interferon-gamma
iNOS	Inducible nitric oxide synthase
IR1	Inverted hexanucleotide repeat separated by 1 nucleotide
KO	Knockout
LBD	Ligand-binding domain
LCA	Lithocholic acid
LRH-1	Liver receptor homologue 1
LPS	Lipopolysaccharide
JNK	JUN *N*-terminal kinase
JUN	Jun proto-oncogene
MAGs	Metagenome-assembled genomes
MCA	Muricholic acid
MCD	Methionine-choline deficient
MCP-1	Monocyte chemoattractant protein-1
MRP2	Multidrug resistance-associated protein 2
mTOR	Mammalian target of rapamycin
NAFLD	Non-alcoholic fatty liver disease
NASH	Non-alcoholic steatohepatitis
NF-kB	Nuclear factor kappa-light chain enhancer of activated B cell
NPY	Neuropeptide Y
NR s	Nuclear receptors
NLRP3	NACHT LRR and PYD domains-containing protein 3
NTCP	Sodium taurocholate co-transporting polypeptide
OATP	Organic anion transporting polypeptide
OCA	Obeticholic acid
OSTα	Organic solute transporter alpha
OSTb	Organic solute transporter beta
P90RSK	p90 ribosomal S6 kinase
PBC	Primary biliary cirrhosis
PEPCK	Phosphoenolpyruvate carboxykinase

PFIC	Progressive familial intrahepatic cholestasis
PGC-1α	Peroxisome proliferator-activated receptor gamma coactivator 1-alpha
PGC-1b	Peroxisome proliferator-activated receptor gamma coactivator 1-beta
PPARγ	Peroxisomal proliferator activated receptor gamma
PXR	Pregnane X receptor
RCT	Reverse cholesterol transport
RXR	Retinoid X receptor
S1PR2	Sphingosine-1-phosphate receptor 2
S6K1	S6 kinase beta-1
SCD1	Stearoyl-CoA desaturase 1
SIBO	Small intestinal bacterial overgrowth
SHP	Small heterodimer partner
SREBP1c	Sterol regulatory element-binding protein 1c
STAT3	Signal transducer and activator of transcription 3
SULT2A1	Sulfotransferase 2A1
TCA	Taurocholic acid
TDCA	Taurodeoxycholic acid
TGR5	Takeda G-protein receptor 5
TLCA	Taurolithocholic acid
TNFα	Tumor necrosis factor alpha
VDR	Vitamin D receptor
VLDL	Very low-density lipoprotein
WAT	White adipose tissue

References

1. Gonzalez, F.J. Nuclear receptor control of enterohepatic circulation. *Compr. Physiol.* **2012**, *2*, 2811–2828. [CrossRef] [PubMed]
2. Matsubara, T.; Li, F.; Gonzalez, F.J. FXR signaling in the enterohepatic system. *Mol. Cell Endocrinol.* **2013**, *368*, 17–29. [CrossRef] [PubMed]
3. Sun, L.; Cai, J.; Gonzalez, F.J. The role of farnesoid X receptor in metabolic diseases, and gastrointestinal and liver cancer. *Nat. Rev. Gastroenterol. Hepatol.* **2021**, *18*, 335–347. [CrossRef]
4. Nguyen, J.T.; Shaw, R.P.H.; Anakk, S. Bile Acids—A Peek Into Their History and Signaling. *Endocrinology* **2022**, *163*, bqac155. [CrossRef]
5. Poland, J.C.; Flynn, C.R. Bile Acids, Their Receptors, and the Gut Microbiota. *Physiology* **2021**, *36*, 235–245. [CrossRef]
6. Sorrentino, G.; Perino, A.; Yildiz, E.; El Alam, G.; Bou Sleiman, M.; Gioiello, A.; Pellicciari, R.; Schoonjans, K. Bile Acids Signal via TGR5 to Activate Intestinal Stem Cells and Epithelial Regeneration. *Gastroenterology* **2020**, *159*, 956–968.e8. [CrossRef]
7. Guillot, A.; Guerri, L.; Feng, D.; Kim, S.J.; Ahmed, Y.A.; Paloczi, J.; He, Y.; Schuebel, K.; Dai, S.; Liu, F.; et al. Bile acid-activated macrophages promote biliary epithelial cell proliferation through integrin alphavbeta6 upregulation following liver injury. *J. Clin. Investig.* **2021**, *131*, e132305. [CrossRef] [PubMed]
8. Hylemon, P.B.; Takabe, K.; Dozmorov, M.; Nagahashi, M.; Zhou, H. Bile acids as global regulators of hepatic nutrient metabolism. *Liver Res.* **2017**, *1*, 10–16. [CrossRef] [PubMed]
9. Jiao, N.; Baker, S.S.; Chapa-Rodriguez, A.; Liu, W.; Nugent, C.A.; Tsompana, M.; Mastrandrea, L.; Buck, M.J.; Baker, R.D.; Genco, R.J.; et al. Suppressed hepatic bile acid signalling despite elevated production of primary and secondary bile acids in NAFLD. *Gut* **2018**, *67*, 1881–1891. [CrossRef]
10. Pavlovic, N.; Stanimirov, B.; Mikov, M. Bile Acids as Novel Pharmacological Agents: The Interplay Between Gene Polymorphisms, Epigenetic Factors and Drug Response. *Curr. Pharm. Des.* **2017**, *23*, 187–215. [CrossRef]
11. Kim, Y.C.; Jung, H.; Seok, S.; Zhang, Y.; Ma, J.; Li, T.; Kemper, B.; Kemper, J.K. MicroRNA-210 Promotes Bile Acid-Induced Cholestatic Liver Injury by Targeting Mixed-Lineage Leukemia-4 Methyltransferase in Mice. *Hepatology* **2020**, *71*, 2118–2134. [CrossRef] [PubMed]
12. Zimny, S.; Koob, D.; Li, J.; Wimmer, R.; Schiergens, T.; Nagel, J.; Reiter, F.P.; Denk, G.; Hohenester, S. Hydrophobic Bile Salts Induce Pro-Fibrogenic Proliferation of Hepatic Stellate Cells through PI3K p110 Alpha Signaling. *Cells* **2022**, *11*, 2344. [CrossRef] [PubMed]
13. Kumari, A.; Pal Pathak, D.; Asthana, S. Bile acids mediated potential functional interaction between FXR and FATP5 in the regulation of Lipid Metabolism. *Int. J. Biol. Sci.* **2020**, *16*, 2308–2322. [CrossRef]
14. McGlone, E.R.; Bloom, S.R. Bile acids and the metabolic syndrome. *Ann. Clin. Biochem.* **2019**, *56*, 326–337. [CrossRef] [PubMed]
15. Ma, K.; Tang, D.; Yu, C.; Zhao, L. Progress in research on the roles of TGR5 receptor in liver diseases. *Scand. J. Gastroenterol.* **2021**, *56*, 717–726. [CrossRef]
16. Wang, Y.; Aoki, H.; Yang, J.; Peng, K.; Liu, R.; Li, X.; Qiang, X.; Sun, L.; Gurley, E.C.; Lai, G.; et al. The role of sphingosine 1-phosphate receptor 2 in bile-acid-induced cholangiocyte proliferation and cholestasis-induced liver injury in mice. *Hepatology* **2017**, *65*, 2005–2018. [CrossRef]

17. Nagahashi, M.; Yuza, K.; Hirose, Y.; Nakajima, M.; Ramanathan, R.; Hait, N.C.; Hylemon, P.B.; Zhou, H.; Takabe, K.; Wakai, T. The roles of bile acids and sphingosine-1-phosphate signaling in the hepatobiliary diseases. *J. Lipid Res.* **2016**, *57*, 1636–1643. [CrossRef]
18. Garruti, G.; Wang, H.H.; Bonfrate, L.; de Bari, O.; Wang, D.Q.; Portincasa, P. A pleiotropic role for the orphan nuclear receptor small heterodimer partner in lipid homeostasis and metabolic pathways. *J. Lipids* **2012**, *2012*, 304292. [CrossRef]
19. Portincasa, P.; Di Ciaula, A.; Garruti, G.; Vacca, M.; De Angelis, M.; Wang, D.Q. Bile Acids and GPBAR-1: Dynamic Interaction Involving Genes, Environment and Gut Microbiome. *Nutrients* **2020**, *12*, 3709. [CrossRef]
20. Zollner, G.; Trauner, M. Mechanisms of cholestasis. *Clin. Liver Dis.* **2008**, *12*, 1–26. [CrossRef]
21. Di Ciaula, A.; Garruti, G.; Lunardi Baccetto, R.; Molina-Molina, E.; Bonfrate, L.; Wang, D.Q.; Portincasa, P. Bile Acid Physiology. *Ann. Hepatol.* **2017**, *16*, s4–s14. [CrossRef] [PubMed]
22. Di Ciaula, A.; Wang, D.Q.; Molina-Molina, E.; Lunardi Baccetto, R.; Calamita, G.; Palmieri, V.O.; Portincasa, P. Bile Acids and Cancer: Direct and Environmental-Dependent Effects. *Ann. Hepatol.* **2017**, *16*, s87–s105. [CrossRef] [PubMed]
23. Panzitt, K.; Zollner, G.; Marschall, H.U.; Wagner, M. Recent advances on FXR-targeting therapeutics. *Mol. Cell. Endocrinol.* **2022**, *552*, 111678. [CrossRef] [PubMed]
24. Shen, T.; Shi, A.; Wei, Y.; Luo, X.; Xi, L. Farnesoid X Receptor as a Promising Therapeutic Target for Nonalcoholic Fatty Liver Disease (NAFLD) and the Current Development of Its Agonists. *Discov. Med.* **2021**, *32*, 113–121.
25. Wang, X.X.; Xie, C.; Libby, A.E.; Ranjit, S.; Levi, J.; Myakala, K.; Bhasin, K.; Jones, B.A.; Orlicky, D.J.; Takahashi, S.; et al. The Role of FXR and TGR5 in Reversing and Preventing Progression of Western Diet-induced Hepatic Steatosis, Inflammation, and Fibrosis in Mice. *J. Biol. Chem.* **2022**, *298*, 102530. [CrossRef] [PubMed]
26. Luo, M.; Yan, J.; Wu, L.; Wu, J.; Chen, Z.; Jiang, J.; Chen, Z.; He, B. Probiotics Alleviated Nonalcoholic Fatty Liver Disease in High-Fat Diet-Fed Rats via Gut Microbiota/FXR/FGF15 Signaling Pathway. *J. Immunol. Res.* **2021**, *2021*, 2264737. [CrossRef] [PubMed]
27. Liu, H.; Pathak, P.; Boehme, S.; Chiang, J.L. Cholesterol 7alpha-hydroxylase protects the liver from inflammation and fibrosis by maintaining cholesterol homeostasis. *J. Lipid Res.* **2016**, *57*, 1831–1844. [CrossRef] [PubMed]
28. Pineda Torra, I.; Claudel, T.; Duval, C.; Kosykh, V.; Fruchart, J.C.; Staels, B. Bile acids induce the expression of the human peroxisome proliferator-activated receptor alpha gene via activation of the farnesoid X receptor. *Mol. Endocrinol.* **2003**, *17*, 259–272. [CrossRef] [PubMed]
29. Sirvent, A.; Claudel, T.; Martin, G.; Brozek, J.; Kosykh, V.; Darteil, R.; Hum, D.W.; Fruchart, J.C.; Staels, B. The farnesoid X receptor induces very low density lipoprotein receptor gene expression. *FEBS Lett.* **2004**, *566*, 173–177. [CrossRef]
30. Claudel, T.; Inoue, Y.; Barbier, O.; Duran-Sandoval, D.; Kosykh, V.; Fruchart, J.; Fruchart, J.C.; Gonzalez, F.J.; Staels, B. Farnesoid X receptor agonists suppress hepatic apolipoprotein CIII expression. *Gastroenterology* **2003**, *125*, 544–555. [CrossRef]
31. Zhao, T.; Wang, J.; He, A.; Wang, S.; Chen, Y.; Lu, J.; Lv, J.; Li, S.; Wang, J.; Qian, M.; et al. Mebhydrolin ameliorates glucose homeostasis in type 2 diabetic mice by functioning as a selective FXR antagonist. *Metabolism* **2021**, *119*, 154771. [CrossRef] [PubMed]
32. Zhang, S.; Wang, J.; Liu, Q.; Harnish, D.C. Farnesoid X receptor agonist WAY-362450 attenuates liver inflammation and fibrosis in murine model of non-alcoholic steatohepatitis. *J. Hepatol.* **2009**, *51*, 380–388. [CrossRef] [PubMed]
33. Roth, J.D.; Veidal, S.S.; Fensholdt, L.K.D.; Rigbolt, K.T.G.; Papazyan, R.; Nielsen, J.C.; Feigh, M.; Vrang, N.; Young, M.; Jelsing, J.; et al. Combined obeticholic acid and elafibranor treatment promotes additive liver histological improvements in a diet-induced ob/ob mouse model of biopsy-confirmed NASH. *Sci. Rep.* **2019**, *9*, 9046. [CrossRef] [PubMed]
34. Stellaard, F.; Lutjohann, D. Dynamics of the enterohepatic circulation of bile acids in healthy humans. *Am. J. Physiol. Gastrointest. Liver Physiol.* **2021**, *321*, G55–G66. [CrossRef] [PubMed]
35. Parks, D.J.; Blanchard, S.G.; Bledsoe, R.K.; Chandra, G.; Consler, T.G.; Kliewer, S.A.; Stimmel, J.B.; Willson, T.M.; Zavacki, A.M.; Moore, D.D.; et al. Bile acids: Natural ligands for an orphan nuclear receptor. *Science* **1999**, *284*, 1365–1368. [CrossRef]
36. Inagaki, T.; Choi, M.; Moschetta, A.; Peng, L.; Cummins, C.L.; McDonald, J.G.; Luo, G.; Jones, S.A.; Goodwin, B.; Richardson, J.A.; et al. Fibroblast growth factor 15 functions as an enterohepatic signal to regulate bile acid homeostasis. *Cell Metab.* **2005**, *2*, 217–225. [CrossRef] [PubMed]
37. Modica, S.; Gofflot, F.; Murzilli, S.; D'Orazio, A.; Salvatore, L.; Pellegrini, F.; Nicolucci, A.; Tognoni, G.; Copetti, M.; Valanzano, R.; et al. The intestinal nuclear receptor signature with epithelial localization patterns and expression modulation in tumors. *Gastroenterology* **2010**, *138*, 636–648, 648.e1-12. [CrossRef]
38. Liu, H.; Hu, C.; Zhang, X.; Jia, W. Role of gut microbiota, bile acids and their cross-talk in the effects of bariatric surgery on obesity and type 2 diabetes. *J. Diabetes Investig.* **2018**, *9*, 13–20. [CrossRef] [PubMed]
39. Kemper, J.K. Regulation of FXR transcriptional activity in health and disease: Emerging roles of FXR cofactors and post-translational modifications. *Biochim. Biophys. Acta* **2011**, *1812*, 842–850. [CrossRef] [PubMed]
40. Goodwin, B.; Jones, S.A.; Price, R.R.; Watson, M.A.; McKee, D.D.; Moore, L.B.; Galardi, C.; Wilson, J.G.; Lewis, M.C.; Roth, M.E.; et al. A regulatory cascade of the nuclear receptors FXR, SHP-1, and LRH-1 represses bile acid biosynthesis. *Mol. Cell* **2000**, *6*, 517–526. [CrossRef]

41. Lu, T.T.; Makishima, M.; Repa, J.J.; Schoonjans, K.; Kerr, T.A.; Auwerx, J.; Mangelsdorf, D.J. Molecular basis for feedback regulation of bile acid synthesis by nuclear receptors. *Mol. Cell* **2000**, *6*, 507–515. [CrossRef]
42. Nishimaki-Mogami, T.; Une, M.; Fujino, T.; Sato, Y.; Tamehiro, N.; Kawahara, Y.; Shudo, K.; Inoue, K. Identification of intermediates in the bile acid synthetic pathway as ligands for the farnesoid X receptor. *J. Lipid Res.* **2004**, *45*, 1538–1545. [CrossRef]
43. Pircher, P.C.; Kitto, J.L.; Petrowski, M.L.; Tangirala, R.K.; Bischoff, E.D.; Schulman, I.G.; Westin, S.K. Farnesoid X receptor regulates bile acid-amino acid conjugation. *J. Biol. Chem.* **2003**, *278*, 27703–27711. [CrossRef]
44. Garcia-Rodriguez, J.L.; Barbier-Torres, L.; Fernandez-Alvarez, S.; Gutierrez-de Juan, V.; Monte, M.J.; Halilbasic, E.; Herranz, D.; Alvarez, L.; Aspichueta, P.; Marin, J.J.; et al. SIRT1 controls liver regeneration by regulating bile acid metabolism through farnesoid X receptor and mammalian target of rapamycin signaling. *Hepatology* **2014**, *59*, 1972–1983. [CrossRef] [PubMed]
45. Schmidt, D.R.; Schmidt, S.; Holmstrom, S.R.; Makishima, M.; Yu, R.T.; Cummins, C.L.; Mangelsdorf, D.J.; Kliewer, S.A. AKR1B7 is induced by the farnesoid X receptor and metabolizes bile acids. *J. Biol. Chem.* **2011**, *286*, 2425–2432. [CrossRef] [PubMed]
46. Modica, S.; Bellafante, E.; Moschetta, A. Master regulation of bile acid and xenobiotic metabolism via the FXR, PXR and CAR trio. *Front. Biosci. (Landmark Ed.)* **2008**, *14*, 4719–4745. [CrossRef]
47. Wagner, M.; Zollner, G.; Trauner, M. New molecular insights into the mechanisms of cholestasis. *J. Hepatol.* **2009**, *51*, 565–580. [CrossRef] [PubMed]
48. Gillard, J.; Clerbaux, L.A.; Nachit, M.; Sempoux, C.; Staels, B.; Bindels, L.B.; Tailleux, A.; Leclercq, I.A. Bile acids contribute to the development of non-alcoholic steatohepatitis in mice. *JHEP Rep.* **2022**, *4*, 100387. [CrossRef] [PubMed]
49. Dommett, R.; Zilbauer, M.; George, J.T.; Bajaj-Elliott, M. Innate immune defence in the human gastrointestinal tract. *Mol. Immunol.* **2005**, *42*, 903–912. [CrossRef] [PubMed]
50. Di Ciaula, A.; Baj, J.; Garruti, G.; Celano, G.; De Angelis, M.; Wang, H.H.; Di Palo, D.M.; Bonfrate, L.; Wang, D.Q.; Portincasa, P. Liver Steatosis, Gut-Liver Axis, Microbiome and Environmental Factors. A Never-Ending Bidirectional Cross-Talk. *J. Clin. Med.* **2020**, *9*, 2648. [CrossRef] [PubMed]
51. Portincasa, P.; Bonfrate, L.; Vacca, M.; De Angelis, M.; Farella, I.; Lanza, E.; Khalil, M.; Wang, D.Q.; Sperandio, M.; Di Ciaula, A. Gut Microbiota and Short Chain Fatty Acids: Implications in Glucose Homeostasis. *Int. J. Mol. Sci.* **2022**, *23*, 1105. [CrossRef]
52. Biedermann, L.; Zeitz, J.; Mwinyi, J.; Sutter-Minder, E.; Rehman, A.; Ott, S.J.; Steurer-Stey, C.; Frei, A.; Frei, P.; Scharl, M.; et al. Smoking cessation induces profound changes in the composition of the intestinal microbiota in humans. *PLoS ONE* **2013**, *8*, e59260. [CrossRef]
53. Mutlu, E.A.; Gillevet, P.M.; Rangwala, H.; Sikaroodi, M.; Naqvi, A.; Engen, P.A.; Kwasny, M.; Lau, C.K.; Keshavarzian, A. Colonic microbiome is altered in alcoholism. *Am. J. Physiol. Gastrointest. Liver Physiol.* **2012**, *302*, G966–G978. [CrossRef]
54. Gao, B.; Chi, L.; Mahbub, R.; Bian, X.; Tu, P.; Ru, H.; Lu, K. Multi-Omics Reveals that Lead Exposure Disturbs Gut Microbiome Development, Key Metabolites, and Metabolic Pathways. *Chem. Res. Toxicol.* **2017**, *30*, 996–1005. [CrossRef]
55. Gao, B.; Bian, X.; Mahbub, R.; Lu, K. Sex-Specific Effects of Organophosphate Diazinon on the Gut Microbiome and Its Metabolic Functions. *Environ. Health Perspect.* **2017**, *125*, 198–206. [CrossRef]
56. Joly, C.; Gay-Queheillard, J.; Leke, A.; Chardon, K.; Delanaud, S.; Bach, V.; Khorsi-Cauet, H. Impact of chronic exposure to low doses of chlorpyrifos on the intestinal microbiota in the Simulator of the Human Intestinal Microbial Ecosystem (SHIME) and in the rat. *Environ. Sci. Pollut. Res. Int.* **2013**, *20*, 2726–2734. [CrossRef]
57. Joly Condette, C.; Bach, V.; Mayeur, C.; Gay-Queheillard, J.; Khorsi-Cauet, H. Chlorpyrifos Exposure During Perinatal Period Affects Intestinal Microbiota Associated With Delay of Maturation of Digestive Tract in Rats. *J. Pediatr. Gastroenterol. Nutr.* **2015**, *61*, 30–40. [CrossRef]
58. Di Ciaula, A.; Bonfrate, L.; Portincasa, P. The role of microbiota in nonalcoholic fatty liver disease. *Eur. J. Clin. Investig.* **2022**, *52*, e13768. [CrossRef]
59. Garruti, G.; Di Ciaula, A.; Wang, H.H.; Wang, D.Q.; Portincasa, P. Cross-Talk Between Bile Acids and Gastro-Intestinal and Thermogenic Hormones: Clues from Bariatric Surgery. *Ann. Hepatol.* **2017**, *16*, s68–s82. [CrossRef]
60. Sayin, S.I.; Wahlstrom, A.; Felin, J.; Jantti, S.; Marschall, H.U.; Bamberg, K.; Angelin, B.; Hyotylainen, T.; Oresic, M.; Backhed, F. Gut microbiota regulates bile acid metabolism by reducing the levels of tauro-beta-muricholic acid, a naturally occurring FXR antagonist. *Cell Metab.* **2013**, *17*, 225–235. [CrossRef]
61. Long, S.L.; Gahan, C.G.M.; Joyce, S.A. Interactions between gut bacteria and bile in health and disease. *Mol. Asp. Med.* **2017**, *56*, 54–65. [CrossRef] [PubMed]
62. Parseus, A.; Sommer, N.; Sommer, F.; Caesar, R.; Molinaro, A.; Stahlman, M.; Greiner, T.U.; Perkins, R.; Backhed, F. Microbiota-induced obesity requires farnesoid X receptor. *Gut* **2017**, *66*, 429–437. [CrossRef] [PubMed]
63. Yokota, A.; Fukiya, S.; Islam, K.B.; Ooka, T.; Ogura, Y.; Hayashi, T.; Hagio, M.; Ishizuka, S. Is bile acid a determinant of the gut microbiota on a high-fat diet? *Gut Microbes* **2012**, *3*, 455–459. [CrossRef] [PubMed]
64. Zhou, H.; Hylemon, P.B. Bile acids are nutrient signaling hormones. *Steroids* **2014**, *86*, 62–68. [CrossRef]
65. Ridlon, J.M.; Wolf, P.G.; Gaskins, H.R. Taurocholic acid metabolism by gut microbes and colon cancer. *Gut Microbes* **2016**, *7*, 201–215. [CrossRef] [PubMed]
66. O'Keefe, S.J.; Li, J.V.; Lahti, L.; Ou, J.; Carbonero, F.; Mohammed, K.; Posma, J.M.; Kinross, J.; Wahl, E.; Ruder, E.; et al. Fat, fibre and cancer risk in African Americans and rural Africans. *Nat. Commun.* **2015**, *6*, 6342. [CrossRef] [PubMed]

67. Jiang, M.; Li, F.; Liu, Y.; Gu, Z.; Zhang, L.; Lee, J.; He, L.; Vatsalya, V.; Zhang, H.G.; Deng, Z.; et al. Probiotic-derived nanoparticles inhibit ALD through intestinal miR194 suppression and subsequent FXR activation. *Hepatology* **2022**. [CrossRef]
68. Boursier, J.; Diehl, A.M. Implication of gut microbiota in nonalcoholic fatty liver disease. *PLoS Pathog.* **2015**, *11*, e1004559. [CrossRef]
69. Lake, A.D.; Novak, P.; Shipkova, P.; Aranibar, N.; Robertson, D.; Reily, M.D.; Lu, Z.; Lehman-McKeeman, L.D.; Cherrington, N.J. Decreased hepatotoxic bile acid composition and altered synthesis in progressive human nonalcoholic fatty liver disease. *Toxicol. Appl. Pharm.* **2013**, *268*, 132–140. [CrossRef]
70. Chavez-Talavera, O.; Tailleux, A.; Lefebvre, P.; Staels, B. Bile Acid Control of Metabolism and Inflammation in Obesity, Type 2 Diabetes, Dyslipidemia, and Nonalcoholic Fatty Liver Disease. *Gastroenterology* **2017**, *152*, 1679–1694.e3. [CrossRef]
71. Mouzaki, M.; Wang, A.Y.; Bandsma, R.; Comelli, E.M.; Arendt, B.M.; Zhang, L.; Fung, S.; Fischer, S.E.; McGilvray, I.G.; Allard, J.P. Bile Acids and Dysbiosis in Non-Alcoholic Fatty Liver Disease. *PLoS ONE* **2016**, *11*, e0151829. [CrossRef] [PubMed]
72. Cao, H.; Xu, M.; Dong, W.; Deng, B.; Wang, S.; Zhang, Y.; Wang, S.; Luo, S.; Wang, W.; Qi, Y.; et al. Secondary bile acid-induced dysbiosis promotes intestinal carcinogenesis. *Int. J. Cancer* **2017**, *140*, 2545–2556. [CrossRef] [PubMed]
73. Liquori, G.E.; Mastrodonato, M.; Rossi, R.; Scillitani, G.; Gena, P.; Portincasa, P.; Calamita, G.; Ferri, D. Altered membrane glycoprotein targeting in cholestatic hepatocytes. *Eur. J. Clin. Investig.* **2010**, *40*, 393–400. [CrossRef] [PubMed]
74. Arab, J.P.; Karpen, S.J.; Dawson, P.A.; Arrese, M.; Trauner, M. Bile acids and nonalcoholic fatty liver disease: Molecular insights and therapeutic perspectives. *Hepatology* **2017**, *65*, 350–362. [CrossRef] [PubMed]
75. Jiang, C.; Xie, C.; Li, F.; Zhang, L.; Nichols, R.G.; Krausz, K.W.; Cai, J.; Qi, Y.; Fang, Z.Z.; Takahashi, S.; et al. Intestinal farnesoid X receptor signaling promotes nonalcoholic fatty liver disease. *J. Clin. Investig.* **2015**, *125*, 386–402. [CrossRef] [PubMed]
76. Li, F.; Jiang, C.; Krausz, K.W.; Li, Y.; Albert, I.; Hao, H.; Fabre, K.M.; Mitchell, J.B.; Patterson, A.D.; Gonzalez, F.J. Microbiome remodelling leads to inhibition of intestinal farnesoid X receptor signalling and decreased obesity. *Nat. Commun.* **2013**, *4*, 2384. [CrossRef]
77. Lorenzo-Zúñiga, V.; Bartolí, R.; Planas, R.; Hofmann, A.F.; Viñado, B.; Hagey, L.R.; Hernández, J.M.; Mañé, J.; Alvarez, M.A.; Ausina, V. Oral bile acids reduce bacterial overgrowth, bacterial translocation, and endotoxemia in cirrhotic rats. *Hepatology* **2003**, *37*, 551–557. [CrossRef]
78. Clements, W.; Parks, R.; Erwin, P.; Halliday, M.; Barr, J.; Rowlands, B. Role of the gut in the pathophysiology of extrahepatic biliary obstruction. *Gut* **1996**, *39*, 587–593. [CrossRef]
79. Inagaki, T.; Moschetta, A.; Lee, Y.K.; Peng, L.; Zhao, G.; Downes, M.; Yu, R.T.; Shelton, J.M.; Richardson, J.A.; Repa, J.J.; et al. Regulation of antibacterial defense in the small intestine by the nuclear bile acid receptor. *Proc. Natl. Acad. Sci. USA* **2006**, *103*, 3920–3925. [CrossRef]
80. Sun, L.; Pang, Y.; Wang, X.; Wu, Q.; Liu, H.; Liu, B.; Liu, G.; Ye, M.; Kong, W.; Jiang, C. Ablation of gut microbiota alleviates obesity-induced hepatic steatosis and glucose intolerance by modulating bile acid metabolism in hamsters. *Acta Pharm. Sin. B* **2019**, *9*, 702–710. [CrossRef]
81. Sun, L.; Xie, C.; Wang, G.; Wu, Y.; Wu, Q.; Wang, X.; Liu, J.; Deng, Y.; Xia, J.; Chen, B.; et al. Gut microbiota and intestinal FXR mediate the clinical benefits of metformin. *Nat. Med.* **2018**, *24*, 1919–1929. [CrossRef] [PubMed]
82. Baron, E.J.; Summanen, P.; Downes, J.; Roberts, M.C.; Wexler, H.; Finegold, S.M. Bilophila wadsworthia, gen. nov. and sp. nov., a unique gram-negative anaerobic rod recovered from appendicitis specimens and human faeces. *J. Gen. Microbiol.* **1989**, *135*, 3405–3411. [CrossRef] [PubMed]
83. Devkota, S.; Wang, Y.W.; Musch, M.W.; Leone, V.; Fehlner-Peach, H.; Nadimpalli, A.; Antonopoulos, D.A.; Jabri, B.; Chang, E.B. Dietary-fat-induced taurocholic acid promotes pathobiont expansion and colitis in Il10($^{-/-}$) mice. *Nature* **2012**, *487*, 104–108. [CrossRef] [PubMed]
84. Loomba, R.; Ling, L.; Dinh, D.M.; DePaoli, A.M.; Lieu, H.D.; Harrison, S.A.; Sanyal, A.J. The Commensal Microbe Veillonella as a Marker for Response to an FGF19 Analog in NASH. *Hepatology* **2021**, *73*, 126–143. [CrossRef]
85. Danneskiold-Samsøe, N.B.; Barros, H.D.d.F.Q.; Santos, R.; Bicas, J.L.; Cazarin, C.B.B.; Madsen, L.; Kristiansen, K.; Pastore, G.M.; Brix, S.; Junior, M.R.M. Interplay between food and gut microbiota in health and disease. *Food Res. Int.* **2019**, *115*, 23–31. [CrossRef]
86. Liu, R.; Zhao, R.; Zhou, X.; Liang, X.; Campbell, D.J.; Zhang, X.; Zhang, L.; Shi, R.; Wang, G.; Pandak, W.M.; et al. Conjugated bile acids promote cholangiocarcinoma cell invasive growth through activation of sphingosine 1-phosphate receptor 2. *Hepatology* **2014**, *60*, 908–918. [CrossRef]
87. Maruyama, T.; Miyamoto, Y.; Nakamura, T.; Tamai, Y.; Okada, H.; Sugiyama, E.; Nakamura, T.; Itadani, H.; Tanaka, K. Identification of membrane-type receptor for bile acids (M-BAR). *Biochem. Biophys. Res. Commun.* **2002**, *298*, 714–719. [CrossRef]
88. Guo, G.L.; Lambert, G.; Negishi, M.; Ward, J.M.; Brewer, H.B., Jr.; Kliewer, S.A.; Gonzalez, F.J.; Sinal, C.J. Complementary roles of farnesoid X receptor, pregnane X receptor, and constitutive androstane receptor in protection against bile acid toxicity. *J. Biol. Chem.* **2003**, *278*, 45062–45071. [CrossRef]
89. Kir, S.; Kliewer, S.A.; Mangelsdorf, D.J. Roles of FGF19 in liver metabolism. *Cold Spring Harb. Symp. Quant. Biol.* **2011**, *76*, 139–144. [CrossRef]

90. Owen, B.M.; Mangelsdorf, D.J.; Kliewer, S.A. Tissue-specific actions of the metabolic hormones FGF15/19 and FGF21. *Trends Endocrinol. Metab.* **2015**, *26*, 22–29. [CrossRef]
91. Zarrinpar, A.; Loomba, R. Review article: The emerging interplay among the gastrointestinal tract, bile acids and incretins in the pathogenesis of diabetes and non-alcoholic fatty liver disease. *Aliment. Pharm.* **2012**, *36*, 909–921. [CrossRef] [PubMed]
92. Henkel, A.S.; Anderson, K.A.; Dewey, A.M.; Kavesh, M.H.; Green, R.M. A chronic high-cholesterol diet paradoxically suppresses hepatic CYP7A1 expression in FVB/NJ mice. *J. Lipid Res.* **2011**, *52*, 289–298. [CrossRef] [PubMed]
93. Miyata, M.; Hata, T.; Yamazoe, Y.; Yoshinari, K. SREBP-2 negatively regulates FXR-dependent transcription of FGF19 in human intestinal cells. *Biochem. Biophys. Res. Commun.* **2014**, *443*, 477–482. [CrossRef] [PubMed]
94. Degirolamo, C.; Sabba, C.; Moschetta, A. Therapeutic potential of the endocrine fibroblast growth factors FGF19, FGF21 and FGF23. *Nat. Rev. Drug Discov.* **2016**, *15*, 51–69. [CrossRef]
95. Beenken, A.; Mohammadi, M. The FGF family: Biology, pathophysiology and therapy. *Nat. Rev. Drug Discov.* **2009**, *8*, 235–253. [CrossRef]
96. Itoh, N.; Ornitz, D.M. Functional evolutionary history of the mouse Fgf gene family. *Dev. Dyn.* **2008**, *237*, 18–27. [CrossRef]
97. Itoh, N. Hormone-like (endocrine) Fgfs: Their evolutionary history and roles in development, metabolism, and disease. *Cell Tissue Res* **2010**, *342*, 1–11. [CrossRef]
98. Goetz, R.; Beenken, A.; Ibrahimi, O.A.; Kalinina, J.; Olsen, S.K.; Eliseenkova, A.V.; Xu, C.; Neubert, T.A.; Zhang, F.; Linhardt, R.J.; et al. Molecular insights into the klotho-dependent, endocrine mode of action of fibroblast growth factor 19 subfamily members. *Mol. Cell. Biol.* **2007**, *27*, 3417–3428. [CrossRef]
99. Nishimura, T.; Utsunomiya, Y.; Hoshikawa, M.; Ohuchi, H.; Itoh, N. Structure and expression of a novel human FGF, FGF-19, expressed in the fetal brain. *Biochim. Biophys. Acta* **1999**, *1444*, 148–151. [CrossRef]
100. Fon Tacer, K.; Bookout, A.L.; Ding, X.; Kurosu, H.; John, G.B.; Wang, L.; Goetz, R.; Mohammadi, M.; Kuro-o, M.; Mangelsdorf, D.J.; et al. Research resource: Comprehensive expression atlas of the fibroblast growth factor system in adult mouse. *Mol. Endocrinol.* **2010**, *24*, 2050–2064. [CrossRef]
101. Xie, M.H.; Holcomb, I.; Deuel, B.; Dowd, P.; Huang, A.; Vagts, A.; Foster, J.; Liang, J.; Brush, J.; Gu, Q.; et al. FGF-19, a novel fibroblast growth factor with unique specificity for FGFR4. *Cytokine* **1999**, *11*, 729–735. [CrossRef] [PubMed]
102. Holt, J.A.; Luo, G.; Billin, A.N.; Bisi, J.; McNeill, Y.Y.; Kozarsky, K.F.; Donahee, M.; Wang, D.Y.; Mansfield, T.A.; Kliewer, S.A.; et al. Definition of a novel growth factor-dependent signal cascade for the suppression of bile acid biosynthesis. *Genes Dev.* **2003**, *17*, 1581–1591. [CrossRef] [PubMed]
103. Kim, I.; Ahn, S.H.; Inagaki, T.; Choi, M.; Ito, S.; Guo, G.L.; Kliewer, S.A.; Gonzalez, F.J. Differential regulation of bile acid homeostasis by the farnesoid X receptor in liver and intestine. *J. Lipid Res.* **2007**, *48*, 2664–2672. [CrossRef] [PubMed]
104. Lundasen, T.; Galman, C.; Angelin, B.; Rudling, M. Circulating intestinal fibroblast growth factor 19 has a pronounced diurnal variation and modulates hepatic bile acid synthesis in man. *J. Intern. Med.* **2006**, *260*, 530–536. [CrossRef]
105. Potthoff, M.J.; Boney-Montoya, J.; Choi, M.; He, T.; Sunny, N.E.; Satapati, S.; Suino-Powell, K.; Xu, H.E.; Gerard, R.D.; Finck, B.N.; et al. FGF15/19 regulates hepatic glucose metabolism by inhibiting the CREB-PGC-1alpha pathway. *Cell Metab.* **2011**, *13*, 729–738. [CrossRef] [PubMed]
106. Potthoff, M.J.; Kliewer, S.A.; Mangelsdorf, D.J. Endocrine fibroblast growth factors 15/19 and 21: From feast to famine. *Genes Dev.* **2012**, *26*, 312–324. [CrossRef]
107. Alvarez-Sola, G.; Uriarte, I.; Latasa, M.U.; Urtasun, R.; Barcena-Varela, M.; Elizalde, M.; Jimenez, M.; Rodriguez-Ortigosa, C.M.; Corrales, F.J.; Fernandez-Barrena, M.G.; et al. Fibroblast Growth Factor 15/19 in Hepatocarcinogenesis. *Dig. Dis.* **2017**, *35*, 158–165. [CrossRef]
108. Uriarte, I.; Latasa, M.U.; Carotti, S.; Fernandez-Barrena, M.G.; Garcia-Irigoyen, O.; Elizalde, M.; Urtasun, R.; Vespasiani-Gentilucci, U.; Morini, S.; de Mingo, A. Ileal FGF 15 contributes to fibrosis-associated hepatocellular carcinoma development. *Int. J. Cancer* **2015**, *136*, 2469–2475. [CrossRef]
109. French, D.M.; Lin, B.C.; Wang, M.; Adams, C.; Shek, T.; Hotzel, K.; Bolon, B.; Ferrando, R.; Blackmore, C.; Schroeder, K.; et al. Targeting FGFR4 inhibits hepatocellular carcinoma in preclinical mouse models. *PLoS ONE* **2012**, *7*, e36713. [CrossRef]
110. Wu, X.; Lemon, B.; Li, X.; Gupte, J.; Weiszmann, J.; Stevens, J.; Hawkins, N.; Shen, W.; Lindberg, R.; Chen, J.L.; et al. C-terminal tail of FGF19 determines its specificity toward Klotho co-receptors. *J. Biol. Chem.* **2008**, *283*, 33304–33309. [CrossRef]
111. Yang, C.; Jin, C.; Li, X.; Wang, F.; McKeehan, W.L.; Luo, Y. Differential specificity of endocrine FGF19 and FGF21 to FGFR1 and FGFR4 in complex with KLB. *PLoS ONE* **2012**, *7*, e33870. [CrossRef] [PubMed]
112. Lin, B.C.; Wang, M.; Blackmore, C.; Desnoyers, L.R. Liver-specific activities of FGF19 require Klotho beta. *J. Biol. Chem.* **2007**, *282*, 27277–27284. [CrossRef] [PubMed]
113. Schumacher, J.D.; Guo, G.L. Pharmacologic Modulation of Bile Acid-FXR-FGF15/FGF19 Pathway for the Treatment of Nonalcoholic Steatohepatitis. *Handb. Exp. Pharm.* **2019**, *256*, 325–357. [CrossRef]
114. Naugler, W.E.; Tarlow, B.D.; Fedorov, L.M.; Taylor, M.; Pelz, C.; Li, B.; Darnell, J.; Grompe, M. Fibroblast Growth Factor Signaling Controls Liver Size in Mice with Humanized Livers. *Gastroenterology* **2015**, *149*, 728–740.e15. [CrossRef] [PubMed]
115. Wang, C.; Yang, C.; Chang, J.Y.; You, P.; Li, Y.; Jin, C.; Luo, Y.; Li, X.; McKeehan, W.L.; Wang, F. Hepatocyte FRS2α is essential for the endocrine fibroblast growth factor to limit the amplitude of bile acid production induced by prandial activity. *Curr. Mol. Med.* **2014**, *14*, 703–711. [CrossRef]

116. Kurosu, H.; Choi, M.; Ogawa, Y.; Dickson, A.S.; Goetz, R.; Eliseenkova, A.V.; Mohammadi, M.; Rosenblatt, K.P.; Kliewer, S.A.; Kuro, O.M. Tissue-specific expression of betaKlotho and fibroblast growth factor (FGF) receptor isoforms determines metabolic activity of FGF19 and FGF21. *J. Biol. Chem.* **2007**, *282*, 26687–26695. [CrossRef] [PubMed]
117. Di Ciaula, A.; Molina-Molina, E.; Bonfrate, L.; Wang, D.Q.; Dumitrascu, D.L.; Portincasa, P. Gastrointestinal defects in gallstone and cholecystectomized patients. *Eur. J. Clin. Investig.* **2019**, *49*, e13066. [CrossRef]
118. Molina-Molina, E.; Lunardi Baccetto, R.; Wang, D.Q.; de Bari, O.; Krawczyk, M.; Portincasa, P. Exercising the hepatobiliary-gut axis. The impact of physical activity performance. *Eur. J. Clin. Investig.* **2018**, *48*, e12958. [CrossRef]
119. Diella, G.; Di Ciaula, A.; Lorusso, M.P.; Summo, C.; Caggiano, G.; Caponio, F.; Montagna, M.T.; Portincasa, P. Distinct Effects of two Almond Cultivars on Agreeability and Gastrointestinal Motility in Healthy Subjects: More than mere Nutraceuticals. *J. Gastrointest. Liver Dis. JGLD* **2018**, *27*, 31–39. [CrossRef]
120. Rizzello, C.G.; Portincasa, P.; Montemurro, M.; Di Palo, D.M.; Lorusso, M.P.; De Angelis, M.; Bonfrate, L.; Genot, B.; Gobbetti, M. Sourdough Fermented Breads are More Digestible than Those Started with Baker's Yeast Alone: An In Vivo Challenge Dissecting Distinct Gastrointestinal Responses. *Nutrients* **2019**, *11*, 2954. [CrossRef] [PubMed]
121. Luiking, Y.C.; Peeters, T.L.; Stolk, M.F.; Nieuwenhuijs, V.B.; Portincasa, P.; Depoortere, I.; van Berge Henegouwen, G.P.; Akkermans, L.M. Motilin induces gall bladder emptying and antral contractions in the fasted state in humans. *Gut* **1998**, *42*, 830–835. [CrossRef] [PubMed]
122. Portincasa, P.; Di Ciaula, A.; Wang, H.H.; Palasciano, G.; van Erpecum, K.J.; Moschetta, A.; Wang, D.Q. Coordinate regulation of gallbladder motor function in the gut-liver axis. *Hepatology* **2008**, *47*, 2112–2126. [CrossRef] [PubMed]
123. Kong, B.; Wang, L.; Chiang, J.Y.; Zhang, Y.; Klaassen, C.D.; Guo, G.L. Mechanism of tissue-specific farnesoid X receptor in suppressing the expression of genes in bile-acid synthesis in mice. *Hepatology* **2012**, *56*, 1034–1043. [CrossRef]
124. Zollner, G.; Wagner, M.; Moustafa, T.; Fickert, P.; Silbert, D.; Gumhold, J.; Fuchsbichler, A.; Halilbasic, E.; Denk, H.; Marschall, H.U.; et al. Coordinated induction of bile acid detoxification and alternative elimination in mice: Role of FXR-regulated organic solute transporter-alpha/beta in the adaptive response to bile acids. *Am. J. Physiol. Gastrointest. Liver Physiol.* **2006**, *290*, G923–G932. [CrossRef] [PubMed]
125. Landrier, J.F.; Eloranta, J.J.; Vavricka, S.R.; Kullak-Ublick, G.A. The nuclear receptor for bile acids, FXR, transactivates human organic solute transporter-alpha and -beta genes. *Am. J. Physiol. Gastrointest. Liver Physiol.* **2006**, *290*, G476–G485. [CrossRef]
126. Chen, F.; Ma, L.; Dawson, P.A.; Sinal, C.J.; Sehayek, E.; Gonzalez, F.J.; Breslow, J.; Ananthanarayanan, M.; Shneider, B.L. Liver receptor homologue-1 mediates species- and cell line-specific bile acid-dependent negative feedback regulation of the apical sodium-dependent bile acid transporter. *J. Biol. Chem.* **2003**, *278*, 19909–19916. [CrossRef]
127. Yu, C.; Wang, F.; Jin, C.; Huang, X.; McKeehan, W.L. Independent repression of bile acid synthesis and activation of c-Jun N-terminal kinase (JNK) by activated hepatocyte fibroblast growth factor receptor 4 (FGFR4) and bile acids. *J. Biol. Chem.* **2005**, *280*, 17707–17714. [CrossRef] [PubMed]
128. Ito, S.; Fujimori, T.; Furuya, A.; Satoh, J.; Nabeshima, Y.; Nabeshima, Y. Impaired negative feedback suppression of bile acid synthesis in mice lacking betaKlotho. *J. Clin. Investig.* **2005**, *115*, 2202–2208. [CrossRef] [PubMed]
129. Tomiyama, K.; Maeda, R.; Urakawa, I.; Yamazaki, Y.; Tanaka, T.; Ito, S.; Nabeshima, Y.; Tomita, T.; Odori, S.; Hosoda, K.; et al. Relevant use of Klotho in FGF19 subfamily signaling system in vivo. *Proc. Natl. Acad. Sci. USA* **2010**, *107*, 1666–1671. [CrossRef] [PubMed]
130. Kir, S.; Zhang, Y.; Gerard, R.D.; Kliewer, S.A.; Mangelsdorf, D.J. Nuclear receptors HNF4alpha and LRH-1 cooperate in regulating Cyp7a1 in vivo. *J. Biol. Chem.* **2012**, *287*, 41334–41341. [CrossRef] [PubMed]
131. Miao, J.; Choi, S.E.; Seok, S.M.; Yang, L.; Zuercher, W.J.; Xu, Y.; Willson, T.M.; Xu, H.E.; Kemper, J.K. Ligand-dependent regulation of the activity of the orphan nuclear receptor, small heterodimer partner (SHP), in the repression of bile acid biosynthetic CYP7A1 and CYP8B1 genes. *Mol. Endocrinol.* **2011**, *25*, 1159–1169. [CrossRef]
132. Pols, T.W.; Noriega, L.G.; Nomura, M.; Auwerx, J.; Schoonjans, K. The bile acid membrane receptor TGR5: A valuable metabolic target. *Dig. Dis.* **2011**, *29*, 37–44. [CrossRef] [PubMed]
133. Broeders, E.P.; Nascimento, E.B.; Havekes, B.; Brans, B.; Roumans, K.H.; Tailleux, A.; Schaart, G.; Kouach, M.; Charton, J.; Deprez, B.; et al. The Bile Acid Chenodeoxycholic Acid Increases Human Brown Adipose Tissue Activity. *Cell Metab.* **2015**, *22*, 418–426. [CrossRef]
134. Thomas, C.; Gioiello, A.; Noriega, L.; Strehle, A.; Oury, J.; Rizzo, G.; Macchiarulo, A.; Yamamoto, H.; Mataki, C.; Pruzanski, M.; et al. TGR5-mediated bile acid sensing controls glucose homeostasis. *Cell Metab.* **2009**, *10*, 167–177. [CrossRef] [PubMed]
135. Perino, A.; Schoonjans, K. TGR5 and Immunometabolism: Insights from Physiology and Pharmacology. *Trends Pharm. Sci.* **2015**, *36*, 847–857. [CrossRef] [PubMed]
136. Schaap, F.G.; Trauner, M.; Jansen, P.L. Bile acid receptors as targets for drug development. *Nat. Rev. Gastroenterol. Hepatol.* **2014**, *11*, 55–67. [CrossRef]
137. Wang, D.Q.H.; Neuschwander-Tetri, B.A.; Portincasa, P. *The Biliary System, Second Edition*; Morgan & Claypool Life Sciences: San Rafael, CA, USA, 2017; Volume 8, p. i-178.
138. Musso, G.; Gambino, R.; Cassader, M. Cholesterol metabolism and the pathogenesis of non-alcoholic steatohepatitis. *Prog. Lipid Res.* **2013**, *52*, 175–191. [CrossRef] [PubMed]

139. Li, T.; Owsley, E.; Matozel, M.; Hsu, P.; Novak, C.M.; Chiang, J.Y. Transgenic expression of cholesterol 7alpha-hydroxylase in the liver prevents high-fat diet-induced obesity and insulin resistance in mice. *Hepatology* **2010**, *52*, 678–690. [CrossRef] [PubMed]
140. Sinal, C.J.; Tohkin, M.; Miyata, M.; Ward, J.M.; Lambert, G.; Gonzalez, F.J. Targeted disruption of the nuclear receptor FXR/BAR impairs bile acid and lipid homeostasis. *Cell* **2000**, *102*, 731–744. [CrossRef]
141. Watanabe, M.; Houten, S.M.; Wang, L.; Moschetta, A.; Mangelsdorf, D.J.; Heyman, R.A.; Moore, D.D.; Auwerx, J. Bile acids lower triglyceride levels via a pathway involving FXR, SHP, and SREBP-1c. *J. Clin. Investig.* **2004**, *113*, 1408–1418. [CrossRef] [PubMed]
142. Zhang, Y.; Lee, F.Y.; Barrera, G.; Lee, H.; Vales, C.; Gonzalez, F.J.; Willson, T.M.; Edwards, P.A. Activation of the nuclear receptor FXR improves hyperglycemia and hyperlipidemia in diabetic mice. *Proc. Natl. Acad. Sci. USA* **2006**, *103*, 1006–1011. [CrossRef] [PubMed]
143. Bhatnagar, S.; Damron, H.A.; Hillgartner, F.B. Fibroblast growth factor-19, a novel factor that inhibits hepatic fatty acid synthesis. *J. Biol. Chem.* **2009**, *284*, 10023–10033. [CrossRef]
144. Kast, H.R.; Nguyen, C.M.; Sinal, C.J.; Jones, S.A.; Laffitte, B.A.; Reue, K.; Gonzalez, F.J.; Willson, T.M.; Edwards, P.A. Farnesoid X-activated receptor induces apolipoprotein C-II transcription: A molecular mechanism linking plasma triglyceride levels to bile acids. *Mol. Endocrinol.* **2001**, *15*, 1720–1728. [CrossRef] [PubMed]
145. Schoenfield, L.J.; Lachin, J.M. Chenodiol (chenodeoxycholic acid) for dissolution of gallstones: The National Cooperative Gallstone Study. A controlled trial of efficacy and safety. *Ann. Intern. Med.* **1981**, *95*, 257–282. [CrossRef] [PubMed]
146. Lu, D.; Liu, Y.; Luo, Y.; Zhao, J.; Feng, C.; Xue, L.; Xu, J.; Wang, Q.; Yan, T.; Xiao, P.; et al. Intestinal farnesoid X receptor signaling controls hepatic fatty acid oxidation. *Biochim. Biophys. Acta Mol. Cell Biol. Lipids* **2022**, *1867*, 159089. [CrossRef] [PubMed]
147. Wang, D.Q.H.; Neuschwander-Tetri, B.A.; Portincasa, P. The Biliary System. *Colloq. Ser. Integr. Syst. Physiol. Mol. Funct.* **2012**, *4*, 1–148. [CrossRef]
148. Wang, D.Q.; Portincasa, P.; Tso, P. Transintestinal cholesterol excretion: A secondary, nonbiliary pathway contributing to reverse cholesterol transport. *Hepatology* **2017**, *66*, 1337–1340. [CrossRef] [PubMed]
149. de Boer, J.F.; Schonewille, M.; Boesjes, M.; Wolters, H.; Bloks, V.W.; Bos, T.; van Dijk, T.H.; Jurdzinski, A.; Boverhof, R.; Wolters, J.C.; et al. Intestinal Farnesoid X Receptor Controls Transintestinal Cholesterol Excretion in Mice. *Gastroenterology* **2017**, *152*, 1126–1138.e6. [CrossRef]
150. Deng, W.; Fan, W.; Tang, T.; Wan, H.; Zhao, S.; Tan, Y.; Oware, K.A.; Tan, J.; Li, J.; Qu, S. Farnesoid X Receptor Deficiency Induces Hepatic Lipid and Glucose Metabolism Disorder via Regulation of Pyruvate Dehydrogenase Kinase 4. *Oxid. Med. Cell Longev.* **2022**, *2022*, 3589525. [CrossRef] [PubMed]
151. Kir, S.; Beddow, S.A.; Samuel, V.T.; Miller, P.; Previs, S.F.; Suino-Powell, K.; Xu, H.E.; Shulman, G.I.; Kliewer, S.A.; Mangelsdorf, D.J. FGF19 as a postprandial, insulin-independent activator of hepatic protein and glycogen synthesis. *Science* **2011**, *331*, 1621–1624. [CrossRef] [PubMed]
152. Jiao, Y.; Lu, Y.; Li, X.Y. Farnesoid X receptor: A master regulator of hepatic triglyceride and glucose homeostasis. *Acta Pharm. Sin.* **2015**, *36*, 44–50. [CrossRef] [PubMed]
153. Yamagata, K.; Daitoku, H.; Shimamoto, Y.; Matsuzaki, H.; Hirota, K.; Ishida, J.; Fukamizu, A. Bile acids regulate gluconeogenic gene expression via small heterodimer partner-mediated repression of hepatocyte nuclear factor 4 and Foxo1. *J. Biol. Chem.* **2004**, *279*, 23158–23165. [CrossRef] [PubMed]
154. Thomas, A.M.; Hart, S.N.; Kong, B.; Fang, J.; Zhong, X.B.; Guo, G.L. Genome-wide tissue-specific farnesoid X receptor binding in mouse liver and intestine. *Hepatology* **2010**, *51*, 1410–1419. [CrossRef]
155. Zhan, L.; Liu, H.X.; Fang, Y.; Kong, B.; He, Y.; Zhong, X.B.; Fang, J.; Wan, Y.J.; Guo, G.L. Genome-wide binding and transcriptome analysis of human farnesoid X receptor in primary human hepatocytes. *PLoS ONE* **2014**, *9*, e105930. [CrossRef]
156. Ma, K.; Saha, P.K.; Chan, L.; Moore, D.D. Farnesoid X receptor is essential for normal glucose homeostasis. *J. Clin. Investig.* **2006**, *116*, 1102–1109. [CrossRef] [PubMed]
157. Kim, K.H.; Choi, S.; Zhou, Y.; Kim, E.Y.; Lee, J.M.; Saha, P.K.; Anakk, S.; Moore, D.D. Hepatic FXR/SHP axis modulates systemic glucose and fatty acid homeostasis in aged mice. *Hepatology* **2017**, *66*, 498–509. [CrossRef] [PubMed]
158. Lee, J.M.; Wagner, M.; Xiao, R.; Kim, K.H.; Feng, D.; Lazar, M.A.; Moore, D.D. Nutrient-sensing nuclear receptors coordinate autophagy. *Nature* **2014**, *516*, 112–115. [CrossRef] [PubMed]
159. Trabelsi, M.S.; Daoudi, M.; Prawitt, J.; Ducastel, S.; Touche, V.; Sayin, S.I.; Perino, A.; Brighton, C.A.; Sebti, Y.; Kluza, J.; et al. Farnesoid X receptor inhibits glucagon-like peptide-1 production by enteroendocrine L cells. *Nat. Commun.* **2015**, *6*, 7629. [CrossRef] [PubMed]
160. Pathak, P.; Xie, C.; Nichols, R.G.; Ferrell, J.M.; Boehme, S.; Krausz, K.W.; Patterson, A.D.; Gonzalez, F.J.; Chiang, J.Y.L. Intestine farnesoid X receptor agonist and the gut microbiota activate G-protein bile acid receptor-1 signaling to improve metabolism. *Hepatology* **2018**, *68*, 1574–1588. [CrossRef]
161. Cully, M. FXR and JAK step up to BAT. *Nat. Rev. Drug Discov.* **2015**, *14*, 91. [CrossRef]
162. Fang, S.; Suh, J.M.; Reilly, S.M.; Yu, E.; Osborn, O.; Lackey, D.; Yoshihara, E.; Perino, A.; Jacinto, S.; Lukasheva, Y.; et al. Intestinal FXR agonism promotes adipose tissue browning and reduces obesity and insulin resistance. *Nat. Med.* **2015**, *21*, 159–165. [CrossRef] [PubMed]

163. Velazquez-Villegas, L.A.; Perino, A.; Lemos, V.; Zietak, M.; Nomura, M.; Pols, T.W.H.; Schoonjans, K. TGR5 signalling promotes mitochondrial fission and beige remodelling of white adipose tissue. *Nat. Commun.* **2018**, *9*, 245. [CrossRef] [PubMed]
164. Stayrook, K.R.; Bramlett, K.S.; Savkur, R.S.; Ficorilli, J.; Cook, T.; Christe, M.E.; Michael, L.F.; Burris, T.P. Regulation of carbohydrate metabolism by the farnesoid X receptor. *Endocrinology* **2005**, *146*, 984–991. [CrossRef] [PubMed]
165. Fu, L.; John, L.M.; Adams, S.H.; Yu, X.X.; Tomlinson, E.; Renz, M.; Williams, P.M.; Soriano, R.; Corpuz, R.; Moffat, B.; et al. Fibroblast growth factor 19 increases metabolic rate and reverses dietary and leptin-deficient diabetes. *Endocrinology* **2004**, *145*, 2594–2603. [CrossRef]
166. Somm, E.; Jornayvaz, F.R. Fibroblast Growth Factor 15/19: From Basic Functions to Therapeutic Perspectives. *Endocr. Rev.* **2018**, *39*, 960–989. [CrossRef]
167. Morton, G.J.; Matsen, M.E.; Bracy, D.P.; Meek, T.H.; Nguyen, H.T.; Stefanovski, D.; Bergman, R.N.; Wasserman, D.H.; Schwartz, M.W. FGF19 action in the brain induces insulin-independent glucose lowering. *J. Clin. Investig.* **2013**, *123*, 4799–4808. [CrossRef]
168. Taylor, R.; Magnusson, I.; Rothman, D.L.; Cline, G.W.; Caumo, A.; Cobelli, C.; Shulman, G.I. Direct assessment of liver glycogen storage by 13C nuclear magnetic resonance spectroscopy and regulation of glucose homeostasis after a mixed meal in normal subjects. *J. Clin. Investig.* **1996**, *97*, 126–132. [CrossRef]
169. Benoit, B.; Meugnier, E.; Castelli, M.; Chanon, S.; Vieille-Marchiset, A.; Durand, C.; Bendridi, N.; Pesenti, S.; Monternier, P.A.; Durieux, A.C.; et al. Fibroblast growth factor 19 regulates skeletal muscle mass and ameliorates muscle wasting in mice. *Nat. Med.* **2017**, *23*, 990–996. [CrossRef]
170. Massafra, V.; Milona, A.; Vos, H.R.; Ramos, R.J.J.; Gerrits, J.; Willemsen, E.C.L.; Ramos Pittol, J.M.; Ijssennagger, N.; Houweling, M.; Prinsen, H.; et al. Farnesoid X Receptor Activation Promotes Hepatic Amino Acid Catabolism and Ammonium Clearance in Mice. *Gastroenterology* **2017**, *152*, 1462–1476.e10. [CrossRef]
171. Bag Soytas, R.; Suzan, V.; Arman, P.; Emiroglu Gedik, T.; Unal, D.; Cengiz, M.; Bolayirli, I.M.; Suna Erdincler, D.; Doventas, A.; Yavuzer, H. Association of FGF-19 and FGF-21 levels with primary sarcopenia. *Geriatr. Gerontol. Int.* **2021**, *21*, 959–962. [CrossRef]
172. Qiu, Y.; Yu, J.; Li, Y.; Yang, F.; Yu, H.; Xue, M.; Zhang, F.; Jiang, X.; Ji, X.; Bao, Z. Depletion of gut microbiota induces skeletal muscle atrophy by FXR-FGF15/19 signalling. *Ann. Med.* **2021**, *53*, 508–522. [CrossRef] [PubMed]
173. Tomlinson, E.; Fu, L.; John, L.; Hultgren, B.; Huang, X.; Renz, M.; Stephan, J.P.; Tsai, S.P.; Powell-Braxton, L.; French, D.; et al. Transgenic mice expressing human fibroblast growth factor-19 display increased metabolic rate and decreased adiposity. *Endocrinology* **2002**, *143*, 1741–1747. [CrossRef] [PubMed]
174. Marcelin, G.; Jo, Y.H.; Li, X.; Schwartz, G.J.; Zhang, Y.; Dun, N.J.; Lyu, R.M.; Blouet, C.; Chang, J.K.; Chua, S., Jr. Central action of FGF19 reduces hypothalamic AGRP/NPY neuron activity and improves glucose metabolism. *Mol. Metab.* **2014**, *3*, 19–28. [CrossRef] [PubMed]
175. Zhou, M.; Learned, R.M.; Rossi, S.J.; DePaoli, A.M.; Tian, H.; Ling, L. Engineered FGF19 eliminates bile acid toxicity and lipotoxicity leading to resolution of steatohepatitis and fibrosis in mice. *Hepatol. Commun.* **2017**, *1*, 1024–1042. [CrossRef] [PubMed]
176. Wang, Y.D.; Chen, W.D.; Wang, M.; Yu, D.; Forman, B.M.; Huang, W. Farnesoid X receptor antagonizes nuclear factor kappaB in hepatic inflammatory response. *Hepatology* **2008**, *48*, 1632–1643. [CrossRef] [PubMed]
177. Daly, C.; Rollins, B.J. Monocyte chemoattractant protein-1 (CCL2) in inflammatory disease and adaptive immunity: Therapeutic opportunities and controversies. *Microcirculation* **2003**, *10*, 247–257. [CrossRef]
178. Liu, T.; Zhang, L.; Joo, D.; Sun, S. NF-kB signaling in inflammation. *Signal Transduct. Target. Ther.* **2017**, *2*, 247–257.
179. Renga, B.; Mencarelli, A.; Migliorati, M.; Cipriani, S.; D'Amore, C.; Distrutti, E.; Fiorucci, S. SHP-dependent and -independent induction of peroxisome proliferator-activated receptor-gamma by the bile acid sensor farnesoid X receptor counter-regulates the pro-inflammatory phenotype of liver myofibroblasts. *Inflamm. Res.* **2011**, *60*, 577–587. [CrossRef]
180. Fiorucci, S.; Rizzo, G.; Antonelli, E.; Renga, B.; Mencarelli, A.; Riccardi, L.; Morelli, A.; Pruzanski, M.; Pellicciari, R. Cross-talk between farnesoid-X-receptor (FXR) and peroxisome proliferator-activated receptor γ contributes to the antifibrotic activity of FXR ligands in rodent models of liver cirrhosis. *J. Pharmacol. Exp. Ther.* **2005**, *315*, 58–68. [CrossRef]
181. El Kasmi, K.C.; Ghosh, S.; Anderson, A.L.; Devereaux, M.W.; Balasubramaniyan, N.; D'Alessandro, A.; Orlicky, D.J.; Suchy, F.J.; Shearn, C.T.; Sokol, R.J. Pharmacologic activation of hepatic farnesoid X receptor prevents parenteral nutrition-associated cholestasis in mice. *Hepatology* **2022**, *75*, 252–265. [CrossRef]
182. Schumacher, J.D.; Kong, B.; Wu, J.; Rizzolo, D.; Armstrong, L.E.; Chow, M.D.; Goedken, M.; Lee, Y.H.; Guo, G.L. Direct and Indirect Effects of Fibroblast Growth Factor (FGF) 15 and FGF19 on Liver Fibrosis Development. *Hepatology* **2020**, *71*, 670–685. [CrossRef] [PubMed]
183. Carulli, N.; Bertolotti, M.; Carubbi, F.; Concari, M.; Martella, P.; Carulli, L.; Loria, P. Review article: Effect of bile salt pool composition on hepatic and biliary functions. *Aliment. Pharm.* **2000**, *14* (Suppl. 2), 14–18. [CrossRef] [PubMed]
184. Merlen, G.; Ursic-Bedoya, J.; Jourdainne, V.; Kahale, N.; Glenisson, M.; Doignon, I.; Rainteau, D.; Tordjmann, T. Bile acids and their receptors during liver regeneration: "Dangerous protectors". *Mol. Asp. Med.* **2017**, *56*, 25–33. [CrossRef]
185. Donkers, J.M.; Roscam Abbing, R.L.P.; van de Graaf, S.F.J. Developments in bile salt based therapies: A critical overview. *Biochem. Pharm.* **2019**, *161*, 1–13. [CrossRef] [PubMed]
186. Dawson, P.A.; Karpen, S.J. Intestinal transport and metabolism of bile acids. *J. Lipid Res.* **2015**, *56*, 1085–1099. [CrossRef] [PubMed]

187. Doignon, I.; Julien, B.; Serriere-Lanneau, V.; Garcin, I.; Alonso, G.; Nicou, A.; Monnet, F.; Gigou, M.; Humbert, L.; Rainteau, D.; et al. Immediate neuroendocrine signaling after partial hepatectomy through acute portal hyperpressure and cholestasis. *J. Hepatol.* **2011**, *54*, 481–488. [CrossRef] [PubMed]
188. Fahrner, R.; Patsenker, E.; de Gottardi, A.; Stickel, F.; Montani, M.; Stroka, D.; Candinas, D.; Beldi, G. Elevated liver regeneration in response to pharmacological reduction of elevated portal venous pressure by terlipressin after partial hepatectomy. *Transplantation* **2014**, *97*, 892–900. [CrossRef]
189. Miura, T.; Kimura, N.; Yamada, T.; Shimizu, T.; Nanashima, N.; Yamana, D.; Hakamada, K.; Tsuchida, S. Sustained repression and translocation of Ntcp and expression of Mrp4 for cholestasis after rat 90% partial hepatectomy. *J. Hepatol.* **2011**, *55*, 407–414. [CrossRef]
190. Bhushan, B.; Borude, P.; Edwards, G.; Walesky, C.; Cleveland, J.; Li, F.; Ma, X.; Apte, U. Role of bile acids in liver injury and regeneration following acetaminophen overdose. *Am. J. Pathol.* **2013**, *183*, 1518–1526. [CrossRef]
191. Meng, Z.; Wang, Y.; Wang, L.; Jin, W.; Liu, N.; Pan, H.; Liu, L.; Wagman, L.; Forman, B.M.; Huang, W. FXR regulates liver repair after CCl4-induced toxic injury. *Mol. Endocrinol.* **2010**, *24*, 886–897. [CrossRef]
192. Huang, W.; Ma, K.; Zhang, J.; Qatanani, M.; Cuvillier, J.; Liu, J.; Dong, B.; Huang, X.; Moore, D.D. Nuclear receptor-dependent bile acid signaling is required for normal liver regeneration. *Science* **2006**, *312*, 233–236. [CrossRef] [PubMed]
193. Pean, N.; Doignon, I.; Garcin, I.; Besnard, A.; Julien, B.; Liu, B.; Branchereau, S.; Spraul, A.; Guettier, C.; Humbert, L.; et al. The receptor TGR5 protects the liver from bile acid overload during liver regeneration in mice. *Hepatology* **2013**, *58*, 1451–1460. [CrossRef] [PubMed]
194. Naugler, W.E. Bile acid flux is necessary for normal liver regeneration. *PLoS ONE* **2014**, *9*, e97426. [CrossRef] [PubMed]
195. Hoekstra, L.T.; Rietkerk, M.; van Lienden, K.P.; van den Esschert, J.W.; Schaap, F.G.; van Gulik, T.M. Bile salts predict liver regeneration in rabbit model of portal vein embolization. *J. Surg. Res.* **2012**, *178*, 773–778. [CrossRef]
196. Hoekstra, L.T.; van Lienden, K.P.; Schaap, F.G.; Chamuleau, R.A.; Bennink, R.J.; van Gulik, T.M. Can plasma bile salt, triglycerides, and apoA-V levels predict liver regeneration? *World J. Surg.* **2012**, *36*, 2901–2908. [CrossRef]
197. Fickert, P.; Fuchsbichler, A.; Marschall, H.U.; Wagner, M.; Zollner, G.; Krause, R.; Zatloukal, K.; Jaeschke, H.; Denk, H.; Trauner, M. Lithocholic acid feeding induces segmental bile duct obstruction and destructive cholangitis in mice. *Am. J. Pathol.* **2006**, *168*, 410–422. [CrossRef]
198. Gilgenkrantz, H.; Tordjmann, T. Bile acids and FGF receptors: Orchestrators of optimal liver regeneration. *Gut* **2015**, *64*, 1351–1352. [CrossRef]
199. Padrissa-Altes, S.; Bachofner, M.; Bogorad, R.L.; Pohlmeier, L.; Rossolini, T.; Bohm, F.; Liebisch, G.; Hellerbrand, C.; Koteliansky, V.; Speicher, T.; et al. Control of hepatocyte proliferation and survival by Fgf receptors is essential for liver regeneration in mice. *Gut* **2015**, *64*, 1444–1453. [CrossRef]
200. Park, Y.J.; Qatanani, M.; Chua, S.S.; LaRey, J.L.; Johnson, S.A.; Watanabe, M.; Moore, D.D.; Lee, Y.K. Loss of orphan receptor small heterodimer partner sensitizes mice to liver injury from obstructive cholestasis. *Hepatology* **2008**, *47*, 1578–1586. [CrossRef]
201. Modica, S.; Petruzzelli, M.; Bellafante, E.; Murzilli, S.; Salvatore, L.; Celli, N.; Di Tullio, G.; Palasciano, G.; Moustafa, T.; Halilbasic, E.; et al. Selective activation of nuclear bile acid receptor FXR in the intestine protects mice against cholestasis. *Gastroenterology* **2012**, *142*, 355–365.e4. [CrossRef]
202. Honda, A.; Miyazaki, T.; Iwamoto, J.; Hirayama, T.; Morishita, Y.; Monma, T.; Ueda, H.; Mizuno, S.; Sugiyama, F.; Takahashi, S.; et al. Regulation of bile acid metabolism in mouse models with hydrophobic bile acid composition. *J. Lipid Res.* **2020**, *61*, 54–69. [CrossRef] [PubMed]
203. de Boer, J.F.; Verkade, E.; Mulder, N.L.; de Vries, H.D.; Huijkman, N.; Koehorst, M.; Boer, T.; Wolters, J.C.; Bloks, V.W.; van de Sluis, B.; et al. A human-like bile acid pool induced by deletion of hepatic Cyp2c70 modulates effects of FXR activation in mice. *J. Lipid Res.* **2020**, *61*, 291–305. [CrossRef] [PubMed]
204. Takahashi, S.; Fukami, T.; Masuo, Y.; Brocker, C.N.; Xie, C.; Krausz, K.W.; Wolf, C.R.; Henderson, C.J.; Gonzalez, F.J. Cyp2c70 is responsible for the species difference in bile acid metabolism between mice and humans. *J. Lipid Res.* **2016**, *57*, 2130–2137. [CrossRef] [PubMed]
205. Hrycay, E.; Forrest, D.; Liu, L.; Wang, R.; Tai, J.; Deo, A.; Ling, V.; Bandiera, S. Hepatic bile acid metabolism and expression of cytochrome P450 and related enzymes are altered in Bsep ($^{-/-}$) mice. *Mol. Cell Biochem.* **2014**, *389*, 119–132. [CrossRef] [PubMed]
206. Srivastava, A. Progressive familial intrahepatic cholestasis. *J. Clin. Exp. Hepatol.* **2014**, *4*, 25–36. [CrossRef]
207. Fuchs, C.D.; Paumgartner, G.; Wahlstrom, A.; Schwabl, P.; Reiberger, T.; Leditznig, N.; Stojakovic, T.; Rohr-Udilova, N.; Chiba, P.; Marschall, H.U.; et al. Metabolic preconditioning protects BSEP/ABCB11($^{-/-}$) mice against cholestatic liver injury. *J. Hepatol.* **2017**, *66*, 95–101. [CrossRef]
208. Lee, C.S.; Kimura, A.; Wu, J.F.; Ni, Y.H.; Hsu, H.Y.; Chang, M.H.; Nittono, H.; Chen, H.L. Prognostic roles of tetrahydroxy bile acids in infantile intrahepatic cholestasis. *J. Lipid Res.* **2017**, *58*, 607–614. [CrossRef]
209. Woolbright, B.L.; Jaeschke, H. Inflammation and Cell Death During Cholestasis: The Evolving Role of Bile Acids. *Gene Expr.* **2019**, *19*, 215–228. [CrossRef]
210. Yerushalmi, B.; Dahl, R.; Devereaux, M.W.; Gumpricht, E.; Sokol, R.J. Bile acid-induced rat hepatocyte apoptosis is inhibited by antioxidants and blockers of the mitochondrial permeability transition. *Hepatology* **2001**, *33*, 616–626. [CrossRef]

211. Grattagliano, I.; Montezinho, L.P.; Oliveira, P.J.; Fruhbeck, G.; Gomez-Ambrosi, J.; Montecucco, F.; Carbone, F.; Wieckowski, M.R.; Wang, D.Q.; Portincasa, P. Targeting mitochondria to oppose the progression of nonalcoholic fatty liver disease. *Biochem. Pharm.* **2019**, *160*, 34–45. [CrossRef]
212. Reinehr, R.; Graf, D.; Haussinger, D. Bile salt-induced hepatocyte apoptosis involves epidermal growth factor receptor-dependent CD95 tyrosine phosphorylation. *Gastroenterology* **2003**, *125*, 839–853. [CrossRef]
213. Miyoshi, H.; Rust, C.; Roberts, P.J.; Burgart, L.J.; Gores, G.J. Hepatocyte apoptosis after bile duct ligation in the mouse involves Fas. *Gastroenterology* **1999**, *117*, 669–677. [CrossRef]
214. Allen, K.; Jaeschke, H.; Copple, B.L. Bile acids induce inflammatory genes in hepatocytes: A novel mechanism of inflammation during obstructive cholestasis. *Am. J. Pathol.* **2011**, *178*, 175–186. [CrossRef]
215. Csanaky, I.L.; Aleksunes, L.M.; Tanaka, Y.; Klaassen, C.D. Role of hepatic transporters in prevention of bile acid toxicity after partial hepatectomy in mice. *Am. J. Physiol. Gastrointest. Liver Physiol.* **2009**, *297*, G419–G433. [CrossRef] [PubMed]
216. Geier, A.; Wagner, M.; Dietrich, C.G.; Trauner, M. Principles of hepatic organic anion transporter regulation during cholestasis, inflammation and liver regeneration. *Biochim. Biophys. Acta* **2007**, *1773*, 283–308. [CrossRef] [PubMed]
217. Kong, B.; Zhang, M.; Huang, M.; Rizzolo, D.; Armstrong, L.E.; Schumacher, J.D.; Chow, M.D.; Lee, Y.H.; Guo, G.L. FXR deficiency alters bile acid pool composition and exacerbates chronic alcohol induced liver injury. *Dig Liver Dis.* **2019**, *51*, 570–576. [CrossRef]
218. Kong, B.; Huang, J.; Zhu, Y.; Li, G.; Williams, J.; Shen, S.; Aleksunes, L.M.; Richardson, J.R.; Apte, U.; Rudnick, D.A.; et al. Fibroblast growth factor 15 deficiency impairs liver regeneration in mice. *Am. J. Physiol. Gastrointest. Liver Physiol.* **2014**, *306*, G893–G902. [CrossRef]
219. Uriarte, I.; Fernandez-Barrena, M.G.; Monte, M.J.; Latasa, M.U.; Chang, H.C.; Carotti, S.; Vespasiani-Gentilucci, U.; Morini, S.; Vicente, E.; Concepcion, A.R.; et al. Identification of fibroblast growth factor 15 as a novel mediator of liver regeneration and its application in the prevention of post-resection liver failure in mice. *Gut* **2013**, *62*, 899–910. [CrossRef]
220. Kong, B.; Sun, R.; Huang, M.; Chow, M.D.; Zhong, X.B.; Xie, W.; Lee, Y.H.; Guo, G.L. Fibroblast Growth Factor 15-Dependent and Bile Acid-Independent Promotion of Liver Regeneration in Mice. *Hepatology* **2018**, *68*, 1961–1976. [CrossRef]
221. Cai, S.Y.; Ouyang, X.; Chen, Y.; Soroka, C.J.; Wang, J.; Mennone, A.; Wang, Y.; Mehal, W.Z.; Jain, D.; Boyer, J.L. Bile acids initiate cholestatic liver injury by triggering a hepatocyte-specific inflammatory response. *JCI Insight* **2017**, *2*, e90780. [CrossRef]
222. Schumacher, J.D.; Kong, B.; Pan, Y.; Zhan, L.; Sun, R.; Aa, J.; Rizzolo, D.; Richardson, J.R.; Chen, A.; Goedken, M.; et al. The effect of fibroblast growth factor 15 deficiency on the development of high fat diet induced non-alcoholic steatohepatitis. *Toxicol. Appl. Pharm.* **2017**, *330*, 1–8. [CrossRef] [PubMed]
223. Li, M.; Cai, S.Y.; Boyer, J.L. Mechanisms of bile acid mediated inflammation in the liver. *Mol. Asp. Med.* **2017**, *56*, 45–53. [CrossRef] [PubMed]
224. Meadows, V.; Kennedy, L.; Ekser, B.; Kyritsi, K.; Kundu, D.; Zhou, T.; Chen, L.; Pham, L.; Wu, N.; Demieville, J.; et al. Mast Cells Regulate Ductular Reaction and Intestinal Inflammation in Cholestasis Through Farnesoid X Receptor Signaling. *Hepatology* **2021**, *74*, 2684–2698. [CrossRef] [PubMed]
225. Wu, A.L.; Coulter, S.; Liddle, C.; Wong, A.; Eastham-Anderson, J.; French, D.M.; Peterson, A.S.; Sonoda, J. FGF19 regulates cell proliferation, glucose and bile acid metabolism via FGFR4-dependent and independent pathways. *PLoS ONE* **2011**, *6*, e17868. [CrossRef]
226. Luo, J.; Ko, B.; Elliott, M.; Zhou, M.; Lindhout, D.A.; Phung, V.; To, C.; Learned, R.M.; Tian, H.; DePaoli, A.M.; et al. A nontumorigenic variant of FGF19 treats cholestatic liver diseases. *Sci. Transl. Med.* **2014**, *6*, 247ra100. [CrossRef] [PubMed]
227. Zhou, M.; Wang, X.; Phung, V.; Lindhout, D.A.; Mondal, K.; Hsu, J.Y.; Yang, H.; Humphrey, M.; Ding, X.; Arora, T.; et al. Separating Tumorigenicity from Bile Acid Regulatory Activity for Endocrine Hormone FGF19. *Cancer Res.* **2014**, *74*, 3306–3316. [CrossRef] [PubMed]
228. Gaich, G.; Chien, J.Y.; Fu, H.; Glass, L.C.; Deeg, M.A.; Holland, W.L.; Kharitonenkov, A.; Bumol, T.; Schilske, H.K.; Moller, D.E. The effects of LY2405319, an FGF21 analog, in obese human subjects with type 2 diabetes. *Cell Metab.* **2013**, *18*, 333–340. [CrossRef]
229. Ogawa, Y.; Kurosu, H.; Yamamoto, M.; Nandi, A.; Rosenblatt, K.P.; Goetz, R.; Eliseenkova, A.V.; Mohammadi, M.; Kuro-o, M. BetaKlotho is required for metabolic activity of fibroblast growth factor 21. *Proc. Natl. Acad. Sci. USA* **2007**, *104*, 7432–7437. [CrossRef]
230. Chen, M.M.; Hale, C.; Stanislaus, S.; Xu, J.; Veniant, M.M. FGF21 acts as a negative regulator of bile acid synthesis. *J. Endocrinol.* **2018**, *237*, 139–152. [CrossRef]
231. Zhang, J.; Gupte, J.; Gong, Y.; Weiszmann, J.; Zhang, Y.; Lee, K.J.; Richards, W.G.; Li, Y. Chronic Over-expression of Fibroblast Growth Factor 21 Increases Bile Acid Biosynthesis by Opposing FGF15/19 Action. *EBioMedicine* **2017**, *15*, 173–183. [CrossRef]
232. Talukdar, S.; Zhou, Y.; Li, D.; Rossulek, M.; Dong, J.; Somayaji, V.; Weng, Y.; Clark, R.; Lanba, A.; Owen, B.M.; et al. A Long-Acting FGF21 Molecule, PF-05231023, Decreases Body Weight and Improves Lipid Profile in Non-human Primates and Type 2 Diabetic Subjects. *Cell Metab.* **2016**, *23*, 427–440. [CrossRef]
233. Kim, A.M.; Somayaji, V.R.; Dong, J.Q.; Rolph, T.P.; Weng, Y.; Chabot, J.R.; Gropp, K.E.; Talukdar, S.; Calle, R.A. Once-weekly administration of a long-acting fibroblast growth factor 21 analogue modulates lipids, bone turnover markers, blood pressure and body weight differently in obese people with hypertriglyceridaemia and in non-human primates. *Diabetes Obes. Metab.* **2017**, *19*, 1762–1772. [CrossRef] [PubMed]

234. Charles, E.D.; Neuschwander-Tetri, B.A.; Pablo Frias, J.; Kundu, S.; Luo, Y.; Tirucherai, G.S.; Christian, R. Pegbelfermin (BMS-986036), PEGylated FGF21, in Patients with Obesity and Type 2 Diabetes: Results from a Randomized Phase 2 Study. *Obesity* **2019**, *27*, 41–49. [CrossRef] [PubMed]
235. Sanyal, A.; Charles, E.D.; Neuschwander-Tetri, B.A.; Loomba, R.; Harrison, S.A.; Abdelmalek, M.F.; Lawitz, E.J.; Halegoua-DeMarzio, D.; Kundu, S.; Noviello, S.; et al. Pegbelfermin (BMS-986036), a PEGylated fibroblast growth factor 21 analogue, in patients with non-alcoholic steatohepatitis: A randomised, double-blind, placebo-controlled, phase 2a trial. *Lancet* **2019**, *392*, 2705–2717. [CrossRef]
236. Luo, Y.; Decato, B.E.; Charles, E.D.; Shevell, D.E.; McNaney, C.; Shipkova, P.; Apfel, A.; Tirucherai, G.S.; Sanyal, A.J. Pegbelfermin selectively reduces secondary bile acid concentrations in patients with non-alcoholic steatohepatitis. *JHEP Rep.* **2022**, *4*, 100392. [CrossRef]
237. Neuschwander-Tetri, B.A.; Loomba, R.; Sanyal, A.J.; Lavine, J.E.; Van Natta, M.L.; Abdelmalek, M.F.; Chalasani, N.; Dasarathy, S.; Diehl, A.M.; Hameed, B.; et al. Farnesoid X nuclear receptor ligand obeticholic acid for non-cirrhotic, non-alcoholic steatohepatitis (FLINT): A multicentre, randomised, placebo-controlled trial. *Lancet* **2015**, *385*, 956–965. [CrossRef]
238. Pellicciari, R.; Costantino, G.; Camaioni, E.; Sadeghpour, B.M.; Entrena, A.; Willson, T.M.; Fiorucci, S.; Clerici, C.; Gioiello, A. Bile acid derivatives as ligands of the farnesoid X receptor. Synthesis, evaluation, and structure-activity relationship of a series of body and side chain modified analogues of chenodeoxycholic acid. *J. Med. Chem.* **2004**, *47*, 4559–4569. [CrossRef]
239. Bowlus, C.L.; Pockros, P.J.; Kremer, A.E.; Pares, A.; Forman, L.M.; Drenth, J.P.H.; Ryder, S.D.; Terracciano, L.; Jin, Y.; Liberman, A.; et al. Long-Term Obeticholic Acid Therapy Improves Histological Endpoints in Patients With Primary Biliary Cholangitis. *Clin. Gastroenterol. Hepatol.* **2020**, *18*, 1170–1178.e6. [CrossRef]
240. Hirschfield, G.M.; Dyson, J.K.; Alexander, G.J.M.; Chapman, M.H.; Collier, J.; Hübscher, S.; Patanwala, I.; Pereira, S.P.; Thain, C.; Thorburn, D.; et al. The British Society of Gastroenterology/UK-PBC primary biliary cholangitis treatment and management guidelines. *Gut* **2018**, *67*, 1568–1594. [CrossRef]
241. Chapman, R.W.; Lynch, K.D. Obeticholic acid-a new therapy in PBC and NASH. *Br. Med. Bull.* **2020**, *133*, 95–104. [CrossRef]
242. Hernandez, E.D.; Zheng, L.; Kim, Y.; Fang, B.; Liu, B.; Valdez, R.A.; Dietrich, W.F.; Rucker, P.V.; Chianelli, D.; Schmeits, J.; et al. Tropifexor-Mediated Abrogation of Steatohepatitis and Fibrosis Is Associated With the Antioxidative Gene Expression Profile in Rodents. *Hepatol. Commun.* **2019**, *3*, 1085–1097. [CrossRef] [PubMed]
243. Badman, M.K.; Chen, J.; Desai, S.; Vaidya, S.; Neelakantham, S.; Zhang, J.; Gan, L.; Danis, K.; Laffitte, B.; Klickstein, L.B. Safety, Tolerability, Pharmacokinetics, and Pharmacodynamics of the Novel Non-Bile Acid FXR Agonist Tropifexor (LJN452) in Healthy Volunteers. *Clin. Pharmacol. Drug Dev.* **2020**, *9*, 395–410. [CrossRef] [PubMed]
244. Camilleri, M.; Nord, S.L.; Burton, D.; Oduyebo, I.; Zhang, Y.; Chen, J.; Im, K.; Bhad, P.; Badman, M.K.; Sanders, D.S.; et al. Randomised clinical trial: Significant biochemical and colonic transit effects of the farnesoid X receptor agonist tropifexor in patients with primary bile acid diarrhoea. *Aliment. Pharm.* **2020**, *52*, 808–820. [CrossRef] [PubMed]
245. Pedrosa, M.; Seyedkazemi, S.; Francque, S.; Sanyal, A.; Rinella, M.; Charlton, M.; Loomba, R.; Ratziu, V.; Kochuparampil, J.; Fischer, L.; et al. A randomized, double-blind, multicenter, phase 2b study to evaluate the safety and efficacy of a combination of tropifexor and cenicriviroc in patients with nonalcoholic steatohepatitis and liver fibrosis: Study design of the TANDEM trial. *Contemp. Clin. Trials* **2020**, *88*, 105889. [CrossRef]
246. Camilleri, M.; Nord, S.L.; Burton, D.; Oduyebo, I.; Zhang, Y.M.; Chen, J.; Bhad, P.; Badman, M.K.; Sanders, D.; Walters, J.R. A Double-Blind, Randomized, Placebo-Controlled, Crossover, Multiple-Dose Study of Tropifexor, a Non Bile Acid Fxr Agonist, in Patients with Primary Bile Acid Diarrhea. *Gastroenterology* **2019**, *156*, S204–S205. [CrossRef]
247. Schramm, C.; Hirschfield, G.; Mason, A.L.; Wedemeyer, H.; Klickstein, L.; Neelakantham, S.; Koo, P.; Sanni, J.; Badman, M.; Jones, D. Early assessment of safety and efficacy of tropifexor, a potent non bile-acid FXR agonist, in patients with primary biliary cholangitis: An interim analysis of an ongoing phase 2 study. *J. Hepatol.* **2018**, *68*, S103. [CrossRef]
248. Sanyal, A.; Lopez, P.; Lawitz, E.; Kim, W.; Huang, J.-F.; Andreone, P.; Goh, G.B.B.; Chen, Y.-C.; Ratziu, V.; Kim, Y.J. Tropifexor, a farnesoid X receptor agonist for the treatment of nonalcoholic steatohepatitis–Interim results based on baseline body mass index from first two parts of Phase 2b study FLIGHT-FXR. *J. Hepatol.* **2019**, *70*, E796–E797. [CrossRef]
249. Schramm, C.; Wedemeyer, H.; Mason, A.; Hirschfield, G.M.; Levy, C.; Kowdley, K.V.; Milkiewicz, P.; Janczewska, E.; Malova, E.S.; Sanni, J.; et al. Farnesoid X receptor agonist tropifexor attenuates cholestasis in a randomised trial in patients with primary biliary cholangitis. *JHEP Rep.* **2022**, *4*, 100544. [CrossRef]
250. Patel, K.; Harrison, S.A.; Elkhashab, M.; Trotter, J.F.; Herring, R.; Rojter, S.E.; Kayali, Z.; Wong, V.W.; Greenbloom, S.; Jayakumar, S.; et al. Cilofexor, a Nonsteroidal FXR Agonist, in Patients With Noncirrhotic NASH: A Phase 2 Randomized Controlled Trial. *Hepatology* **2020**, *72*, 58–71. [CrossRef]
251. Trauner, M.; Bowlus, C.L.; Gulamhusein, A.; Hameed, B.; Caldwell, S.H.; Shiffman, M.L.; Landis, C.; Muir, A.J.; Billin, A.; Xu, J.; et al. Safety and sustained efficacy of the farnesoid X receptor (FXR) agonist cilofexor over a 96-week open-label extension in patients with PSC. *Clin. Gastroenterol. Hepatol.* **2022**, S1542-3565(22)00720-0. [CrossRef]
252. Erken, R.; Andre, P.; Roy, E.; Kootstra, N.; Barzic, N.; Girma, H.; Laveille, C.; Radreau-Pierini, P.; Darteil, R.; Vonderscher, J.; et al. Farnesoid X receptor agonist for the treatment of chronic hepatitis B: A safety study. *J. Viral Hepat.* **2021**, *28*, 1690–1698. [CrossRef] [PubMed]

253. Chianelli, D.; Rucker, P.V.; Roland, J.; Tully, D.C.; Nelson, J.; Liu, X.; Bursulaya, B.; Hernandez, E.D.; Wu, J.; Prashad, M.; et al. Nidufexor (LMB763), a Novel FXR Modulator for the Treatment of Nonalcoholic Steatohepatitis. *J. Med. Chem.* **2020**, *63*, 3868–3880. [CrossRef] [PubMed]
254. Dwivedi, S.K.; Singh, N.; Kumari, R.; Mishra, J.S.; Tripathi, S.; Banerjee, P.; Shah, P.; Kukshal, V.; Tyagi, A.M.; Gaikwad, A.N.; et al. Bile acid receptor agonist GW4064 regulates PPARgamma coactivator-1alpha expression through estrogen receptor-related receptor alpha. *Mol. Endocrinol.* **2011**, *25*, 922–932. [CrossRef] [PubMed]
255. Harrison, S.A.; Bashir, M.R.; Lee, K.J.; Shim-Lopez, J.; Lee, J.; Wagner, B.; Smith, N.D.; Chen, H.C.; Lawitz, E.J. A structurally optimized FXR agonist, MET409, reduced liver fat content over 12 weeks in patients with non-alcoholic steatohepatitis. *J. Hepatol.* **2021**, *75*, 25–33. [CrossRef] [PubMed]
256. Marzano, M.; Fosso, B.; Colliva, C.; Notario, E.; Passeri, D.; Intranuovo, M.; Gioiello, A.; Adorini, L.; Pesole, G.; Pellicciari, R.; et al. Farnesoid X receptor activation by the novel agonist TC-100 (3alpha, 7alpha, 11beta-Trihydroxy-6alpha-ethyl-5beta-cholan-24-oic Acid) preserves the intestinal barrier integrity and promotes intestinal microbial reshaping in a mouse model of obstructed bile acid flow. *Biomed. Pharm.* **2022**, *153*, 113380. [CrossRef]
257. Nara, S.J.; Jogi, S.; Cheruku, S.; Kandhasamy, S.; Jaipuri, F.; Kathi, P.K.; Reddy, S.; Sarodaya, S.; Cook, E.M.; Wang, T.; et al. Discovery of BMS-986339, a Pharmacologically Differentiated Farnesoid X Receptor Agonist for the Treatment of Nonalcoholic Steatohepatitis. *J. Med. Chem.* **2022**, *65*, 8948–8960. [CrossRef]
258. Cao, S.; Yang, X.; Zhang, Z.; Wu, J.; Chi, B.; Chen, H.; Yu, J.; Feng, S.; Xu, Y.; Li, J.; et al. Discovery of a tricyclic farnesoid X receptor agonist HEC96719, a clinical candidate for treatment of non-alcoholic steatohepatitis. *Eur. J. Med. Chem.* **2022**, *230*, 114089. [CrossRef]
259. Gege, C.; Hambruch, E.; Hambruch, N.; Kinzel, O.; Kremoser, C. Nonsteroidal FXR Ligands: Current Status and Clinical Applications. *Handb. Exp. Pharm.* **2019**, *256*, 167–205. [CrossRef]
260. Liu, X.; Xue, R.; Ji, L.; Zhang, X.; Wu, J.; Gu, J.; Zhou, M.; Chen, S. Activation of farnesoid X receptor (FXR) protects against fructose-induced liver steatosis via inflammatory inhibition and ADRP reduction. *Biochem. Biophys. Res. Commun.* **2014**, *450*, 117–123. [CrossRef]
261. Al-Khaifi, A.; Rudling, M.; Angelin, B. An FXR Agonist Reduces Bile Acid Synthesis Independently of Increases in FGF19 in Healthy Volunteers. *Gastroenterology* **2018**, *155*, 1012–1016. [CrossRef]
262. Traussnigg, S.; Halilbasic, E.; Hofer, H.; Munda, P.; Stojakovic, T.; Fauler, G.; Kashofer, K.; Krssak, M.; Wolzt, M.; Trauner, M. Open-label phase II study evaluating safety and efficacy of the non-steroidal farnesoid X receptor agonist PX-104 in non-alcoholic fatty liver disease. *Wien. Klin. Wochenschr.* **2021**, *133*, 441–451. [CrossRef] [PubMed]
263. Wang, Y.; Crittenden, D.B.; Eng, C.; Zhang, Q.; Guo, P.; Chung, D.; Fenaux, M.; Klucher, K.; Jones, C.; Jin, F.; et al. Safety, Pharmacokinetics, Pharmacodynamics, and Formulation of Liver-Distributed Farnesoid X-Receptor Agonist TERN-101 in Healthy Volunteers. *Clin. Pharmacol. Drug Dev.* **2021**, *10*, 1198–1208. [CrossRef] [PubMed]
264. Ratziu, V.; Rinella, M.E.; Neuschwander-Tetri, B.A.; Lawitz, E.; Denham, D.; Kayali, Z.; Sheikh, A.; Kowdley, K.V.; Desta, T.; Elkhashab, M.; et al. EDP-305 in patients with NASH: A phase II double-blind placebo-controlled dose-ranging study. *J. Hepatol.* **2022**, *76*, 506–517. [CrossRef] [PubMed]
265. Miyata, S.; Kawashima, Y.; Sakai, M.; Matsubayashi, M.; Motoki, K.; Miyajima, Y.; Watanabe, Y.; Chikamatsu, N.; Taniguchi, T.; Tokuyama, R. Discovery, optimization, and evaluation of non-bile acid FXR/TGR5 dual agonists. *Sci. Rep.* **2021**, *11*, 9196. [CrossRef] [PubMed]
266. Hu, Y.B.; Liu, X.Y.; Zhan, W. Farnesoid X receptor agonist INT-767 attenuates liver steatosis and inflammation in rat model of nonalcoholic steatohepatitis. *Drug Des. Devel.* **2018**, *12*, 2213–2221. [CrossRef]
267. Cipriani, S.; Mencarelli, A.; Palladino, G.; Fiorucci, S. FXR activation reverses insulin resistance and lipid abnormalities and protects against liver steatosis in Zucker (fa/fa) obese rats. *J. Lipid Res.* **2010**, *51*, 771–784. [CrossRef]
268. Renga, B.; Mencarelli, A.; Vavassori, P.; Brancaleone, V.; Fiorucci, S. The bile acid sensor FXR regulates insulin transcription and secretion. *Biochim. Biophys. Acta* **2010**, *1802*, 363–372. [CrossRef]
269. Mudaliar, S.; Henry, R.R.; Sanyal, A.J.; Morrow, L.; Marschall, H.U.; Kipnes, M.; Adorini, L.; Sciacca, C.I.; Clopton, P.; Castelloe, E.; et al. Efficacy and safety of the farnesoid X receptor agonist obeticholic acid in patients with type 2 diabetes and nonalcoholic fatty liver disease. *Gastroenterology* **2013**, *145*, 574–582.e1. [CrossRef]
270. Chalasani, N.; Younossi, Z.; Lavine, J.E.; Charlton, M.; Cusi, K.; Rinella, M.; Harrison, S.A.; Brunt, E.M.; Sanyal, A.J. The diagnosis and management of nonalcoholic fatty liver disease: Practice guidance from the American Association for the Study of Liver Diseases. *Hepatology* **2018**, *67*, 328–357. [CrossRef]
271. Younossi, Z.M.; Ratziu, V.; Loomba, R.; Rinella, M.; Anstee, Q.M.; Goodman, Z.; Bedossa, P.; Geier, A.; Beckebaum, S.; Newsome, P.N.; et al. Obeticholic acid for the treatment of non-alcoholic steatohepatitis: Interim analysis from a multicentre, randomised, placebo-controlled phase 3 trial. *Lancet* **2019**, *394*, 2184–2196. [CrossRef]
272. Younossi, Z.M.; Stepanova, M.; Nader, F.; Loomba, R.; Anstee, Q.M.; Ratziu, V.; Harrison, S.; Sanyal, A.J.; Schattenberg, J.M.; Barritt, A.S.; et al. Obeticholic Acid Impact on Quality of Life in Patients With Nonalcoholic Steatohepatitis: REGENERATE 18-Month Interim Analysis. *Clin. Gastroenterol. Hepatol.* **2022**, *20*, 2050–2058.e12. [CrossRef] [PubMed]
273. Tully, D.C.; Rucker, P.V.; Chianelli, D.; Williams, J.; Vidal, A.; Alper, P.B.; Mutnick, D.; Bursulaya, B.; Schmeits, J.; Wu, X. *Discovery of tropifexor (LJN452), a Highly Potent Non-Bile Acid FXR agonist for the Treatment of Cholestatic Liver Diseases and Nonalcoholic Steatohepatitis (NASH)*; ACS Publications: Washington, DC, USA, 2017.

274. ClinicalTrials.gov—Studies on Tropifexor. Available online: https://clinicaltrials.gov/ct2/results?cond=&term=tropifexor&cntry=&state=&city=&dist= (accessed on 18 November 2022).
275. ClinicalTrials.gov—Studies on FXF agonists. Available online: https://clinicaltrials.gov/ct2/results?cond=&term=FXR+agonist&cntry=&state=&city=&dist=&Search=Search (accessed on 18 November 2022).
276. Oduyebo, I.; Camilleri, M.; Nelson, A.D.; Khemani, D.; Nord, S.L.; Busciglio, I.; Burton, D.; Rhoten, D.; Ryks, M.; Carlson, P.; et al. Effects of NGM282, an FGF19 variant, on colonic transit and bowel function in functional constipation: A randomized phase 2 trial. *Am. J. Gastroenterol.* **2018**, *113*, 725–734. [CrossRef] [PubMed]
277. Qiao, P.; Jia, Y.; Ma, A.; He, J.; Shao, C.; Li, X.; Wang, S.; Yang, B.; Zhou, H. Dapagliflozin protects against nonalcoholic steatohepatitis in db/db mice. *Front. Pharm.* **2022**, *13*, 934136. [CrossRef] [PubMed]
278. Bellanti, F.; Lo Buglio, A.; Dobrakowski, M.; Kasperczyk, A.; Kasperczyk, S.; Aich, P.; Singh, S.P.; Serviddio, G.; Vendemiale, G. Impact of sodium glucose cotransporter-2 inhibitors on liver steatosis/fibrosis/inflammation and redox balance in non-alcoholic fatty liver disease. *World J. Gastroenterol.* **2022**, *28*, 3243–3257. [CrossRef]
279. Wu, X.; Ge, H.; Gupte, J.; Weiszmann, J.; Shimamoto, G.; Stevens, J.; Hawkins, N.; Lemon, B.; Shen, W.; Xu, J.; et al. Co-receptor requirements for fibroblast growth factor-19 signaling. *J. Biol. Chem.* **2007**, *282*, 29069–29072. [CrossRef]

nutrients

MDPI

Review

Is microRNA-33 an Appropriate Target in the Treatment of Atherosclerosis?

Malgorzata Sidorkiewicz

Department of Medical Biochemistry, Faculty of Health Sciences, Medical University of Lodz, 90-419 Lodz, Poland; malgorzata.sidorkiewicz@umed.lodz.pl

Abstract: The maintenance of cholesterol homeostasis is a complicated process involving regulation of cholesterol synthesis, dietary uptake and bile acid synthesis and excretion. Reverse cholesterol transport, described as the transfer of cholesterol from non-hepatic cells, including foam cells in atherosclerotic plaques, to the liver and then its excretion in the feces is important part of this regulation. High-density lipoproteins are the key mediators of reverse cholesterol transport. On the other hand, microRNA-33 was identified as a key regulator of cholesterol homeostasis. Recent studies indicate the impact of microRNA-33 not only on cellular cholesterol efflux and HDL production but also on bile metabolism in the liver. As proper coordination of cholesterol metabolism is essential to human health, discussion of recent findings in this field may open new perspectives in the microRNA-dependent treatment of a cholesterol imbalance.

Keywords: microRNA-33; HDL; atherosclerosis; cholesterol efflux; reverse cholesterol transport; bile acids

1. Introduction

The loss of cholesterol homeostasis is associated with major cardiometabolic risk factors including atherosclerotic cardiovascular disease and type 2 diabetes [1]. Cholesterol is an important structural component of biological membranes and a substrate for the synthesis of other steroids; thus, both cholesterol intake from the diet as well as its delivery from endogenous synthesis are essential for the maintenance of sufficient concentration of cholesterol in blood. On the other hand, aberrant cholesterol content is linked to coronary atherosclerosis and to other diseases [2,3]. Cholesterol synthesis de novo occurs in the human body through a series of enzymatic reactions from a commonly present substrate, acetyl-CoA. Additionally, cholesterol with triglycerides is delivered to cells from the circulation in the form of lipoproteins [4]. Blood lipoproteins are complex particles that contain cholesterol esters and triglycerides in a central core that is surrounded by free cholesterol, phospholipids and variety of apolipoproteins (Apo) that are responsible for lipoprotein formation and function. Two types of non-high-density lipoproteins, containing apolipoprotein apoB, are involved in lipid delivery to cells (Figure 1). The first one, chylomicrons (CL), equipped with apolipoprotein B-48, are produced from diet-delivered lipids in the intestine. The second, very low-density lipoproteins (VLDL), containing apolipoprotein B-100, are synthetized endogenously in the liver. In peripheral tissues, mostly in muscles and adipose tissue, triglycerides from CL and VLDL are broken down by lipoprotein lipase (LPL), releasing free fatty acids as well as chylomicrons remnants and intermediate-density lipoproteins (IDL), respectively. IDL are then metabolized to low-density lipoproteins (LDL) that are taken up via the LDL receptors both on the surface of both extrahepatic cells as well as on the surface of the liver, the predominant site of LDL uptake. A chronic inflammatory disease, atherosclerosis, is characterized mainly by the LDL-dependent deposition of excess cholesterol in the arterial walls [5]. Circulating LDL can transcytose across the endothelium and accumulate in the arterial wall, leading to the formation of atherosclerotic

Citation: Sidorkiewicz, M. Is microRNA-33 an Appropriate Target in the Treatment of Atherosclerosis? *Nutrients* **2023**, *15*, 902. https://doi.org/10.3390/nu15040902

Academic Editor: Frans Stellaard

Received: 30 December 2022
Revised: 7 February 2023
Accepted: 8 February 2023
Published: 10 February 2023

plaque [6]. The progression of the disease is associated with a number of different cells such as macrophages, endothelial cells and vascular smooth muscle cells [7].

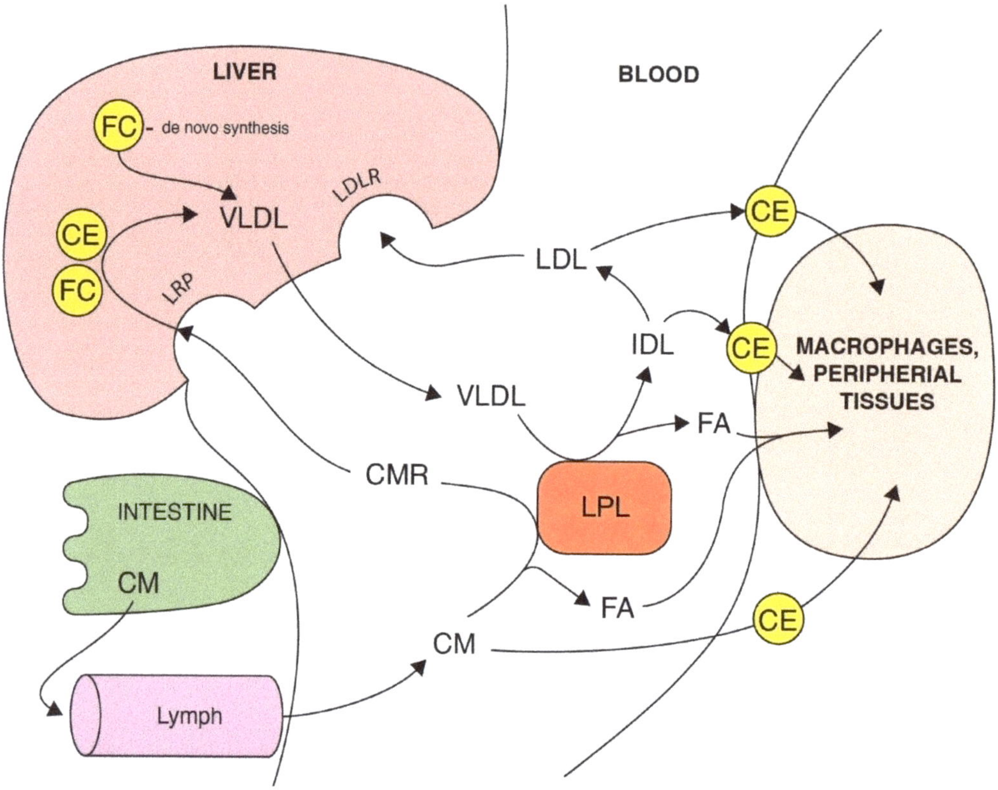

Figure 1. The route of dietary and endogenous cholesterol in an organism. The absorption of dietary triglycerides, free cholesterol (FC), and cholesterol esters (CE) occurs in the small intestine. In enterocytes, dietary lipids are packed in chylomicrons (CM) that are diffused to the bloodstream via lymph. After the digestion of triglycerides by lipoprotein lipase (LPL), fatty acids (FA) are absorbed by target cells and chylomicrons remnant (CMR) uptake to the liver is mediated by LDL receptor-related protein (LRP). The endogenous cholesterol pathway starts from the liver, recirculating triglycerides and cholesterol in the bloodstream, packed in very low-density lipoproteins (VLDL). VLDL through the LPL activity delivers FA to peripheral cells and are metabolized to intermediate-density lipoproteins (IDL) which are enriched in cholesterol. Low-density lipoproteins (LDL) derived from IDL carry the majority of the cholesterol and transports mostly esterified cholesterol (CE) to peripheral tissues, including macrophages in vein walls. LDL are eventually cleared by the liver via the LDL receptor.

An opposite action to LDL is characteristic of the high-density lipoproteins (HDL) fraction containing apolipoprotein A1 (Apo-A1). HDL, by stimulating the efflux of an excess of cellular cholesterol from extrahepatic tissues and transporting it back to the liver (Figure 2), plays a key role in the maintenance of cholesterol homeostasis [8]. The dyslipidemia characterized by an increased concentration of LDL and triglycerides and a decreased level of HDL remains a major factor of cardiovascular diseases [9]. It was only in the second half of the twentieth century when it became clear that HDL concentration and apo-A1 are inversely correlated with cardiovascular risk. This discovery has consequently led to the identification of the reverse cholesterol transport (RCT) pathway as essential for protection against atherosclerosis. RCT promotes cholesterol efflux from artery walls to HDL by activating a series of factors such as: transporters, enzymes and receptors (Figure 2). The ATP-binding membrane cassette transport proteins, ABCA1 and ABCG1, seem to be crucial for both HDL formation and the cholesterol efflux in RCT process [10,11]. Cholesterol homeostasis is closely controlled by feedback mechanisms, and an important part of it depends on the regulation of RCT. At the transcriptional level, the sterol-activated liver X-receptor (LXR) seems to play a key role in the regulation of RCT [12]. In vivo, LXR is activated by oxysterols in cholesterol-loaded cells, and it controls the response to cholesterol excess by regulating target genes *ABCA1* and *ABCG1*. Increased expression of ABCA1 and ABCG1 transporters [13,14] mediates cholesterol efflux in macrophages, reducing cellular sterol content.

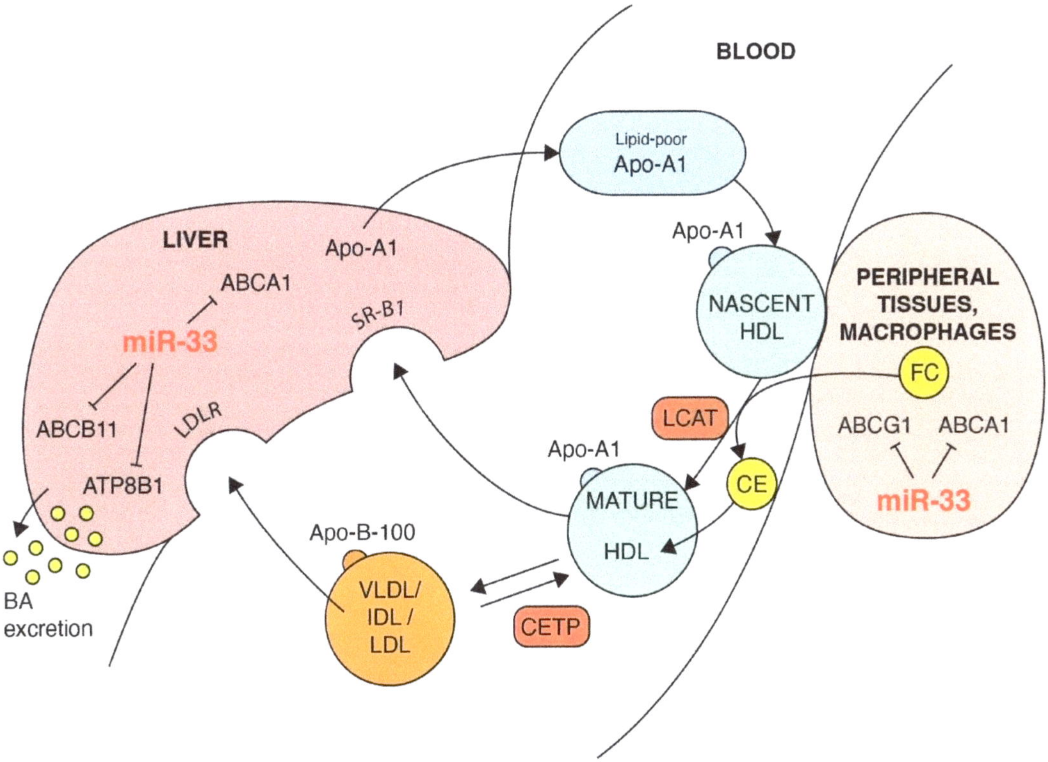

Figure 2. The impact of miR-33 on some steps of reverse cholesterol transport (RCT). Black lines ending with a short perpendicular line indicate target genes that are repressed by miR-33. ABCA1-mediated lipidation of lipid-poor apolipoprotein A-I (Apo-A1) forms nascent HDL. Cholesterol efflux from extrahepatic cells, including macrophages, is mediated by ABCA1 and ABCG1 activity. Through the action of the lecithin:cholesterol acyl transferase (LCAT), free cholesterol (FC) is esterified and cholesterol esters (CE) are absorbed inside of HDL. The progressive action of LCAT on nascent HDL generates a spectrum of mature HDL particles. Part of CE can be transferred by cholesterol ester transfer protein (CETP) from HDL to apolipoprotein B-100 containing lipoproteins (VLDL, IDL, LDL) in exchange for triglycerides (TG), promoting cholesterol clearance by the LDL receptor. Hepatic SR-BI mediates removal of FC and CE from HDL. Excess cholesterol is converted to bile acids excreted from the liver into the bile. Both the ABC11 and ATP8B1 transporters promote hepatic excretion of the bile.

The sterol regulatory element-binding proteins 1 and 2 (SREBP-1 and SREBP-2) are the next important factors controlling the transcription of genes involved in the regulation of cholesterol metabolism [15–17]. While SREBP-1 is induced by hyperinsulinemia and is mostly involved in the synthesis of fatty acids, SREBP-2 controls the intracellular cholesterol status [16]. If cells accumulate an excess of cholesterol or other sterols, inhibition of SREBP processing causes the decrease of LDL receptor expression, limiting the influx of cholesterol into the cell. At the same time, the activity of rate-limiting enzyme for *the novo* cholesterol synthesis, 3-hydroxy-3-methylglutaryl CoA reductase, declines. Both effects, the decrease of cholesterol uptake and the decline of intracellular cholesterol synthesis, are controlled by SREBP. Recently, microRNAs (miRNAs) were identified as important transcription regulators that modulate numerous cellular processes including the cholesterol homeostasis. Several lines of evidence indicate that microRNA-33 (miR-33) regulates cholesterol metabolism in cooperation with the SREBP host genes [18]. On one hand, miR-33 was demonstrated to reduce cholesterol efflux and RCT by targeting genes involved in cholesterol export [11]. On the other hand, it was demonstrated that the inhibition of miR-33 expression improves the blood cholesterol profile and promotes reversing of atheroma [19,20]. Thus, miR-33 appears to plays an important role in the complicated, multifactorial process of maintaining a balance between dietary cholesterol and blood cholesterol. Moreover, it suggests the potential of miR-33 inhibition as a therapeutic target in the treatment of dyslipidemia and atherosclerosis [21]. This review shows how miR-33 expression may have an influence on the homeostasis of cholesterol by indicating the impact of miR-33 on HDL biogenesis, RCT and bile acid synthesis.

2. How Can microRNA-33 Regulate Cholesterol Metabolism?

Careful coordination of cholesterol uptake, biosynthesis and use is essential for human health. That explains why cholesterol metabolism is tightly regulated at a cellular and whole-body level. In addition to the classical regulators of cholesterol expression, sterol regulatory element-binding proteins, numerous noncoding RNAs (including microRNAs) play an essential role in this process. The human genome contains thousands of non-coding RNAs. MicroRNAs are a class of short non-coding RNAs with 18–22 nucleotides [22] that regulate the expression of target genes as sequence-specific inhibitors of messenger RNA (mRNA) by binding to partially complementary regions in target mRNAs. Thus, the effectiveness of this post-transcriptional regulation of target genes depends on a complementarity between microRNA and mRNA. It is noteworthy that multiple microRNAs may cooperate in regulation of the same gene target and, conversely, any mRNA may contain multiple binding sites that are recognized by various miRNAs. Although the most common site for microRNA binding to mRNA is the 3′ UTR regions of target messenger RNA, in some cases, such as when mi122 interacts with hepatitis C virus (HCV), the binding site is the 5′UTR noncoding region of HCV RNA genome [23]. However, regardless of the place of interaction, microRNAs primarily act as mRNA-sequence-specific inhibitors of translation by the induction of mRNA degradation or simply translational repression. MicroRNAs are transcribed by RNA polymerase II that produce a long primary miRNA gene transcript (pri-miRNA). The pri-miRNAs then undergo sequential processing by nuclear and cytoplasmic enzymes. First, these RNAs are cleaved by the catalytic activity of DROSHA and DGCR8 microprocessor complexes into precursor miRNA (pre-miRNA) [24]. After exporting to cytoplasm, pre-miRNA is processed by the RNAse III enzyme Dicer to generate 18–25-nt duplex. The mature miRNA interacts with Argonaute protein and forms a miRNA-induced silencing complex ready for mRNA degradation and translation repression. Many microRNAs are key regulators of lipid metabolism [25]. The indispensable role in the regulation of cholesterol homeostasis has been indicated for miR-33 [21], miR-122 [26], mir-125b [27], miR-148a [28], mir-483 [29] and many others [30,31]. Among hundreds of microRNAs, miRNA-33 has been identified as a major regulator of cholesterol homeostasis and atherosclerosis [18,21]. Both members of the miR-33 family—miR-33a and miR-33b (the second being present only in primates)—are located in intron 16 of SREBP-2 and intron 17 of SREBP-1 gene, respectively and are controlled simultaneously with the SREBP genes [18] to govern cholesterol homeostasis. It was demonstrated that miR-33a and miR-33b are co-transcribed with SREBPF1 and SREBP2 and act to repress genes that oppose

SREBP functions. For example, in the case of a low cholesterol concentration in a cell, SREB transcription factors and miR-33 are activated and cause an increase of cholesterol synthesis and, due to ABCA1 repression, the reduction of cholesterol efflux. The general regulator of several microRNAs involved in cholesterol metabolism, including miR-33, is a long noncoding primate-specific RNA (CHROME) expressed in macrophages and hepatocytes [32]. Elevated levels of CHROME in the plasma and atherosclerotic plaque were found in patients with coronary artery disease. It was observed that the level of CHROME expression depends on the levels of dietary and cellular cholesterol. CHROME becomes unregulated in response to cholesterol overload via LXR. Due to repression of miR-33 expression, CHROME post-transcriptionally regulates ABCA1 expression and increases cholesterol efflux. In turn, cells lacking CHROME exhibit a reduced expression of ABCA1 as a result of increased expression of miR-33. More details on the role of miR-33 in the regulation of cholesterol balance by influencing RCT and HDL biogenesis are presented in the following sections.

3. HDL Formation and microRNA-33 Influence

As the name indicates, high-density lipoproteins are a class of lipoproteins characterized by high density ranges from 1.063 to 1.25 g/mL. This heterogeneity is connected with different contents of apolipoproteins, enzymes and lipids in individual HDL subclasses (HDL2, HDL3 and very high-density HDL). HDL particles consist of divers proportions of phospholipids (PL), free cholesterol (FC), cholesterol esters (CE) and triacylglycerols (TG). Apo-A1 is a major structural protein accounting for approximately 70% of HDL proteins; however, other apolipoproteins, including A-II, A-IV, C-1I, C-II, C-III and E, are also associated with HDL [4]. Continuous remodeling of HDL is responsible not only for changing HDL density and shape but also size, ranging from 7.5 to 15 nm. The main function of HDL is reverse cholesterol transport from peripheral tissues, including macrophages in the vessel walls, to the liver, by which HDL may be anti-atherogenic. The negative correlation between HDL concentration and the risk of cardiovascular diseases strongly suggests cardioprotective properties of HDL [9]. These properties include, beyond the role of HDL in RCT, anti-inflammatory, anti-oxidative, anti-thrombotic and anti-apoptotic effects. Apo-A1, due to the interaction with ATP-binding cassette proteins, SR-B1 receptor and the activation of lecithin:cholesterol acyltransferase (LCAT), has an impact on both HDL biogenesis and reverse cholesterol transport [33]. The first step of HDL formation involves synthesis of Apo-A1 by the liver and intestine. The Apo-A1 mRNA and protein expression was recently demonstrated in samples of kidney renal clear cell carcinoma and in normal renal cells [34], but so far it has been difficult to assess to what extent the expression of Apo-A1 in the kidneys has an effect on the formation of HDL and on RCT. The formation of HDL particles that starts from Apo-A1 production depends on several factors. First, Apo-A1, newly-synthetized in hepatocytes and enterocytes, acquires phospholipid and cholesterol. The ABCA1 is critical for cholesterol and phospholipid efflux from these cells to lipid-poor Apo A1 and accounts for the formation of majority of nascent HDL particles [35]. Additionally, the phospholipid transfer protein (PLTP) promotes phospholipid efflux to HDL from peripheral cells and facilitates the movement of phospholipids between lipoproteins. The expression of ABCA1 is regulated by LXR family and retinoic acid receptors (RXR). When the cellular level of sterols increases, cholesterol is oxygenated to oxysterols that activate ABCA1 expression via activation of LXR and RXR. As a consequence, increased cholesterol efflux from cells is observed. It was demonstrated that mice with targeted knock-out of ABCA1 had a significantly reduced ABCA1 level in both the liver and intestine [36,37]. It was found that ABCA1 expression in the liver determines the HDL level in sera and influences the general cholesterol homeostasis [36]. It is important to note that ABCA1 mRNA with specific long 3′ UTR is highly susceptible to impact of micro RNA. The studies confirmed the mir-33-dependent repression of ABCA1 not only in hepatocytes but also in endothelial cells and macrophages [11,38]. MiR-33, like several other microRNAs, is regulated by CHROME [32], which influences the cholesterol efflux and HDL biogenesis.

The knockdown of CHROME in human hepatocytes and macrophages increases the levels of miR-33 expression, and a result of this is reduced expression of ABCA1, which inhibits cholesterol efflux and HDL particle formation [32].

4. The Impact of microRNA-33 on Reverse Cholesterol Transport

The main part of reverse cholesterol transport is the transfer of cholesterol from peripheral cells, including foam cells in atherosclerotic plaques, to the liver. Peripheral cells accumulate cholesterol through the uptake of cholesterol-rich lipoproteins and de novo synthesis. In the human body, the excess of cholesterol is not simply degraded to acetyl-CoA, as it is in fatty acids. Thus, the only way to remove cholesterol is via the bile acid (BA) synthesis in the liver and BA excretion as bile-related products into feces. In this context, the effectiveness of RCT largely determines the balance between the absorbed cholesterol and the blood cholesterol. RCT begins with the hydrolysis of cholesteryl esters associated with cytoplasmic lipid-droplets by specific hydrolases and lipases [39]. In the early stage of RCT, cholesterol efflux is determined by cholesterol-deficient and phospholipid-depleted apoA-1 particles that are produced by the liver and intestine, as was described in the previous section. The nascent HDL particles acquire cholesterol and phospholipids by ABCA1, indicating the crucial role of ABCA1 expression not only for HDL formation but also for RCT [40]. Additionally, through the lecithin-cholesterol acyltransferase (LCAT) activity, cholesterol in HDL particles is esterified. Cholesteryl esters are moved to the core of HDL particles, forming a steady gradient of free cholesterol and enabling HDLs to accept more cholesterol [41]. Mature HDL particles can acquire additional cholesterol not only by ABCA1 but also via ATP-binding membrane cassette transport protein G1 (ABCG1) and scavenger receptor type B1 (SR-B1). Several studies have confirmed the contribution of the mature HDL particles in mediation of cholesterol efflux to HDL in macrophage foam cells [42–44]. By participating in this essential part of RCT, they contribute to the protection against atherosclerosis [45]. It was observed that overexpression of miR-33 down-regulates ABCA1 and ABCG1 transporters, thus decreasing HDL concentration in plasma. In this context, the impact of miR-33 on the expression of ATP-binding cassette transporters seems to be extremely important, decreasing cholesterol efflux from the atherosclerotic plague to circulating HDL particles. Moreover, it was established that microRNA-33 is responsible for remodeling of membrane microdomains and regulation of the innate immune response through the ABCA1 and ABCG1-dependent mechanism [46], indicating a close connection between cholesterol homeostasis and the immune system. This is another example confirming the importance of maintaining cholesterol homeostasis in the body. In continuing to describe the RCT process, the well-established role of SR-B1 in the selective uptake of cholesteryl esters from mature HDL to the liver should be emphasized. SR-B1, known to modulate HDL metabolism, is a glycoprotein of 509 aa that is involved in bidirectional cholesterol transport at the cell membrane [42]. SR-BI is expressed in the liver and tissues that need cholesterol for steroid synthesis. Unlike cholesterol, Apo-A1, the main protein component of HDL, is catabolized mostly by the kidneys and partially by the liver. Free Apo-A1 and lipid-poor Apo-A1 are filtered in the kidneys and taken up by the renal tubules [4]. The rate of Apo-A1 degradation depends on the degree of lipidation. No less important for the success of reverse cholesterol transport is the reciprocal exchange of cholesteryl ester for triglycerides. This process is mediated by cholesteryl ester transfer protein (CETP). The highest expression of CETP was found in the liver and in adipose tissue. CETP moves the bulk of the cholesteryl esters from HDL to apoB-containing lipoproteins (VLDL, IDL, and LDL). CETP is also responsible for the CE transfer among HDL subclasses. A study [47] has demonstrated that CETP inhibitors could have a significant impact on the management of dyslipidemic coronary heart disease patients. Both free and esterified cholesterol content of apoB-containing lipoproteins are taken up by the liver, predominantly via the low-density lipoprotein receptor (LDLR). In turn, HDL becomes enriched with triglycerides that are then degraded by hepatic lipase forming smaller HDL particles that are recognized by scavenger receptor type B1. Further catabolism of HDL-derived cholesterol occurs in the liver trough conversion of cholesterol into bile components. Improvement of RCT by upregulating of ABCA1 remains one of the potential targets for the development of new therapeutic agents against atherosclerosis. As miR-33 represses expression of ABCA1 transporter, antagonizing miR-33 seems to be an effective strategy for elevation of HDL in blood and for protecting patients from atherosclerosis. The results of many studies demonstrate that silencing of miR-33 in mice, with a variety of methods like targeted deletion [17], anti-sense nucleotides [18,19] or viral delivery inhibitors [21,38], significantly increased ABCA1 expression and the level of HDL in blood. The adverse effect of long-term silencing of miR-33, the increased level of triglyceride, was observed, but only in high-fat-diet mice [48]. In turn, another study [49] of the impact of hepatic miR-33 deficiency in mice demonstrated that loss of hepatic miR-33 improves metabolic homeostasis without any adverse effects. The absence of miR-33b in mice somehow limited the significance of these findings. Thus, a study performed in non-human primates (African green monkeys) with inhibiting of both miR-33a and miR-33b confirmed that inhibition of miR-33 not only increases the concentration of plasma HDL but also reduces VLDL triglycerides, in a model highly related to humans [50]. These results suggest the potential utility of miR-33 inhibition as a novel approach for the maintenance of cholesterol homeostasis.

5. Bile Formation and Secretion Can Be Deregulated by microRNA-33

Modulation of cholesterol metabolism in the human body depends on complex conditions encompassing both anabolic and catabolic processes. As excess cholesterol is not simply degraded, the only catabolic pathway responsible for cleansing the body of excess cholesterol is to convert it in the liver to bile acids (BAs) that are excreted into feces. The mixture of bile acids, cholesterol, phospholipids and ions forms so-called bile. The main functions of bile are the removal of metabolic wastes and the emulsification of dietary lipids. Deregulations of BA transportation as well as the alterations in BA receptor signaling are connected with development of dyslipidemias and atherosclerosis [51]. Thus, BA synthesis and secretion is the next crucial step of RCT that influences the whole-body cholesterol homeostasis. The bile acids level is transcriptionally controlled by the Farnesoid X receptor (FXR) [52]. FXR regulates bile acid synthesis depending on the accumulation of BAs [53]. Cholesterol delivered to the liver via LDLR and SR-BI can be eliminated by two pathways. Firstly, part of cholesterol is directly secreted into bile. Secondly, cholesterol is converted in hepatocytes into bile acids by the catalytic activity of cholesterol 7 alpha-hydroxylase (CYP7A1) [54]. As demonstrated in [55], the enzymatic activity of CYP7A1 can be inhibited by miR-33. After formation, bile acids are secreted across the apical membrane of hepatocytes by the canalicular transporters into the intestine. The most important in this step are transmembrane transporters, ATP-binding cassette: ABCB11, ABCG5, ABCB4, ATP8B1. Mutations that can inactivate the functions of ATP8B1, ABCB11, and ABCB4 are strictly connected with progressive familial intrahepatic cholestasis [56]. In the case of mutations in ABCG5 and ABCG8, the development of sitosterolemia is observed instead [57]. Interestingly, at this step of RCT a detrimental role of miR-33 was also identified. It was found that 3′ UTR of genes coding for ABCB11 and ATP8B1 revealed conserved parts complementary to miR-33 [58]. An experimental study [58] confirmed that both bile salt exporter ABCB11 as well as ATP8B1 are direct targets of miR-33. In normal conditions, both the ABC11 and ATP8B1 transporters promote hepatic clearance, directly via biliary lipid secretion and indirectly via arrangement of adequate canalicular membrane phospholipid asymmetry required for bile salt movement, respectively. Hepatic overexpression of miR-33 causes a significant reduction of ABCB11 and ATP8B1. As an effect of miR-33 overexpression, a decrease of biliary output and thus inhibition of cholesterol elimination in fecal bile acids was observed.

6. Conclusions

This review has been intended to demonstrate the impact of miR-33 on cholesterol homeostasis, and hopefully it has provided a positive answer to the titular question. miR-33 is one of the essential regulatory elements necessary to maintain a balance between dietary and blood cholesterol. This is primarily a result of its critical impact on the reverse transport of cholesterol. miR-33 controls cholesterol metabolism mostly by repressing ABCA1 and ABCG1, responsible for cholesterol efflux. As a result of balanced regulation by miR-33 and SREBP, we can observe a reduction of de novo cholesterol synthesis and an increase of cholesterol efflux from cells in the case of higher dietary intake of cholesterol and an increase of sterol concentration in cells afterwards. In turn, a decrease of cholesterol content in cells is sufficient to increase cholesterol synthesis and to abrogate cholesterol efflux. Besides the repression of ABCA1 and ABCG1, miR-33 influences other steps of RCT by targeting the transformation of cholesterol into bile salts and their transport. By controlling this final step of cholesterol elimination, miR-33 once again influences the homeostasis of cholesterol in an organism. In this context, the possibility of therapeutic use of miR-33 deserves special attention. The most common causes of cardiometabolic diseases are disorders of cholesterol metabolism, associated with an imbalance between cholesterol intake and elimination [59]. The World Health Organization estimates that the annual number of deaths from cardiovascular diseases, including atherosclerosis, will increase to over 22 million by 2030. While LDL levels are directly associated with atherosclerosis, HDL levels and the reverse cholesterol pathway show a strong inverse correlation with cardiovascular diseases. A majority of studies have demonstrated that antagonism of miR-33 in vivo increases circulating HDL and reverse cholesterol transport, thereby reducing the progression and enhancing the regression of atherosclerosis. Taking into account the number of genes that are targeted by mir-33, it can be assumed that miR-33 is a promising target for the treatment of dyslipidemia. A better understanding of the mechanisms that underlie miR-33 mediated regulation of cholesterol homeostasis may improve future therapies in the field.

Funding: The APC was funded by the statutory funds of the Medical University of Lodz (503/6-086-01/503/-66-001).

Institutional Review Board Statement: Not applicable.

Informed Consent Statement: Not applicable.

Data Availability Statement: Not applicable.

Conflicts of Interest: The author declares no conflict of interest.

References

1. Barquera, S.; Pedroza-Tobías, A.; Medina, C.; Hernández-Barrera, L.; Bibbins-Domingo, K.; Lozano, R.; Moran, A.E. Global Overview of the Epidemiology of Atherosclerotic Cardiovascular Disease. *Arch. Med. Res.* **2015**, *46*, 328–338. [CrossRef] [PubMed]
2. Glass, C.K.; Witztum, J.L. Atherosclerosis: The road ahead. *Cell* **2001**, *104*, 503–516. [CrossRef] [PubMed]
3. Maxfield, F.R.; Tabas, I. Role of cholesterol and lipid organization in disease. *Nature* **2005**, *438*, 612–621. [CrossRef] [PubMed]
4. Feingold, K.R. Introduction to Lipids and Lipoproteins. In *Endotext [Internet]*; Feingold, K.R., Anawalt, B., Blackman, M.R., Boyce, A., Chrousos, G., Corpas, E., de Herder, W.W., Dhatariya, K., Hofland, J., Dungan, K., et al., Eds.; MDText.com, Inc.: South Dartmouth, MA, USA, 2021.
5. Zhou, R.; Stouffer, G.A.; Frishman, W.H. Cholesterol Paradigm and Beyond in Atherosclerotic Cardiovascular Disease: Cholesterol, Sterol Regulatory Element-Binding Protein, Inflammation, and Vascular Cell Mobilization in Vasculopathy. *Cardiol. Rev.* **2022**, *30*, 267–273. [CrossRef] [PubMed]
6. Ramírez, C.M.; Zhang, X.; Bandyopadhyay, C.; Rotllan, N.; Sugiyama, M.G.; Aryal, B.; Liu, X.; He, S.; Kraehling, J.R.; Ulrich, V.; et al. Caveolin-1 Regulates Atherogenesis by Attenuating Low-Density Lipoprotein Transcytosis and Vascular Inflammation Independently of Endothelial Nitric Oxide Synthase Activation. *Circulation* **2019**, *140*, 225–239. [CrossRef]
7. Susser, L.I.; Rayner, K.J. Through the layers: How macrophages drive atherosclerosis across the vessel wall. *J. Clin. Investig.* **2022**, *132*, e157011. [CrossRef]
8. Zhang, Y.; Zanotti, I.; Reilly, M.P.; Glick, J.M.; Rothblat, G.H.; Rader, D.J. Overexpression of apolipoprotein A-I promotes reverse transport of cholesterol from macrophages to feces in vivo. *Circulation* **2003**, *108*, 661–663. [CrossRef]
9. Barter, P.; Gotto, A.M.; LaRosa, J.C.; Maroni, J.; Szarek, M.; Grundy, S.M.; Kastelein, J.J.; Bittner, V.; Fruchart, J.C. Treating to New Targets Investigators. HDL cholesterol, very low levels of LDL cholesterol, and cardiovascular events. *N. Engl. J. Med.* **2007**, *357*, 1301–1310. [CrossRef]

10. Tall, A.R.; Yvan-Charvet, L.; Terasaka, N.; Pagler, T.; Wang, N. HDL, ABC transporters, and cholesterol efflux: Implications for the treatment of atherosclerosis. *Cell Metab.* **2008**, *7*, 365–375. [CrossRef]
11. Mao, M.; Lei, H.; Liu, Q.; Chen, Y.; Zhao, L.; Li, Q.; Luo, S.; Zuo, Z.; He, Q.; Huang, W.; et al. Effects of miR-33a-5P on ABCA1/G1-mediated cholesterol efflux under inflammatory stress in THP-1 macrophages. *PLoS ONE* **2014**, *9*, e109722. [CrossRef]
12. Janowski, B.A.; Willy, P.J.; Devi, T.R.; Falck, J.R.; Mangelsdorf, D.J. An oxysterol signalling pathway mediated by the nuclear receptor LXR alpha. *Nature* **1996**, *383*, 728–731. [CrossRef] [PubMed]
13. Costet, P.; Luo, Y.; Wang, N.; Tall, A.R. Sterol-dependent transactivation of the ABC1 promoter by the liver X receptor/retinoid X receptor. *J. Biol. Chem.* **2000**, *275*, 28240–28245. [CrossRef] [PubMed]
14. Kennedy, M.A.; Venkateswaran, A.; Tarr, P.T.; Xenarios, I.; Kudoh, J.; Shimizu, N.; Edwards, P.A. Characterization of the human ABCG1 gene: Liver X receptor activates an internal promoter that produces a novel transcript encoding an alternative form of the protein. *J. Biol. Chem.* **2001**, *276*, 39438–39447, Erratum in *J. Biol. Chem.* **2002**, *277*, 17375. [CrossRef] [PubMed]
15. Brown, M.S.; Goldstein, J.L. The SREBP pathway: Regulation of cholesterol metabolism by proteolysis of a membrane-bound transcription factor. *Cell* **1997**, *89*, 331–340. [CrossRef]
16. Horton, J.D.; Goldstein, J.L.; Brown, M.S. SREBPs: Activators of the complete program of cholesterol and fatty acid synthesis in the liver. *J. Clin. Investig.* **2002**, *109*, 1125–1131. [CrossRef]
17. Horie, T.; Ono, K.; Horiguchi, M.; Nishi, H.; Nakamura, T.; Nagao, K.; Kinoshita, M.; Kuwabara, Y.; Marusawa, H.; Iwanaga, Y.; et al. MicroRNA-33 encoded by an intron of sterol regulatory element-binding protein 2 (Srebp2) regulates HDL in vivo. *Proc. Natl. Acad. Sci. USA* **2010**, *107*, 17321–17326. [CrossRef]
18. Najafi-Shoushtari, S.H.; Kristo, F.; Li, Y.; Shioda, T.; Cohen, D.E.; Gerszten, R.E.; Näär, A.M. MicroRNA-33 and the SREBP host genes cooperate to control cholesterol homeostasis. *Science* **2010**, *328*, 1566–1569. [CrossRef]
19. Rayner, K.J.; Sheedy, F.J.; Esau, C.C.; Hussain, F.N.; Temel, R.E.; Parathath, S.; van Gils, J.M.; Rayner, A.J.; Chang, A.N.; Suarez, Y.; et al. Antagonism of miR-33 in mice promotes reverse cholesterol transport and regression of atherosclerosis. *J. Clin. Investig.* **2011**, *121*, 2921–2931. [CrossRef]
20. Rotllan, N.; Ramírez, C.M.; Aryal, B.; Esau, C.C.; Fernández-Hernando, C. Therapeutic silencing of microRNA-33 inhibits the progression of atherosclerosis in *Ldlr*$^{-/-}$ mice—Brief report. *Arterioscler. Thromb. Vasc. Biol.* **2013**, *33*, 1973–1977. [CrossRef]
21. Rayner, K.J.; Suárez, Y.; Dávalos, A.; Parathath, S.; Fitzgerald, M.L.; Tamehiro, N.; Fisher, E.A.; Moore, K.J.; Fernández-Hernando, C. MiR-33 contributes to the regulation of cholesterol homeostasis. *Science* **2010**, *328*, 1570–1573. [CrossRef]
22. Bartel, D.P. MicroRNAs: Target recognition and regulatory functions. *Cell* **2009**, *136*, 215–233. [CrossRef] [PubMed]
23. Jopling, C.L.; Yi, M.; Lancaster, A.M.; Lemon, S.M.; Sarnow, P. Modulation of hepatitis C virus RNA abundance by a liver-specific MicroRNA. *Science* **2005**, *309*, 1577–1581. [CrossRef] [PubMed]
24. Lee, Y.; Ahn, C.; Han, J.; Choi, H.; Kim, J.; Yim, J.; Lee, J.; Provost, P.; Rådmark, O.; Kim, S.; et al. The nuclear RNase III Drosha initiates microRNA processing. *Nature* **2003**, *425*, 415–419. [CrossRef]
25. Yang, Z.; Cappello, T.; Wang, L. Emerging role of microRNAs in lipid metabolism. *Acta Pharm. Sin. B* **2015**, *5*, 145–150. [CrossRef]
26. Esau, C.; Davis, S.; Murray, S.F.; Yu, X.X.; Pandey, S.K.; Pear, M.; Watts, L.; Booten, S.L.; Graham, M.; McKay, R.; et al. miR-122 regulation of lipid metabolism revealed by in vivo antisense targeting. *Cell Metab.* **2006**, *3*, 87–98. [CrossRef]
27. Hueso, M.; Griñán, R.; Mallen, A.; Navarro, E.; Purqueras, E.; Gomá, M.; Sbraga, F.; Blasco-Lucas, A.; Revilla, G.; Santos, D.; et al. MiR-125b downregulates macrophage scavenger receptor type B1 and reverse cholesterol transport. *Biomed. Pharmacother.* **2022**, *146*, 112596. [CrossRef] [PubMed]
28. Goedeke, L.; Rotllan, N.; Canfrán-Duque, A.; Aranda, J.F.; Ramírez, C.M.; Araldi, E.; Lin, C.S.; Anderson, N.N.; Wagschal, A.; de Cabo, R.; et al. MicroRNA-148a regulates LDL receptor and ABCA1 expression to control circulating lipoprotein levels. *Nat. Med.* **2015**, *21*, 1280–1289. [CrossRef]
29. Dong, J.; He, M.; Li, J.; Pessentheiner, A.; Wang, C.; Zhang, J.; Sun, Y.; Wang, W.T.; Zhang, Y.; Liu, J.; et al. microRNA-483 ameliorates hypercholesterolemia by inhibiting PCSK9 production. *JCI Insight* **2020**, *5*, e143812. [CrossRef]
30. Zhang, X.; Price, N.L.; Fernández-Hernando, C. Non-coding RNAs in lipid metabolism. *Vascul. Pharmacol.* **2019**, *114*, 93–102. [CrossRef]
31. Fernández-Tussy, P.; Ruz-Maldonado, I.; Fernández-Hernando, C. MicroRNAs and Circular RNAs in Lipoprotein Metabolism. *Curr. Atheroscler. Rep.* **2021**, *23*, 33. [CrossRef]
32. Hennessy, E.J.; van Solingen, C.; Scacalossi, K.R.; Ouimet, M.; Afonso, M.S.; Prins, J.; Koelwyn, G.J.; Sharma, M.; Ramkhelawon, B.; Carpenter, S.; et al. The long noncoding RNA CHROME regulates cholesterol homeostasis in primate. *Nat. Metab.* **2019**, *1*, 98–110. [CrossRef] [PubMed]
33. Wang, S.; Gulshan, K.; Brubaker, G.; Hazen, S.L.; Smith, J.D. ABCA1 mediates unfolding of apolipoprotein AI N terminus on the cell surface before lipidation and release of nascent high-density lipoprotein. *Arterioscler. Thromb. Vasc. Biol.* **2013**, *33*, 1197–1205. [CrossRef] [PubMed]
34. Zeng, W.; Xiong, G.; Hua, L.; Hu, Y.; Guo, X.; Peng, X. APOA1 mRNA and protein in kidney renal clear cell carcinoma correlate with the disease outcome. *Sci. Rep.* **2022**, *12*, 12406. [CrossRef] [PubMed]
35. Jacobo-Albavera, L.; Domínguez-Pérez, M.; Medina-Leyte, D.J.; González-Garrido, A.; Villarreal-Molina, T. The Role of the ATP-Binding Cassette A1 (ABCA1) in Human Disease. *Int. J. Mol. Sci.* **2021**, *22*, 1593. [CrossRef] [PubMed]

36. Timmins, J.M.; Lee, J.Y.; Boudyguina, E.; Kluckman, K.D.; Brunham, L.R.; Mulya, A.; Gebre, A.K.; Coutinho, J.M.; Colvin, P.L.; Smith, T.L.; et al. Targeted inactivation of hepatic Abca1 causes profound hypoalphalipoproteinemia and kidney hypercatabolism of apoA-I. *J. Clin. Investig.* **2005**, *115*, 1333–1342. [CrossRef] [PubMed]
37. Brunham, L.R.; Kruit, J.K.; Iqbal, J.; Fievet, C.; Timmins, J.M.; Pape, T.D.; Coburn, B.A.; Bissada, N.; Staels, B.; Groen, A.K.; et al. Intestinal ABCA1 directly contributes to HDL biogenesis in vivo. *J. Clin. Investig.* **2006**, *116*, 1052–1062. [CrossRef]
38. Marquart, T.J.; Allen, R.M.; Ory, D.S.; Baldán, A. miR-33 links SREBP-2 induction to repression of sterol transporters. *Proc. Natl. Acad. Sci. USA* **2010**, *107*, 12228–12232. [CrossRef]
39. Ouimet, M.; Marcel, Y.L. Regulation of lipid droplet cholesterol efflux from macrophage foam cells. *Arterioscler. Thromb. Vasc. Biol.* **2012**, *32*, 575–581. [CrossRef]
40. Zaiou, M.; Bakillah, A. Epigenetic Regulation of ATP-Binding Cassette Protein A1 (ABCA1) Gene Expression: A New Era to Alleviate Atherosclerotic Cardiovascular Disease. *Diseases* **2018**, *6*, 34. [CrossRef]
41. Czarnecka, H.; Yokoyama, S. Regulation of cellular cholesterol efflux by lecithin:cholesterol acyltransferase reaction through nonspecific lipid exchange. *J. Biol. Chem.* **1996**, *271*, 2023–2028. [CrossRef]
42. Ji, Y.; Jian, B.; Wang, N.; Sun, Y.; Moya, M.L.; Phillips, M.C.; Rothblat, G.H.; Swaney, J.B.; Tall, A.R. Scavenger receptor BI promotes high density lipoprotein-mediated cellular cholesterol efflux. *J. Biol. Chem.* **1997**, *272*, 20982–20985. [CrossRef] [PubMed]
43. Jessup, W.; Gelissen, I.C.; Gaus, K.; Kritharides, L. Roles of ATP binding cassette transporters A1 and G1, scavenger receptor BI and membrane lipid domains in cholesterol export from macrophages. *Curr. Opin. Lipidol.* **2006**, *17*, 247–257. [CrossRef] [PubMed]
44. Adorni, M.P.; Zimetti, F.; Billheimer, J.T.; Wang, N.; Rader, D.J.; Phillips, M.C.; Rothblat, G.H. The roles of different pathways in the release of cholesterol from macrophages. *J. Lipid Res.* **2007**, *48*, 2453–2462. [CrossRef] [PubMed]
45. Rosenson, R.S.; Brewer, H.B., Jr.; Davidson, W.S.; Fayad, Z.A.; Fuster, V.; Goldstein, J.; Hellerstein, M.; Jiang, X.C.; Phillips, M.C.; Rader, D.J.; et al. Cholesterol efflux and atheroprotection: Advancing the concept of reverse cholesterol transport. *Circulation* **2012**, *125*, 1905–1919. [CrossRef] [PubMed]
46. Lai, L.; Azzam, K.M.; Lin, W.C.; Rai, P.; Lowe, J.M.; Gabor, K.A.; Madenspacher, J.H.; Aloor, J.J.; Parks, J.S.; Näär, A.M.; et al. MicroRNA-33 Regulates the Innate Immune Response via ATP Binding Cassette Transporter-mediated Remodeling of Membrane Microdomains. *J. Biol. Chem.* **2016**, *291*, 19651–19660. [CrossRef] [PubMed]
47. Mabuchi, H.; Nohara, A.; Inazu, A. Cholesteryl ester transfer protein (CETP) deficiency and CETP inhibitors. *Mol. Cells* **2014**, *37*, 777–784. [CrossRef]
48. Goedeke, L.; Salerno, A.; Ramírez, C.M.; Guo, L.; Allen, R.M.; Yin, X.; Langley, S.R.; Esau, C.; Wanschel, A.; Fisher, E.A.; et al. Long-term therapeutic silencing of miR-33 increases circulating triglyceride levels and hepatic lipid accumulation in mice. *EMBO Mol. Med.* **2014**, *6*, 1133–1141. [CrossRef]
49. Price, N.L.; Zhang, X.; Fernández-Tussy, P.; Singh, A.K.; Burnap, S.A.; Rotllan, N.; Goedeke, L.; Sun, J.; Canfrán-Duque, A.; Aryal, B.; et al. Loss of hepatic miR-33 improves metabolic homeostasis and liver function without altering body weight or atherosclerosis. *Proc. Natl. Acad. Sci. USA* **2021**, *118*, e2006478118. [CrossRef]
50. Rayner, K.J.; Esau, C.C.; Hussain, F.N.; McDaniel, A.L.; Marshall, S.M.; van Gils, J.M.; Ray, T.D.; Sheedy, F.J.; Goedeke, L.; Liu, X.; et al. Inhibition of miR-33a/b in non-human primates raises plasma HDL and lowers VLDL triglycerides. *Nature* **2011**, *478*, 404–407. [CrossRef]
51. Trauner, M.; Claudel, T.; Fickert, P.; Moustafa, T.; Wagner, M. Bile acids as regulators of hepatic lipid and glucose metabolism. *Dig. Dis.* **2010**, *28*, 220–224. [CrossRef]
52. Claudel, T.; Zollner, G.; Wagner, M.; Trauner, M. Role of nuclear receptors for bile acid metabolism, bile secretion, cholestasis, and gallstone disease. *Biochim. Biophys. Acta* **2011**, *1812*, 867–878. [CrossRef] [PubMed]
53. Zhang, Y.; Edwards, P.A. FXR signaling in metabolic disease. *FEBS Lett.* **2008**, *582*, 10–18. [CrossRef] [PubMed]
54. Chiang, J.Y. Regulation of bile acid synthesis: Pathways, nuclear receptors, and mechanisms. *J. Hepatol.* **2004**, *40*, 539–551. [CrossRef]
55. Li, T.; Francl, J.M.; Boehme, S.; Chiang, J.Y. Regulation of cholesterol and bile acid homeostasis by the cholesterol 7α-hydroxylase/steroid response element-binding protein 2/microRNA-33a axis in mice. *Hepatology* **2013**, *58*, 1111–1121. [CrossRef] [PubMed]
56. Morotti, R.A.; Suchy, F.J.; Magid, M.S. Progressive familial intrahepatic cholestasis (PFIC) type 1, 2, and 3: A review of the liver pathology findings. *Semin. Liver Dis.* **2011**, *31*, 3–10. [CrossRef] [PubMed]
57. Yu, L.; Hammer, R.E.; Li-Hawkins, J.; Von Bergmann, K.; Lutjohann, D.; Cohen, J.C.; Hobbs, H.H. Disruption of Abcg5 and Abcg8 in mice reveals their crucial role in biliary cholesterol secretion. *Proc. Natl. Acad. Sci. USA* **2002**, *99*, 16237–16242. [CrossRef]
58. Allen, R.M.; Marquart, T.J.; Albert, C.J.; Suchy, F.J.; Wang, D.Q.; Ananthanarayanan, M.; Ford, D.A.; Baldán, A. miR-33 controls the expression of biliary transporters, and mediates statin- and diet-induced hepatotoxicity. *EMBO Mol. Med.* **2012**, *4*, 882–895. [CrossRef]
59. Bornfeldt, K.E.; Linton, M.F.; Fisher, E.A.; Guyton, J.R. JCL roundtable: Lipids and inflammation in atherosclerosis. *J. Clin. Lipidol.* **2021**, *15*, 3–17. [CrossRef]

nutrients

MDPI

Article

Non-Cholesterol Sterols in Breast Milk and Risk of Allergic Outcomes in the First Two Years of Life

Lieve van Brakel [1], Carel Thijs [2], Ronald P. Mensink [1], Dieter Lütjohann [3] and Jogchum Plat [1,*]

[1] Department of Nutrition and Movement Sciences, NUTRIM School of Translational Research in Metabolism, Maastricht University, 6200 MD Maastricht, The Netherlands; lieve.vanbrakel@maastrichtuniversity.nl (L.v.B.); r.mensink@maastrichtuniversity.nl (R.P.M.)

[2] Department of Epidemiology, CaPHRI (Care and Public Health Research Institute), Maastricht University, 6200 MD Maastricht, The Netherlands; c.thijs@maastrichtuniversity.nl

[3] Institute of Clinical Chemistry and Clinical Pharmacology, University Hospital Bonn, 53127 Bonn, Germany; dieter.luetjohann@ukbonn.de

* Correspondence: j.plat@maastrichtuniversity.nl; Tel.: +31-43-388-1309

Abstract: This study aimed to explore associations between non-cholesterol sterol concentrations in breast milk and allergic outcomes in children aged two. Data from the KOALA Birth Cohort Study, the Netherlands, were used. Non-cholesterol sterols were analyzed by gas–liquid chromatography–mass spectrometry in breast milk sampled one-month postpartum ($N = 311$). Sterols were selected for each allergic outcome, i.e., eczema, wheeze, and allergic sensitization, prior to analyses. Associations between the selected sterols with allergic outcomes were analyzed using multiple logistic regression to calculate odds ratios (ORs). The odds of eczema in the first two years of life were lower with higher concentrations of cholestanol (OR (95%CI): 0.98 (0.95; 1.00), $p = 0.04$), lanosterol (0.97 (0.95; 1.00), $p = 0.02$), lathosterol (0.93 (0.87; 0.99), $p = 0.02$), and stigmasterol (0.51 (0.29; 0.91), $p = 0.02$) in breast milk sampled one-month postpartum. None of the sterols were associated with wheeze in the first two years of life. The odds of allergic sensitization at age two were lower with higher concentrations of campesterol in breast milk (OR (95%CI): 0.81 (0.70; 0.95), $p = 0.01$). In conclusion, our data suggest that exposure to higher non-cholesterol sterol concentrations in breast milk may indeed be associated with the prevention of allergic outcomes in the first two years of life.

Keywords: non-cholesterol sterols; eczema; wheeze; allergic sensitization; breast milk; immune system; infant health

Citation: van Brakel, L.; Thijs, C.; Mensink, R.P.; Lütjohann, D.; Plat, J. Non-Cholesterol Sterols in Breast Milk and Risk of Allergic Outcomes in the First Two Years of Life. *Nutrients* **2022**, *14*, 766. https://doi.org/10.3390/nu14040766

Academic Editor: Michael J. Barratt

Received: 18 January 2022
Accepted: 8 February 2022
Published: 11 February 2022

1. Introduction

Breastfeeding is the preferred nutrition for newborns and infants [1]. The World Health Organization therefore recommends exclusive breastfeeding for the first six months of life and to combine breastfeeding with complementary foods for children aged from six months to two years and beyond [2]. Breastfeeding has several health benefits for infants. For example, breastfeeding has been associated with a decreased risk of child mortality in the first two years of life [3]. In addition, probiotic bacteria in breast milk play an essential role in developing the gut microbiota in early life by seeding the infant gut [4]. Breastfeeding even has health benefits tracking into adulthood. Breastfeeding has been associated with a lower risk of several non-communicable diseases in adults, such as cardiovascular diseases [5,6], obesity [7,8], and type 2 diabetes [7,9]. Moreover, in recent years, there has been increasing interest in the potential role of breastfeeding for the prevention of allergic outcomes in newborns and infants [10–15].

Although results are inconclusive [10], breastfeeding has been associated with a reduced risk of developing asthma [11–13], eczema [13,14], and allergic diseases [15]. It is therefore important to identify compounds within breast milk that could be responsible for the supposed reduced risk of developing allergic diseases. However, identification of these compounds is difficult, since breast milk composition is highly variable, especially during the first month of breastfeeding [16]. Variability in composition is highest in the milk produced during the first three weeks postpartum: colostrum (produced in first 4–7 days) and transitional milk (produced approximately from day 7 to 21 postpartum) [17]. This variability may reflect the infant's needs, e.g., for infant growth [18]. The composition of mature milk (produced from approximately day 21 postpartum onwards) is less variable and contains approximately 3–5% (*w*/*w*) fat, 6.9–7.2% carbohydrates, 0.8–0.9% protein, and 0.2% mineral constituents [17,19]. Lipids in breast milk are the most important energy source for infants [16]. The lipid fraction of breast milk mainly consists of triacylglycerol and for approximately 0.5% of cholesterol [20]. It also contains plant sterols, which surprisingly do not reflect the circulating plasma plant sterol concentrations of the mother [21]. In addition, mRNA expression for sterol transporters ABCG5/G8 was previously observed in bovine mammary glands [22]. When this is also the case in human mammary glands, it could explain the presence of the specific plant sterol concentrations in breast milk [21]. Altogether, these findings suggest a regulated transport process of plant sterols into breast milk.

Today, plant sterols are mainly recognized for their LDL-cholesterol-lowering effects [23]. However, in a paper published by our group, Plat and colleagues have suggested that plant sterols in breast milk may have a perinatal role, e.g., in growth and development of the child [21]. This rationale was, among others, based on that plant sterols have been shown to interact with immune cells [24]. In more detail, plant sterols and stanols (saturated derivatives of plant sterols) may affect T-helper cell behavior, potentially by interacting with regulatory T cells (Tregs) [25–27]. This effect of plant sterols and stanols could be relevant in conditions characterized by a disbalance between T-helper cell subsets Th1 and Th2. For instance, a disbalance in T-helper cell activity towards the Th2 profile has been related to increased immunoglobulin E (IgE) concentrations and allergic diseases, such as allergic asthma [28]. Brüll and colleagues studied the effects of plant stanols on immune cells of allergic asthma patients. Based on their in vitro and in vivo observations, they suggested that plant stanols stimulated Treg and Th1 cell activity, while inhibiting Th2 cell activity [25,26]. Furthermore, plant sterols share a structural similarity with cholesterol precursors, which are intermediates in the endogenous cholesterol synthesis pathways [21]. Previous research has shown that some of these compounds can also interact with immune cells, thereby affecting immune responses. For example, desmosterol was found to inhibit inflammatory cascades within macrophages [29]. Moreover, mevalonate, which is another intermediate in the cholesterol synthesis pathway, was found to be crucial to induce trained immunity [30]. As with plant sterols, cholesterol precursors are also present in breast milk [21]. Together, these sterol compounds can be grouped as non-cholesterol sterols. However, it is important to consider these non-cholesterol sterols in breast milk as nutrients and not as markers for intestinal cholesterol absorption and endogenous synthesis for which their serum concentrations have been validated [31].

It is unknown whether the suggested effects of non-cholesterol sterols on immune cell behavior translate in to a benefit for children when exposed to these compounds in early life. This raises the question whether the amount of non-cholesterol sterols in breast milk could influence immune maturation, alter T-helper cell behavior and immune responses in early life, and thereby prevent allergic diseases. Therefore, this study aimed to determine the association between non-cholesterol sterol concentrations in breast milk and allergic outcomes in breastfed children in the first two years of life.

2. Materials and Methods

2.1. Study Population

The cohort used in this study is part of the "Kind, Ouders en gezondheid: Aandacht voor Leefstijl en Aanleg" (KOALA) Birth Cohort Study, the Netherlands, which has been described in detail elsewhere [32]. Briefly, recruitment of pregnant women started in October 2000. Participants with a conventional (N = 2343) or an 'alternative' lifestyle

(N = 491) with regard to, e.g., child rearing practices or diet (organic or vegetarian) were recruited and enrolled between the 14th and 18th week of gestation. Participants were followed during gestation and up to several years postpartum and completed relevant questionnaires during follow-up. From January 2002, we started collecting biosamples, such as maternal blood at 36 weeks of pregnancy. In this subcohort of the KOALA study (KOALA-SUB), other samples were also obtained, including a breast milk sample from the mother one-month postpartum and a venous blood sample from the child at age two. For this study, we used these two samples and the data collected from questionnaires until the age of two. The KOALA study was approved by the Medical Ethical Committee of Maastricht University Medical Center, Maastricht, the Netherlands (MEC 01-139 and 00-182) and the Central Committee on Research Involving Human Subjects, The Hague, the Netherlands (CCMO P01.1265L). Inclusion criteria for the present study were participation in KOALA-SUB and an available one-month postpartum breast milk sample. Exclusion criteria were prematurity (<37 weeks gestation) and diseases or disorders such as cystic fibrosis, Down's syndrome, and arthritis. The current study included N = 311 children (N = 141 mothers with conventional lifestyle, N = 166 mothers with alternative lifestyle, and N = 4 mothers have missing data on lifestyle).

2.2. Study Outcomes

Allergic outcomes of interest were eczema and wheeze during the first two years of life, and allergic sensitization at age two. At that age, the immune system has had the opportunity to mature while being exposed to different concentrations of non-cholesterol sterols in breast milk [33]. The International Study of Asthma and Allergies in Childhood Questionnaire (ISAAC) was used to determine the presence of eczema and wheeze at 3, 7, 12 and 24 months postpartum, as described previously [34]. In short, if parents ever reported symptoms of eczema (itchy rash that was coming and going) or wheeze (wheezing or whistling in the chest) in this questionnaire, the child was defined as a case of eczema or wheeze, respectively. Children who only had diaper rash, rash around the eyes, or scalp scaling were not considered to have eczema. Allergic sensitization against hen's egg, cow's milk, peanut, birch, grass pollen, cat, dog, or house dust mite was determined in a venous blood sample at age two. Allergic sensitization was defined as having specific serum IgE levels > 0.30 IU/mL against one of the allergens tested. As described earlier, IgE was measured with a detection limit of 0.10 IU/mL [35,36].

2.3. Breast Milk Sampling and Analysis of Non-Cholesterol Sterols

Methods for breast milk sampling and storage have been described elsewhere [37]. Briefly, breast milk was collected in the morning in sterile tubes (Greiner Bio-One, Kremsmuenster, Austria). A sample was collected from the contra-lateral breast since the last feeding, before breastfeeding the child. The milk samples were kept in the refrigerator (4 °C) and picked up by a researcher on the same day. During transport, the milk samples were stored in a cooler (Coleman Company, Inc., Breda, the Netherlands) on packed ice (4 °C) until processing on the same day at the Biobank Maastricht. After measuring the volume of the sample, it was mixed (gently shaking by hand) and five Eppendorf tubes (2 mL) were filled with whole milk for storage. Two Eppendorf tubes were filled for creamatocrit measurement. The remaining sample was centrifuged (400× g, 12 min, no brake, 4 °C) to separate the lipid and aqueous fraction. The lipid layer was trimmed off with a pipette and released in plastic storage vials (Sarstedt, Nümbrecht, Germany) and stored at −80 °C in the Biobank Maastricht until further processing. Creamatocrit was determined as described previously [38,39]. In short, milk samples were centrifuged for 15 min at 12,000× g. The length of the total milk column and of the cream layer were measured directly after centrifuging. Creamatocrit was determined by calculating which percentage of the total length of the milk sample consisted of cream.

For cholesterol and non-cholesterol sterol analysis, the frozen breast milk samples were transported on dry ice and delivered on the same day at the Institute of Clinical

Chemistry and Clinical Pharmacology of the University Hospital Bonn, Germany. Concentrations of cholesterol were measured by gas-liquid chromatography flame ionization detection (GC-FID) on an HP6890 Series GC-System (Agilent Technologies, Waldbronn, Germany), using 5α-cholestane (Serva Electrophoresis GmbH, Heidelberg, Germany) as internal standard. Cholesterol and non-cholesterol standards were applied by Sigma-Aldrich Chemie GmbH, München, Germany. Plant sterol (sitosterol, campesterol, stigmasterol, brassicasterol), cholestanol and cholesterol precursor (lanosterol, lathosterol and desmosterol) concentrations in breast milk were analyzed by gas–liquid chromatography–mass spectrometry-selected ion monitoring (GC-MS-SIM) (Agilent Technologies 6890 Network GC coupled with an Agilent Technologies 5975B inert MSD, Agilent Technologies, Waldbronn, Germany), using epicoprostanol (sigma-Aldrich Chemie GmbH, München, Germany) as internal standard [40]. Sterol concentrations were corrected for the lipid levels (creamatocrit) of the breast milk sample by dividing the sterol concentrations by creamatocrit expressed as fraction.

2.4. Statistical Analysis

Two types of exploratory analyses were first conducted to determine which sterols could be associated with allergic outcomes. First, factor analysis was performed to determine correlations between sterol concentrations in breast milk. A varimax rotation was used to maximize between-subject variance and the minimal eigenvalue was set to 1. The obtained factors were then used in multiple logistic regression analysis to determine factors associated with the allergic outcomes, and to calculate odds ratios (ORs) and their corresponding 95% confidence intervals (95%CIs). Second, independent-sample *t*-tests were performed to determine differences in sterol concentrations between cases and controls for each allergic outcome. Based on these two exploratory analyses, sterols were selected for further analysis with $p < 0.10$ as selection threshold. Sterols were selected when they: (1) were present in factors that were associated with an allergic outcome in the multiple logistic regression analyses (trend [$p < 0.10$] or significant association [$p < 0.05$]), or (2) differed in concentration between cases and controls in the independent-sample *t*-tests (trend [$p < 0.10$] or significant association [$p < 0.05$]). Next, multiple logistic regression models were made for each selected sterol to determine which sterols were significantly associated with allergic outcomes ($p < 0.05$). ORs and corresponding 95%CIs were calculated. A priori, confounders to be used in these regression models were determined by drawing and analyzing causal diagrams (DAGs). Confounders that were tested in the models included study group, smoking, season of milk sampling, gestational age, prepregnancy BMI, maternal age, atopy of parents, maternal education, gender of child, gravidity, duration of breastfeeding, and birthweight. If confounders changed the regression coefficient $\beta 1$ by at least 10%, they were added to regression models. Finally, Spearman correlations were used to explore relations between non-cholesterol sterol and cholesterol concentrations in breast milk.

A result was considered significantly different when $p < 0.05$. All analyses were conducted using IBM SPSS Statistics for Windows, Version 25.0 (Armonk, NY, USA).

3. Results

3.1. Baseline Characteristics and Flow Chart

The selection of participants from the KOALA study is shown in Figure 1. Of the total cohort ($N = 2834$), the women with an available breast milk sample were selected ($N = 315$). The 311 women who fulfilled the criteria for the current study were selected.

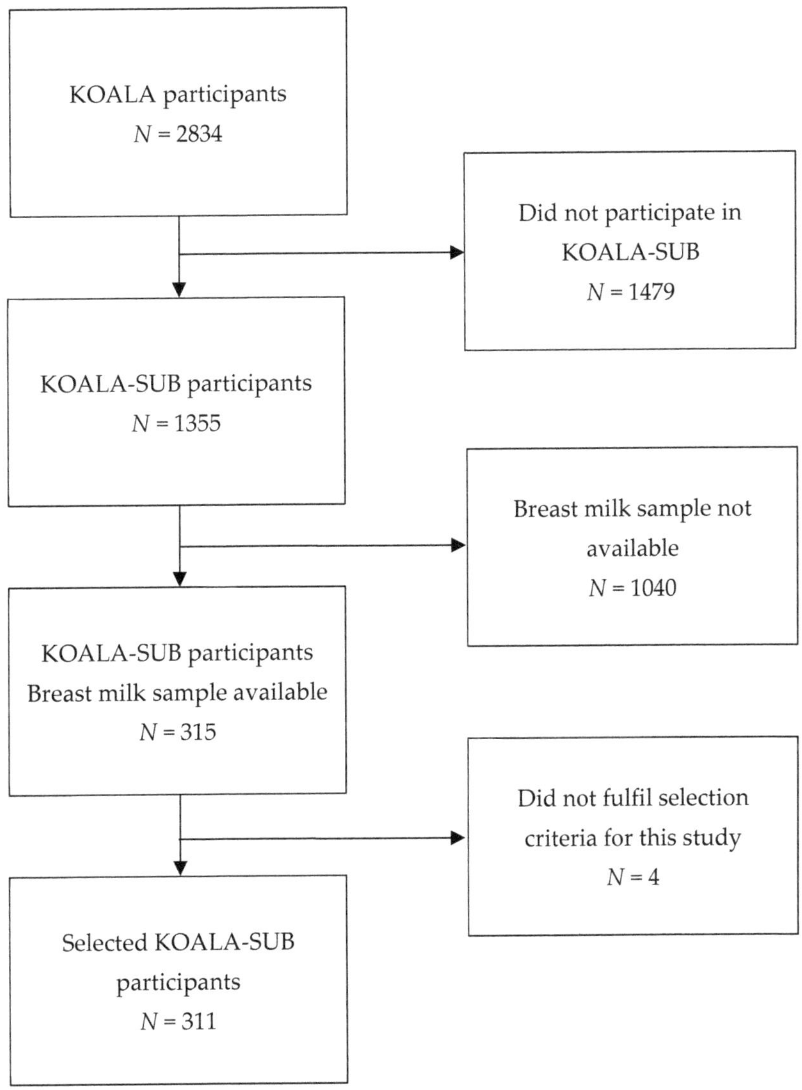

Figure 1. Flow chart of participants included in this study. KOALA = "Kind, Ouders en gezondheid: Aandacht voor Leefstijl en Aanleg"; KOALA-SUB = subcohort of the KOALA study.

Characteristics of the study population are shown in Table 1. The mean (SD) maternal age and median BMI (IQR) of the 311 mothers at the start of pregnancy were 32.4 (3.9) years and 22.4 (20.6–24.5) kg/m^2, respectively. In total, 91 children developed eczema and 79 children developed wheeze in the first two years of life, and 49 children were allergically sensitized against common allergens at age two. The baseline characteristics were comparable between the women with either a conventional or alternative lifestyle. Only sitosterol was higher in the alternative lifestyle group (Table 1).

Table 1. Baseline characteristics of the KOALA-SUB cohort selected for this study. Data are shown as the mean (SD) or the median (Q1–Q3), unless otherwise indicated.

	Total (*N* = 311)	**Conventional Lifestyle** [1] (*N* = 141)	**Alternative Lifestyle** [1] (*N* = 166)
Maternal age, years (SD)	32.4 (3.9)	31.5 (3.4)	33.1 (4.2)
Prepregnancy BMI, kg/m^2 (IQR)	22.4 (20.6–24.5)	23.0 (21.5–25.2)	21.7 (20.1–24.0)
Smoking during pregnancy [2], N (%)	5 (2%)	4 (3%)	1 (1%)
Atopic history parents [1], N (%)			
None	113 (37%)	50 (36%)	62 (38%)
Only father	78 (25%)	37 (26%)	40 (24%)
Only mother	64 (21%)	31 (22%)	32 (20%)
Both	52 (17%)	22 (16%)	30 (18%)
Gender child female, N (%)	161 (52%)	74 (53%)	83 (50%)
Duration breastfeeding, N (%)			
1–3 months	64 (21%)	46 (33%)	17 (10%)
4–6 months	70 (23%)	40 (28%)	29 (18%)
7–9 months	70 (23%)	28 (20%)	41 (25%)
10–12 months	53 (17%)	17 (12%)	36 (22%)
≥13 months	53 (17%)	10 (7%)	43 (26%)
Maternal education, N (%)			
Lower	12 (4%)	7 (5%)	5 (3%)
Middle	96 (31%)	53 (38%)	41 (25%)
Higher vocational	131 (42%)	60 (43%)	69 (42%)
Academic	66 (21%)	19 (14%)	47 (28%)
Other	6 (2%)	2 (1%)	4 (2%)
Season breast milk sampling [1], N (%)			
December 2002–February 2003	112 (37%)	64 (45%)	48 (29%)
March–May 2003	120 (39%)	59 (42%)	61 (37%)
June–September 2003	75 (24%)	18 (13%)	57 (34%)
Gravidity, N (%)			
1	123 (40%)	65 (46%)	58 (34.9)
2	110 (35%)	51 (36%)	57 (34.3)
≥3	78 (25%)	25 (18%)	51 (30.7)
Eczema in first two years [3], N (%)	91 (30%)	41 (30%)	50 (31.1)
Wheeze in first two years [4], N (%)	79 (26%)	37 (27%)	42 (26%)
Allergic sensitization at age 2 [5], N (%)	49 (24%)	25 (28%)	24 (22%)
Creamatocrit value [6], % (IQR)	7 (5–9%)	7 (5–9%)	7 (5–9%)
Cholesterol concentration breast milk, mmol/L (IQR)	0.35 (0.28–0.42)	0.35 (0.27–0.43)	0.36 (0.29–0.41)
Cholesterol concentration breast milk corrected for creamatocrit [7], mmol/Lf (IQR)	4.81 (4.14–5.90)	4.81 (4.17–5.93)	4.83 (5.84–6.87)
Non-cholesterol sterol concentrations breast milk, µmol/L (IQR)			
Brassicasterol	0.23 (0.18–0.27)	0.24 (0.19–0.29)	0.22 (0.17–0.25)
Campesterol [7]	0.32 (0.20–0.52)	0.37 (0.21–0.60)	0.28 (0.19–0.46)
Cholestanol	1.51 (1.28–1.73)	1.50 (1.26–1.75)	1.52 (1.31–1.73)
Desmosterol	52.2 (37.4–70.3)	51.7 (36.3–68.3)	54.3 (38.1–71.8)
Lanosterol	2.03 (1.43–2.89)	2.01 (1.31–2.85)	2.04 (1.50–2.93)
Lathosterol	0.62 (0.40–0.84)	0.62 (0.38–0.87)	0.62 (0.40–0.84)
Sitosterol	0.70 (0.49–1.41)	0.59 (0.45–0.87)	0.82 (0.55–1.48)
Stigmasterol	0.05 (0.04–0.06)	0.05 (0.04–0.06)	0.05 (0.04–0.06)
Non-cholesterol sterol concentrations breast milk corrected for creamatocrit, µmol/Lf (IQR)			

Table 1. *Cont.*

	Total (N = 311)	Conventional Lifestyle [1] (N = 141)	Alternative Lifestyle [1] (N = 166)
Brassicasterol [7]	3.18 (2.44–4.16)	3.32 (2.53–4.25)	2.95 (2.31–4.10)
Campesterol [8]	4.90 (3.01–7.07)	5.34 (3.50–7.87)	4.27 (2.61–6.13)
Cholestanol [7]	21.0 (17.2–27.7)	22.5 (17.3–27.9)	20.5 (17.0–27.9)
Desmosterol [7]	776.5 (592.7–997.9)	716.1(561.6–957.0)	807.2 (633.6–1026.5)
Lanosterol [7]	29.1 (22.7–37.5)	28.4 (22.8–37.3)	29.7 (22.7–37.5)
Lathosterol [7]	8.69 (6.25–11.9)	8.80 (6.47–12.2)	8.31 (5.99–11.6)
Sitosterol [7]	10.3 (7.17–18.7)	8.26 (6.60–12.6)	13.7 (7.88–23.8)
Stigmasterol [7]	0.68 (0.51–0.93)	0.67 (0.51–0.93)	0.69 (0.51–0.94)

BMI = body mass index; IQR = interquartile range; SD = standard deviation; mmol/Lf = mmol per liter milk fat; [1] Missing data N = 4; [2] Missing data N = 1; [3] Missing data N = 12; [4] Missing data N = 9; [5] Missing data N = 110; [6] Missing data N = 37; [7] Missing data N = 8; [8] Missing data N = 45.

Desmosterol was the non-cholesterol sterol with the highest concentration in breast milk (median (IQR): 52.2 (37.4–70.3) µmol/L), which was 25 to 1000 fold higher as compared to the other sterols. Stigmasterol was the sterol with the lowest concentration (0.05 (0.04–0.06) µmol/L). Overall, non-cholesterol concentrations in breast milk were similar in the women with a conventional lifestyle and an alternative lifestyle.

3.2. *Selection Process of Sterols*

3.2.1. Exploratory Factor Analysis

To explore which of the eight non-cholesterol sterols that were analyzed in breast milk were intercorrelated, exploratory factor analysis was performed. Two factors were found based on the sterol concentration in breast milk corrected for creamatocrit (Table 2). The two factors separated brassicasterol, stigmasterol, campesterol, and lathosterol (factor 1); and lanosterol and desmosterol (factor 2). Cholestanol and sitosterol loaded on both factors, although to a higher extent on factor 1 than on factor 2.

Table 2. Factor loadings after varimax rotation. Non-cholesterol sterol concentrations were corrected for creamatocrit.

Non-Cholesterol Sterol (µmol/Lf)	Factor 1	Factor 2
Cholestanol	0.88	0.34
Brassicasterol	0.83	
Stigmasterol	0.78	
Sitosterol	0.65	0.31
Campesterol	0.64	
Lathosterol	0.37	
Lanosterol		0.93
Desmosterol		0.85

µmol/Lf = µmol per liter milk fat; only factor loadings ≥ 0.30 are shown; there were no factor loadings ≤ −0.30.

3.2.2. Multiple Logistic Regression Using the Factors

Multiple logistic regression was performed to explore relations between factors 1 and 2 with the allergic outcomes of interest (i.e., eczema and wheeze in the first two years of life, and allergic sensitization at age two) (Table 3). None of the factors were significantly associated with eczema, wheeze, or allergic sensitization. However, trends were observed for associations between factor 1 and eczema (OR (95%CI): 0.69 (0.46; 1.03), p = 0.07), factor 2 and eczema (0.69 (0.46; 1.04), p = 0.08), and factor 1 and allergic sensitization (0.52 (0.26; 1.07), p = 0.07).

Table 3. Results of the multiple logistic regression analyses to determine relations between the factors and allergic outcomes.

	Factor 1			Factor 2		
Outcome Variable	**OR**	**95% CI**	***p*-Value**	**OR**	**95% CI**	***p*-Value**
Eczema (N = 256) [1]	0.69	0.46; 1.03	0.07 *	0.69	0.46; 1.04	0.08 *
Wheeze (N = 259) [2]	1.04	0.77; 1.40	0.82	0.99	0.69; 1.42	0.95
Allergic sensitization (N = 171) [3]	0.52	0.26; 1.07	0.07 *	1.17	0.79; 1.73	0.43

OR: odds ratio; 95% CI: 95% confidence interval; [1]: adjusted for season, atopy of parents, maternal education, and duration of breastfeeding; [2]: adjusted for smoking, season, gestational age, prepregnancy BMI, atopy of parents, maternal education, gender of child, gravidity, and duration of breastfeeding; [3]: adjusted for study group, smoking, season, prepregnancy BMI, atopy of parents, maternal education, gender of child, gravidity, duration of breastfeeding, and birthweight. * $p < 0.10$.

3.2.3. Independent-Sample *t*-Tests

Independent-sample *t*-tests were performed to explore which of the individual sterols differed between cases and controls for each allergic outcome (Table 4). Lathosterol ($p = 0.06$) and stigmasterol ($p = 0.08$) concentrations in breast milk tended to be lower in eczema cases compared to controls. For wheeze, all sterol concentrations were similar in cases and controls. For allergic sensitization, campesterol concentrations in breast milk were significantly lower in cases compared to controls ($p = 0.03$).

Table 4. Results of the independent-sample *t*-tests to explore differences between non-cholesterol sterol concentrations in breast milk between cases and controls for eczema (N = 80 cases, N = 179 controls), wheeze (N = 67 cases, N = 195 controls), and allergic sensitization (N = 41 cases, N = 130 controls). Non-cholesterol sterol concentrations were adjusted for creamatocrit.

	Eczema			Wheeze			Allergic Sensitization		
Sterol (µmol/Lf)	**Mean Difference**	**95% CI**	***p*-Value**	**Mean Difference**	**95% CI**	***p*-Value**	**Mean Difference**	**95% CI**	***p*-Value**
Brassicasterol	−0.11	−0.76; 0.55	0.75	0.16	−0.52; 0.84	0.64	−0.25	−1.06; 0.55	0.54
Campesterol	−0.76	−1.85; 0.32	0.17	0.11	−1.03; 1.26	0.84	−1.78	−3.40; −0.16	0.03 **
Cholestanol	−3.19	−7.33; 0.94	0.13	0.3	−4.08; 4.67	0.89	−2.68	−8.04; 2.68	0.33
Desmosterol [1]	−0.04	−0.14; 0.06	0.44	0.01	−0.11; 0.13	0.83	−0.05	−0.22; 0.12	0.57
Lanosterol	−2.03	−5.55; 1.48	0.26	−0.36	−4.82; 4.10	0.88	0.94	−4.77; 6.65	0.75
Lathosterol	−1.18	−2.39; 0.03	0.06 *	−0.93	−2.23; 0.36	0.16	−0.53	−2.22; 1.16	0.54
Sitosterol	−0.13	−0.28; 0.03	0.11	−0.1	−0.26; 0.07	0.25	−0.13	−0.31; 0.04	0.14
Stigmasterol	−0.01	−0.02; 0.00	0.08 *	0	−0.01; 0.01	0.89	−0.01	−0.02; 0.01	0.36

95% CI: 95% confidence interval; µmol/Lf = micromoles per liter milk fat; [1] concentration in mmol/Lf; * $p < 0.10$; ** $p < 0.05$.

3.3. *Multiple Logistic Regression Using Selected Sterols*

3.3.1. Eczema

Based on the multiple logistic regression analysis using the obtained factors (Table 3), all sterols included in factors 1 and 2 were selected for eczema. In addition, based on the independent-sample *t*-tests (Table 4), lathosterol and stigmasterol were selected for eczema. Thus, all eight sterols were included in the final multiple logistic regression analyses. Separate models were made for each individual sterol (Table 5). The odds of eczema in the first two years of life were significantly lower with higher concentrations of cholestanol (OR (95%CI): 0.98 (0.95; 1.00), $p = 0.04$), lanosterol (0.97 (0.95; 1.00), $p = 0.02$), lathosterol (0.93 (0.87; 0.99), $p = 0.02$), and stigmasterol (0.51 (0.29; 0.91), $p = 0.02$) in breast milk one-month postpartum. The other sterols did not affect the odds of eczema during the first two years of life.

Table 5. Results of the multiple logistic regression analyses using the selected sterols for eczema. Separate models were made for each individual sterol. Non-cholesterol sterol concentrations were adjusted for creamatocrit.

Sterol (μmol/Lf)	N	OR	95% CI	p-Value
Brassicasterol [1]	264	0.95	0.87; 1.03	0.22
Campesterol [2]	259	0.95	0.88; 1.03	0.25
Cholestanol [3]	264	0.98	0.95; 1.00	0.04 *
Desmosterol [a,1]	264	0.52	0.22; 1.22	0.13
Lanosterol [4]	267	0.97	0.95; 1.00	0.02 *
Lathosterol [5]	267	0.93	0.87; 0.99	0.02 *
Sitosterol [6]	264	0.98	0.95; 1.00	0.09
Stigmasterol [6]	264	0.51	0.29; 0.91	0.02 *

OR: odds ratio; 95% CI: 95% confidence interval; μmol/Lf = micromoles per liter milk fat; [a]: unit is mmol/Lf; [1]: adjusted for season, atopy of parents, maternal education, duration of breastfeeding, and gravidity; [2]: adjusted for season and maternal education; [3]: adjusted for atopy of parents and maternal education; [4]: adjusted for season, maternal education, and duration breastfeeding; [5]: adjusted for maternal education; [6]: adjusted for season, atopy of parents, and maternal education; * $p < 0.05$.

3.3.2. Wheeze

None of the factors from exploratory factor analysis were associated with wheeze, nor were there differences in sterol concentrations in breast milk between cases and controls. Therefore, none of the sterols were evaluated in further analysis for wheeze.

3.3.3. Allergic Sensitization

Based on the multiple logistic regression analysis using the obtained factors (Table 3), the sterols included in factor 1 were selected for allergic sensitization. In addition, based on the independent-sample *t*-tests (Table 4), campesterol was selected for allergic sensitization. Thus, brassicasterol, campesterol, cholestanol, lathosterol, sitosterol, and stigmasterol were included in the final multiple logistic regression analyses. Separate models were made for each individual sterol (Table 6). The odds of allergic sensitization at age 2 were significantly lower with a higher concentration of campesterol in breast milk one-month postpartum (OR (95%CI): 0.81 (0.70; 0.95), $p = 0.01$). The other sterols did not affect the odds of allergic sensitization at age 2.

Table 6. Results of the multiple logistic regression analysis using the selected sterols for allergic sensitization. Non-cholesterol sterol concentrations were adjusted for creamatocrit.

Sterol (μmol/Lf)	N	OR	95% CI	p-Value
Brassicasterol [1]	176	0.93	0.78; 1.12	0.47
Campesterol [2]	171	0.81	0.70; 0.95	0.01 *
Cholestanol [3]	176	0.98	0.95; 1.01	0.26
Lathosterol [4]	176	0.99	0.91; 1.07	0.77
Sitosterol [5]	176	0.97	0.93; 1.01	0.13
Stigmasterol [5]	176	0.77	0.42; 1.40	0.38

OR: odds ratio; 95% CI: 95% confidence interval; μmol/Lf = micromoles per liter milk fat; [1]: adjusted for study group, season, and duration breastfeeding; [2]: adjusted for season and duration breastfeeding; [3]: adjusted for smoking; [4]: adjusted for smoking, season, and duration breastfeeding; [5]: adjusted for smoking and season * $p < 0.05$.

3.4. Cholesterol and Allergic Outcomes

Non-cholesterol sterol concentrations were significantly correlated to cholesterol concentrations (corrected for creamatocrit) in breast milk, except for lanosterol and lathosterol concentrations (Table 7). Therefore, relationships between cholesterol concentrations (corrected for creamatocrit) in breast milk and allergic outcomes were also considered. However, the odds of having eczema or wheeze in the first two years of life were not lower with higher cholesterol concentrations, nor were the odds for allergic sensitization at age two (data not shown). Hence, non-cholesterol sterols did not act as a marker for cholesterol. The reported associations can instead be attributed specifically to the non-cholesterol sterols.

Table 7. Spearman correlations between cholesterol and non-cholesterol sterols, $N = 311$.

Sterol (μmol/Lf)	Spearman's ρ Cholesterol (mmol/Lf)
Brassicasterol	−0.42 **
Campesterol [1]	0.40 **
Cholestanol	−0.45 **
Desmosterol	−0.13 *
Lanosterol	−0.07
Lathosterol	−0.04
Sitosterol	−0.37 **
Stigmasterol	−0.51 **

μmol/Lf = micromoles per liter milk fat; [1] $N = 303$; * $p < 0.05$; ** $p < 0.001$.

4. Discussion

The aim of this study was to determine the association between non-cholesterol sterols in breast milk and allergic outcomes in breastfed children in the first two years of life. We found that the odds of eczema during the first two years of life were significantly lower with higher concentrations of cholestanol, lanosterol, lathosterol, and stigmasterol in breast milk one-month postpartum. We also showed that the odds of allergic sensitization at age 2 were significantly lower with a higher concentration of campesterol in breast milk. None of the sterols were associated with wheeze during the first two years of life. Study groups (women with a conventional or 'alternative' lifestyle with regard to, e.g., child rearing practices) were not further compared, since study group did not seem to influence the reported associations. A priori, we hypothesized that exposure of the immune system to non-cholesterol sterols through breastfeeding early in life influences the maturation of the immune system and thereby prevents allergic outcomes later in life. Our results presented here support this hypothesis and are in line with previous suggestions that non-cholesterol sterols may play a role in infant health [21].

Non-cholesterol sterols in serum are known for their relationship with cholesterol metabolism [21,23,24]. This group of sterols can be divided into two subgroups, i.e., some are diet derived and considered as markers for intestinal cholesterol absorption (brassicasterol, campesterol, cholestanol, sitosterol, stigmasterol), while others are endogenously synthesized and markers for cholesterol synthesis (desmosterol, lanosterol, lathosterol). However, these two subgroups were not identified when exploratory factor analysis was performed using concentrations of these non-cholesterol sterols in breast milk (with or without correction for creamatocrit). This finding is in line with our hypothesis that sterols provided by breast milk should be considered as nutrients (and not as markers for intestinal cholesterol absorption and endogenous synthesis), which may have specific effects in the body in early life, e.g., involvement in the maturation of the immune system. In addition, studies in adults have also reported effects of non-cholesterol sterols on the immune system. Brüll and colleagues [25] used antibody production to a hepatitis A vaccine as a measure for immune function in adult asthma patients that received either plant stanols or placebo. They reported that daily intake of 4 g of plant stanols increased antibody production by 22% compared to placebo [25]. In addition, changes in serum plant stanol concentrations were positively correlated to the Th1/Th2 cytokine balance towards more Th1 activity [25]. These results, together with our current findings, indicate that consuming plant sterols and stanols may not only affect cholesterol metabolism, but may also be related to developing and sustaining immune function throughout life.

Results of studies evaluating the effect of breastfeeding on allergic outcomes in children are inconclusive, and information on the maternal diet during breastfeeding is often missing [41]. Therefore, a clear recommendation for future studies is to include data about maternal diet composition during pregnancy and breastfeeding, and/or breast milk composition. Unfortunately, no studies have related non-cholesterol sterols in breast milk and infant feeding with allergic outcomes in children, which makes it difficult to compare our study results. However, results can be compared with studies evaluating the effect of children's intake of diets or foods rich in these non-cholesterol sterols on allergic outcomes. Our results are in accordance with a review, which stated that plant-based diets and diets similar to the Mediterranean diet, which are generally rich in plant sterols, could reduce inflammation and asthma symptoms in children [42]. Another study reported a negative association between fruit and vegetable intake and allergic symptoms in children [43]. Moreover, the results of our study could be compared to studies evaluating the effects of non-cholesterol sterols on immune-related outcomes in other parts of the body, such as the gut. Van Gorp and colleagues found that intra amniotic administration of β-sitosterol and campesterol prevented gut inflammation in fetal lambs that were intra-amniotically infected with *Ureaplasma parvum* [44]. Plasma IL-6, influx of mucosal myeloperoxidase-positive cells, and intestinal damage were all lowered by the intra amniotic administration of plant sterols [44]. In addition, de Smet and colleagues showed that an acute intake of plant stanols down-regulated genes regulating T-cell functioning in the jejunum of healthy volunteers [45]. These two studies also indicate that non-cholesterol sterols are able to influence immune cell behavior, although the exact mechanisms remain unclear.

Considering our results and the studies described above, it is tempting to suggest that higher intakes of non-cholesterol sterols via breast milk would result in better health outcomes. Infant formulas sometimes also contain high concentrations of plant sterols, especially when produced with vegetable oil as fat source [46], while formula feeding has not been associated to better health outcomes compared to breastfeeding [47–49]. Claumarchirant and colleagues [46] reported that total plant sterol concentrations (sum of brassicasterol, campesterol, β-sitosterol, stigmasterol, and sitostanol) in various infant formulas ranged between 3.1–5.0 mg/100 mL (78.1–132.3 μmol/L). These concentrations are higher than those reported in our study, where the median total plant sterol concentration in breast milk (sum of brassicasterol, campesterol, sitosterol, and stigmasterol) was 1.3 μmol/L. However, desmosterol (0.2–0.4 mg/100 mL (6.2–11.1 μmol/L)) and cholesterol (1.6–5.1 mg/100 mL (0.04–0.1 mmol/L)) concentrations in infant formulas [46] were lower as compared to the concentrations we found in breast milk (desmosterol: 52.2 μmol/L; cholesterol 0.35 mmol/L). Hence, not only absolute concentrations, but also ratios between individual sterols differ between breast milk and infant formula. In more detail, the ratio between cholesterol and sitosterol concentrations in infant formula ranges from approximately 0.6 to 1.8 [46], whereas in our study the median cholesterol concentration was 500 times higher than the sitosterol concentration (350 μmol/L and 0.70 μmol/L, respectively). The lower concentration of cholesterol in infant formulas as compared to breast milk induces higher endogenous cholesterol synthesis in formula fed infants, whereas breastfed infants have a higher intestinal cholesterol absorption [50,51]. To the best of our knowledge, the bioavailability of non-cholesterol in breast milk and infant formula has not been studied. Therefore, it is currently not completely understood how serum non-cholesterol concentrations in children are affected by either breastfeeding or formula feeding.

Additionally, it is currently unknown whether the differences in sterol concentrations between breast milk and infant formula are associated with the difference in the immune responses of breastfed versus formula fed children. However, it should be kept in mind that breast milk and infant formula differ in many more aspects that could affect immune system development than solely non-cholesterol sterol and cholesterol concentrations. For example, breast milk contains human immune factors, which help forming the neonatal immune system [52]. Ultimately, the relation between early infant feeding and allergic outcomes is not fully understood. Future studies should further evaluate: (1) whether the reported effects of non-cholesterol sterols on allergic outcomes in our study can be attributed to their concentration in breast milk, (2) whether the ratio between the different sterols could also play a role, and (3) whether other components in breast milk are potentially involved in this association.

Although the associations between sterols in breast milk, eczema, and allergic sensitization were statistically significant, none of the sterols were significantly associated with wheeze in the first two years of life. There are several wheezing phenotypes, based on the age at which wheezing first occurs [53]. For example, phenotypes such as transient early wheeze and prolonged early wheeze are characterized by wheezing only in the first years of life, while wheezing disappears as the child gets older. Other phenotypes such as intermediate onset wheeze or late onset wheeze are characterized by wheezing occurring at a later age (18–42 months old). The phenotypes characterized by later onset wheeze are strongest associated with allergic outcomes later in life [53]. For this study, it means some of the children could have suffered from the phenotypes characterized by early onset wheeze, which could also be caused by viral infections [54]. It would be interesting to evaluate the association between non-cholesterol sterols in breast milk and wheezing or even asthma at a later age, and to take wheezing phenotypes into account. Unfortunately, the number of late onset wheeze and asthma cases in the subgroup with breast milk samples was insufficient to allow proper statistics. The associations between non-cholesterol sterols, wheeze, and asthma should therefore be evaluated in a larger study.

Another limitation of this study was the extensive selection of sterols and that multiple allergic outcomes were tested. The exploratory nature of this study may have increased the chance of type I errors. Therefore, data should be interpreted with care and additional studies are needed to confirm or refute our findings. In addition, future studies should consider whether there are optimal sterol concentrations in breast milk, whether ratios between different sterols in breast milk play a role in the prevention of allergic outcomes, and how breast milk composition fits into this association.

5. Conclusions

In conclusion, our data suggest that exposure to higher non-cholesterol sterol concentrations (corrected for creamatocrit) in breast milk may indeed contribute to the prevention of allergic outcomes, such as eczema and allergic sensitization at the age of two. Evidence regarding the elaborate role of sterols in human health rapidly grows and should be explored in further detail. Future studies should consider a role for breast milk composition and maternal diet during pregnancy and lactation in the association between breastfeeding and allergic outcomes in children. The effects of sterol intake via breastfeeding versus bottle feeding on allergic disease prevention should also be studied in more detail.

Author Contributions: C.T. collected the samples and data of the KOALA study. D.L. analyzed non-cholesterol sterol concentrations in breast milk samples. L.v.B. performed statistical analyses, supervised by C.T., L.v.B., C.T., R.P.M. and J.P., interpreted the data and wrote the manuscript. D.L. reviewed and edited the manuscript. All authors have read and agreed to the published version of the manuscript.

Funding: The following organizations contributed to the funding of the KOALA study up to 2 years of follow-up: Netherlands Organisation for Health Research and Development (ZonMw) project no. 2100.0090; Netherlands Asthma Foundation (grants 3.2.03.48 and 3.2.07.022); Royal Friesland Foods; Triodos Foundation; Phoenix Foundation; Raphaël Foundation; Iona Foundation; Foundation for the Advancement of Heilpedagogie (all in the Netherlands). The current collaboration project making these analyses possible is co-funded by the PPP Allowance made available by Health~Holland, Top Sector Life Sciences and Health, to stimulate public–private partnerships (project number LSHM18008).

Institutional Review Board Statement: The KOALA study was conducted according to the guidelines of the Declaration of Helsinki, and approved by the Medical Ethical Committee of Maastricht University Medical Center, Maastricht, the Netherlands (MEC 01-139 and 00-182) and the Central Committee on Research Involving Human Subjects, The Hague, the Netherlands (CCMO P01.1265L).

Informed Consent Statement: Informed consent was obtained from all subjects involved in the KOALA study.

Acknowledgments: We thank the parents and children who participated in the KOALA study.

Conflicts of Interest: The authors declare no conflict of interest.

References

1. Victora, C.G.; Bahl, R.; Barros, A.J.D.; França, G.V.A.; Horton, S.; Krasevec, J.; Murch, S.; Sankar, M.J.; Walker, N.; Rollins, N.C.; et al. Breastfeeding in the 21st century: Epidemiology, mechanisms, and lifelong effect. *Lancet* **2016**, *387*, 475–490. [CrossRef]
2. Infant and Young Child Feeding Home Page. Available online: https://www.who.int/news-room/fact-sheets/detail/infant-and-young-child-feeding (accessed on 31 March 2021).
3. Sankar, M.J.; Sinha, B.; Chowdhury, R.; Bhandari, N.; Taneja, S.; Martines, J.C.; Bahl, R. Optimal breastfeeding practices and infant and child mortality: A systematic review and meta-analysis. *Acta Paediatr.* **2015**, *104*, 3–13. [CrossRef] [PubMed]
4. Pannaraj, P.S.; Li, F.; Cerini, C.; Bender, J.M.; Yang, S.; Rollie, A.; Adisetiyo, H.; Zabih, S.; Lincez, P.J.; Bittinger, K. Association between breast milk bacterial communities and establishment and development of the infant gut microbiome. *JAMA Pediatrics* **2017**, *171*, 647–654. [CrossRef] [PubMed]
5. Parikh, N.I.; Hwang, S.-J.; Ingelsson, E.; Benjamin, E.J.; Fox, C.S.; Vasan, R.S.; Murabito, J.M. Breastfeeding in Infancy and Adult Cardiovascular Disease Risk Factors. *Am. J. Med.* **2009**, *122*, 656–663.e1. [CrossRef] [PubMed]
6. Umer, A.; Hamilton, C.; Edwards, R.A.; Cottrell, L.; Giacobbi, P.; Innes, K.; John, C.; Kelley, G.A.; Neal, W.; Lilly, C. Association between breastfeeding and childhood cardiovascular disease risk factors. *Matern. Child Health J.* **2019**, *23*, 228–239. [CrossRef]
7. Horta, B.L.; Loret De Mola, C.; Victora, C.G. Long-term consequences of breastfeeding on cholesterol, obesity, systolic blood pressure and type 2 diabetes: A systematic review and meta-analysis. *Acta Paediatr.* **2015**, *104*, 30–37. [CrossRef]
8. Wang, L.; Collins, C.; Ratliff, M.; Xie, B.; Wang, Y. Breastfeeding Reduces Childhood Obesity Risks. *Child. Obes.* **2017**, *13*, 197–204. [CrossRef]
9. Temples, H.S. Breastfeeding reduces risk of Type 2 Diabetes in the (PETS). *Nurs. Outlook* **2019**, *67*, 115. [CrossRef]
10. Matheson, M.; Allen, K.; Tang, M. Understanding the evidence for and against the role of breastfeeding in allergy prevention. *Clin. Exp. Allergy* **2012**, *42*, 827–851. [CrossRef]
11. Dogaru, C.M.; Nyffenegger, D.; Pescatore, A.M.; Spycher, B.D.; Kuehni, C.E. Breastfeeding and Childhood Asthma: Systematic Review and Meta-Analysis. *Am. J. Epidemiol.* **2014**, *179*, 1153–1167. [CrossRef]
12. Klopp, A.; Vehling, L.; Becker, A.B.; Subbarao, P.; Mandhane, P.J.; Turvey, S.; Lefebvre, D.L.; Sears, M.R.; Azad, M.B.; Daley, D.; et al. Modes of Infant Feeding and the Risk of Childhood Asthma: A Prospective Birth Cohort Study. *J. Pediatr.* **2017**, *190*, 192–199.e2. [CrossRef] [PubMed]
13. Lodge, C.J.; Tan, D.J.; Lau, M.X.Z.; Dai, X.; Tham, R.; Lowe, A.J.; Bowatte, G.; Allen, K.J.; Dharmage, S.C. Breastfeeding and asthma and allergies: A systematic review and meta-analysis. *Acta Paediatr.* **2015**, *104*, 38–53. [CrossRef] [PubMed]
14. Kull, I.; Böhme, M.; Wahlgren, C.-F.; Nordvall, L.; Pershagen, G.; Wickman, M. Breast-feeding reduces the risk for childhood eczema. *J. Allergy Clin. Immunol.* **2005**, *116*, 657–661. [CrossRef] [PubMed]
15. Van Odijk, J.; Kull, I.; Borres, M.P.; Brandtzaeg, P.; Edberg, U.; Hanson, L.Å.; Høst, A.; Kuitunen, M.; Olsen, S.F.; Skerfving, S.; et al. Breastfeeding and allergic disease: A multidisciplinary review of the literature (1966–2001) on the mode of early feeding in infancy and its impact on later atopic manifestations. *Allergy* **2003**, *58*, 833–843. [CrossRef]
16. Lyons, K.E.; Ryan, C.A.; Dempsey, E.M.; Ross, R.P.; Stanton, C. Breast milk, a source of beneficial microbes and asso-ciated benefits for infant health. *Nutrients* **2020**, *12*, 1039. [CrossRef]
17. Institute of Medicine (US) Committee on Nutritional Status During Pregnancy and Lactation. *Nutrition During Lactation*; (DC): National Academies Press: Washington, DC, USA, 1991.
18. Prentice, P.; Ong, K.K.; Schoemaker, M.H.; Van Tol, E.A.F.; Vervoort, J.; Hughes, I.A.; Acerini, C.L.; Dunger, D.B. Breast milk nutrient content and infancy growth. *Acta Paediatr.* **2016**, *105*, 641–647. [CrossRef]
19. Jenness, R. The composition of human milk. *Semin. Perinatol.* **1979**, *3*, 225–239.
20. Albrecht, C.; Huang, X.; Ontsouka, E. Cholesterol transporters in lactating and nonlactating human mammary tissue. In *Handbook of Dietary and Nutritional Aspects of Human Breast Milk*; Wageningen Academic Publishers: Wageningen, The Netherlands, 2013; p. e20895.
21. Plat, J.; Baumgartner, S.; Vanmierlo, T.; Lütjohann, D.; Calkins, K.; Burrin, D.; Guthrie, G.; Thijs, C.; Velde, A.T.; Vreugdenhil, A.; et al. Plant-based sterols and stanols in health & disease: "Consequences of human development in a plant-based environment?" *Prog. Lipid Res.* **2019**, *74*, 87–102. [CrossRef]
22. Farke, C.; Viturro, E.; Meyer, H.H.D.; Albrecht, C. Identification of the bovine cholesterol efflux regulatory protein ABCA1 and its expression in various tissues1. *J. Anim. Sci.* **2006**, *84*, 2887–2894. [CrossRef]
23. Ras, R.T.; Geleijnse, J.M.; Trautwein, E.A. LDL-cholesterol-lowering effect of plant sterols and stanols across different dose ranges: A meta-analysis of randomised controlled studies. *Br. J. Nutr.* **2014**, *112*, 214–219. [CrossRef]
24. de Jong, A.; Plat, J.; Mensink, R.P. Metabolic effects of plant sterols and stanols. *J. Nutr. Biochem.* **2003**, *14*, 362–369. [CrossRef]

25. Brüll, F.; De Smet, E.; Mensink, R.P.; Vreugdenhil, A.; Kerksiek, A.; Lütjohann, D.; Wesseling, G.; Plat, J. Dietary plant stanol ester consumption improves immune function in asthma patients: Results of a randomized, double-blind clinical trial1. *Am. J. Clin. Nutr.* **2016**, *103*, 444–453. [CrossRef] [PubMed]
26. Brüll, F.; Mensink, R.P.; Steinbusch, M.F.; Husche, C.; Lütjohann, D.; Wesseling, G.J.; Plat, J. Beneficial effects of sito-stanol on the attenuated immune function in asthma patients: Results of an in vitro approach. *PLoS ONE* **2012**, *7*, e46895. [CrossRef] [PubMed]
27. Plat, J.; Baumgartner, S.; Mensink, R.P. Mechanisms Underlying the Health Benefits of Plant Sterol and Stanol Ester Consumption. *J. AOAC Int.* **2015**, *98*, 697–700. [CrossRef] [PubMed]
28. Kay, A.B. Allergy and allergic diseases. *N. Engl. J. Med.* **2001**, *344*, 30–37. [CrossRef] [PubMed]
29. Spann, N.J.; Garmire, L.X.; McDonald, J.G.; Myers, D.S.; Milne, S.B.; Shibata, N.; Reichart, D.; Fox, J.N.; Shaked, I.; Heudobler, D.; et al. Regulated Accumulation of Desmosterol Integrates Macrophage Lipid Metabolism and Inflammatory Responses. *Cell* **2012**, *151*, 138–152. [CrossRef]
30. Bekkering, S.; Arts, R.; Novakovic, B.; Kourtzelis, I.; van der Heijden, C.; Li, Y.; Popa, C.; Ter Horst, R.; van Tuijl, J.; Netea-Maier, R.T.; et al. Metabolic Induction of Trained Immunity through the Mevalonate Pathway. *Cell* **2018**, *172*, 135–146. [CrossRef]
31. Miettinen, T.; Gylling, H.; Nissinen, M. The role of serum non-cholesterol sterols as surrogate markers of absolute cholesterol synthesis and absorption. *Nutr. Metab. Cardiovasc. Dis.* **2011**, *21*, 765–769. [CrossRef]
32. Kummeling, I.; Thijs, C.; Penders, J.; Snijders, B.E.P.; Stelma, F.; Reimerink, J.; Koopmans, M.; Dagnelie, P.C.; Huber, M.; Jansen, M.C.J.F.; et al. Etiology of atopy in infancy: The KOALA Birth Cohort Study. *Pediatr. Allergy Immunol.* **2005**, *16*, 679–684. [CrossRef]
33. Ygberg, S.; Nilsson, A. The developing immune system—From foetus to toddler. *Acta Paediatr.* **2012**, *101*, 120–127. [CrossRef]
34. Thijs, C.; Müller, A.; Rist, L.; Kummeling, I.; Snijders, B.; Huber, M.; Van Ree, R.; Simões-Wüst, A.; Dagnelie, P.; Van Den Brandt, P. Fatty acids in breast milk and development of atopic eczema and allergic sensitisation in infancy. *Allergy* **2011**, *66*, 58–67. [CrossRef] [PubMed]
35. Akkerdaas, J.H.; Wensing, M.; Asero, R.; Rivas, M.F.; Knulst, A.C.; Bolhaar, S.; Hefle, S.L.; Aalberse, R.C.; Van Ree, R. IgE Binding to Pepsin-Digested Food Extracts. *Int. Arch. Allergy Immunol.* **2005**, *138*, 203–208. [CrossRef] [PubMed]
36. Kummeling, I.; Thijs, C.; Huber, M.; van de Vijver, L.P.L.; Snijders, B.E.P.; Penders, J.; Stelma, F.; van Ree, R.; Brandt, P.A.V.D.; Dagnelie, P.C. Consumption of organic foods and risk of atopic disease during the first 2 years of life in the Netherlands. *Br. J. Nutr.* **2008**, *99*, 598–605. [CrossRef] [PubMed]
37. Rist, L.; Mueller, A.; Barthel, C.; Snijders, B.; Jansen, M.; Simões-Wüst, A.P.; Huber, M.; Kummeling, I.; von Mandach, U.; Steinhart, H.; et al. Influence of organic diet on the amount of conjugated linoleic acids in breast milk of lactating women in the Netherlands. *Br. J. Nutr.* **2007**, *97*, 735–743. [CrossRef] [PubMed]
38. Lucas, A.; A Gibbs, J.; Lyster, R.L.; Baum, J.D. Creamatocrit: Simple clinical technique for estimating fat concentration and energy value of human milk. *BMJ* **1978**, *1*, 1018–1020. [CrossRef]
39. Chan, M.K.; Huang, Y.C.; Varghese, Z.; Baillod, R.A.; Moorhead, J.F. Predicting plasma concentrations of common biochemical values from residual peritoneal fluid in patients on peritoneal dialysis. *Br. Med. J.* **1978**, *1*, 1670. [CrossRef]
40. Thelen, K.M.; Laaksonen, R.; Päivä, H.; Lehtimäki, T.; Lütjohann, D. High-Dose Statin Treatment Does Not Alter Plasma Marker for Brain Cholesterol Metabolism in Patients with Moderately Elevated Plasma Cholesterol Levels. *J. Clin. Pharmacol.* **2006**, *46*, 812–816. [CrossRef]
41. Ferrante, G.; Carta, M.; Montante, C.; Notarbartolo, V.; Corsello, G.; Giuffrè, M. Current Insights on Early Life Nutrition and Prevention of Allergy. *Front. Pediatrics* **2020**, *8*, 448. [CrossRef]
42. Alwarith, J.; Kahleova, H.; Crosby, L.; Brooks, A.; Brandon, L.; Levin, S.M.; Barnard, N.D. The role of nutrition in asthma prevention and treatment. *Nutr. Rev.* **2020**, *78*, 928–938. [CrossRef]
43. Cepeda, A.M.; The ISAAC Phase III Latin America Group; Thawer, S.; Boyle, R.J.; Villalba, S.; Jaller, R.; Tapias, E.; Segura, A.M.; Villegas, R.; Larsen, V.G.; et al. Diet and Respiratory Health in Children from 11 Latin American Countries: Evidence from ISAAC Phase III. *Lung* **2017**, *195*, 683–692. [CrossRef]
44. van Gorp, C.; de Lange, I.H.; Spiller, O.B.; Dewez, F.; Cillero, P.B.; Heeren, R.; Kessels, L.; Kloosterboer, N.; van Gemert, W.G.; Beeton, M.L. Protection of the ovine fetal gut against ureaplasma-induced chorioamnionitis: A potential role for plant sterols. *Nutrients* **2019**, *11*, 968. [CrossRef] [PubMed]
45. De Smet, E.; Mensink, R.P.; Boekschoten, M.V.; de Ridder, R.; Germeraad, W.T.; Wolfs, T.G.; Plat, J. An acute intake of plant stanol esters alters immune-related pathways in the jejunum of healthy volunteers. *Br. J. Nutr.* **2015**, *113*, 794–802. [CrossRef] [PubMed]
46. Claumarchirant, L.; Matencio, E.; Sanchez-Siles, L.M.; Alegría, A.; Lagarda, M.J. Sterol composition in infant formulas and estimated intake. *J. Agric. Food Chem.* **2015**, *63*, 7245–7251. [CrossRef] [PubMed]
47. Greer, F.R.; Sicherer, S.H.; Burks, A.W.; Committee on Nutrition; Section on Allergy and Immunology. The Effects of Early Nutritional Interventions on the Development of Atopic Disease in Infants and Children: The Role of Maternal Dietary Restriction, Breastfeeding, Hydrolyzed Formulas, and Timing of Introduction of Allergenic Complementary Foods. *Pediatrics* **2019**, *143*, e20190281. [CrossRef] [PubMed]
48. Laubereau, B.; Brockow, I.; Zirngibl, A.; Koletzko, S.; Gruebl, A.; von Berg, A.; Filipiak-Pittroff, B.; Berdel, D.; Bauer, C.P.; Reinhardt, D. Effect of breast-feeding on the development of atopic dermatitis during the first 3 years of life—Results from the GINI-birth cohort study. *J. Pediatrics* **2004**, *144*, 602–607. [CrossRef]

49. Friedman, N.J.; Zeiger, R. The role of breast-feeding in the development of allergies and asthma. *J. Allergy Clin. Immunol.* **2005**, *115*, 1238–1248. [CrossRef]
50. Woollett, L.A.; Heubi, J.E. *Fetal and Neonatal Cholesterol Metabolism*; MDText.com, Inc.: South Dartmouth, MA, USA, 2016.
51. Delplanque, B.; Gibson, R.; Koletzko, B.; Lapillonne, A.; Strandvik, B. Lipid quality in infant nutrition: Current knowledge and future opportunities. *J. Pediatric Gastroenterol. Nutr.* **2015**, *61*, 8. [CrossRef]
52. Le Doare, K.; Holder, B.; Bassett, A.; Pannaraj, P.S. Mother's milk: A purposeful contribution to the development of the infant microbiota and immunity. *Front. Immunol.* **2018**, *9*, 361. [CrossRef]
53. Henderson, J.; Granell, R.; Heron, J.; Sherriff, A.; Simpson, A.; Woodcock, A.; Strachan, D.P.; Shaheen, S.O.; Sterne, J.A. Associations of wheezing phenotypes in the first 6 years of life with atopy, lung function and airway responsiveness in mid-childhood. *Thorax* **2008**, *63*, 974–980. [CrossRef]
54. Granell, R.; Sterne, J.A.; Henderson, J. Associations of different phenotypes of wheezing illness in early childhood with environmental variables implicated in the aetiology of asthma. *PLoS ONE* **2012**, *7*, e48359. [CrossRef]

nutrients MDPI

Article

Effects of Diet-Induced Weight Loss on Plasma Markers for Cholesterol Absorption and Synthesis: Secondary Analysis of a Randomized Trial in Abdominally Obese Men

Sultan Mashnafi [1,2], Jogchum Plat [1], Ronald P. Mensink [1], Peter J. Joris [1], Yvo H. A. M. Kusters [3], Alfons J. H. M. Houben [3], Coen D. A. Stehouwer [3], Casper G. Schalkwijk [3] and Sabine Baumgartner [1,*]

1 Department of Nutrition and Movement Sciences, NUTRIM School of Nutrition and Translational Research in Metabolism, Maastricht University Medical Center, 6200 MD Maastricht, The Netherlands; s.mashnafi@maastrichtuniversity.nl (S.M.); j.plat@maastrichtuniversity.nl (J.P.); r.mensink@maastrichtuniversity.nl (R.P.M.); p.joris@maastrichtuniversity.nl (P.J.J.)
2 Department of Medical Basic Sciences, Faculty of Applied Medical Sciences, AlBaha University, AlBaha 65779-7738, Saudi Arabia
3 Department of Internal Medicine, CARIM School for Cardiovascular Diseases, Maastricht University Medical Center, 6200 MD Maastricht, The Netherlands; ivo.kusters@mumc.nl (Y.H.A.M.K.); b.houben@maastrichtuniversity.nl (A.J.H.M.H.); cda.stehouwer@mumc.nl (C.D.A.S.); c.schalkwijk@maastrichtuniversity.nl (C.G.S.)
* Correspondence: sabine.baumgartner@maastrichtuniversity.nl; Tel.: +31-43-3881305

Abstract: Cross-sectional studies have shown that obesity is associated with lower intestinal cholesterol absorption and higher endogenous cholesterol synthesis. These metabolic characteristics have also been observed in patients with type 2 diabetes, metabolic syndrome, steatosis or cholestasis. The number of intervention studies evaluating the effect of weight loss on these metabolic characteristics is, however, limited, while the role of the different fat compartments has not been studied into detail. In a randomized trial, abdominally obese men (N = 54) followed a 6-week very low caloric (VLCD) diet, followed by a 2 week weight-maintenance period. Non-cholesterol sterols were measured at baseline and after 8 weeks, and compared to levels in lean participants (N = 25). After weight loss, total cholesterol (TC)-standardized cholestanol levels increased by 0.18 μmol/mmol ($p < 0.001$), while those of campesterol and lathosterol decreased by 0.25 μmol/mmol ($p < 0.05$) and 0.39 μmol/mmol ($p < 0.001$), respectively. Moreover, after weight loss, TC-standardized lathosterol and cholestanol levels were comparable to those of lean men. Increases in TC-standardized cholestanol after weight loss were significantly associated with changes in waist circumference ($p < 0.01$), weight ($p < 0.001$), BMI ($p < 0.001$) and visceral fat ($p < 0.01$), but not with subcutaneous and intrahepatic lipids. In addition, cross-sectional analysis showed that visceral fat fully mediated the association between BMI and TC-standardized cholestanol levels. Intrahepatic lipid content was a partial mediator for the association between BMI and TC-standardized lathosterol levels. In conclusion, diet-induced weight loss decreased cholesterol synthesis and increased cholesterol absorption. The increase in TC-standardized cholestanol levels was not only related to weight loss, but also to a decrease in visceral fat volume. Whether these metabolic changes ameliorate other metabolic risk factors needs further study.

Keywords: diet-induced weight loss; cholesterol absorption; cholesterol synthesis; non-cholesterol sterols; visceral fat; subcutaneous fat; intrahepatic lipid; cholesterol precursors; plant sterols

Citation: Mashnafi, S.; Plat, J.; Mensink, R.P.; Joris, P.J.; Kusters, Y.H.A.M.; Houben, A.J.H.M.; Stehouwer, C.D.A.; Schalkwijk, C.G.; Baumgartner, S. Effects of Diet-Induced Weight Loss on Plasma Markers for Cholesterol Absorption and Synthesis: Secondary Analysis of a Randomized Trial in Abdominally Obese Men. *Nutrients* **2022**, *14*, 1546. https://doi.org/10.3390/nu14081546

Academic Editor: Frans Stellaard

Received: 25 February 2022
Accepted: 5 April 2022
Published: 8 April 2022

1. Introduction

Obesity and its associated comorbidities are a major health problem worldwide. An increased visceral fat content, a characteristic of people with abdominal obesity, is clinically the most important form of obesity [1]. Abdominal obesity is strongly associated with insulin resistance, dyslipidemia and hypertension [2], which all contribute to an increased cardiovascular disease risk [1,3]. Recently, we have suggested that overweight and obesity are associated with lower intestinal cholesterol absorption and higher endogenous cholesterol synthesis [4]. These metabolic characteristics have also been observed in patients with type 2 diabetes, metabolic syndrome, steatosis or cholestasis [4]. However, these reported cross-sectional associations do not necessarily imply that weight loss will lead to an increase in cholesterol absorption and a decrease in cholesterol synthesis. To assess whether there is a causal association between weight loss with cholesterol absorption and synthesis, well-controlled intervention studies are needed.

To evaluate changes in cholesterol absorption and synthesis in humans, serum non-cholesterol sterols are frequently used as markers [5]. The cholesterol precursors desmosterol and lathosterol reflect endogenous cholesterol synthesis, while the non-cholesterol sterols sitosterol, campesterol and cholestanol reflect fractional intestinal cholesterol absorption [6]. Using these markers, earlier intervention studies in obese individuals with type 2 diabetes [7,8] or metabolic syndrome [9–11] have indeed suggested that diet-induced weight loss increases cholesterol absorption and decreases cholesterol synthesis. However, relations with fat distribution or the different fat compartments, which behave metabolically different [12–14], were not studied.

So far, studies evaluating the effects of diet-induced weight loss on cholesterol metabolism in apparently healthy individuals with abdominal obesity are limited. In addition, in most studies that did evaluate these effects, a no-weight loss control group was not included [7–9,11]. Furthermore, results have not been compared to those of normal-weight volunteers as a reference population in all previous studies. Finally, in some studies, body weight had not reached a new steady state and participants still had a negative energy balance when serum non-cholesterol sterol concentrations were analyzed after weight loss [8]. Therefore, the aim of this study was to examine the effect of a 6-week diet-induced weight-loss program, followed by a 2-week weight stable period, on markers of cholesterol absorption and synthesis in apparently healthy individuals with abdominal obesity. Results before and after the weight loss in the new steady energy balance were compared to those of normal-weight men. In addition, we examined the relations between changes in markers for cholesterol absorption and synthesis with changes in fat distribution and different fat compartments (visceral fat, subcutaneous fat and intrahepatic lipid) to assess whether changes in aforementioned compartments play a role in cholesterol metabolism characteristics after weight loss. Finally, we used cross-sectional mediation analysis to examine the mediating role of each fat compartment on the relationship between body mass index (BMI) as well as markers for cholesterol absorption and synthesis.

2. Materials and Methods

2.1. Participants and Study Design

Details of this study have been published before [15]. Briefly, Caucasian, apparently healthy male subjects aged between 18–56 years were eligible to participate when they met the following inclusion criteria: stable body weight (±3 kg in the last 3 months); no diabetes; no active cardiovascular diseases; no inflammatory diseases; no use of antihypertensive medication; no drug or alcohol abuse; no use of medication known to affect lipid or glucose metabolism and no participation in another biomedical trial in the previous 30 days. Both normal-weight men and abdominally obese men participated in this study. Normal-weight subjects had a waist circumference below 94 cm, while this was between 102 and 110 cm in the abdominal obese group. Upon inclusion, the abdominally obese men were randomized to the diet-induced weight-loss group or the no-weight-loss control group, as described previously [15]. The participants in the weight-loss group consumed, under strict guidance, a very-low-caloric diet (VLCD; Modifast; Nutrition et Santé Benelux, Breda, The Netherlands) for 4 to 5 weeks. The aim was to achieve a waist circumference below 102 cm, which is the cut-off value used for the diagnosis of metabolic syndrome [16]. Daily caloric intake of the VLCD was 2.1 MJ (500 kcal) and the content of minerals and vitamins met the Dutch dietary guidelines. Participants in the weight-loss group consumed three VLCD formulas, which had to be dissolved in water, on a daily basis. Hereafter, in weeks 5 and 6, participants consumed three meals of a mixed solid caloric-restricted diet providing 4.2 MJ/day (1000 kcal) daily for one to two weeks. Again, the composition of this diet met the Dutch dietary guidelines. In week 7 and 8, weight maintenance was achieved by providing weekly menus which were adjusted to individual energy requirements. Men allocated to the no-weight loss group were asked to maintain their habitual diet, physical activity level and alcohol consumption throughout the entire study duration. A total of 79 men were included; 25 men were a normal weight (waist circumference < 94 cm) and 54 men were abdominally obese (waist circumference 102–110 cm). One man dropped out before randomization and thus, 53 of the abdominally obese men were assigned to the weight-loss group (N = 26) or no-weight-loss control group (N = 27). Written informed consent was obtained from all participants before the start of the study. The study protocol was approved by the Medical Ethical Committee of Maastricht University Medical Center (METC 12-30-40) and registered at ClinicalTrials.gov (NCT01675401).

2.2. Anthropometrics, Fat Distribution and Compartments

Information about overall and abdominal obesity was obtained through measurements of weight, body mass index, waist circumference, hip circumference and waist to hip ratio, as previously described [15]. The volume of the visceral fat and subcutaneous fat compartments, as well as the intrahepatic lipid content, was measured by magnetic resonance imaging (MRI) [17].

2.3. Blood Sampling

Venous blood samples were drawn after an overnight fast at baseline and in week 8. Heparin vacutainer tubes were centrifuged at 1300× g for 15 min at 4 °C to collect plasma samples. Serum tubes were centrifuged at 1300× g for 15 min at 21 °C to collect serum samples. Aliquots were stored at −80 °C until analyzed at the end of the study.

2.4. Serum Lipid Analysis

Serum total cholesterol (TC) (CHOD-PAP method; Roche Diagnostic, Mannheim, Germany), high-density lipoprotein cholesterol (HDL-C) (CHOD-PAP method; Roche Diagnostic, Mannheim, Germany), and triglyceride (TG) concentrations—corrected for glycerol levels—were analyzed enzymatically (GPO-Tinder; Sigma-Aldrich Corp., St. Louis, MO, USA). Serum low-density lipoprotein cholesterol (LDL-C) concentrations were calculated using the Friedewald equation [18].

2.5. Non-Cholesterol Sterol Analysis

Sterols were measured by gas chromatography equipped with a flame ionization detector (GC-FID) (Hewlett Packard 6890 plus), and with a capillary column (DB-XLB 30 m × 0.25 mm i.d. × 0.25 µm; Agilent Technologies, Amstelveen, Netherlands). Extraction of cholesterol and non-cholesterol sterols was performed based on Mackay et al. [19]. Briefly, a 100 µL plasma sample was saponified with 1 ml of 90% ethanolic sodium hydroxide for 1 h at 60 °C. 5α-cholestane and epicoprostanol were used as internal standards. After two rounds of cyclohexane extraction, samples were derivatized with 30 µL of TMS reagent (pyridine, hexamethyldisilazane and trimethylchlorosilane (9:3:1, $v/v/v$)). Samples were injected into GC-FID; cholesterol and non-cholesterol sterol peaks were integrated (OpenLab CDS ChemStation Edition; Agilent Technologies, Santa Clara, CA, USA) and their concentrations were calculated relative to the internal standard 5α-cholestane. Non-cholesterol sterol concentrations were standardized for cholesterol concentrations, as determined within the same GC run and expressed as µmol/mmol cholesterol.

2.6. Statistics Analyses

Data are presented as means ± standard deviations (SD) unless indicated otherwise. Normality of the data was assessed using the Kolmogorov–Smirnov test. The differences at baseline between normal weight and abdominally obese men were compared with an independent t-test. A one-way ANCOVA using baseline concentrations as a covariate was used to examine differences in changes between the diet-induced weight loss and no-weight-loss control treatments. An independent t-test was also used to compare differences between the normal-weight men and the abdominally obese men after weight loss. Linear regression analysis was used to examine cross-sectional relations between cholesterol absorption or synthesis markers with anthropometric measures at baseline and with changes after weight loss. Cross-sectionally, we examined whether relationships between BMI (independent variable) with cholesterol absorption or synthesis markers (dependent variables) were mediated by visceral fat, subcutaneous fat or intrahepatic lipids (potential mediators). For this, the PROGRESS plug-in for SPSS version 4.0 (A.F. Hayes, Ohio State University, Columbus, OH, USA) was used (model 4). A p-value < 0.05 was considered statistically significant. All data were analyzed using SPSS versions 25.0 and 27.0 for Mac (SPSS Inc., Chicago, IL, USA).

3. Results

3.1. Clinical Characteristics of Study Participants

A full consort flow diagram is shown in Figure S1. Plasma samples of 25 normal-weight and 53 abdominally obese men were used for measurements at baseline. One participant in the weight-loss group was excluded due to study protocol violations, and three participants dropped out for reasons as indicated previously [15]. In the end, 23 men in the diet-induced weight loss group and 26 men in the no-weight-loss group completed the study. The characteristics of all participants at baseline have been described previously [15]. Briefly, as shown in Table 1, the median age was comparable between normal weight and abdominally obese men. As expected, BMI, waist circumference, waist to hip ratio, subcutaneous fat, visceral fat and intrahepatic lipid contents were higher in the abdominally obese men compared to those with normal weight. At baseline, serum LDL-C concentrations and plasma TC-standardized levels of the cholesterol synthesis marker lathosterol were higher in the abdominally obese men compared to the normal-weight men. On the other hand, TC-standardized levels of all three cholesterol-absorption markers, campesterol, sitosterol and cholestanol, were lower in the abdominally obese men (all $p < 0.05$).

3.2. Effect of Weight Loss

In the abdominally obese participants allocated to the diet-induced weight-loss group, mean body weight decreased by 10.3 kg (95% CI: −11.4, −9.2 kg; $p < 0.001$), waist circumference by 11.0 cm (−9.9, −12.1 cm; $p < 0.001$), subcutaneous fat by 0.81 L (−0.93, −0.69 L; $p < 0.001$), visceral fat by 0.85 L (−1.0, −0.67 L; $p < 0.001$) and intrahepatic lipid content by −5.80% (−6.58, −5.02%; $p < 0.001$) compared with the no-weight-loss control group.

Serum LDL-C and triglycerides concentrations were significantly reduced (all $p < 0.001$) in abdominally obese men after 8 weeks of diet-induced weight loss as compared with the no-weight loss control treatment group, as shown in Table 1. HDL concentrations did not differ between two treatment groups after 8 weeks. Compared with the normal-weight group, abdominally obese men had comparable values for serum LDL-C and triglycerides and HDL concentrations at the end of the dietary weight loss period.

Table 1. Anthropometric characteristics and plasma cholesterol and non-cholesterol concentrations of normal weight and abdominally obese men at baseline and after 8 weeks with diet-induced weight loss or no-weight-loss control treatment.

	Normal-Weight Group (n = 25)	Weight-Loss Group [1] (n = 23)		Non-Weight-Loss Group [1] (n = 26)		
	Baseline [1,2]	Baseline	After 8 Weeks	Baseline	After 8 Weeks	Treatment Effect [3]
Age (year)	53.7 (25.0–61.6)	52.4 (46.8–61.7)		52.0 (45.4–61.1)		
Body weight (kg)	74.9 ± 8.3 $^{\#\#\#}$	98.2 ± 8.1	88.2 ± 7.6	95.9 ± 8.9	96.4 ± 9.2	−10.3 (−11.4, −9.2) ***
BMI (kg/m^2)	23.3 ± 1.8 $^{\#\#\#}$	30.2 ± 1.5	27.1 ± 1.3	29.9 ± 2.5	30.0 ± 2.5	−3.1 (−3.4, −2.8) ***
Waist circumference (cm)	84.9 ± 6.3 $^{\#\#\#}$	106.8 ± 3.4	95.9 ± 4.2	106.2 ± 3.8	106.3 ± 4.2	−11.0 (−12.1,−9.9) ***
Hip circumference (cm)	96.6 ± 4.2	108.1 ± 4.4	102.3 ± 4.0	107.2 ± 5.9	107.2 ± 6.4	−5.8 (−6.5, −5.0) ***
Waist to hip ratio	0.88 ± 0.05	0.99 ± 0.03	0.94 ± 0.04	0.99 ± 0.05	0.99 ± 0.05	−0.05 (−0.06, −0.04) ***
Visceral fat (L) [4]	0.89 ± 0.42	2.17 ± 0.64	1.44 ± 0.51	2.53 ± 0.75	2.62 ± 0.85	−0.85 (−1.0, −0.67) ***
Subcutaneous fat (L) [4]	1.45 ± 0.51	3.23 ± 0.64	2.44 ± 0.54	2.92 ± 0.81	2.98 ± 0.81	−0.81 (−0.93, −0.69) ***
Intrahepatic lipid (%) [4,5]	3.43 (3.14–3.69)	4.21 (3.59–6.53)	3.54 (3.08–4.19)	5.34 (4.33–8.31)	6.31 (4.56–9.45)	−0.18 (−0.25, −0.12) ***
LDL-cholesterol (mmol/L)	2.80 ± 0.71 $^{\#\#\#}$	3.67 ± 1.03	3.04 ± 0.88	3.70 ± 0.89	3.48 ± 0.77	−0.51 (−0.76, −0.25) ***
HDL-cholesterol (mmol/L)	1.26 ± 0.27 $^{\#}$	1.14 ± 0.16	1.13 ± 0.21	1.09 ± 0.24	1.11 ± 0.26	−0.02 (−0.11, 0.06)
Triglycerides (mmol/L)	1.01 ± 0.48 $^{\#\#\#}$	1.63 ± 0.87	1.19 ± 0.54	1.87 ± 0.77	1.92 ± 0.79	−0.60 (−0.89, −0.30) ***
Total cholesterol (mmol/L) †	4.02 ± 0.69 $^{\#\#\#}$	4.89 ± 0.99	4.15 ± 0.86	5.03 ± 0.78	4.87 ± 0.67	−0.62 (−0.90, −0.35) ***
Campesterol (µmol/mmol cholesterol)	2.39 ± 1.02 $^{\#\#}$	1.70 ± 0.56	1.54 ± 0.38 $^{\#\#}$	1.74 ± 0.64	1.83 ± 0.61	−0.25 (−0.43, −0.07) **
Sitosterol (µmol/mmol cholesterol)	1.55 ± 0.70 $^{\#\#}$	1.08 ± 0.27	1.06 ± 0.19 $^{\#}$	1.12 ± 0.40	1.13 ± 0.35	−0.03 (−0.12, 0.04)
Cholestanol (µmol/mmol cholesterol)	1.53 ± 0.27 $^{\#\#\#}$	1.27 ± 0.21	1.45 ± 0.24	1.27 ± 0.27	1.27 ± 0.27	0.18 (0.19, 0.25) ***
Lathosterol (µmol/mmol cholesterol)	1.13 ± 0.46 $^{\#\#}$	1.47 ± 0.26	1.19 ± 0.24	1.46 ± 0.39	1.59 ± 0.49	−0.39 (−0.55, −0.24) ***

[1] Values expressed as means ± SD or medians (25–75 percentiles). [2] Values are significantly different from abdominally obese participants (n = 49) (independent t-test): $^{\#}$ $p < 0.05$, $^{\#\#}$ $p < 0.01$, $^{\#\#\#}$ $p < 0.001$. Significantly different from normal weight group (independent t-test). [3] Values are differences in changes (95% CIs) between treatment groups obtained from one factor ANCOVA with baseline values as a covariate: ** $p < 0.05$, *** $p < 0.001$. [4] Data available from normal weight participants (n = 24). [5] Log-transformed data. † Obtained from GC-FID run.

TC-standardized plasma campesterol levels were significantly reduced after weight loss (−0.25 µmol/mmol cholesterol (95% CI: −0.43, −0.07 µmol/mmol cholesterol; $p < 0.05$)), while TC-standardized sitosterol levels remained unchanged. In contrast to campesterol, TC-standardized plasma cholestanol levels were significantly increased by 0.18 µmol/mmol cholesterol (95% CI: 0.19, 0.25 µmol/mmol cholesterol; $p < 0.001$). After 8 weeks, TC-standardized campesterol and sitosterol levels remained lower in abdominally obese subjects that lost weight as compared to normal-weight subjects ($p < 0.001$ and $p < 0.05$, respectively), while TC-standardized cholestanol levels were comparable between normal-weight and obese participants after attaining weight loss. Diet-induced weight loss significantly reduced TC-standardized lathosterol levels (−0.39 µmol/mmol cholesterol (95% CI: −0.55, −0.24 µmol/mmol cholesterol; $p < 0.001$)). After weight loss, TC-standardized lathosterol levels were comparable between the normal-weight and obese participants.

3.3. *Associations between Anthropometrics, Fat Distribution and Fat Compartments with Cholesterol Absorption and Synthesis Markers*

Cross-sectional analysis including abdominally obese and normal-weight men at baseline showed significant relationships between markers for cholesterol absorption and synthesis with all anthropometric markers, fat distribution and fat compartments (weight, body mass index, waist circumference, hip circumference, waist to hip ratio, visceral fat, subcutaneous fat and intrahepatic lipid content; all $p < 0.05$) (Table S1). The relation between changes in markers for cholesterol absorption and synthesis with changes in these variables is shown in Table 2. Changes in TC-standardized cholestanol levels after diet-induced weight loss were significantly associated with changes in waist circumference ($p < 0.01$), weight ($p < 0.001$), BMI ($p < 0.001$), hip circumference ($p < 0.05$) and visceral fat ($p < 0.01$). Changes in TC-standardized sitosterol levels were only significantly related to changes in body weight ($p < 0.05$). Changes in TC-standardized campesterol and lathosterol levels with changes in anthropometric measures or intrahepatic lipid were not significantly related.

Table 2. Results of linear regression analyses to investigate the relation between changes in cholesterol absorption and synthesis markers with changes in anthropometric measures, fat distribution and IHL after weight loss intervention ($n = 23$).

	Cholesterol Absorption						Cholesterol Synthesis	
	ΔCholestanol		ΔCampesterol		ΔSitosterol		ΔLathosterol	
	B	95% CI	B	95% CI	B	95% CI	B	95% CI
ΔBW	−0.047	(−0.068, −0.025) ***	0.063	(−0.004, 0.130)	0.030	(0.001, 0.058) *	0.011	(−0.039, 0.060)
ΔBMI	−0.149	(−0.223, −0.074) ***	0.203	(−0.020, 0.427)	0.087	(−0.010, 0.184)	0.044	(−0.119, 0.207)
ΔWaist	−0.036	(−0.069, −0.002) **	0.020	(−0.069, 0.109)	0.009	(−0.029, 0.048)	0.027	(−0.032, 0.086)
ΔHip	−0.043	(−0.085, 0.000) *	0.070	(−0.037, 0.177)	0.029	(−0.018, 0.075)	0.005	(−0.070, 0.081)
ΔWaist:Hip	−1.329	(−5.557, 2.899)	−2.827	(−13.046, 7.391)	−0.999	(−5.437, 3.438)	3.867	(−2.900, 10.634)
ΔVF	−0.246	(−0.422, −0.069) **	0.083	(−0.418, 0.585)	−0.032	(−0.250, 0.185)	0.074	(−0.266, 0.414)
ΔSF	−0.066	(−0.362, 0.229)	−0.194	(−0.906, 0.517)	−0.049	(−0.359, 0.260)	0.101	(−0.383, 0.585)
ΔIHL †	0.252	(−0.252, 0.755)	0.376	(−0.857, 1.609)	0.314	(−0.206, 0.834)	−0.280	(−1.115, 0.555)

ΔBW—changes in body weight; ΔBMI—changes in body mass index; ΔWaist—changes in waist circumference; ΔHip—changes in hip circumference; ΔWaist: Hip—changes in waist to hip ratio; ΔVT—changes in visceral fat; ΔST—changes in subcutaneous fat; ΔIHL—changes in intrahepatic lipid content. Significant relationships: * $p < 0.05$, ** $p < 0.01$, *** $p < 0.001$. † Log transformed.

The effect of BMI on TC-standardized cholestanol levels was fully mediated by visceral fat (percentage of mediated effect: −52.9%; bootstrapped 95% CI: −74.0% to −5.5%) and the direct effect of BMI on TC-standardized cholestanol levels was no longer significant ($p > 0.05$) (Figure 1). In addition, the effect of BMI on TC-standardized lathosterol levels was partially mediated by intrahepatic lipid content (34.9%; bootstrapped 95% CI: 10.0% to 44.1%), and BMI had still a significant effect on TC-standardized lathosterol levels ($p < 0.05$).

Subcutaneous fat neither fully nor partially mediated the associations between BMI and markers of cholesterol absorption and synthesis.

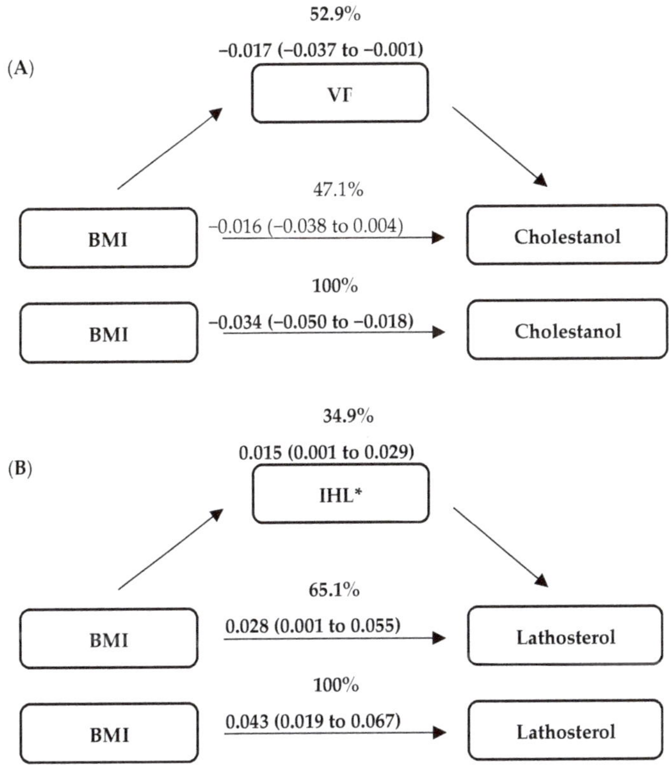

Figure 1. Mediation models of cross-sectional analyses at baseline (n = 73) for effects of each mediator on the relationships between BMI (kg/m^2) and markers of cholesterol absorption (**A**) and synthesis (**B**), expressed in µmol/mmol cholesterol. Data are presented as B (bootstrapped 95% CI). Bold figures indicated for significant effects. VT = visceral fat; IHL = intrahepatic lipid content. * Log-transformed data.

4. Discussion

Diet-induced weight loss reduced levels of TC-standardized campesterol and lathosterol and increased those of TC-standardized cholestanol. After weight loss, TC-standardized lathosterol and cholestanol levels of the (previously) abdominally obese men were comparable to those of normal-weight men. Interestingly, increases in TC-standardized cholestanol levels after weight loss were associated with decreases in waist circumference, BMI, body weight, hip circumference and visceral fat, but not intrahepatic fat and subcutaneous fat volume. Cross-sectionally, visceral fat was a full mediator for the association between BMI and TC-standardized cholestanol levels, while intrahepatic lipid content was a partial mediator for the association between BMI and TC-standardized lathosterol levels.

Our finding of a reduction in endogenous cholesterol synthesis after weight loss (10.3 kg) is in line with earlier studies. A decrease in cholesterol synthesis after weight loss of 6 kg was also observed in a study with apparently healthy obese men, who consumed a hypocaloric diet for 14 weeks followed by a 2 weeks isocaloric diet period [20]. In three studies in obese subjects with metabolic syndrome, cholesterol synthesis also decreased after dietary weight loss of 13 kg, 6 kg and 10 kg, respectively [9–11]. Simonen et al. conducted two weight-loss studies in obese, type 2 diabetic patients. Lathosterol levels tended to decrease after a diet-induced weight loss of 15 kg in 3 months [8], and weight loss of 6 kg resulted in a significant decrease in lathosterol levels after a comparable period immediately followed by a weight-stable period up to 2 years [7].

For cholesterol absorption markers, we observed that after weight loss TC-standardized cholestanol levels increased, TC-standardized campesterol levels decreased and TC-standardized sitosterol levels did not change. The question is how these apparent discrepancies for the three different non-cholesterol sterol markers reflecting intestinal cholesterol absorption can be explained. The major diet-derived plant sterols are campesterol and sitosterol [21]. As the diet of the participants in the weight-loss program was different before and after the intervention period, plasma plant sterol levels may also have changed due to different dietary habits and not only due to changes in intestinal cholesterol absorption. Therefore, it can be debated whether TC-standardized plasma campesterol and sitosterol levels truly reflect intestinal cholesterol absorption when major dietary changes are evident. In this particular situation, TC-standardized plasma cholestanol levels may be a better marker for intestinal cholesterol absorption, as cholestanol levels in the diet are very low [22]. We therefore conclude—based on the increase in TC-standardized cholestanol levels—that diet-induced weight loss increased intestinal cholesterol absorption. This conclusion is in line with the study by Simonen et al. [8] that measured cholestanol concentrations after weight loss in type 2 diabetic subjects.

So far, only a few studies have reported effects of diet-induced weight loss on TC-standardized campesterol levels, and findings are inconsistent. In two studies with type 2 diabetic patients, a decrease of about 6 kg induced by 3 months of very-low-energy diet or low-energy diet increased TC-standardized campesterol levels [7], while a trend for a decrease in these levels was found after a reduction of 15 kg induced by a very-low-energy diet virtually free of cholesterol, cholestanol and plant sterols for 3 months [8]. In a third study, weight loss of nearly 10 kg induced by 20 weeks of a free-living diet with a 500 kcal deficiency in daily energy intake, followed by 5 weeks of Mediterranean diet under an isoenergetic, weight-stabilizing period tended to increase total plant sterols levels (campesterol + sitosterol) in obese men with metabolic syndrome compared with a Mediterranean diet in absence of weight reduction [11]. Chan et al. found that campesterol levels decreased in obese men with insulin resistance after consumption of a hypocaloric diet for 16 weeks followed by a 6-week weight-maintaining period [9]. Taken together, studies on campesterol levels after diet-induced weight loss are conflicting. As discussed above, changes in TC-standardized campesterol levels may have been confounded by changes in dietary composition, and therefore may not truly reflect changes in intestinal cholesterol absorption. Information about dietary intake of plant sterols was only reported in two studies; one reported the total plant sterols content in the Mediterranean diet was higher than the North American control diet [11], while the other study used a diet formula free of cholesterol, cholestanol and plant sterols [8]. The total plant sterol level tended to increase in the former study, while a trend of decreased campesterol and sitosterol levels was demonstrated in the latter study. These observations suggest that circulating sitosterol and campesterol concentrations reflect dietary intake and—in contrast to cholestanol levels—are not valid markers for intestinal cholesterol absorption during weight-loss programs.

To the best of our knowledge, this is the first study in apparently healthy abdominally obese men that examined relationships between changes in cholesterol absorption and synthesis with changes in anthropometric measures, fat distribution as well as the size of different fat compartments after diet-induced weight loss. The relation between changes in TC-standardized cholestanol levels with changes in most anthropometric parameters were consistent, i.e., improvements were seen with increased cholesterol absorption. However, for the different fat compartments, changes in cholestanol were related to changes in visceral fat volume, but not to changes in subcutaneous and intrahepatic lipids. Visceral fat is a metabolically active fat depot and is more strongly associated with CVD risk than subcutaneous fat and IHL [12,23,24]. In addition, the amount of visceral fat is positively associated with cholesterol synthesis in obese subjects [25,26], which has been explained by an increased flux of fatty acids from the visceral fat depot via the portal vein to the liver, thereby stimulating hepatic cholesterol synthesis. However, the current study did not find an association between cholesterol synthesis and intrahepatic fat. In the present study, we demonstrated a positive association between visceral fat and TC-standardized lathosterol levels, but we could not find an association between the changes in visceral fat and cholesterol synthesis. This finding agrees with another controlled dietary intervention study in 26 obese men, in which no association was found between changes in visceral fat and cholesterol synthesis [20].

To examine the associations between BMI and markers of cholesterol absorption and synthesis in more detail, we used mediation analysis to cross-sectionally investigate the impact of several potential mediators (visceral fat, subcutaneous fat or intrahepatic lipid) on the direct association between BMI with cholesterol absorption and synthesis markers. Apparently, visceral fat mediated the link between BMI and cholesterol absorption marker cholestanol, while intrahepatic lipid mediated the link between BMI and cholesterol synthesis marker lathosterol. Due to the altered fatty acid flux from visceral fat to the liver, it can be speculated that there is a link between visceral fat volume with endogenous cholesterol synthesis. However, our findings showed a relation between visceral fat and cholesterol absorption, suggesting that fatty acid fluxes might influence intestinal cholesterol absorption. Although we found significant roles for some fat compartments in the associations between BMI and markers of cholesterol absorption and synthesis in the cross-sectional model, this does not eliminate any other mediators related to determinants or metabolic effects of these fat compartments.

Cholesterol synthesis and absorption clearly show a reciprocal pattern [27–29], which was also evident in the current study as intestinal cholesterol absorption increased and endogenous cholesterol synthesis decreased after weight loss. An important question arises as to whether changes in cholesterol absorption and synthesis after weight loss may reduce the risk for metabolic diseases. Circulating concentrations of desmosterol, a surrogate marker for cholesterol synthesis involved in the Bloch pathway, were associated with the development of non-alcoholic steatohepatitis (NASH) [30]. These findings were confirmed by Plat et al., who described increased serum desmosterol and lathosterol concentrations in patients with NASH [31]. Moreover, a plant sterol and stanol intervention in rodents showed a reduction in hepatic inflammation, which could be linked to changes in cholesterol synthesis and absorption [31]. In the current study, decreased cholesterol absorption and increased cholesterol synthesis in apparently healthy obese men (without diabetes or the metabolic syndrome) were reversed after diet-induced weight-loss intervention. Whether this also suggests a lower risk of developing type 2 diabetes and metabolic syndrome after weight loss cannot be deduced from these data, but definitely deserves further attention.

5. Conclusions

In summary, a 6-week diet-induced weight-loss period followed by a 2-week weight-stable period increases cholesterol absorption and lowers cholesterol synthesis, and resulted in a normalization of cholesterol metabolism characteristics in abdominally obese men as compared to normal-weight men. Moreover, we also showed that changes in cholestanol levels were related not only to weight loss, but also to a decrease in visceral fat volume. Furthermore, mediation analysis results suggest that visceral fat and intrahepatic content play a role in the relationships between BMI and cholesterol absorption and synthesis. Whether this reflects a possible relation with the amelioration of metabolic risk factors needs further study.

Supplementary Materials: The following supporting information can be downloaded at: https://www.mdpi.com/article/10.3390/nu14081546/s1, Figure S1: Consort flow diagram of the study participation; Table S1: Cross-sectional regression analyses to investigate the relations between cholesterol absorption and synthesis markers with anthropometric measures, fat distribution and IHL at baseline in all participants (n = 73).

Author Contributions: S.M.: performed the GC-FID analysis, performed the statistical analyses, interpreted the data, and wrote the manuscript. S.B.: interpreted the data. J.P. and R.P.M.: designed the study, interpreted the data. P.J.J., Y.H.A.M.K., A.J.H.M.H., C.D.A.S. and C.G.S.: designed the study. P.J.J. and Y.H.A.M.K. conducted the study. All authors have read and agreed to the published version of the manuscript.

Funding: The work was supported by a research grant (CH001) from the Top Institute Food and Nutrition, a public–private partnership on precompetitive research in food and nutrition. The public partners were responsible for the study design, data collection and analysis, the decision to publish and preparation of the manuscript. The private partners contributed to the project through regular discussions. Sterol analyses were funded by a scholarship from AlBaha University, Saudi Arabia.

Institutional Review Board Statement: The study was conducted according to the guidelines of the declaration of Helsinki, and approved by the Medical Ethical Committee of Maastricht University Medical Center (METC 12-30-40), and registered at ClinicalTrials.gov (NCT01675401).

Informed Consent Statement: Informed consent was obtained from all participants prior to the study.

Data Availability Statement: The data presented in this work are fully available without restriction.

Acknowledgments: We would like to thank M.C.J.M. Konings for his technical assistance and our participants for taking part in the study.

Conflicts of Interest: The authors declare no conflict of interest.

References

1. Despres, J.P.; Moorjani, S.; Lupien, P.J.; Tremblay, A.; Nadeau, A.; Bouchard, C. Regional distribution of body fat, plasma lipoproteins, and cardiovascular disease. *Arteriosclerosis* **1990**, *10*, 497–511. [CrossRef] [PubMed]
2. Despres, J.P.; Lemieux, I. Abdominal obesity and metabolic syndrome. *Nature* **2006**, *444*, 881–887. [CrossRef] [PubMed]
3. Ritchie, S.A.; Connell, J.M. The link between abdominal obesity, metabolic syndrome and cardiovascular disease. *Nutr. Metab. Cardiovasc. Dis.* **2007**, *17*, 319–326. [CrossRef] [PubMed]
4. Mashnafi, S.; Plat, J.; Mensink, R.P.; Baumgartner, S. Non-Cholesterol Sterol Concentrations as Biomarkers for Cholesterol Absorption and Synthesis in Different Metabolic Disorders: A Systematic Review. *Nutrients* **2019**, *11*, 124. [CrossRef]
5. Miettinen, T.A.; Tilvis, R.S.; Kesaniemi, Y.A. Serum plant sterols and cholesterol precursors reflect cholesterol absorption and synthesis in volunteers of a randomly selected male population. *Am. J. Epidemiol.* **1990**, *131*, 20–31. [CrossRef]
6. Miettinen, T.A.; Gylling, H.; Nissinen, M.J. The role of serum non-cholesterol sterols as surrogate markers of absolute cholesterol synthesis and absorption. *Nutr. Metab. Cardiovasc. Dis.* **2011**, *21*, 765–769. [CrossRef]
7. Simonen, P.; Gylling, H.; Howard, A.N.; Miettinen, T.A. Introducing a new component of the metabolic syndrome: Low cholesterol absorption. *Am. J. Clin. Nutr.* **2000**, *72*, 82–88. [CrossRef]
8. Simonen, P.; Gylling, H.; Miettinen, T.A. Acute effects of weight reduction on cholesterol metabolism in obese type 2 diabetes. *Clin. Chim. Acta* **2002**, *316*, 55–61. [CrossRef]
9. Chan, D.C.; Watts, G.F.; Gan, S.K.; Ooi, E.M.; Barrett, P.H. Effect of ezetimibe on hepatic fat, inflammatory markers, and apolipoprotein B-100 kinetics in insulin-resistant obese subjects on a weight loss diet. *Diabetes Care* **2010**, *33*, 1134–1139. [CrossRef]
10. Chan, D.C.; Watts, G.F.; Ng, T.W.; Yamashita, S.; Barrett, P.H. Effect of weight loss on markers of triglyceride-rich lipoprotein metabolism in the metabolic syndrome. *Eur. J. Clin. Investig.* **2008**, *38*, 743–751. [CrossRef]
11. Richard, C.; Couture, P.; Desroches, S.; Benjannet, S.; Seidah, N.G.; Lichtenstein, A.H.; Lamarche, B. Effect of the Mediterranean diet with and without weight loss on surrogate markers of cholesterol homeostasis in men with the metabolic syndrome. *Br. J. Nutr.* **2012**, *107*, 705–711. [CrossRef] [PubMed]
12. Fox, C.S.; Massaro, J.M.; Hoffmann, U.; Pou, K.M.; Maurovich-Horvat, P.; Liu, C.Y.; Vasan, R.S.; Murabito, J.M.; Meigs, J.B.; Cupples, L.A.; et al. Abdominal visceral and subcutaneous adipose tissue compartments: Association with metabolic risk factors in the Framingham Heart Study. *Circulation* **2007**, *116*, 39–48. [CrossRef] [PubMed]
13. Kershaw, E.E.; Flier, J.S. Adipose tissue as an endocrine organ. *J. Clin. Endocr. Metab.* **2004**, *89*, 2548–2556. [CrossRef] [PubMed]
14. Tchernof, A.; Despres, J.P. Pathophysiology of human visceral obesity: An update. *Physiol. Rev.* **2013**, *93*, 359–404. [CrossRef] [PubMed]
15. Joris, P.J.; Plat, J.; Kusters, Y.H.; Houben, A.J.; Stehouwer, C.D.; Schalkwijk, C.G.; Mensink, R.P. Diet-induced weight loss improves not only cardiometabolic risk markers but also markers of vascular function: A randomized controlled trial in abdominally obese men. *Am. J. Clin. Nutr.* **2017**, *105*, 23–31. [CrossRef]
16. Grundy, S.M.; Becker, D.; Clark, L.T.; Cooper, R.S.; Denke, M.A.; Howard, W.J.; Hunninghake, D.B.; Illingworth, R.; Luepker, R.V.; McBride, P.; et al. Third Report of the National Cholesterol Education Program (NCEP) Expert Panel on Detection, Evaluation, and Treatment of High Blood Cholesterol in Adults (Adult Treatment Panel III) Final Report. *Circulation* **2002**, *106*, 3143–3421. [CrossRef]
17. Kusters, Y.H.; Schalkwijk, C.G.; Houben, A.J.; Kooi, M.E.; Lindeboom, L.; Op't Roodt, J.; Joris, P.J.; Plat, J.; Mensink, R.P.; Barrett, E.J.; et al. Independent tissue contributors to obesity-associated insulin resistance. *JCI Insight* **2017**, *2*, e89695. [CrossRef]

18. Friedewald, W.T.; Levy, R.I.; Fredrickson, D.S. Estimation of the concentration of low-density lipoprotein cholesterol in plasma, without use of the preparative ultracentrifuge. *Clin. Chem.* **1972**, *18*, 499–502. [CrossRef]
19. Mackay, D.S.; Jones, P.J.; Myrie, S.B.; Plat, J.; Lutjohann, D. Methodological considerations for the harmonization of non-cholesterol sterol bio-analysis. *J. Chromatogr. B Analyt. Technol. Biomed. Life Sci.* **2014**, *957*, 116–122. [CrossRef]
20. Riches, F.M.; Watts, G.F.; Hua, J.; Stewart, G.R.; Naoumova, R.P.; Barrett, P.H. Reduction in visceral adipose tissue is associated with improvement in apolipoprotein B-100 metabolism in obese men. *J. Clin. Endocrinol. Metab.* **1999**, *84*, 2854–2861. [CrossRef]
21. Ostlund, R.E., Jr. Phytosterols in human nutrition. *Annu. Rev. Nutr.* **2002**, *22*, 533–549. [CrossRef] [PubMed]
22. Miettinen, T.A.; Tilvis, R.S.; Kesaniemi, Y.A. Serum cholestanol and plant sterol levels in relation to cholesterol metabolism in middle-aged men. *Metabolism* **1989**, *38*, 136–140. [CrossRef]
23. Despres, J.P. Body fat distribution and risk of cardiovascular disease: An update. *Circulation* **2012**, *126*, 1301–1313. [CrossRef] [PubMed]
24. Liu, J.; Fox, C.S.; Hickson, D.; Bidulescu, A.; Carr, J.J.; Taylor, H.A. Fatty liver, abdominal visceral fat, and cardiometabolic risk factors: The Jackson Heart Study. *Arterioscler. Thromb. Vasc. Biol.* **2011**, *31*, 2715–2722. [CrossRef] [PubMed]
25. Adeli, K.; Taghibiglou, C.; Van Iderstine, S.C.; Lewis, G.F. Mechanisms of hepatic very low-density lipoprotein overproduction in insulin resistance. *Trends Cardiovasc. Med.* **2001**, *11*, 170–176. [CrossRef]
26. Tobin, K.A.; Ulven, S.M.; Schuster, G.U.; Steineger, H.H.; Andresen, S.M.; Gustafsson, J.A.; Nebb, H.I. Liver X receptors as insulin-mediating factors in fatty acid and cholesterol biosynthesis. *J. Biol. Chem.* **2002**, *277*, 10691–10697. [CrossRef]
27. Alphonse, P.A.; Jones, P.J. Revisiting Human Cholesterol Synthesis and Absorption: The Reciprocity Paradigm and its Key Regulators. *Lipids* **2016**, *51*, 519–536. [CrossRef]
28. Quintao, E.; Grundy, S.M.; Ahrens, E.H., Jr. Effects of dietary cholesterol on the regulation of total body cholesterol in man. *J. Lipid Res.* **1971**, *12*, 233–247. [CrossRef]
29. Santosa, S.; Varady, K.A.; AbuMweis, S.; Jones, P.J. Physiological and therapeutic factors affecting cholesterol metabolism: Does a reciprocal relationship between cholesterol absorption and synthesis really exist? *Life Sci.* **2007**, *80*, 505–514. [CrossRef]
30. Simonen, M.; Mannisto, V.; Leppanen, J.; Kaminska, D.; Karja, V.; Venesmaa, S.; Kakela, P.; Kuusisto, J.; Gylling, H.; Laakso, M.; et al. Desmosterol in human nonalcoholic steatohepatitis. *Hepatology* **2013**, *58*, 976–982. [CrossRef]
31. Plat, J.; Hendrikx, T.; Bieghs, V.; Jeurissen, M.L.; Walenbergh, S.M.; van Gorp, P.J.; De Smet, E.; Konings, M.; Vreugdenhil, A.C.; Guichot, Y.D.; et al. Protective role of plant sterol and stanol esters in liver inflammation: Insights from mice and humans. *PLoS ONE* **2014**, *9*, e110758. [CrossRef] [PubMed]

nutrients

MDPI

Article

A Systems Analysis of Phenotype Heterogeneity in APOE*3Leiden.CETP Mice Induced by Long-Term High-Fat High-Cholesterol Diet Feeding

Yared Paalvast [1], Enchen Zhou [2,3], Yvonne J. W. Rozendaal [4], Yanan Wang [2,3], Albert Gerding [1], Theo H. van Dijk [1], Jan Freark de Boer [1], Patrick C. N. Rensen [2,4], Ko Willems van Dijk [3,5], Jan A. Kuivenhoven [1], Barbara M. Bakker [1], Natal A. W. van Riel [4,6] and Albert K. Groen [1,6,*]

1 Department of Pediatrics, University of Groningen, University Medical Center Groningen, 9713 AV Groningen, The Netherlands
2 Department of Medicine, Division of Endocrinology, Leiden University Medical Center, 2300 RC Leiden, The Netherlands
3 Einthoven Laboratory for Experimental Vascular Medicine, Leiden University Medical Center, 2300 RC Leiden, The Netherlands
4 Department of Biomedical Engineering, Eindhoven University of Technology, 5600 MB Eindhoven, The Netherlands
5 Department of Human Genetics, Leiden University Medical Center, 2300 RC Leiden, The Netherlands
6 Laboratory of Experimental Vascular Medicine, University of Amsterdam, Amsterdam UMC, Meibergdreef, 1105 AZ Amsterdam, The Netherlands
* Correspondence: a.k.groen@amsterdamumc.nl

Citation: Paalvast, Y.; Zhou, E.; Rozendaal, Y.J.W.; Wang, Y.; Gerding, A.; van Dijk, T.H.; de Boer, J.F.; Rensen, P.C.N.; van Dijk, K.W.; Kuivenhoven, J.A.; et al. A Systems Analysis of Phenotype Heterogeneity in APOE*3Leiden.CETP Mice Induced by Long-Term High-Fat High-Cholesterol Diet Feeding. *Nutrients* **2022**, *14*, 4936. https://doi.org/10.3390/nu14224936

Academic Editor: Frans Stellaard

Received: 28 September 2022
Accepted: 15 November 2022
Published: 21 November 2022

Abstract: Within the human population, considerable variability exists between individuals in their susceptibility to develop obesity and dyslipidemia. In humans, this is thought to be caused by both genetic and environmental variation. APOE*3-Leiden.CETP mice, as part of an inbred mouse model in which mice develop the metabolic syndrome upon being fed a high-fat high-cholesterol diet, show large inter-individual variation in the parameters of the metabolic syndrome, despite a lack of genetic and environmental variation. In the present study, we set out to resolve what mechanisms could underlie this variation. We used measurements of glucose and lipid metabolism from a six-month longitudinal study on the development of the metabolic syndrome. Mice were classified as mice with either high plasma triglyceride (responders) or low plasma triglyceride (non-responders) at the baseline. Subsequently, we fitted the data to a dynamic computational model of whole-body glucose and lipid metabolism (MINGLeD) by making use of a hybrid modelling method called Adaptations in Parameter Trajectories (ADAPT). ADAPT integrates longitudinal data, and predicts how the parameters of the model must change through time in order to comply with the data and model constraints. To explain the phenotypic variation in plasma triglycerides, the ADAPT analysis suggested a decreased cholesterol absorption, higher energy expenditure and increased fecal fatty acid excretion in non-responders. While decreased cholesterol absorption and higher energy expenditure could not be confirmed, the experimental validation demonstrated that the non-responders were indeed characterized by increased fecal fatty acid excretion. Furthermore, the amount of fatty acids excreted strongly correlated with bile acid excretion, in particular deoxycholate. Since bile acids play an important role in the solubilization of lipids in the intestine, these results suggest that variation in bile acid homeostasis may in part drive the phenotypic variation in the APOE*3-Leiden.CETP mice.

Keywords: computational modeling; cholesterol; bile acid; energy expenditure; metabolic syndrome; triglycerides; APOE3; CETP

1. Introduction

Diets characteristic for Western society have spread across the globe, which together with the development of a mostly sedentary lifestyle, have led to an increase in the prevalence of cardiovascular risk factors, such as obesity, insulin resistance and hypertriglyceridemia [1]. It is generally assumed that the vast differences in the genetic make-up between individuals results in some individuals being less prone and some individuals being more prone to developing these risk factors while being subjected to the same environment. However, genome-wide association studies have to date only been able to explain 21% of the variation in body weight [2]. On the other hand, a significant proportion of the variation in body weight is thought to be due to the variation in environmental variables, such as diet and physical activity. Because of the extreme number of putative variables and the complexity of metabolic control, it has been extremely difficult to sort out the interaction between environmental and genetic factors. Nevertheless, when genes are the dominant drivers of the metabolic syndrome, one would expect inbred animal models housed under standardized conditions to show little variation in phenotype when fed a Western diet. Our recent studies have falsified this hypothesis [3].

Using an apolipoprotein E*3-Leiden (APOE*3Leiden).cholesteryl ester transfer protein (CETP) mouse model ([4]), we showed a human-like variation in phenotypic (obesogenic) response when these mice were fed a Western-type diet. This mouse model is heterozygous for the human APOE*3-Leiden variant, conferring a reduced hepatic uptake of triglyceride-rich lipoprotein remnants from the circulation. Furthermore, the mouse model is heterozygous for human CETP under its endogenous promotor. The combination of these genes results in a 'humanized' lipid metabolism with more cholesterol in apoB-containing lipoproteins and a relatively low HDL cholesterol level [4].

A striking observation in our previous studies was the major variation in important parameters of the metabolic syndrome, including body weight, plasma triglyceride and cholesterol as well as insulin resistance, not only in time, but also amongst individual mice [3]. Given the fact that the mice are inbred and maintained under identical conditions, this model seems promising to elucidate the mechanism underlying the observed phenotypic variation. Instead of using the usual gene-focused mechanism, we reasoned that, whatever the major source of variation in these animals, the effect must take place through processes involved in energy metabolism. To help us direct our search, we made use of a mixed approach in which computational modeling exploited experimentally obtained longitudinal data to identify the processes most likely involved in explaining the phenotypic differences [5]. Recently, we have published an ordinary differential equation based a computational model called Model INtegrating GLucose and Lipid Dynamics (MINGLeD), which encompasses the main metabolic pathways involved in energy metabolism [6]. In the current study, we concentrated on plasma triglyceride in combination with obesity and used MINGLeD to analyse which processes are altered during the progression of the metabolic syndrome in male APOE*3-Leiden.CETP mice on a high-fat diet.

2. Methods

2.1. Animals, Diet and Housing

Experimental conditions have been described previously ([3]). In brief, male APOE*3-Leiden.CETP mice were housed individually and fed a synthetic high-fat and -cholesterol diet (HFCD) containing 60% of energy from fat and 0.25% of weight from cholesterol (D12429, Research Diets) in a light- (lights on 7:00 a.m.–7:00 p.m.) and temperature-controlled (21 °C) facility. At the start of the experimental period, the mice were at the age of 4 months. Prior to the start of the experimental period, mice were co-housed with siblings and fed chow ad libitum. At least one week prior to the start of the experiment, animals were housed individually to acclimatize. Experimental procedures were approved by the Ethics Committees for Animal Experiments of the University of Groningen (Protocol Code 6903).

2.2. Experimental Setup

As previously described [3], four groups of mice were fed HFCDs ad libitum for 4 ($n = 20$), 9 ($n = 19$), 13 ($n = 20$) and 28 weeks ($n = 30$), respectively. At the end of the dietary intervention, mice in the respective cohorts were distributed over two groups, to either measure VLDL-TG production or to measure hepatic de novo lipogenesis, measure bile production and collect tissues. In addition, a cohort (n = 16) was used to measure endogenous glucose production at weeks 3, 9, 15 and 27 and energy expenditure at weeks 1 and 19. In all groups, blood samples were obtained by tail bleeding at 4- to 6-week intervals, to determine plasma TG, plasma total cholesterol (TC), HDL-C and glucose [3]. In addition, 24 h feces were collected from all groups at 4- to 6-week intervals. Plasma bile acid concentrations were determined by liquid chromatography–tandem mass spectrometry (LC–MS/MS). For quantification, internal standard solution containing D4-cholate, D4-chenodeoxycholate, D4-glycocholate, D4-taurocholate, D4-glycochenodeoxycholate and D4-taurochenodeoxycholate was added to the plasma. Bile salt composition of prepared fecal samples was determined by capillary gas chromatography on an Agilent gas chromatograph (HP 6890). All flux measurements and blood sample collections were started at 1 PM under fasting conditions, with food removed at 9 AM.

2.3. The ADAPT Method

ADAPT is a hybrid modelling method combining data assimilation and machine learning to discover differential equations describing the long-term dynamics of a diet intervention. ADAPT separates the multiscale behavior of metabolic physiology in time. The fast metabolic dynamics (minutes to hours, for example, associated with food intake) are decoupled from the slow dynamics (weeks to months) related to long-term, high-fat and high-cholesterol diet feeding. ADAPT starts from measurements taken at the fast time scale; the slow scale system can be assumed to be in steady state. ADAPT first takes a longitudinal data set and fits a series of polynomial curves and data splines through points sampled from the normal distribution of the data at the respective time points [7]. Subsequently, ADAPT minimizes the error between the model output and the data splines from time point to time point using a least-squares algorithm. Since a penalty is put on changes in the parameter values, the algorithm favors gradual changes in parameters through time over abrupt changes. By studying predicted changes in parameters that were not constrained, ADAPT may assist in identifying processes that are likely to be changed as well, and thus may play an important role in explaining the phenomenon of interest. For the current work, we used 200 time steps and applied a regularization parameter (λ) of 0.01. Further details on how we arrived at the settings used for ADAPT are described in Supplemental S1 (Figures S1 and S2).

2.4. Experimental Data and Modeling Constraints

The experimental data have been described in detail previously [3]. Some parameters for the assessment of VLDL-TG production, de novo lipogenesis, biliary sterol secretion and liver lipids were obtained cross-sectionally. The number of non-responders in the VLDL-TG production cohort (n = 2, n = 0, n = 0, n = 4) and in the cohort undergoing bile cannulation (n = 3, n = 1, n = 1, n = 3) for the respective time points of 4, 9, 13 and 28 weeks were too small for reliable differentiation with the responders (n = 8, n = 9, n = 9, n = 13 and n = 7, n = 9, n = 8, n = 14) in the respective cohorts. Therefore, while constraints for food intake, body weight, plasma parameters and fecal samples were directly taken from the data of responders and non-responders, constraints for liver lipids, biliary secretion, hepatic de novo lipogenesis and VLDL-TG production were taken as group averages and therefore were the same for responder and non-responder groups. Further details as to how experimental data were translated to model constraints may be found in Supplemental S2.

2.5. Choices Concerning Model Design

We designed a model that includes all fluxes relevant for whole body fat and cholesterol metabolism. Overall, the computational model may be described as a three-link chain in which the metabolic network within a module is more connected than the number of interactions between modules (Figure 1). The model was named Model INtegrating GLucose and Lipid Dynamics (MINGLeD), emphasizing that we have both glucose and lipid metabolism integrated into one model [6]. In general, we considered reactions to be first order. ADAPT uses a data-driven approach to discover a dynamic model. The rate equations reflect the pathway/network structure (connectivity) and stoichiometry and are not based on actual enzyme kinetics. The data-based constraints are used, since ADAPT will change the parameter values to comply with the constraints regardless of the rate equation. All model equations may be found in the Supplemental Information (Supplemental S3). We highlight some of the relations here because they require explanation. The rate equation for CETP (j34 in Figure 1) was chosen to be dependent on plasma triglyceride concentration, since this is generally considered to be the driver behind CETP action [8]. The trans-intestinal cholesterol excretion (TICE) rate equation (j37 in Figure 1) was chosen to be dependent on the VLDL-C pool. TICE is the flux of cholesterol that enters the intestine directly from the plasma. The plasma compartments contributing to TICE are not completely clear, and may be both coming from apoB-containing lipoproteins as well as from erythrocytes. Therefore, it was decided to make it dependent on VLDL-C only, since erythrocytes were out of the scope of this study [9].

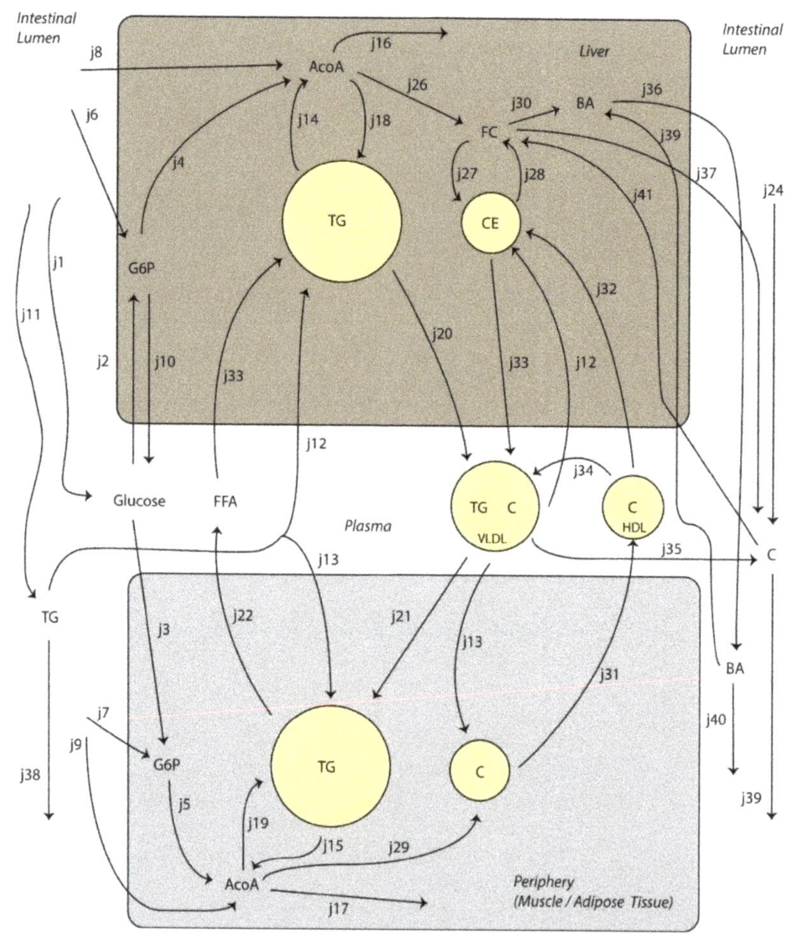

Figure 1. Schematic of the MINGLeD model. The MINGLeD model consists of four compartments (liver, plasma, periphery and intestinal lumen), 18 states and 41 fluxes. Food intake is modeled as glucose entering the plasma (j1), triglyceride (TG, j11) and cholesterol (C, j24) entering the intestinal

lumen, whereas amino acids from protein are distributed to liver and periphery at the level of glucose-6-phosphate (G6P, gluconeogenic) or acetyl-CoA (AcoA, ketogenic) (j6, j7, j8, j9). Glucose in the plasma is absorbed by liver (j2) and periphery (j3) to enter the Krebs cycle (j16, j17) or to be used for biosynthetic processes, such as de novo lipogenesis (j19) or cholesterol (j26, j29) and bile acid synthesis (j30). TG from the intestinal lumen can be absorbed by the liver (j12) or periphery (j13) and be used for beta-oxidation (j14, j15) or redistribution as VLDL (j20) or free fatty acids (FFA, j22). Absorbed dietary cholesterol first enters the liver (j41) where it can be used for bile acid synthesis (j30) or redistributed to the periphery in the form of VLDL-C (j33, j13). Peripheral cholesterol pools can return to the liver through HDL-C (j31, j32) or VLDL-C after action of cholesteryl ester transfer protein (j34, j12). Cholesterol can be cleared from the body through biliary cholesterol secretion (j37) or trans-intestinal cholesterol secretion (j35).

2.6. Validation Experiment

In two institutions (UMCG and LUMC), 13 male APOE*3-L.CETP mice were fed the same HFCD with 60% of energy from fat and 0.25% of weight from cholesterol for 8 weeks and fractional cholesterol absorption was measured as described previously [10]. Animal experiments were approved by the responsible ethics committees. Fecal FFAs were measured as described previously [11].

3. Results

3.1. Stratification to Responder and Non-Responder Phenotypes

We have shown previously that the APOE*3L.CETP mice show great variability in response to treatments with an HFCD [3]. Data on basic parameters, such as body weight and food intake, are presented in [3]. To be able to differentiate the response of the mice to the HFCD, we stratified the individual mice into the responder and non-responder groups by classifying the animals with a plasma TG < 1.0 mM at the baseline (chow diet) as non-responders. Since non-responders have lower body weights, this also selected mice with lower body weights. This stratification procedure yielded 36 responders and 11 non-responders. As shown in Figure 2, the plasma TG in the non-responding group remained low during the full course of the experiment. In contrast, the plasma TG in the responding mice started to increase at week 4, reached a maximum at week 20 and then decreased sharply to reach a new steady state at week 24. The responding group showed a similar response in the plasma total cholesterol (TC) levels. Interestingly, much less of a difference was observed in the peripheral fat content as well as the plasma HDL-C levels. In agreement with the observed differences in plasma TG and body weight, insulin levels were lower in non-responders as well. (Figures S3 and S4).

3.2. Application of ADAPT

To address the question which processes are responsible for the observed kinetics in Figure 2, we used the newly developed MINGLeD model of lipid and carbohydrate metabolism [6]. The model contains 18 state variables, 39 parameters and 41 fluxes. In order to find the underlying mechanisms by which the differences between the responders and non-responders can be explained, ADAPT was applied to the MINGLeD model. As depicted in Figure 2, Figures S5 and S6, the modeling is able to describe the complex evolution in the measured parameters involved in lipid, glucose and energy metabolism.

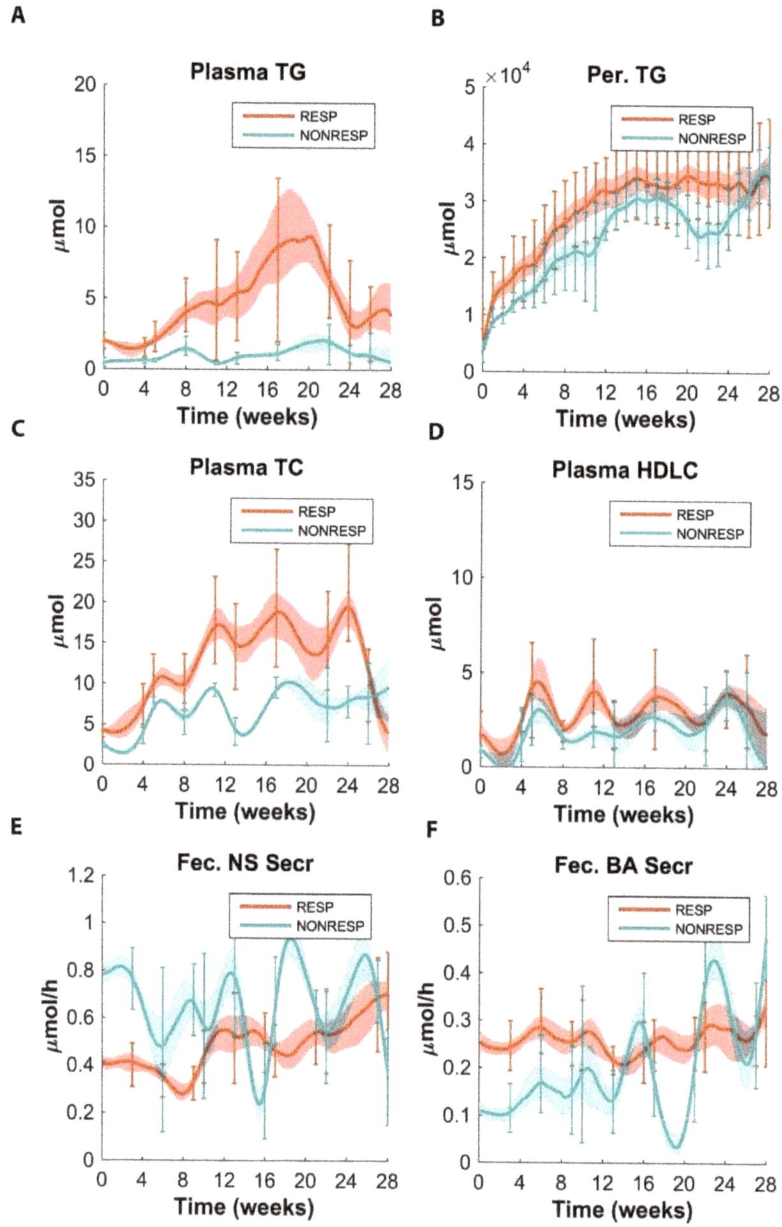

Figure 2. Plasma TG (**A**) and peripheral fat (Per. TG (**B**)), plasma total cholesterol (TC) (**C**), HDL cholesterol (HDLC) (**D**), fecal neutral sterol secretion (**E**) (Fec. NS Secr) and fecal bile acid secretion (**F**) (Fec. BA Secr) for responders (RESP) and non-responders (NONRESP), respectively, with their respective fits in the ADAPT model simulation. Note that non-responders are marked by lower plasma TG and less peripheral fat. Error bars represent data with standard deviation, bold lines represent the median solution of all ADAPT simulations and the areas represent 30% around the median solution.

3.3. ADAPT Predicts Decreased Cholesterol Absorption for Non-Responders

Since non-responders presented with both lower plasma cholesterol values and a higher level of fecal sterol excretion (Figure 2), we inspected which fluxes ADAPT predicted pertaining to cholesterol homeostasis. We then found that ADAPT predicted slightly lower cholesterol absorption for non-responders as compared to responders (Figure 3A), which makes sense in light of the higher observed level of fecal cholesterol excretion and lower plasma TC for the non-responders. While cholesterol synthesis is expected to be decreased with increases in cholesterol absorption [12], no differences in prediction were found for de novo hepatic or peripheral cholesterol syntheses (data not shown).

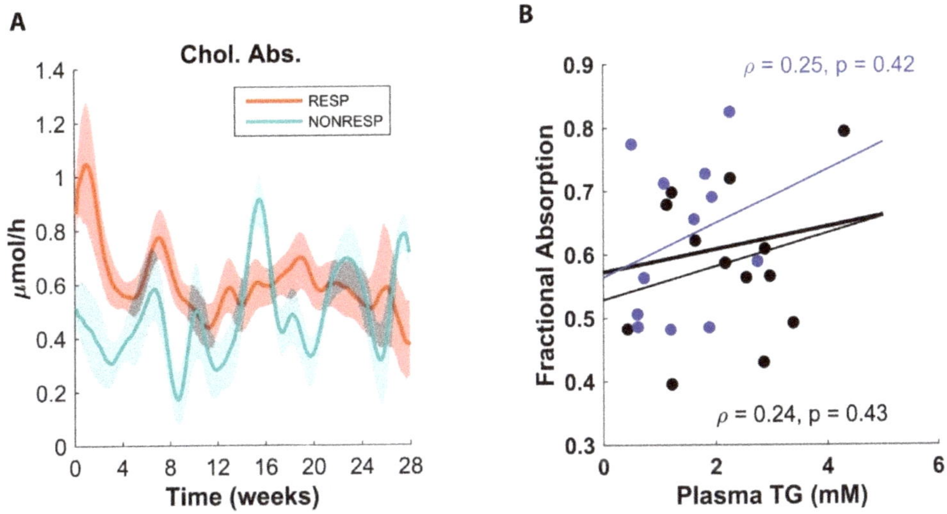

Figure 3. Cholesterol absorption (Chol. Abs.) as predicted by ADAPT (**A**) for responders (RESP) and non-responders (NONRESP); note that ADAPT predicts lower level of cholesterol absorption for non-responders. Fractional cholesterol absorption (**B**) in two cohorts (black and blue) of mice after 8 weeks of HFCD. Note there is no correlation between fractional cholesterol absorption and plasma TG.

3.4. Validation Experiment of Decreased Cholesterol Absorption

Since ADAPT predicted a lower level of cholesterol absorption in the non-responders compared to responders, we performed a validation experiment in which we measured the cholesterol absorption after feeding the APOE*3-Leiden.CETP mice HFCDs for eight weeks. We reasoned that if the prediction of ADAPT was correct, we would find a positive correlation between the cholesterol absorption and plasma TG. To make sure any effect found would not be site- or cohort-dependent, two independent experiments were performed with different cohorts of mice at two different facilities. As depicted in Figure 3B, this prediction was falsified; no correlation between the cholesterol absorption and plasma TG levels was observed. Furthermore, there was no clear negative correlation between plasma TG and fecal neutral sterol excretion (Figure S7). These findings indicate that cholesterol absorption is not consistently decreased in the non-responder animals.

3.5. ADAPT Predicts Higher Glucose Oxidation Rates in Non-responders

Next, we looked for parameter and flux trajectories that may explain the lower body weight observed in the non-responders compared to responders (Figures 2 and S3). Since body weight is the result of the balance between energy absorption and expenditure, any differences must be explained by either. Looking at energy expenditure, we found that ADAPT predicted that non-responders would have higher glucose oxidation rates, while the fat oxidation rates were predicted to be equal between the two groups (Figure 4A,B). This model prediction implies both a higher total energy expenditure for the non-responders, and more glucose stored as fat and enhanced peripheral de novo lipogenesis in the responders. Indeed, if we look at the flux trajectories for these processes, we see that this is also predicted (Figure S8). Since we had put mice in metabolic cages that were monitored at least during the initial period of the experiment, we could compare how the energy expenditure as measured by indirect calorimetry was connected to having a responder or non-responder status. Interestingly, we observed no difference in the energy expenditure between the groups (Figure 4C,D). Of note, even when adjusting for body weight [13], the non-responder animals had a near-average energy expenditure compared to that of the responders.

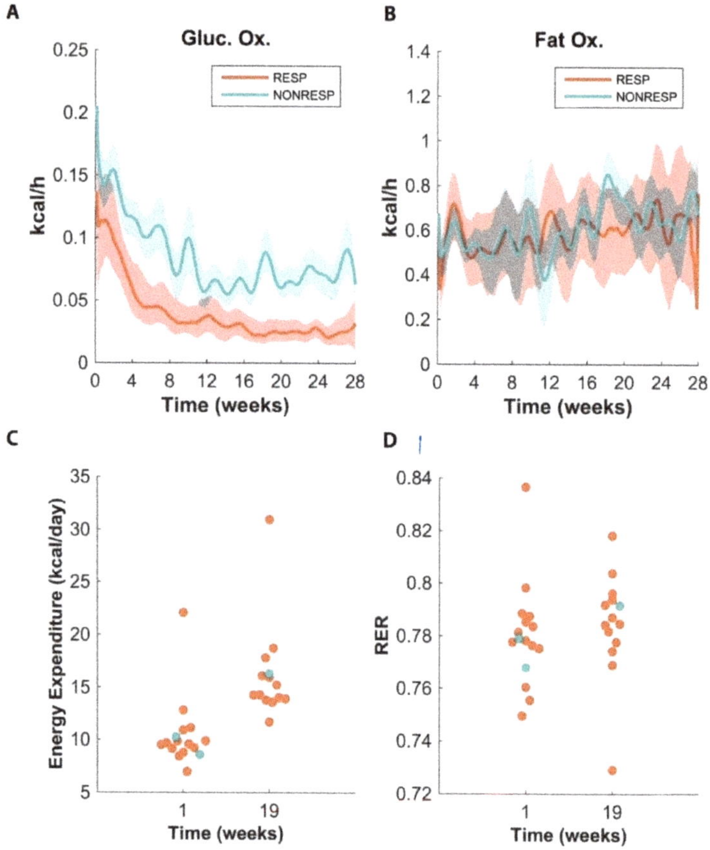

Figure 4. Predictions for glucose oxidation (Gluc. Ox.) (**A**) and fat oxidation (Fat Ox.) rate (**B**) in responders and non-responders, respectively. The line represents the median values, whereas the area around the line denotes 30% of solutions around the median. Energy expenditure (**C**) and respiratory exchange ratio (RER) (**D**) for animals after 1 and 19 weeks of HFCDs. Note how the non-responders (blue) are not necessarily marked by increased energy expenditure.

3.6. ADAPT Predicts Lower Fat Absorption in Non-responders

Interestingly, we found that ADAPT predicted a higher level of fat excretion in the non-responders (Figure 5C). This was further highlighted by predictions of a higher TG content in the intestinal lumen and lower parameter values for fat absorption (Figure 5B,D). This prediction was validated by measuring the amount of fatty acids (FFA) still contained in the feces. The FFA content in the feces from the non-responders was indeed higher than that of the responders, suggesting impaired fat absorption in these mice (Figure 6A). While the cumulative fecal fat excretion was significantly different between the groups, not all the non-responders presented with an increased level of fecal FFA excretion, suggesting that, in these animals, the plasma TG is low for another reason. Interestingly, the fecal FFA excretion also negatively correlated with the body weight and plasma TG (Figure 6E,F). However, the correlation between the body weight and fecal FFA excretion was much more evident than that for the plasma TG, whose additional variation obviously must be from another factor.

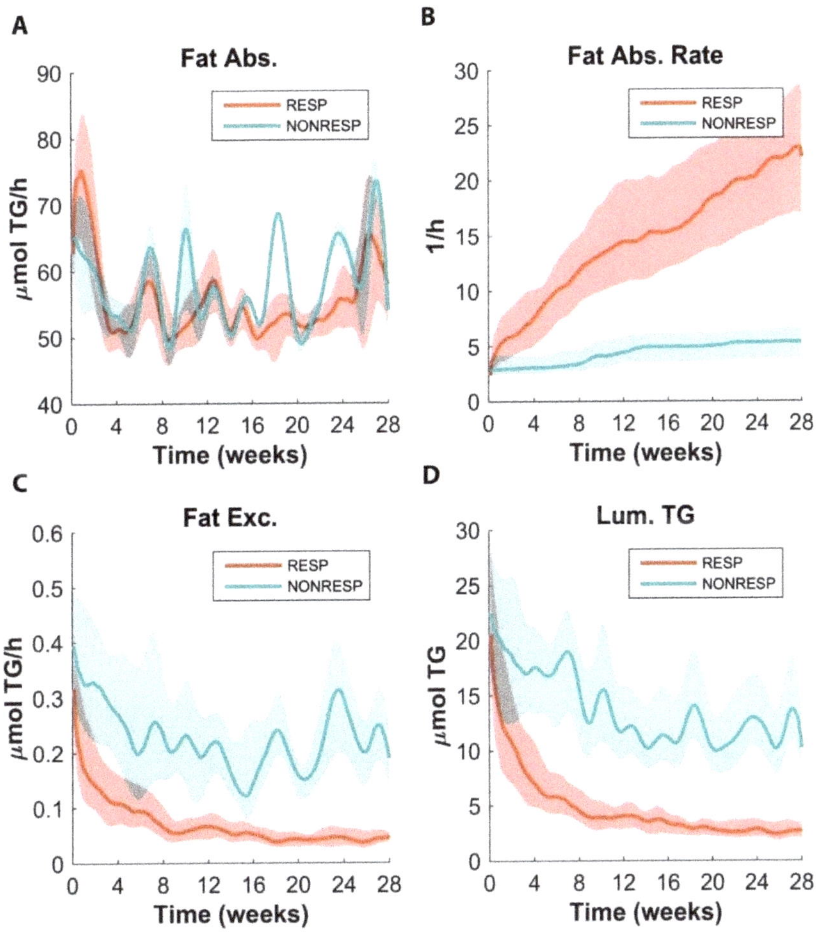

Figure 5. Predictions for fat absorption (Fat Abs.) (**A**), fat absorption rate (Fat Abs. Rate) (**B**), fecal fat excretion (Fat Exc.) (**C**) and intestinal lumen fat content (Lum. TG) (**D**) in responders (RESP) and non-responders (NONRESP), respectively. The bold line represents the median values, whereas the area around the line denotes 30% of solutions around the median. Note how the fat absorption rate is predicted to be slower, while fat excretion and intestinal fat content are predicted to be increased in non-responders.

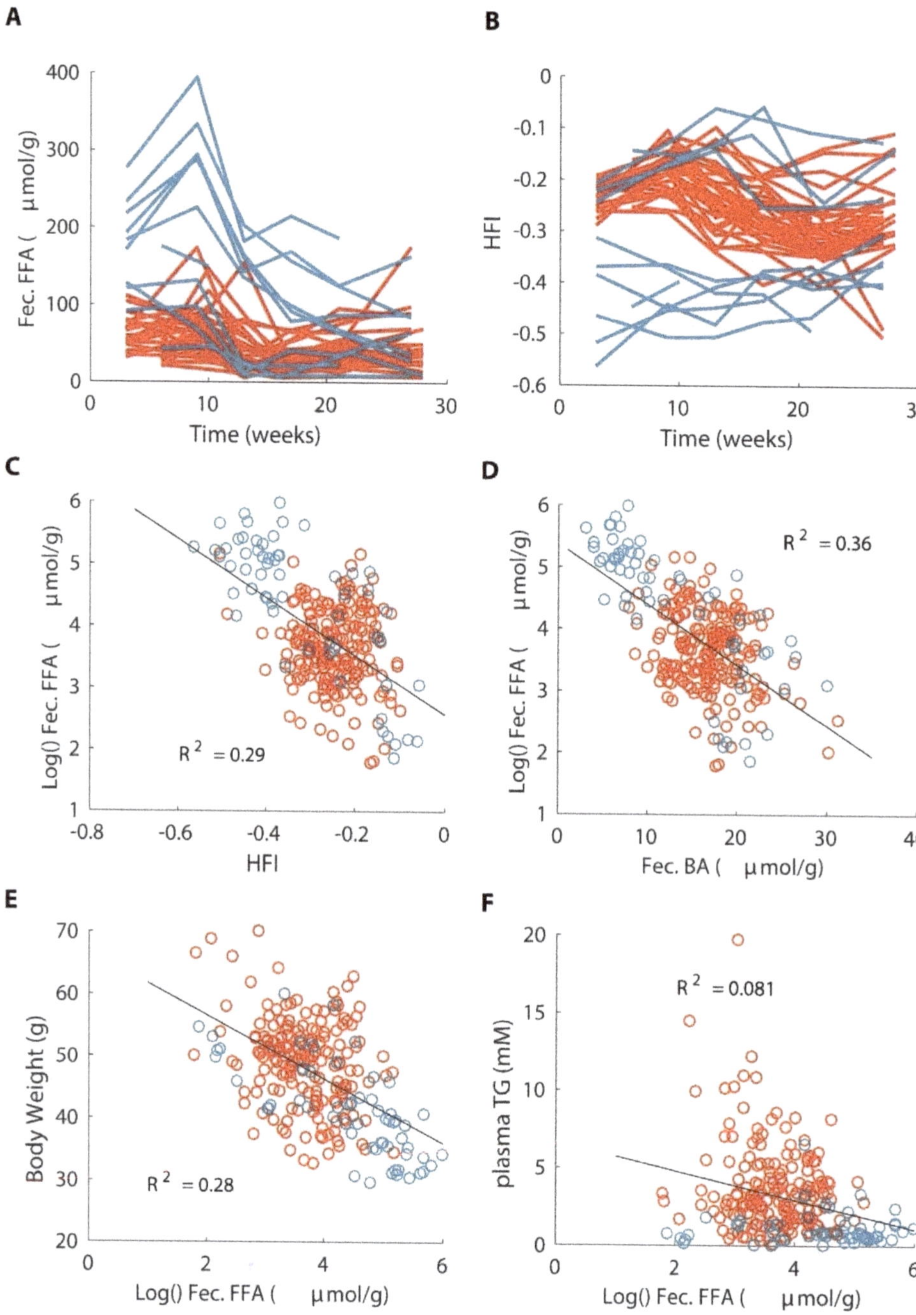

Figure 6. Fecal fatty acid excretion (Fec. FFA) (**A**) and hydrophobicity index (HFI) of fecal bile acids (**B**) over time. Correlations between fecal fatty acid excretion and HFI (**C**), fecal bile acids (Fec. BA) (**D**), body weight (**E**) and plasma TG (**F**). Responders are marked in red and non-responders are marked in blue.

3.7. Decreased Fat Absorption Is Associated with a Lower Hydrophobicity Index of Fecal Bile Acids

Since cholesterol absorption and fat absorption are more promoted by hydrophobic than hydrophilic bile acids [13–15], we compared the bile acid composition profiles in the feces, plasma and bile between the responders and non-responders. We reasoned that the higher observed level of fecal FFA excretion may be related to the hydrophobicity of bile acids. Indeed, regardless of responder or non-responder statuses, the hydrophobicity index of the fecal bile acids was correlated with the fecal FFA excretion (Figure 6C). Furthermore, the fecal hydrophobicity index was positively associated with the fecal bile acid excretion as well. In fact, we found fecal FFA excretion to be more strongly correlated with fecal bile acid excretion than with the hydrophobicity index (Figure 6D). Interestingly, we found that the fecal deoxycholic acid was especially highly correlated with the fecal FFA (Figure S9). Surprisingly, comparing the biliary bile acid profiles of the responders from all time points with those of the non-responders, neither a difference in the total biliary bile acid secretion nor in the individual bile acids was found (Figures S10–S13). However, there was a trend of a lower hydrophobicity index of the biliary bile acids for the non-responders ($p = 0.12$). Interestingly, when only the biliary bile acid profiles of the first three months were compared, when the level of fat excretion was the highest, the hydrophobicity index of the biliary bile acids was indeed lower in the non-responders ($p = 0.001$). Moreover, the mice with low plasma TG, in which the fractional cholesterol absorption was measured after eight weeks of the HFCDs, also showed biliary bile acid profiles with lower hydrophobic indexes than those with higher plasma TG ($p = 0.01$). Furthermore, both cholic acid- and chenodeoxycholic acid-derived bile acids were higher in the plasma of the non-responders (Figures S14–S16). Together, these data suggest that the difference in the fecal FFA excretion in non-responders is driven by changes in the bile acid metabolism.

4. Discussion

The major result of this study is that by using the computational modeling method ADAPT, it is possible to analyze in detail the pathways that induce the progression of comorbidities in complex diseases. In this study, we applied the method to explain the phenotypic variation induced by long-term HFCDs in APOE*3-Leiden.CETP mice. The application of ADAPT for adjusting parameters in the MINGleD model allowed an accurate modelling of the phenotypic changes in long-term experiments. The ADAPT analysis suggested different pathways, such as decreased cholesterol absorption, increased energy expenditure and an increased level of fecal fat excretion to explain the phenotypic variability in body, hepatic and plasma TG. Subsequent validation experiments failed to confirm a decreased cholesterol absorption and an increased energy expenditure to explain the lack of response in a subset of the mice. In contrast, an increased level of fecal fat excretion in the non-responders could be confirmed. Furthermore, we found that the increased level of fecal fat excretion was associated with a decreased level of fecal bile acid excretion, suggesting that a decrease in bile acid production may, at least in part, drive the lower body weight and plasma TG in non-responders. A similar relation between deoxycholic acid and body weight loss was reported by [16] However, they observed the interaction of intestinal fat with GPR119, causing an increased satiety signal. In our study, we did not observe a decreased level of food intake in the non-responder mice.

4.1. Hybrid Modeling Using ADAPT

ADAPT is a hybrid modelling method that employs the power of data-driven and mechanistic modelling techniques to estimate the long-term dynamics of metabolic physiology. ADAPT uses concepts somewhat similar to physics-informed neural networks (PINNs) and other data assimilation methods emerging in cardiovascular modeling [17]. The network structure of the metabolic system imposes strong constraints on the solution space of the mathematical model, which ADAPT combines into equations for kinetics and fluxes with time-series data. The trade-off between the bias variance is controlled by the hyperparameter λ. Lambda was tuned to provide enough flexibility to discover possible explanations underlying the phenotype heterogeneity in APOE*3-Leiden.CETP mice induced by long-term HFCDs (Supplemental S1). For higher values of λ, the fluctuations in the model dynamics could be dampened, but with too high values of λ, the goodness of fit decreases and a bias emerges. The stochastic data model and spline interpolation to take into account experimental and biological uncertainties also contribute to the variation in model predictions. Other types of data, e.g., transcriptomics, can also be included in ADAPT to further constrain the model predictions. Gene expression data have been incorporated by expanding the regularization function in ADAPT [18].

4.2. Fat Absorption and Energy Expenditure

In line with the major impact of fat absorption in the model, we found that the non-responders are marked by an increased level of fecal fat excretion. However, while the cumulative fat excretion varied between 0.5 and 5 g, the body weight difference amounted up to 36 g, and is thus roughly ten times as large. This indicates that any differences in absorption must be accompanied by a difference in energy expenditure. In fact, we predicted that the energy expenditure in the non-responders is increased compared to the responders, which ADAPT mainly attributed to a difference in glucose oxidation. Indirect calorimetry, however, showed no statistical significant differences in the respiratory ratio (RER) or increases in energy expenditure. Our results are in agreement with findings of Tarasco et al., who failed to find differences in energy expenditure between responder and non-responder APOE*3-Leiden.CETP mice as well [19].

A possible explanation is that indirect calorimetry may not be sensitive enough to detect the difference in energy expenditure between the responders and non-responders. The mean difference in the weight increase between non-responders and responders in the first 12 weeks was 5 g. This difference in weight would amount to an energy imbalance, presuming the weight difference is on account of fat (0.6 kcal/day), which, assuming a daily energy expenditure of 12 kcal/day, would be 5% of the energy expenditure. Coincidentally, 5% of the energy expenditure is about the threshold to detect differences in energy expenditure using indirect calorimetry [20,21]. Furthermore, while a two-fold higher glucose oxidation rate was predicted for the non-responders compared to the responders, the RER data from the non-responders are not in agreement with this prediction. However, it should also be considered that, given the number of responders and non-responders and the expected difference in RER if the glucose utilization was 10% for the responders and 20% for the non-responders, the power to detect this difference would be less than 50%. All in all, despite a lack of observed differences using indirect calorimetry, the non-responders likely have an increased level of energy expenditure in addition to a decreased level of fat absorption.

It is tempting to speculate that a decrease in fat absorption also leads to a higher level of energy expenditure. While the mechanism behind this is unclear, it has been proposed that this effect may be due to a shift towards absorption more distally in the small intestine, leading to less chylomicron production and with a smaller size. These smaller chylomicrons are then believed to tip the scale more towards utilization than storage, explaining the higher energy expenditure [22]. In humans, bariatric surgery also leads to increased nutrient availability in the distal small intestine. The weight loss associated with bariatric surgery, however, is neither due to the increased energy expenditure nor malabsorption per se, but attributable to the decrease in food intake in response to the increased production of incretins, such as GLP-1 [23,24]. Since in this study no decrease in food intake was observed (Figure S5), such a mechanism is likely not relevant here.

4.3. Bile Acid Metabolism

An important result of this study was that apart from the tight association in male APOE*3-Leiden.CETP mice between body weight and fecal fat excretion, there is also a tight connection between fecal fat excretion and bile acid metabolism. However, what drives the observed differences in bile acid homeostasis remains unclear. Tarasco et al.

reported that the livers of non-responder APOE*3-Leiden.CETP mice were found to have more inflammation and less steatosis, and occasionally formed neoplasms [19]. In the current study, though no significantly increased inflammation was found, we did observe less steatosis in non-responder mice, and found (pre)neoplastic deformations in two out of the nine non-responder mice whose histology was available. However, the difference in the response to the HFCDs was present from the moment the TG levels rose in the responding mice, probably in part caused by the difference in bile acid excretion which was much lower in the non-responding mice (Figure 2) at the onset of the experiments. Together this makes liver injury as an explanation for the difference in bile acid homeostasis not very likely. We conclude that the altered fecal bile acid profile and concurrent changes in the fat uptake dynamics may contribute substantially to the observed decrease in the plasma lipids and the lower body weight found in the non-responders. While the correlation between the fecal deoxycholic acid level and body weight was strong, it should be noted that deoxycholic acid is mainly produced in the colon and is therefore not likely to significantly contribute to fat uptake in the small intestine. Rather, fecal deoxycholic acid is likely a sensitive marker for the availability of cholic acid in the small intestine, through the interaction with the microbiome [25,26]. Cholic acid in the small intestine then drives the observed fat uptake dynamics. The question arises whether our results in a mouse with a humanized lipid profile can be translated to development of the metabolic syndrome in humans. The efficiency of fat absorption in humans is comparable to that in mice, hence a small decrease may also substantially affect the progression of obesity in humans. Although bile acid homeostasis differs substantially between mice and humans, in both species the level of the excretion of deoxycholic acid is high, hence similar mechanisms could play a role. The pharmacological perturbation of bile acid metabolism may prove successful in combating obesity and dyslipidemia in humans.

5. Conclusions

This study demonstrates how a systems analysis may be used to explore heterogeneity in the propensity to develop the metabolic syndrome. Using ADAPT, we show that there is increased level of fecal fat excretion and that there must be an increased level of energy expenditure in APOE*3-Leiden.CETP-mice that do not respond to a HFCD. Finally, we show that these differences appear to be coupled to a decreased production of bile acids and a decrease in the fecal excretion of deoxycholic acid. Further studies should address whether similar mechanisms may be responsible for the differences in susceptibility in developing dyslipidemia and obesity in the human population.

Supplementary Materials: The following supporting information can be downloaded at: https://www.mdpi.com/article/10.3390/nu14224936/s1, S1: Finding appropriate settings within the ADAPT procedure for the model; S2: Translation of Experimental Data to Modeling Constraints; Table S1: Description of model states; Table S2: Description of model parameters; Table S3: Description of model fluxes; Figure S1: The data fit error and regularization error for different lambdas (logarithmic scale); Figure S2: The Data fit error and regularization error for different amount of time steps; Figure S3: Metabolic parameters; Figure S4: Evolution of insulin sensitivity of responders (red) and non-responders (blue); Figure S5: Simulation results for food intake (FI), hepatic TG (HepTG), total cholesterol (HepTC) and free cholesterol (HepFC); Figure S6: Simulation results; Figure S7: Correlations between plasma TG and sterol parameters; Figure S8: Predictions for hepatic and peripheral glucose oxidation; Figure S9: Correlations between individual bile acid species and fecal FFA concentrations; Figure S10: Biliary bile acid secretion rate and hydrohobicity index of biliary bile acids; Figure S11: Individual bile acids in responders and non responders; Figure S12: Effect of HFCD on bile acids; Figure S13: Bile acid production; Figure S14: Total plasma bile acds; Figure S15: Total plasma bile acids used in calcula-tions; Figure S16: Total bile acids in for all animals used in the longitudinal (6 months) cohort and validation study.

Author Contributions: Conceptualization, Y.P. and A.K.G.; Methodology, Y.P. and Y.J.W.R.; Formal Analysis, Y.P.; Investigation, Y.P., E.Z., J.F.d.B., A.G. and T.H.v.D.; Writing—Original Draft, Y.P.; Writing—Review and Editing, E.Z., Y.W., Y.J.W.R., P.C.N.R., K.W.v.D., J.A.K., B.M.B., N.A.W.v.R. and A.K.G.; Funding Acquisition, A.K.G., J.A.K. and N.A.W.v.R.; Supervision, A.K.G., J.A.K., B.M.B., N.A.W.v.R. and Y.W. All authors have read and agreed to the published version of the manuscript.

Funding: This work was funded by grants from the European Union, grant FP7-HEALTH n 305707; acronym RESOLVE (YP, JAK, YJWR, NAWR and AKG) and FP7-603091; Acronym TransCard as well as the Netherlands CardioVascular Research Initiative: "the Dutch Heart Foundation, Dutch Federation of University Medical Centers, the Netherlands Organization for Health Research and Development and the Royal Netherlands Academy of Sciences" (CVON2017-2020; Acronym Genius2 to JAK and PCNR). JAK and PCNR are Established Investigators of the Netherlands Heart Foundation (2015T068 and 2009T038).

Institutional Review Board Statement: Experimental procedures were approved by the Ethics Committees for Animal Ex-periments of the University of Groningen (Protocol Code 6903).

Data Availability Statement: Data have been published in part in [3] and Supplemental Report.

Conflicts of Interest: The authors declare no conflict of interest.

References

1. Grundy, S.M. Hypertriglyceridemia, Atherogenic Dyslipidemia, and the Metabolic Syndrome. *Am. J. Cardiol.* **1998**, *81*, 18B–25B. [CrossRef]
2. Locke, A.E.; Kahali, B.; Berndt, S.I.; Justice, A.E.; Pers, T.H.; Day, F.R.; Powell, C.; Vedantam, S.; Buchkovich, M.L.; Yang, J.; et al. Genetic studies of body mass index yield new insights for obesity biology. *Nature* **2015**, *518*, 197–206. [CrossRef] [PubMed]
3. Paalvast, Y.; Gerding, A.; Wang, Y.; Bloks, V.W.; van Dijk, T.H.; Havinga, R.; Willems van Dijk, K.; Rensen, P.C.; Bakker, B.M.; Kuivenhoven, J.A.; et al. Male apoE*3-Leiden.CETP mice on high-fat high-cholesterol diet exhibit a biphasic dyslipidemic response, mimicking the changes in plasma lipids observed through life in men. *Physiol. Rep.* **2017**, *5*, e13376. [CrossRef]
4. Westerterp, M.; van der Hoogt, C.C.; de Haan, W.; Offerman, E.H.; Dallinga-Thie, G.M.; Jukema, J.W.; Havekes, L.M.; Rensen, P.C. Cholesteryl ester transfer protein decreases high-density lipoprotein and severely aggravates atheroscler-sis in APOE*3-Leiden mice. *Arterioscler. Thromb. Vasc. Biol.* **2006**, *26*, 2552–2559. [CrossRef]
5. Tiemann, C.A.; Vanlier, J.; Hilbers, P.A.; van Riel, N.A. Parameter adaptations during phenotype transitions in progressive diseases. *BMC Syst. Biol.* **2011**, *5*, 174. [CrossRef] [PubMed]
6. Rozendaal, Y.J.W.; Wang, Y.; Paalvast, Y.; Tambyrajah, L.L.; Li, Z.; Van Dijk, K.W.; Rensen, P.C.N.; Kuivenhoven, J.A.; Groen, A.K.; Hilbers, P.A.J.; et al. In vivo and in silico dynamics of the development of Metabolic Syndrome. *PLOS Comput. Biol.* **2018**, *14*, e1006145. [CrossRef] [PubMed]
7. Tiemann, C.A.; Vanlier, J.; Oosterveer, M.H.; Groen, A.K.; Hilbers, P.A.J.; van Riel, N.A.W. Parameter Trajectory Analysis to Identify Treatment Effects of Pharmacological Interventions. *PLoS Comput. Biol.* **2013**, *9*, e1003166. [CrossRef]
8. Rye, K.-A.; Barter, P.J. Regulation of High-Density Lipoprotein Metabolism. *Circ. Res.* **2014**, *114*, 143–156. [CrossRef]
9. Vrins, C.L.; Ottenhoff, R.; van den Oever, K.; de Waart, D.R.; Kruyt, J.K.; Zhao, Y.; van Berkel, T.J.; Havekes, L.M.; Aerts, J.M.; van Eck, M.; et al. Trans-intestinal cholesterol efflux is not mediated through high density lipo-protein. *J. Lipid Res.* **2012**, *53*, 2017–2023. [CrossRef]
10. van der Veen, J.N.; van Dijk, T.H.; Vrins, C.L.; van Meer, H.; Havinga, R.; Bijsterveld, K.; Tietge, U.J.; Groen, A.K.; Kuipers, F. Faculty Opinions recommendation of Activation of the liver X receptor stimulates trans-intestinal excretion of plasma cholesterol. *J. Biol. Chem.* **2009**, *284*, 19211–19219. [CrossRef]
11. Van den Kamer, J.H.; Huinink, B.O.; Weyers, H.A. Method for the determination of fat in feces. *J. Biol. Chem.* **1949**, *177*, 347–355. [CrossRef]
12. Siperstein, M.; Guest, M. Studies on the site of the feedback control of cholesterol synthesis. *J. Biol. Chem.* **1960**, *39*, 642–652. [CrossRef] [PubMed]
13. Heuman, D.M. Quantitative estimation of the hydrophilic-hydrophobic balance of mixed bile salt solutions. *J. Lipid Res.* **1989**, *30*, 719–730. [CrossRef]
14. Wang, D.Q.H.; Tazuma, S.; Cohen, D.E.; Carey, M.C. Feeding natural hydrophilic bile acids inhibits intestinal choles-terol absorption: Studies in the gallstone-susceptible mouse. *Am. J. Physiol. Gastrointest. Liver Physiol.* **2003**, *285*, G494–G502. [CrossRef]
15. Chevre, R.; Trigueros-Motos, L.; Castaño, D.; Chua, T.; Corlianò, M.; Patankar, J.V.; Sng, L.; Sim, L.; Juin, T.L.; Carissimo, G.; et al. Therapeutic modulation of the bile acid pool by *Cyp8b1* knockdown protects against nonalcoholic fatty liver disease in mice. *FASEB J.* **2018**, *32*, 3792–3802. [CrossRef] [PubMed]
16. Higuchi, S.; Ahmad, T.R.; Argueta, D.A.; Perez, P.A.; Zhao, C.; Schwartz, G.J.; DiPatrizio, N.V.; Haeusler, R.A. Bile acid composition regulates GPR119-dependent intestinal lipid sensing and food intake regulation in mice. *Gut* **2020**, *69*, 1620–1628. [CrossRef]

17. Regazzoni, F.; Chapelle, D.; Moireau, P. Combining data assimilation and machine learning to build data-driven models for unknown long time dynamics—Applications in cardiovascular modeling. *Int. J. Numer. Methods Biomed. Eng.* **2021**, *37*, e3471. [CrossRef]
18. van Riel, N.A.W.; Tiemann, C.A.; Hilbers, P.A.J.; Groen, A.K. Metabolic Modeling Combined with Machine Learning Integrates Longitudinal Data and Identifies the Origin of LXR-Induced Hepatic Steatosis. *Front. Bioeng. Biotechnol.* **2021**, *8*, 536957. [CrossRef]
19. Tarasco, E.; Pellegrini, G.; Whiting, L.; Lutz, T.A. Phenotypical heterogeneity in responder and nonresponder male ApoE*3Leiden.CETP mice. *Am. J. Physiol. Liver Physiol.* **2018**, *315*, G602–G617. [CrossRef]
20. Butler, A.A.; Kozak, L.P. A recurring problem with the analysis of energy expenditure in genetic models express-ing lean and obese phenotypes. *Diabetes* **2010**, *59*, 323–329. [CrossRef]
21. Even, P.C.; Nadkarni, N.A. Indirect calorimetry in laboratory mice and rats: Principles, practical considerations, interpretation and perspectives. *Am. J. Physiol. Regul. Integr. Comp. Physiol.* **2012**, *303*, R459–R476. [CrossRef] [PubMed]
22. Buhman, K.K.; Smith, S.J.; Stone, S.J.; Repa, J.J.; Wong, J.S.; Knapp, F.; Burri, B.J.; Hamilton, R.L.; Abumrad, N.A.; Farese, R.V., Jr. DGAT1 Is Not Essential for Intestinal Triacylglycerol Absorption or Chylomicron Synthesis. *J. Biol. Chem.* **2002**, *277*, 25474–25479. [CrossRef] [PubMed]
23. Carswell, K.A.; Vincent, R.P.; Belgaumkar, A.P.; Sherwood, R.A.; Amiel, S.A.; Patel, A.G.; Le Roux, C.W. The effect of bari-atric surgery on intestinal absorption and transit time. *Obes. Surg.* **2014**, *24*, 796–805. [CrossRef]
24. Gribble, F.M.; Reimann, F. Enteroendocrine Cells: Chemosensors in the Intestinal Epithelium. *Annu. Rev. Physiol.* **2016**, *78*, 277–299. [CrossRef] [PubMed]
25. Ridlon, J.M.; Alves, J.M.; Hylemon, P.B.; Bajaj, J.S. Cirrhosis, bile acids and gut microbiota: Unraveling a complex relationship. *Gut Microbes* **2013**, *4*, 382–387. [CrossRef] [PubMed]
26. Ridlon, J.M.; Kang, D.-J.; Hylemon, P.B.; Bajaj, J.S. Gut Microbiota, Cirrhosis, and Alcohol Regulate Bile Acid Metabolism in the Gut. *Dig. Dis.* **2015**, *33*, 338–345. [CrossRef]

nutrients

MDPI

Article

Serum Low Density Lipoprotein Cholesterol Concentration Is Not Dependent on Cholesterol Synthesis and Absorption in Healthy Humans

Frans Stellaard [1,2,*], Sabine Baumgartner [1], Ronald Mensink [1], Bjorn Winkens [3], Jogchum Plat [1] and Dieter Lütjohann [2]

1 Department of Nutrition and Movement Sciences, NUTRIM (School of Nutrition and Translational Research in Metabolism), Maastricht University Medical Center, 6200 MD Maastricht, The Netherlands
2 Institute of Clinical Chemistry and Clinical Pharmacology, University Hospital Bonn, Venusberg-Campus 1, 53127 Bonn, Germany
3 Department of Methodology and Statistics, CAPHRI (Care and Public Health Research Institute), Maastricht University, 6200 MD Maastricht, The Netherlands
* Correspondence: f.stellaard@maastrichtuniversity.nl; Tel.: +31-652579392

Abstract: Introduction. Pharmacological reduction of cholesterol (C) synthesis and C absorption lowers serum low-density lipoprotein C (LDL-C) concentrations. We questioned whether high baseline C synthesis or C absorption translates into high serum LDL-C concentrations or if there was no connection. Therefore, we studied the association between serum LDL-C and C synthesis or C absorption in healthy subjects. Methods. Three published data sets of young subjects on different diets (study 1), mildly hypercholesterolemic subjects without cardiovascular disease (study 2) and healthy controls of the Framingham study (study 3) were used. The three study populations varied in sex, age, and weight. C synthesis and C fractional absorption rate (FAR) were measured with fecal sterol balance and stable isotope techniques (studies 1 and 2). Additionally, serum lathosterol and campesterol concentrations corrected for the serum total C concentration (R_lathosterol and R_campesterol) were used as markers for hepatic C synthesis and C FAR, respectively (studies 1–3). Linear regression analysis was applied to evaluate associations between LDL-C, C synthesis, and C absorption. Results. Seventy-three, 37, and 175 subjects were included in studies 1, 2, and 3, respectively. No statistically significant associations were found between LDL-C and the measured C synthesis and C FAR, nor for R_lathosterol and R_campesterol in any of the study groups. This lack of associations was confirmed by comparing the male subjects of studies 1 and 2. Study 1 subjects had a 50% lower serum LDL-C than the study 2 subjects ($p < 0.01$), but not a lower C synthesis, C FAR, R-lathosterol, or R_campesterol. Conclusions. Under physiological conditions, C synthesis and C FAR are not major determinants of circulating serum LDL-C concentrations in healthy subjects. The results need to be confirmed in large-scale studies in healthy subjects and patients at risk for cardiovascular disease.

Keywords: LDL cholesterol; synthesis; absorption; lathosterol; campesterol

Citation: Stellaard, F.; Baumgartner, S.; Mensink, R.; Winkens, B.; Plat, J.; Lütjohann, D. Serum Low Density Lipoprotein Cholesterol Concentration Is Not Dependent on Cholesterol Synthesis and Absorption in Healthy Humans. *Nutrients* **2022**, *14*, 5370. https://doi.org/10.3390/nu14245370

Academic Editor: Maria Luz Fernandez

Received: 26 November 2022
Accepted: 15 December 2022
Published: 17 December 2022

1. Introduction

Patients with elevated total serum cholesterol (TC) and in particular low-density lipoprotein C (LDL-C) in combination with additional risk factors for cardiovascular disease (CVD), such as obesity, diabetes mellitus type 2 (DM2), hypertension (HT), and family history with cardiovascular events, are treated by cholesterol-lowering therapies. First, patients need to achieve sufficient reductions in serum LDL-C concentrations and patients with high-risk scores require even more extensive LDL-C lowering to reach LDL-C levels below 1.5 mmol/L or 55 mg/dL [1,2]. LDL-C lowering requires upregulation of the hepatic LDL-receptor, which can be established by a reduction of the hepatic free C pool. The first treatment of choice is a statin treatment, which aims to reduce cholesterol synthesis in the liver by blocking 3-hydroxy-3-methylglutaryl coenzyme A (HMG-CoA) reductase activity in the mevalonate pathway. In the case of insufficient serum LDL-C reduction, even after a dosage increase and/or change in statin, it is generally concluded that the patient's C synthesis is too low to react sufficiently to statin treatment. As an alternative, ezetimibe treatment may be started to inhibit C absorption, as a high C absorption rate is a second potential cause of elevated serum C. Ezetimibe inhibits the Niemann-Pick C1-Like 1 (NPC1L1) protein that transports sterols into the intestinal cell and reduces the absorption of dietary and biliary C and therewith the hepatic C influx [1]. To maximize the C-lowering effect, a statin and ezetimibe may be combined [3]. It is generally known that C-lowering drugs, and in particular statins, may have significant side effects. Increasing the dose using single drug treatment enhances the side effects. During combination treatment, low dosages of both drugs can be used. Twenty mg of statin is combined with 10 mg of ezetimibe daily. A low-dose combination treatment is more effective than a high-dose single treatment of statin or ezetimibe acid [4,5]. In the case of statin intolerance, statins can be replaced by bempedoic acid [6]. Patients with familial hypercholesterolemia have high serum LDL-C levels, i.e., >5 mmol/L or >190 mg/dL [7]. This may be caused by mutations encoding the LDL-receptor leading to very low receptor activity [8,9] or by hyperactivity or a higher number of proprotein convertase subtilisin/kexin type 9 (PCSK9) [10], which catabolizes the LDL-receptor [11]. In this case, statin and ezetimibe treatments are not sufficient. PCSK9 inhibition is then added [12–14].

An interesting observation was made in the Framingham study as described by Matthan et al. [15]. Cases with defined CVD and not taking lipid-lowering drugs were compared with control subjects. The serum LDL-C concentrations were normal to mildly increased and identical in both groups. Thus, in this study, the cardiovascular events were not related to elevated LDL-C concentrations. The authors found that the cases had enhanced serum concentrations of campesterol, sitosterol, and cholestanol corrected for the TC concentration (R_campesterol, R_sitosterol, R_cholestanol), which represent surrogate markers for C absorption [15]. The serum lathosterol concentration corrected for the TC concentration (R_lathosterol), as a surrogate marker for C synthesis was found to be reduced. These results [12] suggest that increased C absorption possibly initiates cardiovascular events without increasing serum LDL-C. This leads to the question of whether LDL-C is related to C synthesis and C absorption. Pharmacological treatment demonstrates that inhibition of synthesis and particular hepatic synthesis by statins and inhibition of absorption by ezetimibe reduces serum LDL-C dose-dependently [3]. However, does the relationship also apply under physiological conditions? The result of the Framingham study may also have been affected by the fact that surrogate serum markers for absorption and synthesis were used instead of a direct measurement of whole-body C synthesis and C absorption. The association between absorption markers and directly measured fractional absorption was questioned under physiological conditions [16]. Do the markers reflect the actual function in the physiological situation or only during pharmacological treatment?

It must be realized that hepatic C synthesis and C absorption affect the hepatic C pool. Reduction of this pool by means of statin and ezetimibe treatment initiates serum LDL-C lowering. However, C synthesis and C absorption are not the only fluxes affecting the hepatic C pool. Other C influxes such as HDL-C and effluxes such as very low density lipoprotein C (VLDL-C), biliary C secretion, and bile acid synthesis, play important roles. Therefore, hepatic C homeostasis is a complex multifactorial process, and, as a consequence, the establishment of the serum LDL-C concentration is also complex. This accentuates the question of whether C synthesis and C absorption play dominant roles in the establishment of the serum LDL-C concentration. Therefore, we aimed to determine the associations between serum LDL-C and C synthesis and C absorption. To study the effects of physiological parameters such as age, sex, and BMI on the potential relationships, we restricted our study population to healthy subjects in order to exclude confounding effects introduced by pathological events. Furthermore, we needed to differentiate between the methods of measuring C synthesis and C absorption. At first, we used the classical fecal sterol balance method [17] and the stable isotope technique [3], respectively. These techniques have been applied only in a limited number of small studies. Secondly, surrogate marker techniques [18,19] have been developed that can be used on a larger scale. In this study, we validated the marker technology for association studies.

2. Materials and Methods

Data from three earlier published studies were re-evaluated for these new research questions. In the first study, serum lipids as well as C absorption and C synthesis were measured with original stable isotope tracer techniques and side-by-side with surrogate marker concentrations in young omnivores, lacto-ovo vegetarians, lacto vegetarians, and vegans [20]. The diet compositions have been described in detail in the original publication [17]. Shortly, omnivores ate all kinds of foods. Lacto-ovo vegetarians consumed no meat, fish, or dairy products. Lacto vegetarians did not eat meat, fish, or eggs. Vegans did not consume meat, fish, eggs, dairy products, or honey. Subjects under any medication or intake of dietary supplements fortified with cholesterol-lowering agents such as plant sterol or stanol esters were excluded. In the second study, the same parameters were measured in older mildly hypercholesterolemic subjects under placebo and cholesterol-lowering conditions [21]. Only the data obtained under placebo conditions were included in this study. Both studies were performed at the Institute of Clinical Chemistry and Clinical Pharmacology, University Hospital Bonn, Bonn, Germany. All serum lipids, whole-body cholesterol synthesis, and the fractional cholesterol absorption rate, as well as surrogate plasma markers, were measured in the Laboratory for Special Lipid Analysis at the same institute. Fecal neutral and acidic sterol excretion was measured by gas chromatography-flame ionization detection (GC-FID) [17,18]. C absorption was measured with stable isotope methodology applying GC-mass spectrometry (MS) [17,18,22]. Synthesis and absorption markers in serum were measured with GC-MS [23,24]. The third study was part of the Framingham study [15]. From this study, only the data of controls who did not suffer from diabetes or hypertension were selected. Serum lipids and C synthesis and absorption marker concentrations and their ratios to enzymatically determined cholesterol were analyzed at the Lipid Metabolism Laboratories of the Tufts University School of Medicine, Boston, USA. Synthesis and absorption markers were measured by GC-FID. Studies 1 and 2 allow us to study the associations between LDL-C and synthesis and absorption measured with the original stable isotope tracer methods and with marker technology (lathosterol, campesterol, R_Lathosterol and R_Campesterol), while in study 3, only the synthesis and absorption markers could be used to relate to serum LDL-C.

3. Statistical Analysis

Within each study group, the association between LDL-C (dependent variable) and the individual parameters C synthesis, C FAR, and the marker concentrations and ratios (independent variables) were determined by applying linear regression analysis. In study 1, associations were tested in all four diet groups separately: omnivores, lacto-ovo vegetarians, lacto vegetarians, and pure vegans. In studies 1 and 3, associations were also tested in males and females separately. In study 3, the effects of age and BMI on the associations were studied. At first, it was determined whether the slope was significantly different from zero ($p \leq 0.05$) and whether the slope was positive or negative. Secondly, the R square was determined in order to measure the goodness of fit and to express the potential overall contribution of the tested parameter to the establishment of the height of serum LDL-C concentration. In studies 1 and 2, the mean measured C FAR, synthesis rates, and marker concentrations and ratios were calculated in all subjects. As both studies were performed at the same institute, the data of the males of study 1 were compared with the data of study 2 which was composed of males only. These groups were compared to check whether there was a difference in LDL-C between the groups, which might be accompanied by a difference in C synthesis, C FAR, their marker concentrations and/or ratios to TC. Group comparison was performed using the Mann Whitney U test. $p \leq 0.05$ was considered significant. GraphPad Prism 8.0.2 was used for all statistical analyses.

4. Results

Subject characteristics including serum lipid and lipoprotein concentrations are presented in Table 1. Differences in age, weight, BMI, and serum lipids are observed. Most strikingly, the subjects in study 1 were the youngest, had the lowest weight and BMI, as well as the lowest serum lipid concentrations. Otherwise, subjects in study 3 were the oldest and had the highest BMI and serum triglyceride levels. In Table 2, the results from linear regression analyses are summarized. In a number of study groups, but not all, the lathosterol and/or campesterol concentrations were positively and significantly associated with LDL-C. However, the marker ratios R_lathosterol and R_campesterol, as well as the C FAR and C synthesis rates, were not significantly associated with LDL-C. These lacking associations were not affected by gender, age, and BMI, as subgroup analyses based on these variables showed similar results. The highest R square value was 0.22, found for C FAR. This indicates that neither C FAR nor C synthesis relevantly contributes to serum LDL-C concentrations.

Table 1. Subject characteristics of the three study populations. The data are expressed as number of subjects or as mean ± standard deviation. TC = total cholesterol, LDL-C =low density lipoprotein cholesterol, HDL-C =high density lipoprotein C, TG = triglycerides).

	Study 1	Study 2	Study 3
N	73	37	175
Sex	37 F, 36 M	M	50 F, 125 M
Age (years)	25 ± 3	41 ± 8	64 ± 8
Weight (kg)	67 ± 13	84 ± 10	82 ± 16
BMI (kg/m^2)	22 ± 3	25 ± 2	28 ± 5
TC (mg/dL)	179 ± 28	233 ± 28	204 ± 34
LDL-C (mg/dL)	105 ± 22	157 ± 22	129 ± 31
HDL-C (g/dL)	57 ± 14	53 ± 13	48 ± 15
TG (mg/dL)	89 ± 28	118 ± 43	141 ± 101

Table 2. R_square data for the linear regression evaluating associations between serum LDL-C concentrations and various parameters expressing C synthesis and C absorption.

	Lathosterol	Campesterol	R_Lathosterol	R_Campesterol	Synthesis	FAR
Study 1 Omnivores (n = 19)	**0.32 (+)**	0.13	0.04	0.03	0.06	0.01
Study 1 Lacto-ovo vegetarians (n = 18)	**0.27 (+)**	0.20	0.01	0.04	0.02	0.01
Study 1 Lacto vegetarians (n = 17)	0.03	0.03	0.12	0.09	0.02	0.22
Study 1 Pure vegans (19)	0.09	0.03	0.03	0.04	0.18	0.04
Study 1 Females (37)	0.03	**0.12 (+)**	0.01	0.01	0.06	0.05
Study 1 Males (36)	0.03	0.00	0.03	0.01	0.04	0.03
Study 2 All subjects (n = 37)	0.00	**0.20 (+)**	0.04	0.06	0.02	0.10
Study 3 All subjects (n = 171)	**0.05 (+)**	**0.13 (+)**	0.00	0.00		
Study 3 Females (n = 49)	0.00	0.01	0.07	0.05		
Study 3 Males (n = 171)	**0.10 (+)**	**0.19 (+)**	0.00	0.00		
Study 3 Age 30 to 59 years (n = 48)	**0.20 (+)**	**0.16 (+)**	0.00	0.02		
Study 3 Age 70 to 82 years (n = 53)	0.06	**0.26 (+)**	0.00	0.03		
Study 3 BMI 20 to 25 (n = 48)	0.01	0.06	0.01	0.04		
Study 3 BMI 30 to 47 (n = 54)	0.04	**0.10 (+)**	0.03	0.00		

Bold numbers indicate that the slope was significantly different from zero. (+) indicates a positive association with LDL-C. R square varies from 0 (no fit between line and points) to 1 (perfect fit). FAR = fractional absorption rate (%). Due to the large number of participants in study 3, we were able to test the effects of age, sex and BMI on the linear regression data.

The measured C synthesis and C FAR values, as well as the absolute marker concentrations and ratios to the TC concentration in studies 1 and 2, are included in Table 3 where the males of study 1 are compared with the males of study 2. Significant differences were found for age, weight, BMI, and serum lipids with the exception of high density lipoprotein C (HDL-C). The participants in study 2 had a 50% higher serum LDL-C concentration than the male participants in study 1, and both lathosterol and campesterol concentrations were higher. However, the marker ratios R_lathosterol and R_campesterol, C FAR and C synthesis were not significantly different between both groups. These observations are in agreement with the lack of associations found with linear regression analysis in the individual study groups.

Table 3. Comparison of subject characteristics, serum lipids, marker concentrations, marker ratios, FAR and synthesis for male subjects of study 1 and 2. Data are expressed as number of subjects or as median and interquartile range (25th and 75th percentile). The statistical difference was tested with the Mann Whitney U Test. TC = total cholesterol, LDL-C =low density lipoprotein cholesterol, HDL-C =high density lipoprotein C, TG = triglycerides), Lath = lathosterol, Camp = campesterol, FAR = fractional absorption rate (%).

	Study 1	Study 1 Males	Study 2	*p*
N	73	36	37	
Sex	37 F, 36 M	36M	37M	
Age (years)	25 (23 to 27)	25 (24 to 26)	41 (37 to 47)	<0.01
Weight (kg)	64 (58 to 74)	73 (65 to 81)	84 (75 to 91)	<0.01
BMI (kg/m^2)	21 (20 to 23)	22 (20 to 24)	25 (24 to2 7)	<0.01
TC (mg/dL)	180 (159 to 200)	171 (149 to 191)	231 (213 to 252)	<0.01
LDL-C (mg/dL)	100 (89 to 124)	101 (84 to 121)	155 (139 to 174)	<0.01
HDL-C (g/dL)	54 (47 to 67)	50 (43 to 56)	50 (46 to 54)	0.98
TG (mg/dL)	91 (67 to 107)	93 (64 to 114)	109 (88 to 144)	<0.01
Lath (mg/dL)	0.26 (0.18 to 0.35)	0.28 (0.19 to 0.36)	0.37 (0.29 to 0.44)	<0.01
Camp (mg/dL)	0.34 (0.24 to 0.42)	0.34 (0.24 to 0.43)	0.58 (0.33 to 0.70)	<0.01
R_Lath (mg/g)	1.5 (1.1 to 1.9)	1.6 (1.3 to 2.2)	1.7 (1.1 to 2.1)	0.56
R_Camp (mg/g)	1.9 (1.5 to 2.6)	2.2 (1.6 to 2.7)	2.5 (1.4 to 2.9)	0.79
FAR (%)	46 (40 to 57)	47 (41 to 58)	52 (43 to 58)	0.59
Synthesis (mg/d)	917 (752 to 1126)	1049 (832 to 1360)	854 (686 to 1328)	0.14

5. Discussion

In this study, we tested the associations between serum LDL-C and C synthesis and C absorption under physiological conditions in healthy subjects. In studies 1 and 2, the associations could be studied using measured C synthesis and C absorption, as well as using the marker ratios R_lathosterol and R_campesterol. Applying linear regression, no significant associations could be detected, not for the measured values, nor for the marker ratios. Additionally, a comparison of the male population of study 1 with the subjects of study 2 showed a 50% higher serum LDL-C in study 2. However, the measured C synthesis and C absorption data as well as the marker ratio data were not significantly different in both studies. Studies 1 and 2 were performed in the same hospital and also the same analytical techniques were used. Study 3 was included in order to test the hypothesis in a larger and independent study population. Study 3 was performed in different clinical and laboratory settings. Therefore, we did not perform a group comparison between studies 3 and 1 or 2. It was important to find similar low associations in studies 1 and 2 for the measured synthesis and absorption data and the marker ratios. This process proved the validity of the marker ratios for testing the hypothesis and allowed the interpretation of the regression data for the marker ratios in study 3. Furthermore, in study 3, no significant associations were found between serum LDL-C and the marker ratios R_lathosterol and R_campesterol. The associations were assessed in the whole study group and in males and females, as well as in subjects with different ages and BMIs. The lack of associations was not affected by sex, age, and BMI. The subjects in studies 1 and 2 were all Caucasian.

Unfortunately, no data were available for the ethnic background of the subjects of the Framingham study (study 3). This could potentially affect the results. As indicated in Table 2, in some groups, significant positive associations were found between serum LDL-C and the serum lathosterol and/or campesterol concentrations. These associations may be explained by the fact that the markers and C are transported and comparably distributed in the same lipoprotein particles and are similarly affected by the hepatic LDL-receptor activity. Linear regression statistics for the marker ratios need special attention. The marker ratio (Y/X) expresses the ratio between the serum marker concentration (Y) and the serum TC concentration (X). The serum TC concentration is strongly positively associated with the LDL-C concentration as the majority of TC consists of LDL-C. Thus, LDL-C and 1/TC exhibit a negative spurious correlation. As a consequence, a negative relationship between the marker ratio and LDL-C may be expected, even if synthesis and absorption are not associated with LDL-C. However, a lacking association between marker ratio and LDL-C directly proves that C synthesis or C absorption are not determinants of the LDL-C concentration. The R square value obtained with linear regression statistics permits testing whether the X-value adds a relevant contribution to the establishment of the (variation in) Y-value. R square varies from 0 (no fit between the regression line and individual points) to 1 (complete fit). As shown in Table 2, the R square values of all parameters expressing C absorption and C synthesis were low and indicate that FAR and synthesis are not relevant determinants of serum LDL-C under physiological conditions. The lathosterol and campesterol concentrations scored higher R squares, but the highest value of 0.31 for lathosterol is still too low to ascribe a relevant contribution to the size of the serum LDL-C concentration. The relationship between synthesis or FAR with LDL-C has been frequently documented under treatment with a C synthesis or C FAR reducing agent [21,25]. Under physiological conditions and particularly in healthy humans, relationships have not been studied systematically. Surprisingly, our study appears the first to specifically address this important question. Miettinen et al. studied the plant sterols and cholesterol precursors as markers for C FAR and C synthesis in volunteers of a randomly selected Finnish male population [26]. Applying their ratios to total C, no associations were found between the ratios and LDL-C. In another study, Miettinen et al. related the serum cholestanol/cholesterol ratio as a marker for C absorption to lipoprotein C concentrations in 50-year-old-men and found no correlation with LDL-C [27]. Silbernagel et al. described the relationships between C absorption markers and the ATP-binding cassette sub-family G member 5 and G member 8 (ABCG5/G8) alleles in the Ludwigshafen Risk and Cardiovascular health study (LURIC) and the Young Finns Study (YFS) cohorts [28]. In both cohorts, strong associations between frequencies of alleles and the serum marker ratios were found ($p < 0.0001$). However, associations with serum LDL-C were much weaker, being $p = 0.02$ for LURIC and 0.28 for YFS. This indicates a missing link between absorption and serum LDL-C concentrations. Kesäniemi et al. studied the effect of C FAR on serum C in the Finnish population [29] and found a high LDL-C to be associated with a high C FAR and a low C synthesis. C FAR was measured with stable isotope technology. These results were confirmed by Miettinen et al. [26] in 50-year-old-men. The results of the last two studies appear to contradict our results as well as those of the other Finnish studies by the same research group. In vegetarians, reduced cholesterol intake results in only small changes in C FAR, enhanced C synthesis, and reduction of LDL-C [20,30]. This identifies the intestinal C flux as an additional parameter. We considered it important to test the associations using measured whole-body synthesis and FAR values first and in a second line with non-cholesterol sterols as markers for synthesis and FAR. The validity of markers may be dependent on many experimental conditions, as shown by Quintao [31].

Whole body C synthesis and C absorption are the two influxes into the endogenous C pool. They are considered to be determinants of serum LDL-C concentration. The serum TC concentration is the result of the C homeostasis process [32]. Absorbed C enters the liver via chylomicron remnant particles formed from chylomicrons secreted by the intestinal cells. Also, C extracted by the liver from VLDL remnants, LDL and HDL, enters the liver. Thirdly, the liver is the main contributor to whole-body C synthesis. The hepatic C effluxes are biliary C secretion and bile acid synthesis. The liver must coordinate the influxes and effluxes to keep the whole-body C pool constant. C synthesis is controlled in order to balance variation in C absorption. This statement has been proven by pharmacological data showing increased synthesis under ezetimibe treatment [19] and under extreme situations under physiological conditions. Excessive high dietary C intake and absorption may be compensated by reduced C absorption and C synthesis, as well as an increased bile acid synthesis, as shown elegantly in a man consuming 25 eggs every day [33] not suffering from hypercholesterolemia. A very low dietary C intake as in pure vegans leads to only a moderately lower serum LDL-C caused by an increased compensatory synthesis [20]. The mechanism by which the regulation takes place has still not been clarified and may not function efficiently in all subjects. Enhanced C absorption results in an increased hepatic C pool, which may be compensated by decreased hepatic C synthesis. However, also hepatic LDL-C uptake may be decreased, and VLDL-C secretion, biliary C secretion, and bile acid synthesis increased. In this extreme situation, hepatic C synthesis may remain unaltered. The sequence and extent of responsive events are unknown (for a general review of the hepatic C homeostasis, see reference [32]). The serum LDL-C concentration is established by a number of C fluxes, such as hepatic VLDL secretion, conversion of VLDL to LDL, exchange of C ester with HDL, and hepatic and extrahepatic uptake of LDL by the LDL-receptor. The hepatic C pool may be the regulator in hepatic LDL uptake [23]. Statin treatment and ezetimibe treatment share the fact that the hepatic C influx is reduced and therewith the hepatic C pool. Under statin treatment, two compensating effects to restore this pool is to lower hepatic VLDL-C secretion and enhance C extraction from the blood by upregulation of the LDL-receptor activity. Ezetimibe treatment reduces the hepatic influx of chylomicron-derived absorbed C. This effect is compensated by enhanced LDL uptake and enhanced hepatic C synthesis [23]. The original publication of study 2 [18] and the follow-up studies show that the effects of 20 mg simvastatin daily and 10 mg ezetimibe [23,34] daily have different effects on whole-body C synthesis and on R_lathosterol. R_lathosterol has been defined as a marker of hepatic C synthesis, as it reflects the expected effects of simvastatin and ezetimibe on hepatic C synthesis more closely. This also predicts that hepatic and extrahepatic synthesis are affected differently. Extrahepatic C synthesis appears increased under both simvastatin and ezetimibe treatment. It is unclear whether this hepatic balancing plays any role under physiological conditions in healthy subjects. In our study, we did not observe conflicting data for whole-body C synthesis and R_lathosterol in the relationship with serum LDL-C. The Framingham data indicated enhanced C absorption and reduced hepatic C synthesis in the case group [15]. This combination of data might explain the normal serum LDL-C concentration. However, on average, the absorption markers were only 6 to 16 % higher in the cases compared with controls, and R_lathosterol was only 15% lower. It remains unclear whether such small changes in C absorption can introduce cardiovascular events. The cases group had a statistically significantly higher number of patients with diabetes and beta-blocker users. These uninvestigated factors may have played a potentially causal role in the increased development of cardiovascular disease. Diabetes is a known risk factor for the development of atherosclerosis. Beta-blockers have been developed to treat abnormal heart rhythms and are also effective in the treatment of high blood pressure, which is also a known risk factor for cardiovascular disease. The direct effects of beta-blockers on cardiovascular disease are not clearly defined [35].

Our study aimed to test the hypothesis that synthesis and absorption are determinants of LDL-C concentration under physiological conditions. Our results do not support this notion. A high C synthesis or a high C absorption were not linked to a high serum LDL-C concentration. The lacking associations may be explained by the inversed relationship between synthesis and absorption. Our findings may be restricted to the healthy condition of our study populations. A drawback of our study is also the small sample size. Repetition of our calculations in a much larger group of healthy subjects is necessary. However, the associations must also be tested in less healthy subjects with an increased risk to develop cardiovascular diseases in whom the balance between synthesis and absorption is potentially disturbed.

It may be expected that patients with a high C synthesis are more able to reduce their synthesis rate under statin therapy, while patients with high C absorption may more strongly reduce their FAR under ezetimibe treatment. Whether these enhanced reductions directly lead to enhanced serum LDL-C reductions cannot be predicted. Using literature data from studies involving statin and ezetimibe treatment, Descamp et al. could not prove this concept [36]. As indicated by the Framingham data, cardiovascular events may be caused by a factor independent of the serum LDL-C level [15]. Thus, serum LDL-C-lowering and the prevention of CVD development may be two different goals that can be reached in parallel in many, but not all patients.

6. Conclusions

This study indicates that C synthesis and C absorption are not major determinants of the serum LDL-C concentration in healthy subjects, which means that either a high C synthesis or high C absorption does not automatically translate into a high serum LDL-C concentration. The results need to be confirmed in large-scale studies in healthy subjects and patients at risk for cardiovascular disease.

Author Contributions: Conceptualization, F.S., J.P. and D.L.; methodology, F.S. and B.W.; validation, S.B., D.L., J.P. and R.M.; formal analysis, B.W.; investigation, F.S.; resources, D.L. and J.P.; writing—original draft preparation, F.S.; writing—review and editing, S.B., R.M., B.W., J.P. and D.L.; visualization, F.S., S.B.; supervision, J.P. and D.L. All authors have read and agreed to the published version of the manuscript.

Funding: This research received no external funding.

Institutional Review Board Statement: Not applicable since only data from previously published human studies were used. Statements have been made in the original publications. These are listed in the reference list.

Informed Consent Statement: Not applicable since only data from previously published human studies were used. Statements have been made in the original publications. These are listed in the reference list.

Data Availability Statement: No publicly archived datasets have been used. The applied internal data sets may be requested from the authors of this study.

Conflicts of Interest: The authors declare no conflict of interest.

References

1. Arnett, D.K.; Blumenthal, R.S.; Albert, M.A.; Buroker, A.B.; Goldberger, Z.D.; Hahn, E.J.; Himmelfarb, C.D.; Khera, A.; Lloyd-Jones, D.; McEvoy, J.W.; et al. 2019 ACC/AHA Guideline on the Primary Prevention of Cardiovascular Disease: A Report of the American College of Cardiology/American Heart Association Task Force on Clinical Practice Guidelines. *Circulation* **2019**, *140*, e596–e646. [CrossRef] [PubMed]
2. Gu, J.; Sanchez, R.; Chauhan, A.; Fazio, S.; Wong, N. Lipid treatment status and goal attainment among patients with atherosclerotic cardiovascular disease in the United States: A 2019 update. *Am. J. Prev. Cardiol.* **2022**, *10*, 100336. [CrossRef] [PubMed]
3. Lütjohann, D.; Stellaard, F.; Mulder, M.T.; Sijbrands, E.J.G.; Weingartner, O. The emerging concept of "individualized cholesterol-lowering therapy": A change in paradigm. *Pharmacol. Ther.* **2019**, *199*, 111–116. [CrossRef] [PubMed]
4. Grigore, L.; Norata, G.D.; Catapano, A.L. Combination therapy in cholesterol reduction: Focus on ezetimibe and statins. *Vasc. Health Risk Manag.* **2008**, *4*, 267–278. [CrossRef]
5. Westerink, J.; Deanfield, J.E.; Imholz, B.P.; Spiering, W.; Basart, D.C.; Coll, B.; Kastelein, J.J.; Visseren, F.L. High-dose statin monotherapy versus low-dose statin/ezetimibe combination on fasting and postprandial lipids and endothelial function in obese patients with the metabolic syndrome: The PANACEA study. *Atherosclerosis* **2013**, *227*, 118–124. [CrossRef]

6. Tummala, R.; Gupta, M.; Devanabanda, A.R.; Bandyopadhyay, D.; Aronow, W.S.; Ray, K.K.; Mamas, M.; Ghosh, R.K. Bempedoic acid and its role in contemporary management of hyperlipidemia in atherosclerosis. *Ann. Med.* **2022**, *54*, 1287–1296. [CrossRef]
7. Jasani, R.; Ahmad, Z.; Schneider, R.; Tujardon, C.; Basit, M.; Khera, A. Applying an LDL-C threshold-based approach to identify individuals with familial hypercholesterolemia. *J. Clin. Lipidol.* **2022**, *16*, 508–515. [CrossRef]
8. Hobbs, H.H.; Brown, M.S.; Goldstein, J.L. Molecular genetics of the LDL receptor gene in familial hypercholesterolemia. *Hum. Mutat.* **1992**, *1*, 445–466. [CrossRef]
9. Hobbs, H.H.; Russell, D.W.; Brown, M.S.; Goldstein, J.L. The LDL receptor locus in familial hypercholesterolemia: Mutational analysis of a membrane protein. *Annu. Rev. Genet.* **1990**, *24*, 133–170. [CrossRef]
10. Stoekenbroek, R.M.; Lambert, G.; Cariou, B.; Hovingh, G.K. Inhibiting PCSK9—Biology beyond LDL control. *Nat. Rev. Endocrinol.* **2018**, *15*, 52–62. [CrossRef]
11. Horton, J.D.; Cohen, J.C.; Hobbs, H.H. Molecular biology of PCSK9: Its role in LDL metabolism. *Trends Biochem. Sci.* **2007**, *32*, 71–77. [CrossRef] [PubMed]
12. Nordestgaard, B.G.; Nicholls, S.J.; Langsted, A.; Ray, K.K.; Tybjaerg-Hansen, A. Advances in lipid-lowering therapy through gene-silencing technologies. *Nat. Rev. Cardiol.* **2018**, *15*, 261–272. [CrossRef] [PubMed]
13. Catapano, A.L.; Pirillo, A.; Norata, G.D. New Pharmacological Approaches to Target PCSK9. *Curr. Atheroscler. Rep.* **2020**, *22*, 24. [CrossRef] [PubMed]
14. Pirillo, A.; Norata, G.D.; Catapano, A.L. LDL-Cholesterol-Lowering Therapy. *Handb. Exp. Pharmacol.* **2022**, *270*, 73–101. [CrossRef]
15. Matthan, N.R.; Pencina, M.; LaRocque, J.M.; Jacques, P.F.; D'Agostino, R.B.; Schaefer, E.J.; Lichtenstein, A.H. Alterations in cholesterol absorption/synthesis markers characterize Framingham offspring study participants with CHD. *J. Lipid Res.* **2009**, *50*, 1927–1935. [CrossRef]
16. Jakulj, L.; Mohammed, H.; van Dijk, T.H.; Boer, T.; Turner, S.; Groen, A.K.; Vissers, M.N.; Stroes, E.S. Plasma plant sterols serve as poor markers of cholesterol absorption in man. *J. Lipid Res.* **2013**, *54*, 1144–1150. [CrossRef]
17. Czubayko, F.; Beumers, B.; Lammsfuss, S.; Lütjohann, D.; von Bergmann, K. A simplified micro-method for quantification of fecal excretion of neutral and acidic sterols for outpatient studies in humans. *J. Lipid Res.* **1991**, *32*, 1861–1867. [CrossRef]
18. Sudhop, T.; Lütjohann, D.; Kodal, A.; Igel, M.; Tribble, D.L.; Shah, S.; Perevozskaya, I.; von Bergmann, K. Inhibition of intestinal cholesterol absorption by ezetimibe in humans. *Circulation* **2002**, *106*, 1943–1948. [CrossRef]
19. Sudhop, T.; Lütjohann, D.; von Bergmann, K. Sterol transporters: Targets of natural sterols and new lipid lowering drugs. *Pharmacol. Ther.* **2005**, *105*, 333–341. [CrossRef]
20. Lütjohann, D.; Meyer, S.; von Bergmann, K.; Stellaard, F. Cholesterol Absorption and Synthesis in Vegetarians and Omnivores. *Mol. Nutr. Food Res.* **2018**, *62*, e1700689. [CrossRef]
21. Sudhop, T.; Reber, M.; Tribble, D.; Sapre, A.; Taggart, W.; Gibbons, P.; Musliner, T.; von Bergmann, K.; Lütjohann, D. Changes in cholesterol absorption and cholesterol synthesis caused by ezetimibe and/or simvastatin in men. *J. Lipid Res.* **2009**, *50*, 2117–2123. [CrossRef] [PubMed]
22. Lütjohann, D.; Meese, C.O.; Crouse, J.R., 3rd; von Bergmann, K. Evaluation of deuterated cholesterol and deuterated sitostanol for measurement of cholesterol absorption in humans. *J. Lipid Res.* **1993**, *34*, 1039–1046. [CrossRef] [PubMed]
23. Stellaard, F.; Lütjohann, D. The Interpretation of Cholesterol Balance Derived Synthesis Data and Surrogate Noncholesterol Plasma Markers for Cholesterol Synthesis under Lipid Lowering Therapies. *Cholesterol* **2017**, *2017*, 5046294. [CrossRef] [PubMed]
24. Mackay, D.S.; Jones, P.J.; Myrie, S.B.; Plat, J.; Lütjohann, D. Methodological considerations for the harmonization of non-cholesterol sterol bio-analysis. *J. Chromatogr. B Analyt. Technol. Biomed. Life Sci.* **2014**, *957*, 116–122. [CrossRef]
25. Gylling, H.; Miettinen, T.A. The effect of cholesterol absorption inhibition on low density lipoprotein cholesterol level. *Atherosclerosis* **1995**, *117*, 305–308. [CrossRef]
26. Miettinen, T.A.; Kesaniemi, Y.A. Cholesterol absorption: Regulation of cholesterol synthesis and elimination and within-population variations of serum cholesterol levels. *Am. J. Clin. Nutr.* **1989**, *49*, 629–635. [CrossRef]
27. Miettinen, T.A.; Tilvis, R.S.; Kesaniemi, Y.A. Serum cholestanol and plant sterol levels in relation to cholesterol metabolism in middle-aged men. *Metabolism* **1989**, *38*, 136–140. [CrossRef]
28. Silbernagel, G.; Chapman, M.J.; Genser, B.; Kleber, M.E.; Fauler, G.; Scharnagl, H.; Grammer, T.B.; Boehm, B.O.; Makela, K.M.; Kahonen, M.; et al. High intestinal cholesterol absorption is associated with cardiovascular disease and risk alleles in ABCG8 and ABO: Evidence from the LURIC and YFS cohorts and from a meta-analysis. *J. Am. Coll. Cardiol.* **2013**, *62*, 291–299. [CrossRef]
29. Kesaniemi, Y.A.; Miettinen, T.A. Cholesterol absorption efficiency regulates plasma cholesterol level in the Finnish population. *Eur. J. Clin. Investig.* **1987**, *17*, 391–395. [CrossRef]
30. Vuoristo, M.; Miettinen, T.A. Absorption, metabolism, and serum concentrations of cholesterol in vegetarians: Effects of cholesterol feeding. *Am. J. Clin. Nutr.* **1994**, *59*, 1325–1331. [CrossRef]
31. Quintao, E.C.R. Plasma Non-cholesterol Sterols as Markers of Cholesterol Synthesis and Intestinal Absorption: A Critical Review. *Curr. Pharm. Des.* **2020**, *26*, 5152–5162. [CrossRef] [PubMed]
32. Stellaard, F. From Dietary Cholesterol to Blood Cholesterol, Physiological Lipid Fluxes, and Cholesterol Homeostasis. *Nutrients* **2022**, *14*, 1643. [CrossRef] [PubMed]

33. Kern, F., Jr. Normal plasma cholesterol in an 88-year-old man who eats 25 eggs a day: Mechanisms of adaptation. *N. Engl. J. Med.* **1991**, *324*, 896–899. [CrossRef]
34. Stellaard, F.; von Bergmann, K.; Sudhop, T.; Lütjohann, D. The value of surrogate markers to monitor cholesterol absorption, synthesis and bioconversion to bile acids under lipid lowering therapies. *J. Steroid. Biochem. Mol. Biol.* **2017**, *169*, 111–122. [CrossRef] [PubMed]
35. Vrablik, M.; Corsini, A.; Tumova, E. Beta-blockers for Atherosclerosis Prevention: A Missed Opportunity? *Curr. Atheroscler. Rep.* **2022**, *24*, 161–169. [CrossRef]
36. Descamps, O.S.; De Sutter, J.; Guillaume, M.; Missault, L. Where does the interplay between cholesterol absorption and synthesis in the context of statin and/or ezetimibe treatment stand today? *Atherosclerosis* **2011**, *217*, 308–321. [CrossRef]

nutrients

MDPI

Article

Consuming High-Fat and Low-Fat Ground Beef Depresses High-Density and Low-Density Lipoprotein Cholesterol Concentrations, and Reduces Small, Dense Low-Density Lipoprotein Particle Abundance

Jason R. Lytle [1,†], Tara Price [2,†], Stephen F. Crouse [1], Dana R. Smith [3], Rosemary L. Walzem [4,5] and Stephen B. Smith [5,6,*]

1 Department of Health & Kinesiology, Texas A&M University, College Station, TX 77843, USA
2 Department of Nutrition and Food Science, College Station, TX 77843, USA
3 Independent Nutrition Consultant, College Station, TX 77845, USA
4 Poultry Science, Texas A&M University, College Station, TX 77843, USA
5 Graduate Faculty of Nutrition, Texas A&M University, College Station, TX 77843, USA
6 Department of Animal Science, Texas A&M University, College Station, TX 77843, USA
* Correspondence: sbsmith@tamu.edu; Tel.: +1-979-845-3939
† These authors contributed equally to this work.

Abstract: We hypothesized that consumption of high-fat (HF) ground beef (24% fat) would not affect plasma concentrations of high-density lipoprotein cholesterol (HDL-C) or low-density lipoprotein (LDL-C), whereas low-fat (LF) ground beef (5% fat) would decrease HDL-C and LDL-C concentrations. In a randomized 2-period crossover, controlled feeding trial, 25 men (mean age and body mass index, 40 years and 31.2) consumed 115-g HF or LF patties, 5/week for 5 weeks with a 4-week washout. The HF treatment increased % energy from fat ($p = 0.006$) and saturated fat ($p = 0.004$) and tended ($p = 0.060$) to depress % energy from carbohydrates. The HF and LF treatments decreased the plasma concentrations of HDL-C ($p = 0.001$) and LDL-C ($p = 0.011$). Both ground beef treatments decreased the abundance of HDL_{3a} and increased the abundance of HDL_3 ($p \leq 0.003$); the LF treatment also decreased the abundance of HDL_{2b} and HDL_{2a} ($p \leq 0.012$). The HF and LF treatments decreased the abundance of LDL_3 and LDL_4 ($p \leq 0.024$) and the HF treatment also decreased LDL_5 ($p = 0.041$). Contrary to our hypothesis, the HF treatment decreased plasma HDL-C and LDL-C concentrations despite increased saturated fat intake, and both treatments decreased the abundance of smaller, denser LDL subfractions.

Keywords: ground beef consumption; saturated fat intake; lipoprotein cholesterol

Citation: Lytle, R.J.; Price, T.; Crouse, S.F.; Smith, D.R.; Walzem, R.L.; Smith, S.B. Consuming High-Fat and Low-Fat Ground Beef Depresses High-Density and Low-Density Lipoprotein Cholesterol Concentrations, and Reduces Small, Dense Low-Density Lipoprotein Particle Abundance. *Nutrients* **2023**, *15*, 337. https://doi.org/10.3390/nu15020337

Academic Editor: Frans Stellaard

Received: 22 November 2022
Revised: 5 January 2023
Accepted: 7 January 2023
Published: 10 January 2023

1. Introduction

Beef is a popular food worldwide, and the United States (U.S.) consumes 21% of the world's beef production, with China, the European Union, and Brazil consuming 16%, 13%, and 13%, respectively, of the world's production (https://www.fas.usda.gov/) (accessed on 2 December 2022). It is estimated that ground beef constitutes 40–45% of the beef consumed in the U.S.; when beef is prepared for meals at home, ground beef is used 60% of the time (https://beef2live.com/story-ground-beef-united-states-128-104332) (accessed on 2 December 2022). Ground beef containing less than 5–9% fat, 10% fat, and 20% fat accounts for 20, 42, and 20% of retail sales, respectively; the remaining 18% is comprised primarily of ground beef containing 30% fat (https://www.beefitswhatsfordinner.com/retail/sales-data-shopper-insights/ground-beef-at-retail-and-foodservice) (accessed on 2 December 2022). Palmitic acid (16:0) and stearic acid (18:0) are the most abundant saturated fatty acids (SFA) in beef, and oleic acid (18:1n-9) is the most abundant monounsaturated fatty acid (MUFA) in ground beef [1–5]. Beef also contains significant amounts of naturally occurring *trans*-fatty acid (TFA), primarily eladic acid (18:1trans-9) and *trans*-vaccenic acid (18:1trans-11), and conventional, 20% fat ground beef (22.8 g fat/114-g patty) contains 5 g palmitic acid, 3 g stearic acid, 8 g oleic acid, and 1.7 g TFA [1]. However, TFA derived from ruminal sources (e.g., dairy products and beef) does not affect the risk for cardiovascular disease (CVD) [6].

As indicated above, ground beef is a popular component in U.S. diet, and as such, ground beef constitutes one of the primary dietary sources of saturated fatty acids. In addition, the total fat and saturated fatty acid content can be changed readily and accurately during the formulation of ground beef preparations. Therefore, we conducted several randomized, controlled trials with men and/or postmenopausal women in which we tested the effects of ground beef differing in fatty acid composition on risk factors for CVD. Adams et al. [1] reported that high-density lipoprotein cholesterol (HDL-C) and low-density lipoprotein cholesterol (LDL-C) concentrations decreased from baseline when mildly hypercholesterolemic men were fed 35% fat, 114-g ground beef patties (40 g fat/patty; 15.7 g SFA/patty) (5 patties/week for 5 weeks). In a randomized controlled trial with normocholesterolemic men, consumption of 24% fat, 114-g patties (27 g fat/patty; 14.4 g SFA/patty) (5 patties/week for 5 weeks) had no effect on HDL-C or LDL-C concentration [2]. In a subsequent randomized controlled trial with postmenopausal women, consumption of 21% fat, 114-g patties (24 g fat/patty; 10.5 g SFA/patty) (5 patties/week for 6 weeks) had no effect on HDL-C or LDL-C concentration [3]. Choi et al. [4] reported that in a randomized controlled trial including postmenopausal women and older men, consumption of 22% fat, 114 g patties (25 g fat/patty; 10.5 g SFA/patty) (5 patties/week for 5 weeks) had no effect on HDL-C or LDL-C concentration. When data were pooled across the four trials (n = 42 men, 24 women), we established that intakes of 114-g high-fat/high-SFA ground beef patties 5 times per week for 5 or 6 weeks did not significantly affect HDL-C or LDL-C concentration [5].

Our previous studies also demonstrated that ground beef interventions elicit differences in lipoprotein particle sizes. LDL particle diameter decreased in men following a 35% fat ground beef intervention [1], and HDL_2 and HDL_3 particle diameter decreased in men following a 24% fat ground beef intervention [2]. In contrast to [1], LDL particle diameter increased in men and women following consumption of 25 g fat/patty ground beef, which was reflected in a greater concentration of cholesterol occurring in the LDL_1 and LDL_2 subfractions [4]. Because our previous studies indicated that ground beef consumption could affect lipoprotein particle size, a feature that varies inversely with flotation density, a secondary outcome of the current trial was to document the effects of LF (5% fat) and HF (25% fat) ground beef consumption on the density distributions for LDL and HDL subfractions. We previously determined particle density distributions using isopycnic density profiling of lipoproteins pre-stained with a lipophilic fluorescent probe [7]. In that method, the density distribution of labeled lipoproteins was analyzed as the area under the curve (AUC), where the image area was measured as pixels (i.e., number of pixels within a lipoprotein density interval). This study demonstrated that in a randomized controlled crossover study, both LF and HF ground beef consumption depressed plasma HDL-C and LDL-C concentrations as well as AUC for most HDL and LDL subfractions, providing new insight into the effects of beef consumption on the risk for CVD.

The primary outcome of the current randomized controlled trial was to document the effects of high-fat (HF) ground beef (27 g/patty, 24% fat by weight) and low-fat (LF) ground beef (6 g/patty, 5% fat by weight) on voluntary nutrient intake, and to establish if changes in major macronutrient intake were responsible for any changes we observed in lipoprotein cholesterol concentrations. We hypothesized that consumption of HF ground beef for 5 weeks would reduce the voluntary intake of carbohydrates but have no effect on HDL-C or LDL-C concentrations. Conversely, consumption of LF ground beef for 5 weeks would depress HDL-C and LDL-C concentrations.

2. Materials and Methods

2.1. Ethics Statements and Participant Recruitment

This randomized, controlled, 2-period crossover trial was conducted in accordance with the Declaration of Helsinki guidelines [8]. The trial was registered at www.clinicaltrials.gov as NCT04841460 accessed on 12 April 2021. All procedures involving human participants were approved by the Texas A&M University Institutional Review Board for use of human participants in research (Protocol number IRB2018-0755). Participants were recruited in October and November 2019, and the ground beef treatments were initiated in February 2020. The final blood samples were collected in July 2020. The study staff were not blinded, but the statistician was blinded to treatment during the initial analyses by identifying the diet conditions as A and B. All subjects were provided with detailed instructions, including potential risks of participation.

Seventy-five healthy males between the ages of 25 and 60 years participated in one of two informational meetings (Figure 1). Four individuals did not meet the inclusion criteria, and 25 men declined to participate. Forty-six men signed Informed Consent forms, and 14 men later declined to participate. Thirty-two men were assigned at random to two treatment groups, LF and HF ground beef, and were provided test ground beef patties. Seven men who left the study were excluded due to inability to comply (did not provide all diet records, $n = 3$; did not provide all blood samples, $n = 4$), and 25 men completed all phases of the study.

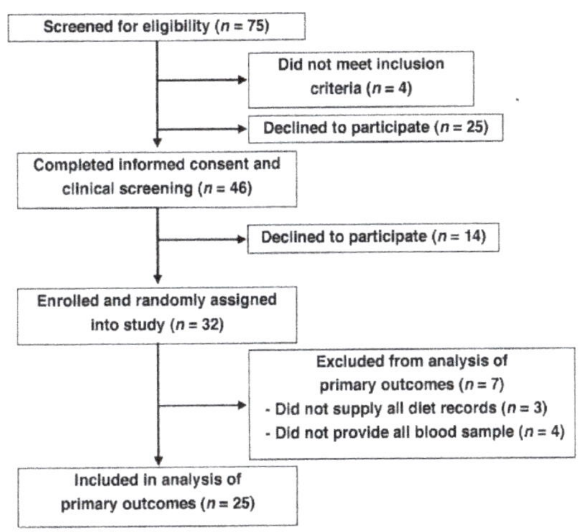

Figure 1. Recruitment flow diagram.

2.2. Inclusion Criteria

The participants had to be non-smoking males, not be on any restrictive diets or cholesterol-lowering medications, and not have a total cholesterol (TC) above 350 mg/dL. The participants were requested not to change their habitual diet or level of physical activity. Due to limited funding, we chose to recruit men only, and not women only or a combination of men and women. Statistical power calculations were based on changes in HDL-C concentration in normocholesterolemic men in response to a ground beef intervention ([2], described below). Unpublished data from previous trials in our laboratory [1–4] indicated that women had much greater variation in HDL-C concentration at entry (47–120 mg/dL) than men (36–76 mg/dL). The range of HDL-C concentrations at entry in the current study (39–73 mg/dL; Figure 2) was similar to the variation in our previous studies with men. The lesser variation in HDL-C concentration for men improved the power of our statistical analyses, and for this reason, men only were chosen for this trial.

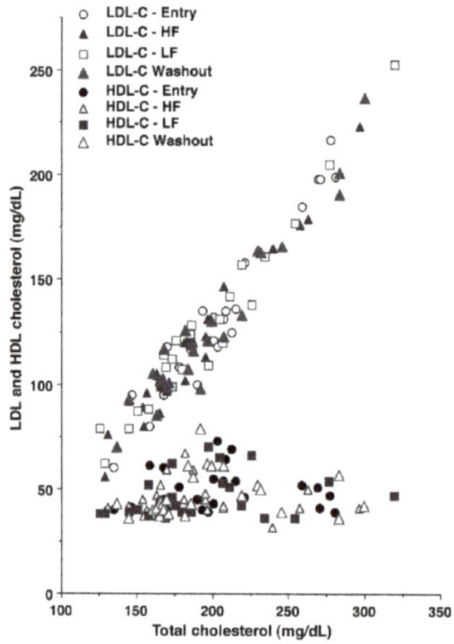

Figure 2. Plasma LDL-C and HDL-C concentrations as a function of plasma total cholesterol concentration. LDL-C, $r^2 = 0.947$; HDL-C, $r^2 = 0.022$.

2.3. Study Design

The study design was a two-period, randomized crossover design in which each participant completed two 5-week ground beef interventions in a randomly assigned order, with a 4-week washout period [9] between the test periods. All blood samples were taken from the fasting subjects. Four blood samples were drawn at baseline immediately before treatment assignment (entry), immediately after the ground beef interventions, and following the 4-week washout period, immediately before the second ground beef intervention. During the 5-week intervention, the men consumed 5 ground beef patties per week for 5 weeks for each ground beef type, LF and HF (total of 25 patties for each type). The participants were assigned to one of two groups, which were balanced based on plasma HDL-C concentrations measured at the initial screening. Before the first ground beef intervention, the men assigned to the LF group had a mean HDL-C concentration of 51 ± 3 mg/dL, and the men assigned to the HF group had a mean HDL-C concentration of 48 ± 3 mg/dL. Of the 25 men who completed the study, 12 men consumed LF ground beef, and 13 men consumed HF ground beef during the first intervention phase. After the washout period, the groups of 12 and 13 men were rotated to the other test ground beef.

2.4. Source of Ground Beef

The source of raw materials for the production of LF and HF patties were the beef pectoralis muscle and 75:25 coarse ground beef, respectively, purchased from a local supplier (Readfield Meats, Bryan, TX, USA). The muscle raw materials were ground, and 4-ounce (115 g) patties were formed in a patty maker, individually vacuum-packaged, and stored at −20 °C. Prior to the initiation of each phase of the ground beef interventions, each participant received an unlabeled box containing 25 frozen, vacuum-packaged patties.

Chemical analysis of the ground beef after patty formation indicated that raw LF patties contained 5% fat (6 g fat/patty) and HF patties contained 24% fat (27 g fat/patty) (Table 1). Diet records from previous studies [1–4] indicated that most study participants pan-broiled the ground beef patties intact, so samples of the LF and HF patties were pan-broiled [10], and total fat and fatty acid composition of the cooked patties were measured. Cooking losses for LF and HF patties were 3% and 41%, respectively. The total lipid and fatty acid composition of the drained pan-broiled patties were used to calculate the daily intake of dietary fats.

Table 1. Fatty acid composition and lipid content of raw and pan-broiled low-fat and high-fat ground beef patties [1].

Fatty Acid	Low-Fat		High-Fat	
	Raw	Pan-Broiled	Raw	Pan-Broiled
g fatty acid/114-g beef patty				
Myristic, 14:0	0.16 ± 0.02	0.15 ± 0.01	0.80 ± 0.04	0.46 ± 0.02
Palmitic, 16:0	1.49 ± 0.16	1.45 ± 0.06	6.37 ± 0.35	3.74 ± 0.48
Palmitoleic, 16:1n-7	0.21 ± 0.02	0.20 ± 0.01	0.98 ± 0.05	0.56 ± 0.02
Stearic, 18:0	0.79 ± 0.08	0.77 ± 0.08	3.53 ± 0.19	2.11 ± 0.08
Oleic, 18:1n-9	2.52 ± 0.27	2.17 ± 0.01	9.70 ± 0.53	5.55 ± 0.20
cis-Vaccenic, 18:1n-7	0.11 ± 0.01	0.13 ± 0.01	0.52 ± 0.03	0.33 ± 0.01
Linoleic, 18:2n-6	0.27 ± 0.03	0.27 ± 0.03	0.65 ± 0.04	0.41 ± 0.02
α-Linolenic, 18:3n-3	0.01 ± 0.01	0.01 ± 0.01	0.05 ± 0.01	0.02 ± 0.01
Total SFA [2]	2.44 ± 0.26	2.37 ± 0.23	10.70 ± 0.58	6.32 ± 0.22
Total MUFA [2]	2.84 ± 0.30	2.50 ± 0.11	11.29 ± 0.71	6.44 ± 0.23
Total PUFA [2]	0.28 ± 0.03	0.28 ± 0.01	0.71 ± 0.04	0.43 ± 0.02
MUFA:SFA	1.16 ± 0.01	1.06 ± 0.01	1.05 ± 0.01	1.02 ± 0.01
PUFA:SFA	0.11 ± 0.01	0.12 ± 0.01	0.07 ± 0.01	0.07 ± 0.01
Total *trans*-fatty acids [3]	0.17 ± 0.02	0.16 ± 0.02	1.44 ± 0.08	0.84 ± 0.03
Total lipid per patty [4]	6.4 ± 0.7	6.2 ± 0.3	26.9 ± 1.5	15.9 ± 0.6

[1] Values are means ± SE, n = 3 batches of ground beef for each study. [2] Total SFA (saturated fatty acids), sum of myristic, palmitic, and stearic acid. Total MUFA (monounsaturated fatty acids), sum of palmitoleic, oleic acid, and cis-vaccenic acid. Total PUFA (polyunsaturated fatty acids), sum of linoleic and α-linolenic acid. Eicosapentaenoic acid (20:5n-3) and docosahexaenoic acid (22:6n-3) were too low to quantify in the ground beef patties. [3] Sum of eladic acid (18:1trans-9) and trans-vaccenic acid (18:1trans-11). [4] Determined gravimetrically before and after cooking. Includes additional minor fatty acids not included in the table.

2.5. Diet Records

The participants were required to complete a 3-day diet record before the diet interventions and once during each intervention to establish nutrient intakes and encourage compliance. Daily intake of major nutrients and dietary exchanges were analyzed by a registered dietitian nutritionist (RDN) using ESHA's Food Processor Nutrition Analysis software (ESHA, Salem, OR, USA). The participants were trained in the use of myfitnesspal (myfitnesspal.com) to record daily intakes, which were forwarded to the RDN. All participants received instructions from the RDN for the preparation of the ground beef, including recipes, but the participants were not restricted to specific cooking methods for the test ground beef. The RDN also contacted the participants at regular intervals to encourage compliance and provide information about completing diet records.

2.6. Body Composition

The body composition of all subjects was assessed at the beginning and at the end of the study using dual-energy X-ray absorptiometry (DXA) (General Electric Lunar Prodigy Advance, Madison, WI, USA). Derived variables of interest from the DXA scans were total body mass, lean body mass, android fat, gynoid fat (all in kg), and percent fat mass. Body mass index (BMI) was calculated for each individual (Table 2).

Table 2. Physiological demographics and DXA-measured body composition for men at baseline and after consumption of ground beef patties (5 patties/week for 5 weeks) initially containing 6 g fat/patty and 27 g fat/patty [1].

Item	Entry (Range)	Final	*p*-Values
Age, years	39.9 ± 2.2 (24–58)	40.0 ± 2.2	0.489
Height, cm	177.4 ± 1.4 (168–191)	177.8 ± 1.3	0.409
Weight, kg	97.3 ± 5.0 (76–178)	97.4 ± 5.0	0.499
BMI	31.2 ± 1.8 (23–58)	30.9 ± 1.7	0.461
DXA measurements			
Fat, kg	30.6 ± 3.9 (14–83)	29.6 ± 3.8	0.425
Lean, kg	64.5 ± 1.9 (52–95)	66.9 ± 1.9	0.200
Body fat, %	30.6 ± 3.8 (18–59)	29.6 ± 3.8	0.425
Android fat, %	35.8 ± 2.8 (20–68)	32.0 ± 2.3	0.160
Gynoid fat, %	31.0 ± 2.0 (16–60)	29.7 ± 1.5	0.247

[1] Values are means ± SEM, n = 25. Range of values for each item is indicated for entry.

2.7. General Blood Sampling and Analyses

Blood sampling and assay procedures were published previously [11]. On the day of blood sampling (at entry and immediately following the 5-week ground beef interventions and 4-week washout), the subjects were asked to report to the laboratory after an overnight fast (approximately 10 h) restricted to water only. The evening meal was not standardized prior to the study visits; rather, throughout this trial, participants consumed their habitual diets, except for the inclusion of the ground beef patties during the ground beef interventions. Blood was collected after 5 min of seated rest via venipuncture from the antecubital fossa region into serum separator tubes using standard sterile phlebotomy procedures. After collection, the blood was allowed to clot at room temperature for 2 h or chilled at 4 °C for serum and plasma separation, respectively, prior to centrifugation in a refrigerated centrifuge for 20 min (2000× *g*). One serum separator vacutainer was couriered the same day to a commercial Clinical Laboratory Improvement Amendments-certified laboratory for determination of TC, HDL-C, LDL-C, and triglyceride (TG) using standard clinical chemistry analyses. Plasma LDL-C concentration was calculated using the Friedwald equation, which is based primarily on TC (LDL-C = TC − HDL-C − TG/5). Aliquots of serum and plasma from additional vacutainers were transferred into separate 2 mL freezer vials, and the vials were stored frozen at −80 °C until analyzed.

2.8. Lipoprotein Density Profiles

Density profiles for circulating lipoproteins were determined by imaging 6 μL serum following NBD-C6-ceramide labeling of lipoproteins, as described [7]. The overall lipoprotein density profile was analyzed as absolute AUC where image area was measured as pixels (i.e., number of pixels within a density interval). Eleven lipoprotein subclasses were identified by their density intervals and quantified by pixel values. The major lipoprotein subclasses were triacylglycerol-rich lipoproteins (TRL; d < 1.019 g/mL), LDL_1 (d = 1.019–1.023 g/mL), LDL_2 (d = 1.023–1.034 g/mL), LDL_3 (d = 1.034–1.044 g/mL), LDL_4 (d = 1.044–1.055 g/mL), LDL_5 (d = 1.055–1.063 g/mL), HDL_{2b} (d = 1.063–1.091 g/mL), HDL_{2a} (d = 1.091–1.110 g/mL), HDL_{3a} (d = 1.110–1.133 g/mL), HDL_{3b} (d = 1.133–1.156 g/mL) and HDL_{3c} (d = 1.156–1.179 g/mL) [12]. Lipoprotein density profiles for the participants (not indicated) were essentially identical to those reported previously for men by the co-author Walzem, R.L. [7].

The average percent relative standard deviation in AUC for different lipoprotein subfractions was 4.45% (within-day) and 7.37% (day-to-day). Data were also used to express HDL subfractions as percentages of total HDL AUC:

$$\%LDL_X \text{ AUC} = 100 * \%LDL_X \text{ AUC/total LDL AUC}$$

$$\%HDL_X \text{ AUC} = 100 * \%HDL_X \text{ AUC/total HDL AUC}$$

2.9. Statistics

Power calculations were conducted to estimate the required sample size based on HDL-C concentrations from our previous studies with normocholesterolemic men [2]. Plasma HDL-C concentrations increased by 2.8 mg/dL on consumption of 24% fat ground beef [2] compared to habitual diets. Analyses used the following assumptions: power was set at 0.8 and $\alpha = 0.05$, 2-sided. It was estimated that a sample size of 18 was sufficient to test the hypothesis that HF ground beef would change HDL-C concentrations.

Ground beef effects were analyzed using a repeated measures mixed model to assess the effects of diet (LF vs. HF), sequence (entry, first LF/HF intervention, washout, and second LF/HF intervention), and the diet-by-sequence interaction. Entry BMI and age were included as covariates in the initial model but were insignificant for all dependent variables and were dropped from the final model. The NORM.DIST model of Excel (Microsoft Excel of Mac version 16.16.27) was used to test for normality, and the data were normally distributed. Pairwise comparisons were assessed by Fisher's Protected LSD method when there was a significant effect of diet or sequence. Associations among plasma lipids were assessed using Pearson's correlations. Absolute change from baseline was calculated by subtracting measurements taken at entry from post-dietary intervention period values. Data are reported as means ± standard error of the mean (SEM) ($n = 25$ men who completed all phases of the study). Differences among means were considered significant at $p \leq 0.05$, but tendencies among treatments ($p \leq 0.08$) will be noted.

3. Results

3.1. Ground Beef Composition and Participant Nutrient Intake

The LF patties did not lose a detectable amount of fat following pan frying, but the HF patties lost nearly 41% fat after frying. The drained, pan-fried HF patties contained more total SFA, MUFA, and polyunsaturated fatty acids (PUFA) than the pan-fried LF patties (Table 1). Eicosapentaenoic acid (20:5n-3) and docosahexaenoic acid (22:6n-3) were not detectable in the raw or cooked ground beef patties. MUFA:SFA and PUFA:SFA ratios of the LF and HF ground beef were unaffected by pan broiling.

Energy intake did not differ among treatment phases, although energy intake during the LF intervention tended ($p = 0.071$) to be lower than at entry (Table 3). The LF treatment increased % energy from protein relative to entry, washout, and HF treatment ($p = 0.002$). There was a tendency ($p = 0.062$) of the HF treatment to decrease % energy from carbohydrates. The HF treatment increased % energy from fat and SFA relative to entry, washout, and LF treatment ($p \leq 0.006$).

Table 3. Intakes of major nutrients for men at entry, following washout, and during consumption of ground beef patties initially containing 6 g fat/patty (Low-fat) or 27 g fat/patty (High-fat) [1].

Item [2]	Entry	Washout	Low-Fat	High-Fat	*p*-Value
Total, MJ/d	8.7 ± 0.4	8.1 ± 0.3	7.9 ± 0.4	8.6 ± 0.4	0.071
%MJ/d					
Protein	18.2 ± 0.7 [b]	19.1 ± 0.9 [b]	21.9 ± 1.0 [a]	18.2 ± 1.1 [b]	0.002
Carbohydrate	42.3 ± 1.7	41.9 ± 1.5	40.9 ± 1.6	38.5 ± 1.7	0.060
Fat	38.4 ± 1.1 [b]	36.7 ± 1.4 [b]	36.5 ± 1.4 [b]	42.4 ± 1.7 [a]	0.006
SFA [3]	12.6 ± 0.6 [b]	12.3 ± 0.5 [b]	12.6 ± 0.7 [b]	14.7 ± 0.7 [a]	0.004
MUFA	6.5 ± 0.6 [b]	6.7 ± 0.9 [b]	7.1 ± 0.5 [b]	9.8 ± 0.9 [a]	0.003
PUFA	3.6 ± 0.3	3.5 ± 0.4	3.5 ± 0.4	3.4 ± 0.3	0.376
Intake, g/d					
Protein	93.8 ± 5.5	91.3 ± 4.2	102.2 ± 6.2	93.2 ± 7.5	0.078
Carbohydrate	217.4 ± 12.2 [a]	203.4 ± 9.9 [ab]	190.5 ± 9.8 [b]	197.0 ± 12.6 [ab]	0.048
Dietary fiber	17.8 ± 1.2	17.4 ± 1.5	15.4 ± 0.9	15.3 ± 1.2	0.062
Soluble fiber	0.8 ± 0.2	0.9 ± 0.1	0.9 ± 0.1	0.8 ± 0.1	0.334

Table 3. *Cont.*

Item [2]	Entry	Washout	Low-Fat	High-Fat	p-Value
Insoluble fiber	2.0 ± 0.4	2.6 ± 0.6	2.5 ± 0.4	2.4 ± 0.4	0.175
Total sugars	68.6 ± 7.0	58.9 ± 5.7	65.6 ± 5.5	64.5 ± 7.9	0.140
Added sugars	8.6 ± 3.1	10.5 ± 3.8	14.6 ± 3.9	14.0 ± 4.3	0.114
Total fat	89.2 ± 5.1 [ab]	80.2 ± 4.9 [bc]	76.7 ± 4.4 [c]	96.6 ± 6.0 [a]	0.006
SFA	29.2 ± 1.8 [ab]	26.9 ± 1.8 [b]	26.5 ± 1.7 [b]	33.7 ± 2.5 [a]	0.013
MUFA	15.4 ± 1.6 [b]	14.1 ± 1.7 [b]	15.2 ± 1.4 [b]	22.2 ± 2.2 [a]	0.002
PUFA	8.1 ± 0.8	7.6 ± 0.9	7.4 ± 1.0	7.6 ± 0.7	0.293
trans-Vaccenic acid	0.7 ± 0.2 [a]	0.5 ± 0.1 [ab]	0.3 ± 0.1 [b]	0.6 ± 0.2 [ab]	0.011
Linoleic acid	5.9 ± 0.8	5.4 ± 0.7	5.1 ± 0.9	4.9 ± 0.5	0.152
α-Linolenic acid	0.7 ± 0.1	0.7 ± 0.1	0.6 ± 0.1	0.5 ± 0.1	0.053
Cholesterol, mg/d	415 ± 57	335 ± 34	328 ± 39	314 ± 33	0.066

[1] Values are means ± SEM, n = 25. [2] Data were derived from 3-day diet records collected during each test period, to include one weekend day. [3] SFA = saturated fatty acids, primarily palmitic and stearic acid. MUFA = monounsaturated fatty acids, primarily oleic acid. PUFA = polyunsaturated fatty acids, primarily linoleic acid and α-linolenic acid. A repeated measures mixed model was used to assess the effects of diet, sequence, and the diet-by-sequence interaction. [abc] Pairwise comparisons were assessed by Fisher's Protected Least Squares Difference method when there was a significant diet effect.

Protein intake (g/d) tended ($p = 0.078$) to be greatest during the LF treatment (Table 3). Carbohydrate intake was less during the LF intervention than at entry ($p = 0.048$), and dietary fiber intake tended ($p = 0.062$) to be greater at entry and during washout than during the LF and HF treatments. The intakes of soluble fiber, insoluble fiber, total sugars, and added sugars were not affected by the ground beef treatments ($p \geq 0.114$). Total fat and SFA intakes were greater during the HF treatment than during the washout or LF treatment ($p \leq 0.013$), but not different from entry. MUFA intake was greatest during the HF treatment ($p = 0.002$) and TFA intake was least during the LF treatment. Intakes of PUFA and linoleic acid were not affected by ground beef consumption ($p \geq 0.152$), but intake of α-linolenic acid tended ($p = 0.053$) to be depressed during the HF treatment. Cholesterol intake tended ($p = 0.066$) to be lower during washout and during the LF and HF treatments than at entry.

There were significant ($p \leq 0.05$) absolute changes from entry for % energy from total fat, SFA, MUFA (which increased), and carbohydrate (which decreased) during the HF intervention (Figure 3). The increase from entry for % energy from protein during the LF intervention also was significant ($p < 0.05$). The mean absolute decreases from entry for cholesterol intake during washout, LF treatment, and HF treatment were 72 mg/d ($p = 0.119$), 95 mg/d ($p = 0.067$), and 101 mg/d ($p = 0.024$), respectively.

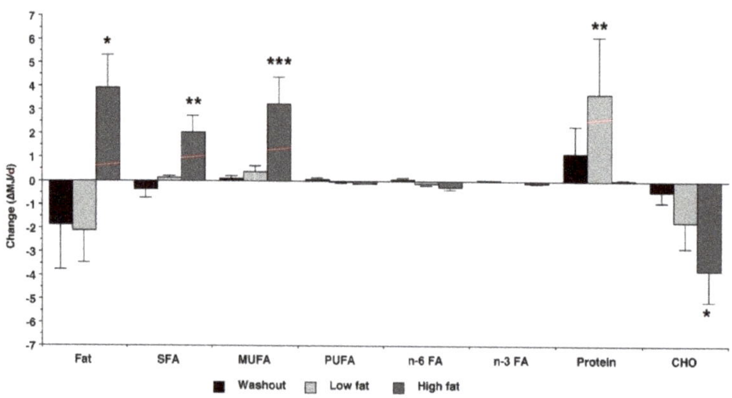

Figure 3. Absolute change (final – entry) in %MJ/d during consumption of ground beef patties initially containing 6 g fat/patty (Low fat) and 27 g fat/patty (High fat). Fat, total fat; SFA, saturated fat; MUFA, monounsaturated fat; PUFA, polyunsaturated fat; CHO, carbohydrate. Bars are means ($n = 25$) with pooled SEM attached. Change statistically different from 0, * $p \leq 0.05$; ** $p \leq 0.01$; *** $p \leq 0.001$.

3.2. Plasma Lipid Concentrations

As indicated above, plasma LDL-C concentration was calculated by the Friedwald equation, which is based primarily on TC (LDL-C = TC − HDL-C − TG/5). For this reason, there was a high correlation between TC and LDL-C concentration (r^2 = 0.941) (Figure 2). There was greater variation in plasma concentrations of TC (126–319 mg/dL) and LDL-C (56–223 mg/dL) than HDL-C (36–73 mg/dL). At entry, 20 out of 25 participants had TC concentrations < 225 mg/dL. There was no correlation between TC and HDL-C concentration (r^2 = 0.022) ($p > 0.25$).

Plasma TG concentration tended ($p = 0.058$) to be greatest following the HF treatment (Table 4). The LF treatment decreased TC, HDL-C, and LDL-C concentrations relative to entry, and the HF treatment decreased TC, HDL-C, and LDL-C concentrations relative to entry and washout. Absolute changes from entry were significant ($p < 0.05$) for TC and HDL-C following the LF treatment, and absolute changes from entry were significant ($p < 0.05$) for TG, TC, HDL-C, and LDL-C following the HF treatment (Figure 4).

Table 4. Lipoprotein cholesterol and triglyceride concentrations and area under the curve for lipoprotein subfractions for men at entry and after consumption of ground beef patties initially containing 6 g fat/patty (Low-fat) and 27 g fat/patty (High-fat) [1].

Item	Entry	Washout	Low-Fat	High-Fat	p-Value
TG, mg/dL [2]	111.4 ± 9.6	114.6 ± 10.9	110.1 ± 8.6	123.0 ± 11.1	0.058
TC, mg/dL	203.0 ± 8.2 [a]	197.4 ± 8.5 [ab]	191.5 ± 8.8 [bc]	188.5 ± 8.1 [c]	0.008
HDL cholesterol, mg/dL	49.3 ± 2.0 [a]	47.6 ± 2.2 [ab]	46.0 ± 1.9 [bc]	45.4 ± 1.8 [c]	0.001
LDL cholesterol, mg/dL	131.2 ± 8.1 [a]	126.9 ± 7.9 [ab]	123.6 ± 8.5 [bc]	118.5 ± 7.5 [c]	0.011
Total AUC	1920 ± 46 [a]	1844 ± 44 [b]	1815 ± 51 [b]	1811 ± 48 [b]	0.003
TRL AUC	150 ± 14 [b]	165 ± 18 [ab]	173 ± 17 [a]	159 ± 15 [ab]	0.050
Total LDL AUC	827 ± 34 [a]	772 ± 31 [b]	752 ± 34 [b]	746 ± 32 [b]	0.001
LDL_1 AUC	29 ± 1	28 ± 2	29 ± 2	30 ± 2	0.225
LDL_2 AUC	50 ± 3	48 ± 28	48 ± 3	48 ± 3	0.450
LDL_3 AUC	170 ± 11 [a]	159 ± 9 [b]	155 ± 12 [b]	152 ± 9 [b]	0.024
LDL_4 AUC	387 ± 25 [a]	354 ± 22 [b]	341 ± 24 [b]	341 ± 22 [b]	0.002
LDL_5 AUC	189 ± 15 [a]	181 ± 15 [ab]	178 ± 10 [ab]	173 ± 13 [b]	0.041
Total HDL AUC	942 ± 29 [a]	906 ± 25 [b]	884 ± 301 [b]	905 ± 28 [b]	0.003
HDL_{2b} AUC	225 ± 18 [a]	209 ± 174 [b]	211 ± 20 [b]	212 ± 18 [ab]	0.012
HDL_{2a} AUC	245 ± 11 [a]	230 ± 8 [b]	225 ± 11 [b]	234 ± 10 [ab]	0.007
HDL_{3a} AUC	278 ± 8 [a]	268 ± 5 [ab]	259 ± 7 [b]	264 ± 6 [b]	0.002
HDL_{3b} AUC	135 ± 4	133 ± 4	128 ± 4	130 ± 5	0.058
HDL_{3c} AUC	58 ± 1 [b]	64 ± 1 [a]	62 ± 1 [a]	64 ± 2 [a]	0.003

[1] Values are means ± SEM, n = 25 in a crossover design. [2] AUC, area under the curve (unitless) of all lipid-rich fractions; TRL, triglyceride-rich lipids; LDL, low-density lipoprotein; HDL, high-density lipoprotein. A repeated measures mixed model was used to assess the effects of diet and sequence, and the diet-by-sequence interaction. [abc] Pairwise comparisons were assessed by Fisher's Protected Least Squares Difference method when there was a significant diet effect.

3.3. Lipoprotein Density Distributions

TRL AUC was greater following the LF intervention than at entry ($p = 0.050$) (Table 4). LDL_4 AUC represented the greatest proportion of AUC, indicating that LDL was predominantly small and dense. AUC for HDL_{3c} was very low, indicating that HDL_{3c} particles are less numerous than for other HDL subfractions. LDL_1 and LDL_2 AUC were not affected by the ground beef treatments, but total AUC and LDL AUC, LDL_3 AUC, LDL_4 AUC, and total HDL AUC were lower following washout and the LF and HF treatments than at entry ($p \leq 0.024$). LDL_5 AUC was less following the HF intervention than at entry ($p = 0.041$) and HDL_{2b} AUC and HDL_{2a} AUC were less following washout and the LF intervention than at entry ($p \leq 0.012$). HDL_{3a} AUC was less following the LF and HF interventions than at entry ($p = 0.002$), and HDL_{3b} AUC tended ($p = 0.058$) to be less following the LF treatment than at entry. HDL_{3c} AUC was greater following washout and the LF and HF treatments than at entry ($p = 0.003$). The HF treatment increased $\%LDL_1$ AUC ($p = 0.021$) and $\%LDL_2$ AUC ($p = 0.023$), and the LF and HF treatments increased $\%HDL_{3c}$ AUC ($p = 0.001$) (Table 5).

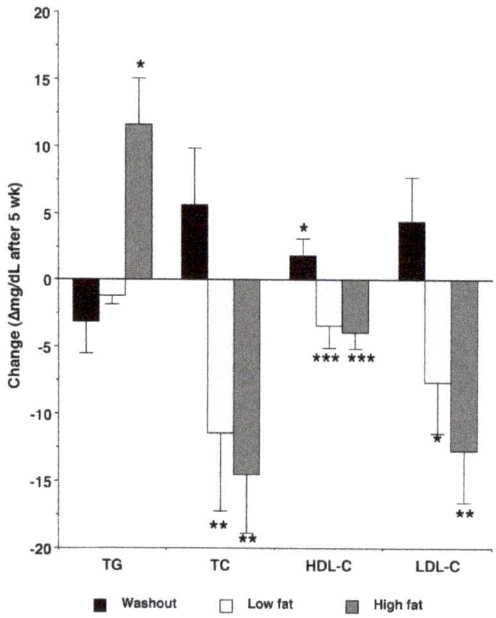

Figure 4. Absolute change (final – entry) in triglyceride and lipoprotein cholesterol following consumption of ground beef patties initially containing 6 g fat/patty (Low fat) and 27 g fat/patty (High fat). TG, triglyceride; TC, total cholesterol; HDL-C, HDL cholesterol; LDL-C, LDL cholesterol. Bars are means ($n = 25$) with pooled SEM attached. Change statistically different from 0, * $p \leq 0.05$; ** $p \leq 0.01$; *** $p \leq 0.001$.

Table 5. Percent area under the curve for lipoprotein fractions for men at entry and after consumption of ground beef patties initially containing 6 g fat/patty (Low-fat) and 27 g fat/patty (High-fat) [1].

Item	Entry	Washout	Low-Fat	High-Fat	*p*-Value
$\%LDL_1$ AUC [2]	3.7 ± 0.2^{b}	3.7 ± 0.2^{b}	4.0 ± 0.2^{ab}	4.1 ± 0.2^{a}	0.021
$\%LDL_2$ AUC	6.1 ± 0.3^{b}	6.3 ± 0.3^{ab}	6.4 ± 0.3^{ab}	6.5 ± 0.2^{a}	0.023
$\%LDL_3$ AUC	20.6 ± 0.9	20.5 ± 0.8	20.5 ± 0.9	20.2 ± 0.7	0.534
$\%LDL_4$ AUC	46.3 ± 1.8	45.7 ± 1.7	45.0 ± 1.9	45.4 ± 1.6	0.229
$\%LDL_5$ AUC	23.3 ± 1.8	23.7 ± 1.8	24.1 ± 1.7	23.7 ± 1.8	0.325
$\%HDL_{2b}$ AUC	23.3 ± 1.2	22.5 ± 1.2	23.0 ± 1.3	22.8 ± 1.2	0.164
$\%HDL_{2a}$ AUC	25.9 ± 0.5	25.4 ± 0.4	25.3 ± 0.5	25.7 ± 0.5	0.125
$\%HDL_{3a}$ AUC	29.8 ± 0.7	29.9 ± 0.7	29.8 ± 0.9	29.6 ± 0.7	0.256
$\%HDL_{3b}$ AUC	14.7 ± 0.6	15.0 ± 0.5	14.8 ± 0.6	14.7 ± 0.6	0.472
$\%HDL_{3c}$ AUC	6.3 ± 0.2^{b}	7.2 ± 0.3^{a}	7.1 ± 0.3^{a}	7.3 ± 0.3^{a}	0.001

[1] Values are means ± SEM, $n = 25$ in a crossover design. [2] AUC, area under the curve of lipid-rich subfractions; LDL, low-density lipoprotein; HDL, high-density lipoprotein. $\%LDL_x$ AUC = 100 ∗ ($\%LDL_x$ AUC/total LDL AUC); $\%HDL_x$ AUC = 100 ∗ ($\%HDL_x$ AUC/total HDL AUC). A repeated measures mixed model was used to assess the effects of diet and sequence, and the diet-by-sequence interaction. [ab] Pairwise comparisons were assessed by Fisher's Protected Least Squares Difference method when there was a significant diet effect.

3.4. Pearson Correlation Coefficients

Plasma HDL-C concentration was negatively correlated with TRL AUC ($r = -0.383$; $p < 0.001$) and HDL_{3b} AUC ($r = -0.277$; $p < 0.01$) and was highly, positively correlated with HDL_{2b} AUC ($r = 0.914$; $p < 0.0001$), HDL_{2a} AUC ($r = 0.905$; $p < 0.0001$), and total HDL AUC ($r = 0.658$; $p < 0.0001$) (Table 6). The correlation of HDL-C concentration with HDL_{3c} AUC was not significant (r = −0.046; $p > 0.25$). LDL-C concentration was not correlated with TRL AUC (r = −0.208; $p > 0.05$) but was positively correlated with LDL_1 AUC (r = 0.322; $p < 0.001$), LDL_2 AUC (r = 0.600; $p < 0.0001$), LDL_3 AUC (r = 0.731; $p < 0.0001$), LDL_4 AUC (r = 0.736; $p < 0.0001$), LDL_5 AUC (r = 0.251; $p < 0.05$), and total LDL AUC (r = 0.911; $p < 0.0001$).

Table 6. Pearson correlations among plasma HDL-C and LDL concentrations and area under the curve for HDL and LDL subfractions [1,2].

Item	Area under the Curve						
	TRL	**HDL_{2b}**	**HDL_{2a}**	**HDL_{3a}**	**HDL_{3b}**	**HDL_{3c}**	**Total HDL**
HDL-C	−0.383 ***	0.914 ****	0.905 ****	0.128	−0.277 **	−0.046	0.658 ****
	TRL	**LDL_1**	**LDL_2**	**LDL_3**	**LDL_4**	**LDL_5**	**Total LDL**
LDL-C	−0.208	0.332 ***	0.600 ****	0.731 ****	0.736 ****	0.251 *	0.911 ****

[1] Values are simple correlation coefficients, n = 100 (entry, washout, low-fat and high-fat values). [2] HDL-C, high-density lipoprotein cholesterol; LDL, low-density lipoprotein cholesterol; Total HDL and LDL, total area under the curve for HDL and LDL subfractions, respectively. * $p < 0.05$; ** $p < 0.01$; *** $p < 0.001$; **** $p < 0.0001$.

4. Discussion

4.1. Energy and Macronutrient Intake and Composition of Patties

The primary outcome of this study was to document which voluntary changes in macronutrient intake were responsible for any observed changes in lipoprotein cholesterol concentrations. The current study confirmed significant increases in % energy from total fat, saturated fatty acids, and monounsaturated fatty acids during the HF treatment. We also calculated absolute change from baseline as previously reported by others [13–16], and there was an absolute increase from entry for % energy from total fat, saturated fatty acids, and monounsaturated fatty acids during the HF treatment and an increase in % energy from protein during the LF treatment. These findings indicate that the participants voluntarily altered macronutrient intake in response to the fat content of the ground beef treatments.

The calculations of daily fat intake during the LF and HF interventions assumed that, regardless of the method of cooking, drippings were not included in the final food product. We previously reported that pan-broiled, 25% fat ground beef lost 44–49% fat, depending on the degree of doneness [17], and the 41% fat loss of the 24% fat, HF ground beef following pan broiling was similar to our previous results. We also assumed that the fatty acid composition (g fatty acid/100 g total fatty acids) of the pan-fried ground beef patties (drippings not included) would be similar to the composition of raw patties, based on earlier studies from this laboratory [17,18]. As indicated above, the MUFA:SFA and PUFA:SFA ratios were similar in raw and pan-broiled LF and HF patties, indicating that neither monounsaturated fatty acids nor polyunsaturated fatty acids were preferentially lost during cooking. Additionally, polyunsaturated fatty acid intake during ground beef intake was unchanged, indicating that changes observed in lipoprotein cholesterol subfractions were not due to changes in polyunsaturated fatty acid intake.

Many randomized studies have provided set guidelines for nutrient intake during the treatment phases and/or provided baseline diets before initiating dietary interventions [13–16]. The design of previous studies [1–4] and the current study differ in that we tested the effects of ground beef interventions on lipoprotein cholesterol concentration following free choice consumption of habitual diets at entry, during the washout period, and during the ground beef interventions. Simple regression over all data (entry, washout, LF, and HF, n = 100) indicated that % energy from saturated fatty acids increased as % energy from fat increased (r = 0.810; $p < 0.0001$) and % energy from carbohydrate decreased as % energy from fat increased (r = −0.544; $p < 0.0001$) (data not shown in tabular form). We conclude from these correlations that participants consuming HF ground beef voluntarily reduced carbohydrates in the diet. In addition, increasing beef fat intake markedly increased saturated fatty acid intake, reflecting the fatty acid composition and content in the HF ground beef.

Saturated fatty acid intake by participants in this study exceeded dietary recommendations of less than 10% energy intake from saturated fatty acids; approximately 73% of U.S. males exceed this recommendation [19]. The men in the current study on average consumed 12–13% energy as saturated fatty acids during entry, washout, and LF treatment, and 15% energy as saturated fatty acids during the HF treatment. Maki et al. [20] recently concluded that although the effects of saturated fatty acids on risk factors for CVD remain uncertain, it is prudent to accept the current recommendation [19] of less 10% energy from saturated fatty acids. Despite the reduction in LDL-C concentration caused by the HF intervention in the current study, we also do not recommend exceeding the recommendations [19] for saturated fatty acid intake. We also do not recommend this high level of ground beef intake (i.e., 5 patties/week for 5 weeks). As indicated above, we have chosen to test the effects of ground beef on lipoprotein cholesterol metabolism because total fat and saturated fatty acid composition can be altered accurately during the formulation of ground beef preparations.

4.2. *High-Density Lipoprotein Cholesterol Concentrations*

HDL particles carry out reverse cholesterol transport and possess antioxidative and anti-inflammatory activities through associated proteins and bioactive lipids [21]. These functional properties are variable and can be ascribed to particles of specific diameter ranges [22]. HDL_3 are small and vary in diameter from approximately 6 to 9.5 nm; HDL_2 vary in diameter from approximately 9.5 to 13 nm [22,23]. Some have concluded that HDL_3 includes both the most beneficial and the most detrimental species of HDL [24,25]. Increased amounts of sphingosine-1-P (S1P) and Apo-A1 in HDL_3 are associated with a robust ability to stabilize LDL against oxidation and attenuate apoptosis in endothelial cells [26]. However, greater plasma concentrations of HDL_{3c}-C are associated with greater probability of mortality [27].

Scott et al. [13] reported that a diet containing lean beef (diet = 31% energy as fat; 8% energy as saturated fatty acids) depressed HDL-C concentration in hypercholesterolemic men by 2.4 mg/dL, relative to a high-fat/high saturated fatty acid stabilization diet (diet = 40% energy as fat; 18% energy as saturated fatty acids). Roussell et al. [14] demonstrated that a Beef in an Optimal Lean Diet (BOLD) and a BOLD+ diet (which contained extra beef; both diets = 28% energy as fat; 6% energy as saturated fatty acids) depressed HDL-C concentration in men and women by 3.2 mg/dL, relative to a Healthy American Diet (HAD; diet = 33% energy as fat; 12% energy as saturated fatty acids). In the current study, the LF treatment reduced HDL-C concentration by 3.3 mg/dL, relative to the entry diet, similar to the results of Scott et al. [13] and Roussell et al. [14]. Furthermore, the HF treatment depressed HDL-C concentration by 3.9 mg/dL in spite of the greater total dietary % energy from fat (42%) and saturated fatty acids (15%) relative to the entry diet. It is difficult to understand why the HF treatment elicited results similar to the LF treatment and previous studies [13,14].

4.3. *Low-Density Lipoprotein Cholesterol Abundance*

Each LDL particle contains one apolipoprotein B (apoB), and the larger, less dense LDL_1 and LDL_2 are comprised of a greater proportion of cholesterol, whereas the smaller, denser LDL_3, LDL_4 and LDL_5 carry less cholesterol per particle [28]. Reductions in apoB are thought to be as or more beneficial for reducing atherosclerotic CVD risk than reductions in LDL-C concentration [29], and a predominance of small, dense LDL particles ($LDL_{3\text{-}5}$) is associated with a greater risk of CVD [29–31], in part because apoB comprises a greater proportion of these particles. The reduction in LDL-C concentration reported in the current trial might not be beneficial if caused by a decrease in cholesterol content of the larger, more buoyant LDL particles (LDL_1 and LDL_2).

Wang et al. [16] reported that a lower-fat diet decreased TC, LDL-C, and HDL-C concentration relative to an Average American Diet. Moreover, the lower-fat diet caused reductions for cholesterol in the LDL_1-C and LDL_2-C subfractions, had no effect on cholesterol content of the LDL_3-C subfraction, and increased cholesterol content of the LDL_4-C subfraction [16]. We previously demonstrated that approximately 75% of the total LDL-C is contained in the LDL_1 plus LDL_2 subfractions [4]. In addition, ground beef treatments containing 18% or 25% fat increased the cholesterol content of LDL_1 plus LDL_2-C subfractions but did not increase the cholesterol content of LDL_3-C or LDL_4-C subfractions [4]. We did not measure cholesterol content or apoB amounts in the LDL subfractions in the current study, but we have demonstrated that our ground beef treatments did not affect LDL_1 or LDL_2 AUC, but depressed $LDL_{3\text{-}5}$ AUC (described below), suggesting that more LDL cholesterol was carried in larger, less dense LDL particles.

4.4. Lipoprotein Area under the Curve

NBD-C6-ceramide labels the surface of lipoprotein particles and can be used to indicate particle diameter associated with a specific density interval [32], and can indicate, within that discrete density interval, relative particle abundance in treatment comparisons. Both ground beef interventions depressed HDL_{3a} AUC (hereafter referred to as abundance) and increased HDL_{3c} abundance. The LF treatment also depressed HDL_{2b} and HDL_{2a} abundance. HDL-C concentration was highly correlated with HDL_{2b} abundance (r = 914; $p < 0.0001$) and HDL_{2a} abundance (r = 0.905; $p < 0.001$). A recent study by the co-author Walzem, R.L. reported similar correlations between HDL_2 abundance and HDL-C concentration [12].

Wang et al. [16] reported that a lower-fat diet depressed the number of small HDL particles in addition to decreasing HDL-C concentration, and we conclude that the LF and HF treatments in the current study also decreased HDL particle abundance. Because HDL_3 abundance increased from entry, the data suggest that HDL_{3a} and HDL_{3b} particles became denser following HF and LF treatments, shifting their particle densities into the HDL_{3c} density interval. $\%HDL_{3c}$ (calculated as a percentage of total HDL AUC) also increased from entry, corroborating a shift from less dense to more dense HDL_3 particles following the ground beef interventions. As described above, diet analysis during the LF and HF interventions indicated that our participants consumed a much greater % energy from fat/SFA and less % energy from carbohydrates than the BOLD/BOLD+ studies [12,14], which likely explains many of the differences in responses to our dietary interventions.

Small, dense LDL particles have the strongest association with the risk for CVD [33]. Wu et al. [12] reported a reduction of $\%LDL_4$ (a small, dense LDL subfraction) following the BOLD+ intervention, suggesting that the BOLD+ diet improved the LDL density profile. These results were similar to the current study, in which the LF treatment decreased abundance for small, dense LDL_4 and the HF treatment decreased abundance for small, dense LDL_4 and LDL_5. Wu et al. [12] also reported a decrease in LDL_2 abundance following the BOLD diet, whereas we observed no change LDL_1 and LDL_2 abundance and increased $\%LDL_1$ and $\%LDL_2$ abundance following the HF treatment.

4.5. Nutrient Intake and Lipoprotein Cholesterol

The LF treatment increased % energy from protein, and the HF treatment increased % energy from fat and saturated fat, and decreased % energy from carbohydrate. However, the LF and HF treatments caused similar reductions in HDL-C and LDL-C concentrations from entry. This seemingly would rule out % energy from protein, fat, or carbohydrate as causative for the reductions in HDL-C and LDL-C concentrations. However, as indicated above, cholesterol intake during washout and the LF and HF treatments was lower than at entry, especially for the HF treatment. Dietary interventions designed to reduce saturated fat intake typically reduced cholesterol intake [13,14,16], and these studies uniformly decreased HDL-C and LDL-C concentrations. We propose that the reduction in cholesterol intake during washout and both ground beef interventions in the current study may have contributed to a general decline in circulating cholesterol concentrations and HDL and LDL subfraction abundance. It is not known why cholesterol intake declined following entry, but we are currently analyzing dietary sources of cholesterol during entry, washout, and ground beef interventions to provide a basis for this observation.

4.6. Limitations

Several limitations of the present study should be considered. Because of the nature of the study treatments, the study staff were not blinded, which leads to the potential for bias associated with lack of blinding. The short duration of the intervention, 35 days, may also be considered a limitation, but previous research conducted by the authors has shown that changes in HDL-C and LDL-C concentrations were evident at 5 weeks of making dietary changes [1,2]. The demonstration of a significant carryover effect for some of the lipid AUC values and cholesterol intake should be considered a limitation of not prescribing diets preceding entry or during the washout period. Another limitation is that the participants were men only. Previous trials in our laboratory that included postmenopausal women [3,4] indicated that their responses to ground beef interventions were more variable than for men [1,2]. Moreover, women have greater HDL-C concentration than men, and restricting the study to one gender increased the power of our analyses by reducing the variability inherent between men and women. Finally, we do not recommend the high level of ground beef intake used in this study for the general population.

5. Conclusions

In summary, the LF and HF ground beef interventions differently affected voluntary nutrient intake, but the LF and HF treatments similarly depressed lipoprotein cholesterol concentrations. The LF and HF treatments improved the LDL density profile by decreasing abundance for small, dense LDL (LDL_4 and LDL_5), and increasing %AUC for large, less dense LDL (LDL_1 and LDL_2), relative to entry levels. However, both ground beef interventions increased abundance and %AUC for HDL_{3c}, potentially increasing the risk of CVD. Despite frequent moderate ground beef intake (114 g/d), cholesterol intake declined, especially during the HF intervention. For this reason, care should be taken in the interpretation of studies that compare low-fat and low-cholesterol diets to higher-fat and higher-cholesterol diets.

Author Contributions: Conceptualization, S.B.S., R.L.W. and S.F.C.; methodology, J.R.L., T.P. and D.R.S.; formal analysis, S.B.S.; writing, S.B.S., R.L.W., J.R.L., T.P. and D.R.S.; funding acquisition, S.B.S. and R.L.W. All authors have read and agreed to the published version of the manuscript.

Funding: This research was funded by the BEEF CHECKOFF PROGRAM, grant number 1803058.

Institutional Review Board Statement: This study was conducted in accordance with the Guidelines of the Declaration of Helsinki guidelines and approved by the Texas A&M University Institutional Review Board for use of human participants in research (Protocol number IRB2018-0755). This trial was registered at www.clinicaltrials.gov as NCT04841460 accessed on 12 April 2021.

Informed Consent Statement: Informed consent was obtained from all subjects involved in the registered trial NCT04841460.

Data Availability Statement: The data presented in this study are available on request from the corresponding author. The data are not publicly available as no appropriate public database is available.

Conflicts of Interest: The authors declare no conflict of interest.

Abbreviations

AUC, area under the curve; BMI, body mass index; CVD, cardiovascular disease; DXA, dual-energy X-ray adsorptiometry; HDL-C, high-density lipoprotein cholesterol; HF, high-fat; LDL-C, low-density lipoprotein cholesterol; LF, low-fat; MUFA, monounsaturated fatty acids; PUFA, polyunsaturated fatty acids; RDN, registered dietitian nutritionist; SFA, saturated fatty acids; TG, triglycerides; TFA, *trans*-fatty acids.

References

1. Adams, T.H.; Walzem, R.L.; Smith, D.R.; Tseng, S.; Smith, S.B. Hamburger high in total, saturated and *trans*-fatty acids decreases HDL cholesterol and LDL particle diameter, and increases TAG, in mildly hypercholesterolaemic men. *Br. J. Nutr.* **2010**, *103*, 91–98. [CrossRef]
2. Gilmore, L.A.; Walzem, R.L.; Crouse, S.F.; Smith, D.R.; Adams, T.H.; Vaidyanathan, V.; Cao, X.; Smith, S.B. Consumption of high-oleic acid ground beef increases HDL-cholesterol concentration but both high- and low-oleic acid ground beef decrease HDL particle diameter in normocholesterolemic men. *J. Nutr.* **2011**, *141*, 1188–1194. [CrossRef]
3. Gilmore, L.A.; Crouse, S.F.; Carbuhn, A.; Klooster, J.; Calles, J.A.E.; Meade, T.; Smith, S.B. Exercise attenuates the increase in plasma monounsaturated fatty acids and high-density lipoprotein cholesterol but not high-density lipoprotein 2b cholesterol caused by high-oleic ground beef in women. *Nutr. Res.* **2013**, *33*, 1003–1011. [CrossRef]
4. Choi, S.H.; Gharahmany, G.; Walzem, R.L.; Meade, T.H.; Smith, S.B. Ground beef high in total fat and saturated fatty acids decreases X Receptor Signaling Targets in peripheral blood mononuclear cells of men and women. *Lipids* **2018**, *53*, 279–290. [CrossRef] [PubMed]
5. Smith, S.B.; Lunt, D.K.; Smith, D.R.; Walzem, R.L. Producing high-oleic acid beef and the impact of ground beef consumption on risk Factors for cardiovascular disease: A review. *Meat Sci.* **2020**, *163*, 108076. [CrossRef] [PubMed]
6. Stender, S.; Astrup, A.; Dyerberg, J. Ruminant and industrially produced *trans* fatty acids: Health aspects. *Food Nutr. Res.* **2008**, *52*, 1651. [CrossRef] [PubMed]
7. Wu, X.; Roussell, M.A.; Hill, A.M.; Kris-Etherton, P.M.; Walzem, R.L. Baseline insulin resistance is a determinant of the small, dense low-density lipoprotein response to diets differing in saturated fat, protein, and carbohydrate contents. *Nutrients* **2021**, *13*, 4328. [CrossRef]
8. World Medical Association. Declaration of Helsinki: Ethical principles for medical research involving human subjects. *JAMA* **2000**, *284*, 3043–3045. [CrossRef]
9. Clevidence, B.A.; Fong, A.K.H.; Todd, K.; Brinkley, L.J.; Heiser, C.R.; Swain, J.F.; Rasmussen, H.; Tsay, R.; Oexmann, M.J.; Salbe, A.D.; et al. Planning diet studies. In *Well-Controlled Diet Studies in Humans*; Dennis, B.H., Ershow, A.G., Obarzanek, E., Clevidence, B.A., Eds.; The American Dietetic Association: Chicago, IL, USA, 1997; pp. 144–145.
10. Blackmon, T.; Miller, R.K.; Kerth, C.; Smith, S.B. Ground beef patties prepared from brisket, flank and plate have unique fatty acid and sensory characteristics. *Meat Sci.* **2015**, *103*, 46–53. [CrossRef]
11. Oliver, J.M.; Joubert, D.P.; Caldwell, A.; Martin, S.E.; Crouse, S.F. A longitudinal study examining the effects of a season of American football on lipids and lipoproteins. *Lipids Health Dis.* **2015**, *14*, 35. [CrossRef] [PubMed]
12. Orsoni, A.; Saheb, S.; Levels, J.H.M.; Dallinga-Thie, G.; Atassi, M.; Bittar, R.; Robillard, P.; Bruckert, E.; Kontush, A.; Carrie, A.; et al. LDL-apheresis depletes apoE-HDL and pre-ß1-HDL in familial hypercholesterolemia: Relevance to atheroprotection. *J. Lipid Res.* **2011**, *52*, 2304–2313. [CrossRef]
13. Scott, L.W.; Dunn, J.K.; Pownall, H.J.; Brauchi, D.J.; McMann, M.C.; Herd, J.A.; Harris, K.B.; Savell, J.W.; Cross, H.R. Effects of beef and chicken consumption on plasma lipid levels in hypercholesterolemic men. *Arch. Inter. Med.* **1994**, *154*, 1261–1267. [CrossRef]
14. Roussell, M.A.; Hill, A.M.; Gaugler, T.L.; West, S.G.; Heuvel, J.P.; Alaupovic, P.; Gillies, P.J.; Kris-Etherton, P.M. Beef in an Optimal Lean Diet study: Effects on lipids, lipoproteins, and apolipoproteins. *Am. J. Clin. Nutr.* **2012**, *95*, 9–16. [CrossRef] [PubMed]
15. Bowen, K.J.; Kris-Etherton, P.M.; West, S.G.; Fleming, J.A.; Connelly, P.W.; Lamarche, B.; Courture, P.; Jenkins, D.J.A.; Taylor, C.G.; Zahradka, P.; et al. Diets enriched with conventional or high-oleic acid canola oil lower atherogenic lipids and lipoproteins compared to a diet with a Western fatty acid profile in adults with central adiposity. *J. Nutr.* **2019**, *149*, 471–478. [CrossRef]
16. Wang, L.; Bordi, P.L.; Fleming, J.A.; Hill, A.M.; Kris-Etherton, P.M. Effect of a moderate fat diet with and without avocados on lipoprotein particle number, size and subclasses in overweight and obese adults: A randomized, controlled trial. *J. Am. Heart Assoc.* **2015**, *4*, e001355. [CrossRef]
17. Cannell, L.E.; Savell, J.W.; Smith, S.B.; Cross, H.R.; St John, L.C. Fatty acid composition and caloric value of ground beef containing low levels of fat. *J. Food Sci.* **1989**, *54*, 1163–1168. [CrossRef]
18. Smith, D.R.; Savell, J.W.; Smith, S.B.; Cross, H.R. Fatty acid and proximate composition of raw and cooked retail cuts of beef trimmed to different external fat levels. *Meat Sci.* **1989**, *26*, 295–311. [CrossRef] [PubMed]
19. U.S. Department of Agriculture; U.S. Department of Health and Human Services. *Dietary Guidelines for Americans, 2020–2025*, 9th ed.; U.S. Department of Health and Human Services: Washington, DC, USA, 2020. Available online: DietaryGuidelines.gov (accessed on 2 December 2022).
20. Maki, K.C.; Dicklin, M.R.; Kirkpatrick, C.F. Saturated fats and cardiovascular health: Current evidence and contraversies. *J. Clin. Lipidol.* **2021**, *15*, 765–772. [CrossRef]
21. Ouimet, M.; Barrett, T.J.; Fisher, E.A. HDL and reverse cholesterol transport: Basic mechanisms and their roles in vascular health and disease. *Circ. Res.* **2019**, *124*, 1505–1518. [CrossRef] [PubMed]
22. Sean Davidson, W.; Silva, R.A.G.D.; Chantepie, S.; Lagor, W.R.; Chapman, M.J.; Kontush, A. Proteomic analysis of defined HDL subpopulations reveals particle-specific protein clusters: Relevance to antioxidative function. *Arterioscler. Thromb. Vasc. Biol.* **2009**, *29*, 870–876. [CrossRef] [PubMed]
23. Nichols, A.V.; Krauss, R.M.; Musliner, T.A. Nondenaturing polyacrylamide gradient gel electrophoresis. *Methods Enzymol.* **1986**, *128*, 417–431. [CrossRef] [PubMed]

24. Patsch, J.; Gotto, A., Jr. Metabolism of high density lipoproteins. In *New Comprehensive Biochemistry*; Gotto, J.A.M., Ed.; Elsevier Science Publishers: New York, NY, USA, 1987; pp. 221–259.
25. Kontush, A.; Chapman, M.J. Antiatherogenic small, dense HDL—Guardian angel of the arterial wall? *Nat. Clin. Pract. Cardiovasc. Med.* **2006**, *3*, 144–153. [CrossRef] [PubMed]
26. Vaisar, T.; Pennathur, S.; Green, P.S.; Gharib, S.A.; Hoofnagle, A.N.; Cheung, M.C.; Byun, J.; Vuletic, S.; Kassim, S.; Singh, P.; et al. Shotgun proteomics implicates protease inhibition and complement activation in the antiinflammatory properties of HDL. *J. Clin. Investig.* **2007**, *117*, 746–756. [CrossRef] [PubMed]
27. Kontush, A.; Therond, P.; Zerrad, A.; Couturier, M.; Negre-Salvayre, A.; de Souza, J.A.; Chantepie, S.; Chapman, M.J. Preferential sphingosine-1-phosphate enrichment and sphingomyelin depletion are key features of small dense HDL3 particles: Relevance to antiapoptotic and antioxidative activities. *Arterioscler. Thromb. Vasc. Biol.* **2007**, *27*, 1843–1849. [CrossRef]
28. Ai, M.; Otokozawa, S.; Asztalos, B.F.; Ito, Y.; Nakajima, K.; White, C.C.; Cupples, L.A.; Wilson, P.W.; Schaefer, E.J. Small dense low density lipoprotein cholesterol and coronary heart disease: Results from the Framingham Offspring Study. *Clin. Chem.* **2010**, *56*, 967–976. [CrossRef]
29. Véniant, M.M.; Sullivan, M.A.; Kim, S.K.; Ambroziak, P.; Chu, A.; Wilson, M.D.; Mellerstein, M.K.; Rudel, L.L.; Walzem, R.L.; Young, S.G. Defining the atherogenicity of large and small lipoproteins containing apolipoprotein B100. *J. Clin. Investig.* **2000**, *106*, 1501–1510. [CrossRef]
30. Sniderman, A.D.; Thanassoulis, G.; Glavinovic, T.; Navar, A.M.; Pencina, M.; Catapano, A.; Ference, B.A. Apolipoprotein B particles and cardiovascular disease: A Narrative Review. *JAMA Cardiol.* **2019**, *4*, 1287–1295. [CrossRef]
31. Campos, H.; Genest, J.J., Jr.; Blijlevens, E.; McNamara, J.R.; Jenner, J.L.; Ordovas, J.M.; Wilson, P.W.; Schaefer, E.J. Low density lipoprotein size and coronary artery disease. *Arterioscler. Thromb.* **1992**, *12*, 187–195. [CrossRef]
32. Austin, M.A.; Breslow, J.L.; Hennekens, C.H.; Buring, J.E.; Willet, W.C.; Krauss, R.M. Low-density lipoprotein subclass patterns and risk of myocardial infarction. *JAMA* **1988**, *260*, 1917–1921. [CrossRef]
33. Shore, V.G. Nondenaturing electrophoresis of lipoproteins in agarose and polyacrylamide gradient gels. In *Analyses of Fats, Oils and Lipoproteins*; Perkins, E.G., Ed.; AOAC Press: Rockville, MD, USA, 1991; pp. 573–588.

nutrients

MDPI

Article

Phytosterols and Cardiovascular Risk Evaluated against the Background of Phytosterolemia Cases—A German Expert Panel Statement †

Eberhard Windler [1], Frank-Ulrich Beil [2], Heiner K. Berthold [3], Ioanna Gouni-Berthold [4], Ursula Kassner [5], Gerald Klose [6], Stefan Lorkowski [7], Winfried März [8,9,10,*], Klaus G. Parhofer [11], Jogchum Plat [12], Günter Silbernagel [13], Elisabeth Steinhagen-Thiessen [14], Oliver Weingärtner [15], Birgit-Christiane Zyriax [16] and Dieter Lütjohann [17]

1 Preventive Medicine, University Heart Center, University Hospital Hamburg-Eppendorf, Hamburg-Eppendorf, Martinistr. 52-Bldg. N26, 20246 Hamburg, Germany
2 Ambulanzzentrum, Universitätsklinikum Hamburg-Eppendorf, Martinistr. 52, 20246 Hamburg, Germany
3 Department of Internal Medicine and Geriatrics, Bethel Clinic, 33611 Bielefeld, Germany
4 Center for Endocrinology, Diabetes and Preventive Medicine, Faculty of Medicine and University Hospital Cologne, University of Cologne, Kerpener Str. 62, 50937 Cologne, Germany
5 Lipid Clinic at the Interdisciplinary Metabolism Center, Charite-Universitätsmedizin Berlin, Augustenburger Platz 1, 13353 Berlin, Germany
6 Praxen Dres. T. Beckenbauer & S. Maierhof, Am Markt 11, 28195 Bremen und Dres. I. van de Loo & K. Spieker, Gerold Janssen Straße 2 A, 28359 Bremen, Germany
7 Institute of Nutritional Science and Competence Cluster for Nutrition and Cardiovascular Health (nutriCARD), Halle-Jena-Leipzig, Friedrich Schiller University Jena, Dornburger Str. 25, 07743 Jena, Germany
8 SYNLAB Akademie für Ärztliche Fortbildung, SYNLAB Holding Deutschland GmbH, P5,7, 68161 Mannheim, Germany
9 Medical Clinic V, Medical Faculty Mannheim, University of Heidelberg, Theodor-Kutzer-Ufer 1-3, 68167 Mannheim, Germany
10 Clinical Institute of Medical and Chemical Laboratory Diagnostics, Medical University of Graz, 8010 Graz, Austria
11 Medizinische Klinik IV, Klinikum der Universität München, Grosshadern, Marchioninistr. 15, 81377 München, Germany
12 Department of Nutrition and Movement Sciences, School of Nutrition and Translational Research in Metabolism (NUTRIM), Maastricht University, 6211 LK Maastricht, The Netherlands
13 Division of Vascular Medicine, Department of Internal Medicine, Medical University of Graz, 8010 Graz, Austria
14 Arbeitsbereich Lipidstoffwechsel der Medizinischen Klinik für Endokrinologie und Stoffwechselmedizin, Charité—Universitätsmedizin Berlin, Augustenburger Platz 1, 13353 Berlin, Germany
15 Klinik für Innere Medizin I, Universitätskliniken Jena, Friedrich-Schiller-Universität Jena, 07743 Jena, Germany
16 Midwifery Science—Health Care Research and Prevention, Research Group, Preventive Medicine and Nutrition, Institute for Health Services Research in Dermatology and Nursing (IVDP), University Medical Center Hamburg-Eppendorf, Martinistr. 52, 20246 Hamburg, Germany
17 Institute of Clinical Chemistry and Clinical Pharmacology, University Clinics Bonn, 53127 Bonn, Germany
* Correspondence: winfried.maerz@synlab.com

† Dedication. This article is dedicated to the memory of Prof Klaus von Bergmann, who passed away on 26 September 2022. Klaus von Bergmann was a gastroenterologist and clinical pharmacologist who held the Chair of Clinical Pharmacology at the University of Bonn, Germany, until 2005. He made substantial scientific contributions in the field of intestinal cholesterol metabolism and absorption. Eberhard Windler, Frank-Ulrich Beil, Heiner K Berthold, Ioanna Gouni-Berthold, Ursula Kassner, Gerald Klose, Stefan Lorkowski, Winfried März *, Klaus G Parhofer, Jogchum Plat, Günter Silbernagel, Elisabeth Steinhagen-Thiessen, Birgit-Christiane Zyriax, Oliver Weingärtner, Dieter Lütjohann.

Citation: Windler, E.; Beil, F.-U.; Berthold, H.K.; Gouni-Berthold, I.; Kassner, U.; Klose, G.; Lorkowski, S.; März, W.; Parhofer, K.G.; Plat, J.; et al. Phytosterols and Cardiovascular Risk Evaluated against the Background of Phytosterolemia Cases—A German Expert Panel Statement. *Nutrients* **2023**, *15*, 828. https://doi.org/10.3390/nu15040828

Academic Editor: Stephen J. Ives

Received: 4 December 2022
Revised: 23 January 2023
Accepted: 25 January 2023
Published: 6 February 2023

Abstract: Phytosterols (PSs) have been proposed as dietary means to lower plasma LDL-C. However, concerns are raised that PSs may exert atherogenic effects, which would offset this benefit. Phytosterolemia was thought to mimic increased plasma PSs observed after the consumption of PS-enriched foods. This expert statement examines the possibility of specific atherogenicity of PSs based on sterol metabolism, experimental, animal, and human data. Observational studies show no evidence that plasma PS concentrations would be associated with an increased risk of atherosclerosis or cardiovascular (CV) events. Since variants of the ABCG5/8 transporter affect the absorption of cholesterol and non-cholesterol sterols, Mendelian randomization studies examining the effects of ABCG5/8 polymorphisms cannot support or refute the potential atherogenic effects of PSs due to pleiotropy. In homozygous patients with phytosterolemia, total PS concentrations are ~4000% higher than under physiological conditions. The prevalence of atherosclerosis in these individuals is variable and may mainly relate to concomitant elevated LDL-C. Consuming PS-enriched foods increases PS concentrations by ~35%. Hence, PSs, on a molar basis, would need to have 20–40 times higher atherogenicity than cholesterol to offset their cholesterol reduction benefit. Based on their LDL-C lowering and absence of adverse safety signals, PSs offer a dietary approach to cholesterol management. However, their clinical benefits have not been established in long-term CV endpoint studies.

Keywords: plant sterols; phytosterols; phytosterolemia; atherosclerosis; cardiovascular disease; cholesterol absorption

1. Introduction

Phytosterolemia, formerly also known as sitosterolemia, is a rare, autosomal-recessively transmitted inborn disorder of metabolism caused by mutations of the genes encoding the adenosine triphosphate (ATP)-binding cassette (ABC) G5/8 co-transporter [1,2]. The disease is often advocated to illustrate the possible consequences of an increased plasma concentration of phytosterols (PSs), also known as plant sterols and stanols, due to an increased dietary intake from foods with added PSs. Typically, daily PS intake is about 300 mg with a habitual diet and can be as high as 600 mg/d in vegetarians [3]. The recommended intake of PSs to lower LDL-C is around 2 g/d. The two most common PSs are sitosterol and campesterol.

Selective observations of individual cases with severe atherosclerosis, especially in children and adolescents affected by phytosterolemia, have provided reasonable grounds to suspect that PSs may advance atherosclerosis as much as or even more severely than LDL-cholesterol (LDL-C), a causal risk factor for atherosclerosis and cardiovascular disease (CVD) [4]. While PSs reduce LDL-C concentrations [3], concerns remain that their atherogenic effects may outweigh their benefits.

In fact, homozygous defects in the genes encoding the ABCG5/G8 tandem transporter, which causes phytosterolemia, modulate the absorption of about forty sterols, including cholesterol and non-cholesterol sterols (also known as xenosterols) with various complex effects on sterol and lipid metabolism and other tissue and organ functions. All these effects are unlikely to be associated with only a single sterol [5]. Furthermore, since the phenotype of phytosterolemia is highly variable, modifying influences must be considered.

This expert statement discusses the hypothesis of a specific atherogenicity of PSs based on the analysis of the current understanding of sterol metabolism, experimental and animal studies, epidemiological data from observational studies, and clinical experience in patients with phytosterolemia.

2. Regulation of Intestinal Absorption and Biliary Excretion of Phytosterols

The absorption and excretion of PSs, as well as of cholesterol (both from dietary and biliary sources), involves several membrane transporters and intracellular proteins. In principle, the organism is protected from absorption and accumulation of xenosterols such as PSs by redundant mechanisms [6]. Plasma concentrations of total PSs (mainly the sum of sitosterol and campesterol) are generally between 7–24 μmol/L (0.3–1.0 mg/dL), which is approximately 500-fold lower than circulating concentrations of total cholesterol; subsequently, PSs account for less than 1% of total plasma sterols [3].

The reason for the considerable difference between the plasma concentrations of cholesterol and PSs is twofold. First, PSs—in contrast to cholesterol—cannot be synthesized by mammalian cells, and second the selectivity of at least five successive absorption and excretion processes mediated via complex and not yet fully understood mechanisms prefer PSs over cholesterol [7]. As an example, the Niemann-Pick C1-like1 (NPC1L1) protein located on the membrane of the enterocytes binds cholesterol with a higher affinity than PSs [8–10]. PSs that are absorbed despite this first selection process are inhibited from incorporation into chylomicrons due to the specificity of two subsequent processes. Acyl-CoA cholesterol acyl transferase 2 (ACAT2) prefers cholesterol over PSs so that cholesteryl esters are predominantly formed and subsequently preferentially incorporated into chylomicrons [11,12]. Thus, the majority of PS remains unesterified for transfer back from the enterocytes into the intestinal lumen by the ABCG5/G8 transporter, which prefers PS over cholesterol [13,14]. This results in a total absorption rate of PSs of less than 5% [14]. In contrast, the average cholesterol absorption is about 54%, with a range from 20 to 80% [15,16].

The PSs remaining in the enterocytes are esterified with fatty acids in the same way as cholesterol by ACAT2 and then incorporated into chylomicrons. Next, PSs in chylomicrons and their remnants or in HDL particles are predominantly taken up into the liver. Here, PSs once again encounter the ABCG5/G8 transporter, which transfers them into the bile ducts for secretion with the bile. Some of the secreted PSs can be reabsorbed back into the liver through the NPC1L1 protein of the bile ducts. However, as in the intestine, there is a higher affinity for cholesterol than PS reabsorption [17]. These redundant protective selection mechanisms are so efficient that even heterozygous mutations in one of the heterodimers of the ABCG5/G8 transporter will result in only a modest increase in plasma PS concentrations of around 0.034 mmol/L (approx. 1.4 mg/dL) [18,19]. Even in the case of homozygous or combined heterozygous mutations of the ABCG5/G8 transporter, such as in phytosterolemia, PS plasma concentrations remain at about 0.3–1.2 mmol/L (approx. 12–48 mg/dL) [20,21]. Hence, PS concentrations never reach the level of plasma cholesterol concentrations despite a comparable dietary intake. The main steps involved in the intestinal absorption and biliary secretion of sterols (cholesterol and PSs) are depicted in Figure 1.

PSs are found in free and, more so, esterified forms in all serum lipoprotein fractions and in the bile [22–24]. Approximately two-thirds of plasma PSs are found in LDL particles, while HDL seems to make the largest contribution to the transport of PSs from the intestine to the liver [25].

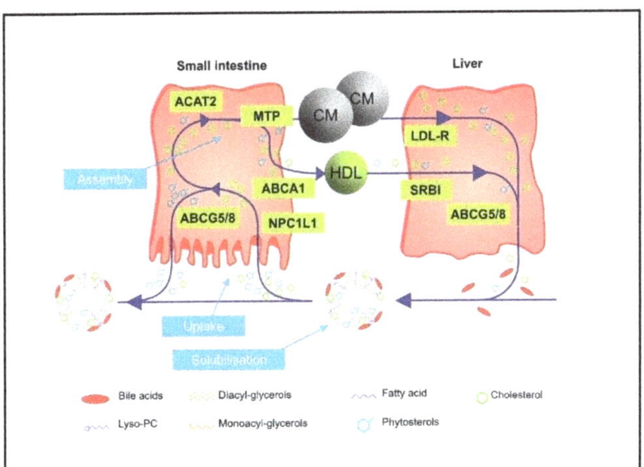

Figure 1. Flux and homeostasis of sterols (cholesterol and non-cholesterol sterols) with a focus on intestinal absorption and biliary secretion (adopted and modified from [26]. ABCA1, ATP binding cassette transporter A1; ABCG5/8, ATP binding cassette co-transporters G5 and G8; ACAT2, acetylcoenzyme-A

cholesterol acyl transferase 2; CM, chylomicron; HDL, high-density lipoprotein; LDL-R, low-density lipoprotein receptor; MTP, microsomal triglyceride transfer protein; NPC1L1, Niemann Pick C1-like 1 protein; SRBI, scavenger receptor class B type 1.

3. Phytosterolemia—An Inherited Disorder of Sterol Transport

Clarification of the pathophysiology of phytosterolemia has improved our understanding of phytosterol metabolism [2]. Phytosterolemia is caused by homozygous or compound heterozygous defects in the genes ABCG5/sterolin-1 and ABCG8/sterolin-2 on chromosome 2p21 [27–32]. These genes encode the obligate heterodimeric ABC transporters, ABCG5 and ABCG8, with the result that a defect in one of the two monomers may cause a malfunction of the entire heterodimer. The ABCG5/G8 transporter pumps sterols, such as cholesterol, but preferably PSs, from the apical surface of the enterocytes back into the small intestine, as well as cholesterol and PSs from hepatocytes into the bile ducts [5,33]. The excretion of PSs from the hepatocytes into the bile canaliculi is necessary since, in mammalians, most PSs cannot be converted into bile acids and are subsequently excreted [34,35].

Phytosterolemia has been considered a rare inherited disease. Worldwide, only some 100 phytosterolemia cases have been described limiting possible correlations to clinical manifestations. However, based on recent advances in genetics, the prevalence of individuals with detrimental mutations in ABCG5/G8 genes may be more than 1 in 200,000 individuals [1]. Phytosterolemia may also be underdiagnosed because clinically it is difficult to distinguish clinically from familial hypercholesterolemia (FH).

3.1. *Plasma Sterol Concentrations and Cardiovascular Disease in Phytosterolemia*

Characteristic symptoms of phytosterolemia, including severe hypercholesterolemia, xanthomas, aortic stenosis, as well as fatal and non-fatal cardiac events, have been reported predominantly in infants/children up to the age of adolescence, whereas these complications are apparently not as frequent with increasing age. In adults, mostly low to average plasma cholesterol concentrations have been reported, but LDL-C levels similar to those in heterozygous FH have also been observed in some cases [36]. The pronounced cardiovascular (CV) manifestations of homozygous phytosterolemia in children and adolescents are possibly the result of extremely high plasma cholesterol concentrations of up to more than 20 mmol/L (770 mg/dL). It has been suggested that the very high cholesterol concentrations in childhood are caused by the immaturity of the mechanisms that limit sterol intake [37]. Apparently, the effect is accentuated by a higher cholesterol burden in breast-fed infants and high consumption of dairy products in childhood [38]. High LDL-C during childhood has been described in several case reports as reviewed previously [14]. For example, in an 11-year-old with phytosterolemia, the plasma sitosterol concentration was 0.7 mmol/L (approx. 28 mg/dL), but the tendinous and tuberous xanthomas were similar to homozygous FH. The total plasma sterol concentration was 555 mg/dL and was reduced to 221 mg/dL with a cholesterol-lowering diet [39]. Treatment with cholestyramine resulted in total sterols of 173 mg/dL, the tuberous xanthomas disappeared, and the tendinous xanthomas decreased, while plasma PSs remained elevated. A similar experience was reported in two Japanese sisters with phytosterolemia whose xanthomas disappeared after a cholesterol-lowering diet alone [40]. These cases underpin that cholesterol concentrations can be extremely high in childhood phytosterolemia and are most likely responsible for the atherosclerosis and CVD encountered in these patients [29,41].

The hypothesis of temporarily high blood cholesterol levels in phytosterolemia is supported by the observation of several adult patients with phytosterolemia without symptoms of atherosclerotic CVD whose plasma concentrations of cholesterol could be well controlled, while their plasma PS concentrations were reduced by ezetimibe but not normalized [42]. In a German outpatient lipid clinic, two clinically inconspicuous brothers with phytosterolemia were monitored until their ninth decade, whereas another brother died at the age of 20 of suspected FH [*personal communication Frank-Ulrich Beil*].

After the first description in 1974 [22], more and more reports have been published on patients with xanthomas and macrothrombocytopenia, homozygous or compound heterozygous for mutations in the ABCG5 or ABCG8 gene, and thus biochemically and genetically diagnosed phytosterolemia, but without evidence of clinically relevant atherosclerosis [33,43–54]. Even in a family of a 13-year-old boy with phytosterolemia and severe atherosclerotic disease who died from sudden cardiac death, some, but not all, affected relatives developed premature ischemic heart disease [55]. In two sisters and three unrelated patients with suspected homozygous FH at the ages of 4–8 years, phytosterolemia was diagnosed. Plasma concentrations of total cholesterol were between 409 and 805 mg/dL (10.6 and 20.8 mmol/L), and LDL-C was between 372 and 630 mg/dL (9.6 and 16.3 mmol/L) [56]. All these patients had xanthomas, and some had thrombocytopenia. While changes in the diet and treatment with cholestyramine and ezetimibe resulted in LDL-C concentrations of 58 and 96 mg/dL [1.5 and 2.5 mmol/L), sitosterol and campesterol remained approximately 10 times higher than normal, ranging between 12 and 29 mg/dL (0.48 and 1.9 mmol/L). However, up to the ages of 17–33 years, no clinical signs of atherosclerosis were found by ultrasound examinations of the carotids, by coronary angiography or by cardiac computed tomography (CT) calcium score. The carotid intima-media thickness (IMT) of the carotids remained approximately constant at 0.45–0.54 mm [56]. This may suggest that any damaging potential of plant sterols may be harnessed once LDL-C cholesterol is sufficiently controlled by drug treatment, but obviously, systematic and challengeable evidence for this is lacking.

A case of a 33-year-old female with chronic tendosynovitis, xanthoma, and elevated LDL-C had campesterol and sitosterol concentrations of 7–12-fold and 20–38-fold higher, respectively, above the upper limit of normal but showed no clinical evidence of premature atherosclerosis [54]. The absence of atherosclerosis may again suggest that phytosterolemia is not necessarily accompanied by premature vascular disease.

The mechanism underlying the high LDL-C concentrations in phytosterolemia has not been understood. Possibly, the inability to limit cholesterol absorption due to defective re-secretion of absorbed cholesterol into the gut lumen may increase the flux of intestinal cholesterol into the liver, which may, in turn, down-regulate hepatic LDL receptor expression. Typical for phytosterolemia, plasma cholesterol concentrations decrease drastically with a cholesterol-lowering diet. This was regarded as a diagnostic sign at the time before PSs were measured, and phytosterolemia was named pseudo-homozygous FH [40,55,57]. A Korean infant, exclusively breast-fed, presented with phytosterolemia and developed extensive intertriginous xanthomas in the third month of life. The infant's serum total cholesterol (675 mg/dL or 17.5 mmol/L) and LDL-C (540 mg/dL or 14 mmol/L) decreased dramatically following a cholesterol-lowering diet and cholestyramine treatment. At three years of age, LDL-C was 118 mg/dL (3.1 mmol/L), but the plasma sitosterol concentration remained increased at 19 mg/dL (0.76 mmol/L, while the xanthomas had completely disappeared [58]. The infant's five-year-old sister with the same PS phenotype was asymptomatic, suggesting that mechanisms other than the PS concentration should be responsible for the clinical manifestation of the disease.

The observed extreme plasma cholesterol elevations seen in childhood may not be encountered when the initial diagnosis is made later in life. It is conceivable that the development of atherosclerosis is initiated during childhood, resulting in CV events in childhood and young adulthood, at a time at which plasma PSs and cholesterol concentrations have already declined so that a definite correlation between events and plasma sterol concentrations is not apparent [42]. At an adult age, the regulatory mechanisms of sterol intake will become effective, resulting in a decrease of both PS and cholesterol intake and enabling the normalization of the plasma cholesterol concentration through dietary changes and targeted therapy. For example, a normocholesterolemic phytosterolemia patient with a total cholesterol concentration of 192 mg/dL (5.0 mmol/L) had to undergo bypass surgery at the age of 29. He was primarily diagnosed with FH, suggesting that he had very high LDL-C concentrations in his younger years and that during that time, major vascular damage had occurred [45].

Clinically, infants heterozygous for phytosterolemia are not affected, while homozygous infants may develop xanthomas of the patella and Achilles tendons, the extensor tendons of the hand, and tubero-eruptive xanthomas at the elbows, or other pressure-bearing areas. These xanthomas, however, may disappear with increasing age in response to diet and, if necessary, medical therapy [6,48,59]. Next, fatal and non-fatal coronary events, as well as aortic stenoses, have been reported, mainly in children from the age of five up to young adulthood [14,22,55,60].

3.2. Heterozygous Phytosterolemia and Plasma Phytosterol Concentrations

Due to the redundant defense mechanisms outlined above, heterozygotes for mutations in the genes of the ABCG5/G8 transporter are clinically asymptomatic, despite slightly elevated plasma PSs [6]. Whether plasma PS concentrations increase in affected individuals, who consume PS-enriched foods with their diet, has been repeatedly examined. For example, two studies with obligate heterozygotes for phytosterolemia consuming a low-fat diet (American Heart Association Step 1 Diet) containing 2.2 g/day of PSs reported an increase in average plasma PS concentrations from 0.9 mg/dL to 1.7 mg/dL (0.04 to 0.07 mmol/L), whereas LDL-C decreased by about 16 mg/dL (0.41 mmol/L) [61]. The individually highest levels for campesterol plus sitosterol of about 2 mg/dL (0.08 mmol/L) increased after intake of PS-enriched margarine to about 3.4 mg/dL (0.14 mmol/L). On average, campesterol concentrations roughly doubled, while sitosterol concentrations increased by about 50%. Similar results were also found in other studies with individuals heterozygous for phytosterolemia [62,63]. These changes are in the range of those reported for individuals without evidence of heterozygous phytosterolemia (an increase in sitosterol of 17–50% and in campesterol of 72–114%) [61].

At the same time, the effect of dietary PS intake on plasma LDL-C lowering in heterozygotes for phytosterolemia is comparable to that of non-genetically affected individuals and can even be significantly higher in the presence of certain mutations in the ABCG5/G8 and NPC1L1 genes [63]. For instance, in the ABCG8 1289 C > A (T400 K) polymorphism, the A allele carriers with high basal plasma PS concentrations demonstrated a 3.9-fold greater reduction in LDL-C than their low basal plasma PS counterparts [63]. It can therefore be speculated that individuals with increased intestinal sterol absorption due to heterozygous mutations in the ABCG5/G8 genes may benefit from a higher dietary PS intake at the extent of any unaffected individual (or even more) due to the reduction of cholesterol absorption. This is supported by the finding that individuals with high cholesterol absorption may benefit more from the inhibition of cholesterol absorption by PS supplementation [18,64,65].

In a recent study, rare heterozygous loss-of-function (LoF) variants in ABCG5 or ABCG8 were associated with blood lipids and CVD risk [66]. Compared to non-carriers, carriers of heterozygous LoF variants in ABCG5 had higher plasma sitosterol and ~25 mg/dL higher LDL-C and were at a two-fold higher risk of CAD [66]. Moreover, the impact of heterozygous LoF carrier status on CVD risk was proportional to the effect on LDL-C elevation. Therefore, LDL-C rather than sitosterol may have been the key driver of accelerated atherosclerosis.

Taken together, accumulating evidence, including cases of infants with phytosterolemia and elevated LDL-C as well as progression/regression of xanthomas, reveal that the elevation of LDL-C may be the major cause for the development of atherosclerosis, and not elevated PS concentrations [1].

3.3. Phytosterol Concentrations in Tissues and Risk of Atherosclerosis

Interestingly, an increase in plasma PSs does not lead to specific enrichment of PSs in tissues, as the PS-to-cholesterol ratio is identical in plasma and in all tissues examined, including xanthomas, arterial walls, and aortic valves [22,67]. An exception is the human brain since it is not or only marginally accessible for PS [47]. Feeding mice with a PS-enriched diet over 6 weeks resulted in an increase in the concentration mainly of campesterol in the brain [68]. However, this does not correspond to the results in humans. In autopsies of patients with phytosterolemia, the brain was explicitly excluded from PS accumulation [47]. In line with these observations, there are no neurological dysfunctions reported in patients with phytosterolemia.

Plasma and tissue from elective endarterectomies of patients, who had consumed foods enriched with PSs, showed, as expected, a slightly higher PS content, while the PS-to-cholesterol ratio corresponded to that of plasma [69]. This observation holds even for individuals with homozygous phytosterolemia. While plasma PS concentrations are significantly increased in these patients, cholesterol still accounts for 80% of sterols. Hence, PSs are not over-represented in xanthomas of phytosterolemia patients; rather, they contain about 80% of sterols as cholesterol, in parallel to their plasma concentrations [22,47,60].

These findings indicate that the tissue PS concentrations merely passively follow the plasma levels, as is also the case for other lipids like fatty acids. Interestingly, despite high circulating levels, at least at a young age, sterols do not accumulate in parenchymal organs in patients with phytosterolemia but only in vessel walls, cardiac valves, tendons, and the skin [47].

The study by Horenstein et al. [70] valuated the association of being a carrier of the heterozygous G574R in ABCG8 with sub-clinical atherosclerosis as assessed by ultrasound measurement of carotid IMT. ABCG8 G574R carriers have moderately elevated plasma PS, but compared with non-carriers, decreased carotid IMT, suggesting that a moderate, life-long elevation in plasma PS is not associated with accelerated atherosclerosis and may even be protective.

In summary, there is no evidence that PSs are deposited preferentially of cholesterol in the vessel wall, nor is there evidence for a stronger atherogenicity of PSs compared to cholesterol. Epidemiologic and genetic studies in humans without phytosterolemia or heterozygous variant carriers revealed no evidence of an increased risk of atherosclerosis in individuals with moderately elevated plasma PS concentrations.

3.4. Phytosterolemia to Evaluate Effects of Dietary Phytosterol Intake

As outlined above, phytosterolemia has often been advocated as a model to infer the possible effects of dietary PSs. The changes in blood lipids seen in affected patients, however, mainly appear to be due to a compromised biliary secretion of PSs and cholesterol rather than an increased intestinal uptake of PS. This has been learned from studies following liver transplantation as a treatment of phytosterolemia [71]. Increased cholesterol absorption and impaired biliary cholesterol secretion in phytosterolemia may be compensated by a markedly diminished to non-measurable endogenous cholesterol biosynthesis with unchanged bile acid synthesis, which counteracts hypercholesterolemia at least in adults [45].

In individuals with phytosterolemia, plasma PS concentrations increase relatively more than plasma cholesterol concentrations. So, what can be learned from cases with phytosterolemia, except that the possible effects of dietary PSs are debatable? First, plasma PS concentrations of 10–65 mg/dL (0.3–1.2 mmol/L) have never been reached by the consumption of foods enriched with PS. Next, phytosterolemia leads to the accumulation not only of the two main PS sitosterol and campesterol but also of other xenosterols such as stigmasterol, brassicasterol, and avenasterol, which may have various effects on their own, even at low concentrations [19].

In phytosterolemia, the circulating concentration of cholesterol quantitatively preponderates and can reach considerable levels. This phenomenon is especially prominent in infant and child cases, as discussed above. In children with homozygous phytosterolemia, extremely high cholesterol levels have been observed in the range similar to those with severe homozygous FH, which can lead to fatal myocardial infarction as early as 5 years of age [10,14,29,36,72]. Therefore, phytosterolemia has also been referred to as pseudo-homozygous FH [40,57].

In a study of untreated children and adolescents with homozygous phytosterolemia aged six months to eleven years, the mean plasma sitosterol concentration was 25.5 mg/dL (0.61 mmol/L) and that of campesterol 12.8 mg/dL (0.32 mmol/L) [14]. In contrast, total cholesterol was about 490 mg/dL (12.7 mmol/L), hence a considerably stronger relative increase than that of PSs. In adults with homozygous phytosterolemia, total cholesterol concentrations were significantly lower than those of younger patients [73].

More recently, cases of phytosterolemia have increasingly been diagnosed in adults without manifest vascular disease due to other consequences of PS accumulation, mostly hematological changes. Rare exceptional cases with arthralgia, endocrine and hepatic dysfunctions have been described as well as one case of a xanthoma in the spinal canal [73,74].

3.5. Can Thrombocytopenia Be Used to Define a Threshold of Phytosterol Toxicity?

Hematological manifestations of phytosterolemia, such as macrothrombocytopenia with bleeding tendency, hemolytic anemia with stomatocytes and splenomegaly, have been described in children and adults because of symptoms or incidental findings [22,33,43–49,51]. Apparently, these symptoms are direct consequences of the alteration of the cell membrane by very high PS concentrations, as has been demonstrated experimentally [50]. In plants, PSs are the equivalent of cholesterol, from which they differ only by modifications in the aliphatic side chain. The incorporation of PS into cell membranes may lead to increased rigidity, which may explain particularly the hematological sequels of pathologically high PS concentrations, but not necessarily all their effects [13,52,72,75–84].

Macrothrombocytopenia appears to be a sensitive marker for increased PS concentrations. In 13 Chinese adult patients with phytosterolemia and a mean age of 44 years (23 to 61 years) from eight unrelated families, common leading symptoms included macrothrombocytopenia and hemolytic anemia with stomatocytes. In eight of these patients, xanthomas were found, and only two had coronary heart disease, although serum sitosterol and stigmasterol concentrations averaged 58 and 55 mg/dL, respectively. Also, five of these patients had LDL-C concentrations of 200 mg/dL and higher, i.e., in the range of heterozygous FH [36,53]. These symptoms improved following cholesterol reduction with cholestyramine and a diet with reduced PS intake.

A reduction of circulating PSs with ezetimibe can increase platelet count and decrease platelet volume. Based on clinical experience and experimental studies, macrothrombocytopenia normalizes around a PS concentration of $\leq$15 mg/dL, while pathological changes are observed at levels of >15 mg/dL [50,85]. The pharmacological reduction of PSs, in addition to the reduction of cholesterol, is certainly useful once plasma levels reach the toxic range >15 mg/dL [86,87]. However, ezetimibe may not fully normalize thrombocytopenia in every case. In addition, ezetimibe does not appear to be effective in infants and young children, possibly due to insufficient glucuronidation at this age [59].

The threshold of >15 mg/dL plasma concentration is so high that an intake of PSs with enriched foods is not likely to have an influence on thrombocytes or on the osmotic fragility of erythrocytes [88]. In homozygous phytosterolemia, the toxic concentration of PSs between 15 and 30 mg/dL, according to clinical criteria, can usually be undercut by appropriate nutrition and other adequate therapy, thanks to the apparent change in the response to the genetic predisposition in adulthood [86]. In heterozygous individuals, this toxic range is never reached, just as in the case of non-affected individuals, even if the consume a PS-enriched diet [20,27,28,61,62].

4. Phytosterols in Animal Models for Atherosclerosis

A possible relationship between circulating PS concentrations and CV risk was also investigated in various animal models [3,89]. Mice with a defect of the ABCG5/G8 transporter do not develop atherosclerosis [90]. Feeding of very high doses of PS to LDL receptor knockout mice (LDR +/−) and apolipoprotein E knockout mice (ApoE−/−) was associated with reduced development of atherosclerotic plaques [91–93]. The experimental PS application in this mouse model corresponded to an intake in humans of 200–250 g per day, which makes it difficult to compare with the effects possibly achievable with PS intakes of 2–3 g daily in humans. Severe toxic effects are reported only in mice with a double knockout of both the ABCG5 and the ABCG8 transporter, in which plasma PS levels rose to 250 mg/dL by feeding a PS-rich chow diet and in which PSs constituted 75% of the total sterols [83]. Toxic reactions under those artificial conditions are not surprising since, otherwise, nature would not have designed multiple mechanisms for the elimination of PSs from mammals.

Nevertheless, the effectiveness of PS consumption for reducing atherosclerosis is supported by the results of animal experiments [89], which would not occur if PSs were more atherogenic than cholesterol. Still, PSs may be as atherogenic as cholesterol on a molar basis since ezetimibe administration which lowers PSs to some extent, appears to be more effective than PS feeding in reducing the development of atherosclerotic plaques at equal total cholesterol concentrations in mice [92]. However, the concentrations of atherogenic lipoproteins have not been determined in those experiments. Furthermore, no additional effect of ezetimibe beyond that mediated via LDL-C reduction has been observed in clinical trials in humans [94,95].

There are yet other reasons that the results from animal experiments cannot be extrapolated to humans; in rodents, most xenosterols and cholesterol are transported in HDL rather than in LDL, and LDL may not be the major atherogenic particles in these animals. Further, not all aspects of murine atherosclerosis are identical to humans [96], and the regulation of lipoprotein metabolism differs vastly from that in humans in that, e.g., mice respond conversely to drugs like statins and fibrates [96,97]. Apart from that, many experiments have been carried out with excessively high PS doses.

5. Are plasma Phytosterol Concentrations Associated with Cardiovascular Risk?

5.1. Evidence from Observational Studies

Based on the suspicion of increased CV risk in individuals with phytosterolemia [47], several observational studies utilizing different cohorts assessing the association between plasma PS concentration and CV risk were carried out. The PROCAM (PROspective CArdiovascular Münster) study revealed a positive correlation between plasma sitosterol but not campesterol concentrations and CV risk [98]. In contrast, in the LASA (Longitudinal Ageing Amsterdam) study [99], the CORA (The COronary Risk factors for Atherosclerosis in women study) study [100], a Finnish study [101], and the Spanish EPIC cohort study [102], increasing plasma PS concentrations were correlated with a low CV risk. In a study with younger participants, an increasing PS-to-cholesterol ratio was associated with lower IMT of the carotid arteries as a surrogate marker of atherosclerosis [103]. The LURIC (Ludwigshafen Risk and Cardiovascular Health) study [104,105] and the Framingham Offspring Study [106] both found a weak positive correlation between plasma PS concentrations and CV risk. In contrast, in the Finland Helsinki Businessmen cohort study, higher PS concentrations in middle-aged men predicted a longer life expectancy [101]. A meta-analysis including 17 studies (sample size ranging from 40 to 2440, with 11,182 study participants in total) revealed no consistent correlation between plasma PS concentrations and CV risk [107].

The heterogeneity of these observational findings may be explained by differences in the prevalence of numerous factors that influence the physiological plasma concentrations of cholesterol and PS. In healthy individuals, plasma concentrations of campesterol may vary between 0.28 and 1.19 mg/dL and those of sitosterol between 0.12 and 0.66 mg/dL. Thus, the total of these two quantitatively dominating PS varies between 0.39 and 1.85 mg/dL [16]. Based on published data, individuals heterozygous for phytosterolemia reached maximum PS concentrations of up to 3.0 mg/dL, while in homozygotes or combined heterozygotes, PS concentrations were up to 65 mg/dL (Figure 2). Factors that influence the plasma concentrations of cholesterol and PS are, for example, the apolipoprotein E-phenotype and polymorphisms in the ABCG5/G8 transporters as well as in NPC1L1, but above all, different non-standardized analytical methods for the measurement of PSs in plasma or serum [16]. Such factors may impact the reported concentrations of cholesterol and PSs as well as their ratio. The differences in PS intake in omnivores, vegetarians, or vegans have only a minimal influence on the plasma PS concentrations, while cholesterol may substantially differ [108]. Individuals with diabetes mellitus and metabolic syndrome have lower plasma PS concentrations, possibly due to lower intestinal cholesterol absorption, while the effects of these conditions on CVD risk are the opposite [16]. Statins and ezetimibe lower cholesterol, while ezetimibe lowers PSs as well but does not decrease CV risk more than statins when normalized to the magnitude of LDL-C reduction [94,109].

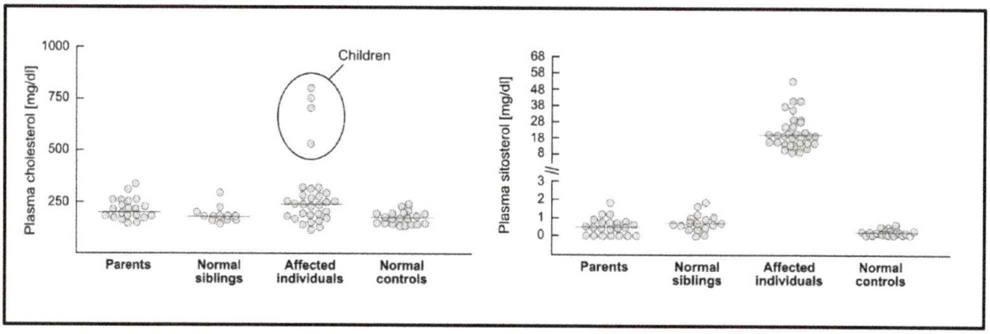

Figure 2. Plasma concentrations of total cholesterol and sitosterol in heterozygous (parents) and homozygous phytosterolemia cases versus unaffected siblings and unrelated controls (according to [28]).

In fact, the positive associations between PSs and CV risk, which were observed in some studies, may be because PSs are surrogate markers of the uptake of any sterols, including cholesterol, from the intestine [103,104,106]. A cross-sectional study with 270 asymptomatic individuals found a direct association of campesterol with carotid atherosclerotic plaques and an inverse association with the rates of cholesterol synthesis/absorption, suggesting that increased absorption and decreased synthesis of cholesterol, in general, are associated with carotid atherosclerosis [110].

5.2. *Evidence from Genetic Epidemiology*

Genetic variations in sterol transporters, which cause increased plasma concentrations of PSs, not only result in higher uptake and higher plasma concentrations of PSs but also in pleiotropically increased intestinal cholesterol absorption and subsequently higher plasma concentrations of LDL-C [111,112]. This may explain the correlation between increased plasma concentrations of PSs and CV risk in some Mendelian randomization studies [113]. The differences in the concentration of PS-transporting lipoproteins can have a fundamental impact on CVD patients compared to healthy controls as well. For example, higher HDL-cholesterol concentrations in healthy study participants may result in higher PS concentrations compared to patients with CVD and low HDL cholesterol [100]. So far, none of the observational studies distinguished between phytosterols in HDL and LDL particles.

Results from Mendelian randomization studies of mutations in the genes of the ABCG5/G8 transporter have been repeatedly interpreted as showing a causal relationship between increased phytosterol absorption and coronary heart disease, even though these transporters regulate the uptake of all xenosterols, i.e., dietary non-cholesterol sterols, and of dietary cholesterol [111]. In the case of such pleiotropy, the principle of Mendelian randomization only allows us to derive correlations with the gene of interest but not with one of several gene products or one of their several functions. In recognition of this limitation, the interpretation of the Mendelian randomization studies needs to be revised in that they do not prove a causal link between phytosterol concentrations and CVD [114].

Utilizing data from three studies of individuals of European origin from Iceland, Denmark, and the UK Biobank, Helgadottir et al. [115,116] investigated whether variability in cholesterol and PS absorption impacts the risk of CVD, using sequence variants of the ABCG5/8 genes. They concluded that the impact of ABCG5/8 variants on CVD cannot be fully explained by non-HDL-cholesterol and that other sterols, e.g., PSs, may contribute directly to atherogenesis. This conclusion, namely that the unexplained CVD risk is caused by absorbed PSs, has been questioned [117]. Moreover, plasma concentrations of cholestanol, a validated biomarker of intestinal cholesterol absorption, have not been reported. This would allow us to differentiate the impact of elevated PS concentrations versus increased intestinal overall sterol absorption on CVD risk.

A recent genome-wide meta-analysis of 32 PS traits reflecting cholesterol absorption, synthesis and esterification confirmed established associations between CVD and ABCG5/8 and ABO (blood group gene) next to showing an extended locus heterogeneity at ABCG5/8 with different functional mechanisms [118]. Subsequent Mendelian randomization analyses revealed a risk-increasing causal relationship between serum sitosterol concentrations and CVD, which was partly mediated by cholesterol. A direct and indirect causal effect of sitosterol on CVD risk was concluded [118].

In contrast, the results of the Copenhagen City Heart Study (CCHS) and the Copenhagen Ischemic Heart Disease Study (CIHDS) support the hypothesis of a relationship between absorbed cholesterol and CV risk [119]. It was shown that single nucleotide polymorphisms (SNPs), which increase sterol uptake and thus also intestinal cholesterol absorption, also result in reduced biliary cholesterol excretion and increased plasma LDL-C concentrations, which in turn are associated with an increased risk of coronary events and a reduced risk of gallstone formation [119]. It is noteworthy in this context that the increase in LDL-C was sufficient to explain the increased CV risk [119].

High cholesterol absorption has also been associated with increased CV risk in hemodialysis patients [120] and with the occurrence of in-stent restenosis [121].

Taken together, the findings from Mendelian randomization studies do not provide convincing evidence that the metabolic effects on CVD risk of variants in ABCG5/8 and others (ABO) are mediated by plasma PSs but rather by LDL-C or, more specifically, by an increased flux of intestinal cholesterol at the expense of hepatic de novo cholesterol biosynthesis even at a given steady state LDL-C concentration.

Essentially, three non-cholesterol surrogate markers have been suggested for the intestinal input of cholesterol into the body, namely the ratios of sitosterol, campesterol or cholestanol to cholesterol. None of them reflects cholesterol excretion into the bile. While sitosterol and campesterol are poorly absorbed plant sterols (phytosterols), cholestanol is a cholesterol degradation metabolite. While ideally, all of them should be reported, the cholestanol to cholesterol ratio is preferable since it is not affected by the supply of plant sterols with the food.

So far, none of the Mendelian randomization studies has so far reported cholestanol concentrations, which would have helped to distinguish between the risks conferred by PSs or by a high overall rate of sterol absorption. Thus, any PS-CVD association may also be explained by higher cholesterol absorption and reciprocally lower cholesterol synthesis rates which become measurable by surrogate markers such as sitosterol, campesterol and cholestanol for cholesterol absorption and plasma/serum lathosterol and desmosterol for cholesterol synthesis [122].

6. Do Phytosterol-Enriched Foods Increase the Risk of Atherosclerosis?

An important question is whether PSs can increase the risk of CVD when the average intake of about 300 mg per day with a habitual diet is increased to 2.5–3 g per day through PS supplements or PS-enriched foods. Several decades ago, the safety of the administration of up to 45 g/day of PSs (in non-esterified formulation) was tested over 16 weeks, and doses of up to 18 g/day were used particularly in pediatric medicine for decades to treat hypercholesterolemia [123–125]. However, this clinical experience can certainly not rule out an atherogenic effect of slightly increased plasma PS concentrations. In recent years, a daily PS intake of up to 9 g/day with fortified foods was tested. Daily PS doses of 2–4 g/day were administered in clinical studies for up to 85 weeks without clinically apparent adverse effects. [126,127]. Yet, the duration of these trials was certainly too short to allow conclusions about long-term atherogenic effects.

Through the redundant defense mechanisms described above, the plasma concentrations of PSs increase only slightly following consumption of PS-containing supplements or fortified foods and never reach levels as high as seen in patients with homozygous phytosterolemia (Figure 3). A meta-analysis of 41 randomized controlled trials with 55 treatment groups in a total of 2084 participants showed that an average PS intake of 1.6 g/day in the form of fortified foods increases plasma concentrations of sitosterol and campesterol by on average 31% and 37%, respectively [128]. In absolute terms, this corresponds to an increase in plasma concentrations of sitosterol of 2.24 μmol/L (0.092 mg/dL) and of campesterol of 5.00 μmol/L (0.20 mg/dL). This compares to a reduction of total cholesterol or LDL-C of 5.9% or 8.5%, respectively, corresponding to an absolute decrease of 0.36 mmol/L (13.9 mg/dL) and 0.33 mmol/L (12.8 mg/dL), respectively. [128]. In the highest PS dose range (2.0–3.2 g/day), sitosterol and campesterol were raised by an average of 3.56 μmol/L (0.15 mg/dL) and 7.64 μmol/L (0.31 mg/dL), respectively. Thus, the proportion of PSs remained clearly below 1% of the total sterol content in the blood.

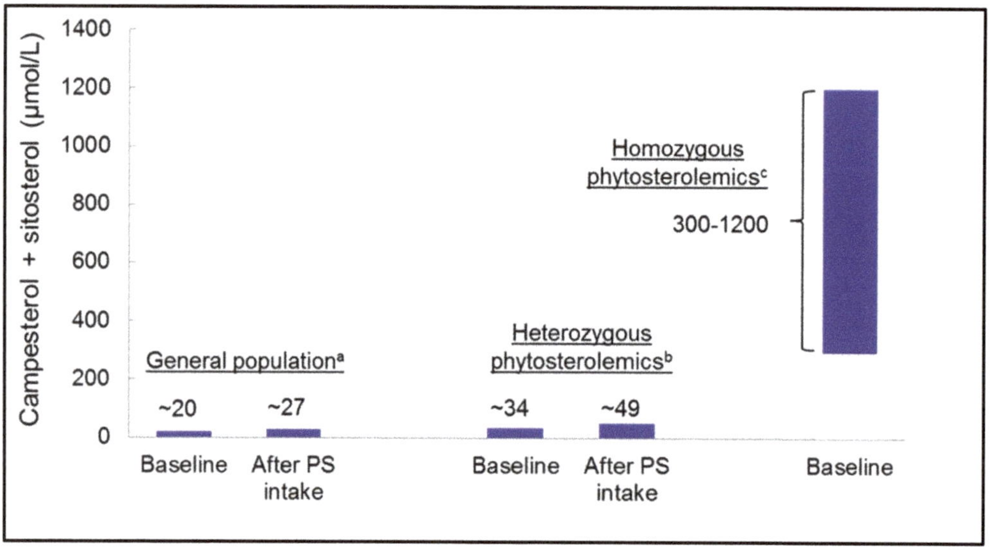

Figure 3. Plasma concentrations (baseline and after phytosterol (PS) intake) of campesterol and sitosterol in the general population and in individuals heterozygous for phytosterolemia versus in patients homozygous for phytosterolemia. [a] based on data from 41 human intervention studies [128]. [b] based on four studies regarding individuals with heterozygous phytosterolemia [18,61,62,129]. [c] based on cases with homozygous phytosterolemia [20,21].

Taken together, the LDL-C lowering effect of 6–12% achieved by the consumption of PS-enriched foods reflects a decrease of plasma cholesterol that is 20–40 times higher than the parallel occurring increase in the PS concentration. Thus, PSs would need to have a 20–40 times higher atherogenicity than cholesterol to cancel the positive effect of cholesterol reduction. Yet, there is no apparent evidence for this. Based on the knowledge of the large variability of plasma PS concentrations, ranging from 0.50 to 2 mg/dL, a higher intake or higher plasma concentrations of PSs, do not necessarily indicate a higher risk for atherosclerotic CVD [130]. This is in line with the findings of a meta-analysis of observational studies, which failed to show any association between plasma PS levels concentrations and CV events [107]. Average concentrations of circulating PS for the first vs. third tertiles were 0.17 and 0.57 mg/dL (4 and 12 μmol/L) for campesterol and 0.13 and 0.38 mg/dL (3 and 10 μmol/L) for sitosterol, which is about a three-fold difference [107]. For comparison, plasma PS concentrations at baseline and after several weeks of intake of foods with added PS increased from 0.52 to 0.71 mg/dL (13–18 μmol/L) for campesterol and from 0.29 to 0.38 mg/dL (7–9 μmol/L) for sitosterol [128]. This magnitude of variation in circulating PS concentrations was covered by the meta-analysis of Genser et al. [107], which found no association.

The increases in dietary PS intake or in the plasma PS-to-cholesterol ratio, respectively, caused by vegetarian diets or by taking statins, are associated with reduced CV risk and support the concept that minor increases in plasma PS concentrations do not trigger atherosclerosis. This is in line with the experience that no increased cardiac risk is known for the heterozygosity for phytosterolemia despite doubling plasma PS concentrations [6]. A lack of effect on CV risk is further reflected in intervention studies with ezetimibe, in particular the IMPROVE-IT and the HIJ-PROPER trials [94,131]. These studies showed no further reduction of CV events that goes beyond what can be expected by the LDL-C reduction alone despite the additional PS reduction. Interestingly, statin intake has been found to be associated with increased PS concentrations in plasma and muscle tissue despite the reduction in CV risk [109,132,133].

7. Conclusions

The synopsis of published data shows that phytosterolemia is primarily a pediatric disease that manifests in early infancy and may extend into late adolescence. In homozygous familial phytosterolemia, atherosclerosis appears to be mainly due to extremely high plasma LDL-C concentrations during childhood, of which the CV consequences may still become manifest later in life. In some cases, hematologic changes may persist into old age. No pathological changes have been described with plasma PS concentrations less than 15 mg/dL. Therefore, phytosterolemia provides no apparent evidence that the more modest increases in plasma PS concentrations induced by the consumption of PS-enriched foods might have undesirable effects that may outweigh the 20–40-fold greater reduction of LDL-C.

Evidence from CV endpoint studies that PS consumption also lowers the risk of CVD beyond the known LDL-C lowering effect is, however, not available. Such a large-scale outcome study of foods with added PS for CVD prevention in the setting of low to intermediate risk is likely not practically feasible, given the large number of study participants (>50,000) and the long duration (>5 years) that would be required for adequate power [3].

Author Contributions: E.W. and W.M. conceived the work. E.W. and D.L. drafted the manuscript. All authors contributed to the compilation and interpretation of the data and have approved the submitted version of this statement. Conceptualization, E.W. and W.M.; methodology, E.W. and D.L.; software, n.a.; validation, E.W., F.-U.B., H.K.B., I.G.-B., U.K., G.K., S.L., W.M., K.G.P., J.P., G.S., E.S.-T., O.W., B.-C.Z. and D.L., formal analysis, n.a.; investigation, n.a.; resources, n.a.; data curation, n.a.; writing—original draft preparation, E.W. and D.L.; writing—review and editing, E.W., F.-U.B., H.K.B., I.G.-B., U.K., G.K., S.L., W.M., K.G.P., J.P., G.S., E.S.-T., O.W., B.-C.Z. and D.L.; visualization, E.W., W.M. and D.L.; supervision, E.W. and W.M.; project administration, E.W.; funding acquisition, n.a. All authors have read and agreed to the published version of the manuscript.

Funding: This work received no external funding.

Institutional Review Board Statement: Not applicable.

Informed Consent Statement: Not applicable.

Data Availability Statement: All presented data are available in the cited literature.

Conflicts of Interest: E.W. reports honoraria for consulting and lectures from Amedes, AMGEN, AstraZeneca, Berlin Chemie, Daiichi-Sankyo, Fresenius, Hexal, MSD Sharp & Dohme, Novartis,

Upfield, Pfizer, Sanofi, Synlab, and an unrestricted investigational grant from Unilever and BASF. I.G.-B. reports honoraria for consulting from AMGEN, Regeneron, Sanofi, Aegerion, Akcea, Novartis, Daiichi-Sankyo, Ultragenyx, and Amarin. U.K. reports honoraria for consulting and lectures from AMGEN, Daiichi-Sankyo, Fresenius Medical Care, and Sanofi. MSD Sharp & Dohme, and Alexion. G.K. reports honoraria for consulting and lectures from Aegerion, Akcea, AMGEN, Bristol-Myers Squibb, Daiichi-Sankyo, Esanum, Novartis, Sanofi, Sobi, and Ultragenyx. S.L. reports research funding and/or honoraria for consultancy and lectures from Akcea, AMARIN, Amedes, AMGEN, Berlin-Chemie, Boehringer Ingelheim, Daiichi-Sankyo, Danone, DSM, Janssen-Cilag, Lilly, Novartis, Novo Nordisk, Roche, Sanofi, Sobi, Synlab, and Upfield. W.M. reports grants and personal fees from AMRYT, AMGEN, Sanofi Abbott, Sobi, Novartis, and employment with Synlab. K.G.P. has received research funding and/or honoraria for consultancy and/or lectures from Akcea, Amarin, AMGEN, Boehringer-Ingelheim, Schär, Daiichi-Sankyo, MSD Sharp & Dohme, Novartis, Regeneron, Sanofi, and Silence Therapeutics. J.P. received research funding from the Dutch Research Council (NWO), The Netherlands Organization for Health Research and Development (ZonMW), Applied and Engineering Sciences (STW), The Dutch Topsector for Life Sciences and Health (TKI-LSH), The Californian Almond Foundation, EzCol, Newtricious, BASF, RAISIO, Upfield, and Unilever, and has received honoraria from Unilever, Raisio, BASF, and Upfield, and is secretary of the board from the foundation Nutrition in Transition (NIT), is chair of the supervisory board of the FIVES foundation and is chair of the Department Nutrition and Movement Sciences of the Maastricht University. G.S. reports grants and honoraria from Bayer, Sanofi, Numares, Daiichi-Sankyo, and AMGEN. EST reports honoraria for consultancy and lectures from Fresenius Medical Care, Daiichi-Sankyo, Novartis, Pfizer, Sanofi, AMGEN, and Sobi. O.W. received honoraria and/or travel support from AMGEN, Daiichi-Sankyo, Amarin, Novo Nordisk, Novartis, Hexal, Sanofi, Fresenius Medical Care, Akcea, TAD Pharma, Pfizer, and from the German Society of Cardiology. B.-C.Z. reports an unrestricted investigational grant from Unilever and BASF. D.L. reports honoraria for consulting and/or lectures from Actelion, Daiichi-Sankyo, MSD Sharp & Dohme, and Unilever. F.-U.B., H.K.B. have nothing to declare.

References

1. Tada, H.; Nohara, A.; Inazu, A.; Sakuma, N.; Mabuchi, H.; Kawashiri, M.A. Sitosterolaemia, Hypercholesterolemia, and Coronary Artery Disease. *J. Atheroscler. Thromb.* **2018**, *25*, 783–789. [CrossRef] [PubMed]
2. Williams, K.; Segard, A.; Graf, G.A. Sitosterolemia: Twenty Years of Discovery of the Function of ABCG5 ABCG8. *Int. J. Mol. Sci.* **2021**, *22*, 2641. [CrossRef]
3. Gylling, H.; Plat, J.; Turley, S.; Ginsberg, H.N.; Ellegård, L.; Jessup, W.; Jones, P.J.; Lütjohann, D.; Maerz, W.; Masana, L.; et al. Plant sterols and plant stanols in the management of dyslipidaemia and prevention of cardiovascular disease. *Atherosclerosis* **2014**, *232*, 346–360. [CrossRef]
4. Ference, B.A.; Ginsberg, H.N.; Graham, I.; Ray, K.K.; Packard, C.J.; Bruckert, E.; Hegele, R.A.; Krauss, R.M.; Raal, F.J.; Schunkert, H.; et al. Low density lipoproteins cause atherosclerotic cardiovascular disease. Evidence from genetic, epidemiologic, and clinical studies. A consensus statement from the European Atherosclerosis Society Consensus Panel. *Eur. Heart J.* **2017**, *38*, 2459–2472. [CrossRef] [PubMed]
5. Hazard, S.E.; Patel, S.B. Sterolins ABCG5 and ABCG8: Regulators of whole body dietary sterols. *Pflügers Arch.* **2007**, *453*, 745–752. [CrossRef] [PubMed]
6. Salen, G.; Tint, G.S.; Shefer, S.; Shore, V.; Nguyen, L. Increased sitosterol absorption is offset by rapid elimination to prevent accumulation in heterozygotes with sitosterolemia. *Arterioscler. Thromb.* **1992**, *12*, 563–568. [CrossRef] [PubMed]
7. Calpe-Berdiel, L.; Escolà-Gil, J.C.; Blanco-Vaca, F. New insights into the molecular actions of plant sterols and stanols in cholesterol metabolism. *Atherosclerosis* **2009**, *203*, 18–31. [CrossRef]
8. Davis, H.R., Jr.; Zhu, L.J.; Hoos, L.M.; Tetzloff, G.; Maguire, M.; Liu, J.; Yao, X.; Iyer, S.P.; Lam, M.H.; Lund, E.G.; et al. Niemann-Pick C1 Like 1 (NPC1L1) is the intestinal phytosterol and cholesterol transporter and a key modulator of whole-body cholesterol homeostasis. *J. Biol. Chem.* **2004**, *279*, 33586–33592. [CrossRef]
9. Yamanashi, Y.; Takada, T.; Suzuki, H. Niemann-Pick C1-like 1 overexpression facilitates ezetimibe-sensitive cholesterol and β-sitosterol uptake in CaCo-2 cells. *J. Pharmacol. Exp. Ther.* **2007**, *320*, 559–564. [CrossRef]
10. Lütjohann, D.; Bjorkhem, I.; Beil, U.F.; von Bergmann, K. Sterol absorption and sterol balance in phytosterolemia evaluated by deuterium-labeled sterols: Effect of sitostanol treatment. *J. Lipid Res.* **1995**, *36*, 1763–1773. [CrossRef]
11. Temel, R.E.; Gebre, A.K.; Parks, J.S.; Rudel, L.L. Compared with acyl- CoA:cholesterol O-acyltransferase (ACAT) 1 and lecithin:cholesterol acyltransferase, ACAT2 displays the greatest capacity to differentiate cholesterol from sitosterol. *J. Biol. Chem.* **2003**, *278*, 47594–47601. [CrossRef]
12. Nguyen, T.M.; Sawyer, J.K.; Kelley, K.L.; Davis, M.A.; Kent, C.R.; Rudel, L.L. ACAT2 and ABCG5/G8 are both required for efficient cholesterol absorption in mice: Evidence from thoracic lymph duct cannulation. *J. Lipid Res.* **2012**, *53*, 1598–1609. [CrossRef] [PubMed]

13. Yu, L.; von Bergmann, K.; Lütjohann, D.; Hobbs, H.H.; Cohen, J.C. Selective sterol accumulation in ABCG5/ABCG8-deficient mice. *J. Lipid Res.* **2004**, *45*, 301–307. [CrossRef]
14. Othman, R.A.; Myrie, S.B.; Jones, P.J. Non-cholesterol sterols and cholesterol metabolism in sitosterolemia. *Atherosclerosis* **2013**, *231*, 291–299. [CrossRef]
15. Bosner, M.S.; Lange, L.G.; Stenson, W.F.; Ostlund, R.E., Jr. Percent cholesterol absorption in normal women and men quantified with dual stable isotopic tracers and negative ion mass spectrometry. *J. Lipid Res.* **1999**, *40*, 302–308. [CrossRef]
16. Chan, Y.M.; Varady, K.A.; Lin, Y.; Trautwein, E.; Mensink, R.P.; Plat, J.; Jones, P.J. Plasma concentrations of plant sterols: Physiology and relationship with coronary heart disease. *Nutr. Rev.* **2006**, *64*, 385–402. [CrossRef]
17. Temel, R.E.; Tang, W.; Ma, Y.; Rudel, L.L.; Willingham, M.C.; Ioannou, Y.A.; Davies, J.P.; Nilsson, L.M.; Yu, L. Hepatic Niemann-Pick C1-like 1 regulates biliary cholesterol concentration and is a target of ezetimibe. *J. Clin. Investig.* **2007**, *117*, 1968–1978. [CrossRef] [PubMed]
18. Myrie, S.B.; Mymin, D.; Triggs-Raine, B.; Jones, P.J. Serum lipids, plant sterols, and cholesterol kinetic responses to plant sterol supplementation in phytosterolemia heterozygotes and control individuals. *Am. J. Clin. Nutr.* **2012**, *95*, 837–844. [CrossRef]
19. Escolà-Gil, J.C.; Quesada, H.; Julve, J.; Martín-Campos, J.M.; Cedó, L.; Blanco-Vaca, F. Sitosterolemia: Diagnosis, investigation, and management. *Curr. Atheroscler. Rep.* **2014**, *16*, 424. [CrossRef] [PubMed]
20. Sudhop, T.; von Bergmann, K. Sitosterolemia—A rare disease. Are Elev. Plant Sterols Addit. Risk Factor? *Z. Kardiol.* **2004**, *93*, 921–928. [CrossRef]
21. Jessup, W.; Herman, W.; Chapman, M.J. Phytosterols in cardiovascular disease: Innocuous dietary components, or accelerators of atherosclerosis? *Future Lipidol.* **2008**, *3*, 301–310. [CrossRef]
22. Bhattacharyya, A.K.; Connor, W.E. Beta-sitosterolemia and xanthomatosis. A newly described lipid storage disease in two sisters. *J. Clin. Investig.* **1974**, *53*, 1033–1043. [CrossRef]
23. Simonen, P.P.; Gylling, H.; Miettinen, T.A. The distribution of squalene and non-cholesterol sterols in lipoproteins in type 2 diabetes. *Atherosclerosis* **2007**, *194*, 222–229. [CrossRef] [PubMed]
24. Gylling, H.; Simonen, P. Phytosterols, phytostanols, and lipoprotein metabolism. *Nutrients* **2015**, *7*, 7965–7977. [CrossRef] [PubMed]
25. Robins, S.J.; Fasulo, J.M. High density lipoproteins, but not other lipoproteins, provide a vehicle for sterol transport to bile. *J. Clin. Investig.* **1997**, *99*, 380–384. [CrossRef]
26. Silbernagel, G.; Baumgartner, I.; Wanner, C.; März, W. Toward individualized cholesterol-lowering treatment in end-stage renal disease. *J. Ren. Nutr.* **2014**, *24*, 65–71. [CrossRef]
27. Patel, S.B.; Honda, A.; Salen, G. Sitosterolemia: Exclusion of genes involved in reduced cholesterol biosynthesis. *J. Lipid Res.* **1998**, *39*, 1055–1056. [CrossRef]
28. Patel, S.B.; Salen, G.; Hidaka, H.; Kwiterovich, P.O.; Stalenhoef, A.F.; Miettinen, T.A.; Grundy, S.M.; Lee, M.H.; Rubenstein, J.S.; Polymeropoulos, M.H.; et al. Mapping a gene involved in regulating dietary cholesterol absorption. The sitosterolemia locus is found at chromosome 2p21. *J. Clin. Investig.* **1998**, *102*, 1041–1044. [CrossRef] [PubMed]
29. Berge, K.E.; Tian, H.; Graf, G.A.; Yu, L.; Grishin, N.V.; Schultz, J.; Kwiterovich, P.; Shan, B.; Barnes, R.; Hobbs, H.H. Accumulation of dietary cholesterol in sitosterolemia caused by mutations in adjacent ABC transporters. *Science* **2000**, *290*, 1771–1775. [CrossRef]
30. Lee, M.H.; Gordon, D.; Ott, J.; Lu, K.; Ose, L.; Miettinen, T.; Gylling, H.; Stalenhoef, A.F.; Pandya, A.; Hidaka, H.; et al. Fine mapping of a gene responsible for regulating dietary cholesterol absorption; founder effects underlie cases of phytosterolaemia in multiple communities. *Eur. J. Hum. Genet.* **2001**, *9*, 375–384. [CrossRef]
31. Lee, M.H.; Lu, K.; Hazard, S.; Yu, H.; Shulenin, S.; Hidaka, H.; Kojima, H.; Allikmets, R.; Sakuma, N.; Pegoraro, R.; et al. Identification of a gene, ABCG5, important in the regulation of dietary cholesterol absorption. *Nat. Genet.* **2001**, *27*, 79–83. [CrossRef] [PubMed]
32. Lu, K.; Lee, M.H.; Hazard, S.; Brooks-Wilson, A.; Hidaka, H.; Kojima, H.; Ose, L.; Stalenhoef, A.F.; Mietinnen, T.; Bjorkhem, I.; et al. Two genes that map to the STSL locus cause sitosterolemia: Genomic structure and spectrum of mutations involving sterolin-1 and sterolin-2, encoded by ABCG5 and ABCG8, respectively. *Am. J. Hum. Genet.* **2001**, *69*, 278–290. [CrossRef] [PubMed]
33. Patel, S.B.; Salen, G. Sitosterolemia: Xenophobia for the Body. In *Evidence-Based Management of Lipid Disorders*; Vissers, M.N., Kastelein, J.J.P., Stroes, E.S., Eds.; TFM Publishing: Harley, UK, 2010; pp. 217–230.
34. Bjorkhem, I.; Danielsson, H. Biosynthesis and metabolism of bile acids in man. *Prog. Liver Dis.* **1976**, *5*, 215–231. [PubMed]
35. Bjorkhem, I. Mechanism of degradation of the steroid side chain in the formation of bile acids. *J. Lipid Res.* **1992**, *33*, 455–471. [CrossRef]
36. Calandra, S.; Tarugi, P.; Speedy, H.E.; Dean, A.F.; Bertolini, S.; Shoulders, C.C. Mechanisms and genetic determinants regulating sterol absorption, circulating LDL levels, and sterol elimination: Implications for classification and disease risk. *J. Lipid Res.* **2011**, *52*, 1885–1926. [CrossRef]
37. Drozdowski, L.A.; Clandinin, T.; Thomson, A.B. Ontogeny, growth and development of the small intestine: Understanding pediatric gastroenterology. *World J. Gastroenterol.* **2010**, *16*, 787–799.
38. Rios, J.; Stein, E.; Shendure, J.; Hobbs, H.H.; Cohen, J.C. Identification by whole-genome resequencing of gene defect responsible for severe hypercholesterolemia. *Hum. Mol. Genet.* **2010**, *19*, 4313–4318. [CrossRef]
39. Belamarich, P.F.; Deckelbaum, R.J.; Starc, T.J.; Dobrin, B.E.; Tint, G.S.; Salen, G. Response to diet and cholestyramine in a patient with sitosterolemia. *Pediatrics* **1990**, *86*, 977–981. [CrossRef]

40. Yoshida, A.; Naito, M.; Miyazaki, K. Japanese sisters associated with pseudohomozygous familial hypercholesterolemia and sitosterolemia. *J. Atheroscler. Thromb.* **2000**, *7*, 33–38. [CrossRef]
41. Rajaratnam, R.A.; Gylling, H.; Miettinen, T.A. Independent association of serum squalene and noncholesterol sterols with coronary artery disease in postmenopausal women. *J. Am. Coll. Cardiol.* **2000**, *35*, 1185–1191. [CrossRef]
42. Tsubakio-Yamamoto, K.; Nishida, M.; Nakagawa-Toyama, Y.; Masuda, D.; Ohama, T.; Yamashita, S. Current therapy for patients with sitosterolemia–effect of ezetimibe on plant sterol metabolism. *J. Atheroscler. Thromb.* **2010**, *17*, 891–900. [CrossRef]
43. Shulman, R.S.; Bhattacharyya, A.K.; Connor, W.E.; Fredrickson, D.S. Beta-sitosterolemia and xanthomatosis. *N. Engl. J. Med.* **1976**, *294*, 482–483. [CrossRef] [PubMed]
44. Khachadurian, A.K.; Clancy, K.F. Familial phytosterolemia (beta-sitosterolemia): Report of five cases and studies in cultured skin fibroblasts. *Clin. Res.* **1978**, *26*, 379A.
45. Miettinen, T.A. Phytosterolaemia, xanthomatosis and premature atherosclerotic arterial disease: A case with high plant sterol absorption, impaired sterol elimination and low cholesterol synthesis. *Eur. J. Clin. Investig.* **1980**, *10*, 27–35. [CrossRef] [PubMed]
46. Wang, C.; Lin, H.J.; Chan, T.K.; Salen, G.; Chan, W.C.; Tse, T.F. A unique patient with coexisting cerebrotendinous xanthomatosis and beta-sitosterolemia. *Am. J. Med.* **1981**, *71*, 313–319. [CrossRef]
47. Salen, G.; Horak, I.; Rothkopf, M.; Cohen, J.L.; Speck, J.; Tint, G.S.; Shore, V.; Dayal, B.; Chen, T.; Shefer, S. Lethal atherosclerosis associated with abnormal plasma and tissue sterol composition in sitosterolemia with xanthomatosis. *J. Lipid Res.* **1985**, *26*, 1126–1133. [CrossRef] [PubMed]
48. Berger, G.M.; Deppe, W.M.; Marais, A.D.; Biggs, M. Phytosterolaemia in three unrelated South African families. *Postgrad. Med. J.* **1994**, *70*, 631–637. [CrossRef]
49. Rees, D.C.; Iolascon, A.; Carella, M.; O'marcaigh, A.S.; Kendra, J.R.; Jowitt, S.N.; Wales, J.K.; Vora, A.; Makris, M.; Manning, N.; et al. Stomatocytic haemolysis and macrothrombocytopenia (Mediterranean stomatocytosis/macrothrombocytopenia) is the haematological presentation of phytosterolaemia. *Br. J. Haematol.* **2005**, *130*, 297–309. [CrossRef]
50. Kruit, J.K.; Drayer, A.L.; Bloks, V.W.; Blom, N.; Olthof, S.G.; Sauer, P.J.; de Haan, G.; Kema, I.P.; Vellenga, E.; Kuipers, F. Plant sterols cause macrothrombocytopenia in a mouse model of sitosterolemia. *J. Biol. Chem.* **2008**, *283*, 6281–6287. [CrossRef]
51. Neff, A.T. Sitosterolemia's stomatocytosis and macrothrombocytopenia. *Blood* **2012**, *120*, 4283. [CrossRef]
52. Kanaji, T.; Kanaji, S.; Montgomery, R.R.; Patel, S.B.; Newman, P.J. Platelet hyperreactivity explains the bleeding abnormality and macrothrombocytopenia in a murine model of sitosterolemia. *Blood* **2013**, *122*, 2732–2742. [CrossRef] [PubMed]
53. Wang, Z.; Cao, L.; Su, Y.; Wang, G.; Wang, R.; Yu, Z.; Bai, X.; Ruan, C. Specific macrothrombocytopenia/hemolytic anemia associated with sitosterolemia. *Am. J. Hematol.* **2014**, *89*, 320–324. [CrossRef]
54. Wadsack, D.; Sattler, H.; Kleber, M.E.; Eirich, K.; Scharnagl, H.; Fauler, G.; März, W. Grammer TBRecurrent tendosynovitis as a rare manifestation of a lipid disorder. *Clin. Lipidol.* **2019**, *13*, 54–61. [CrossRef] [PubMed]
55. Kwiterovich, P.O.; Bachorik, P.S., Jr.; Smith, H.H.; McKusick, V.A.; Connor, W.E.; Teng, B.; Sniderman, A.D. Hyperapobetalipoproteinaemia in two families with xanthomas and phytosterolaemia. *Lancet* **1981**, *1*, 466–469. [CrossRef] [PubMed]
56. Hansel, B.; Carrié, A.; Brun-Druc, N.; Leclert, G.; Chantepie, S.; Coiffard, A.S.; Kahn, J.F.; Chapman, M.J.; Bruckert, E. Premature atherosclerosis is not systematic in phytosterolemic patients: Severe hypercholesterolemia as a confounding factor in five subjects. *Atherosclerosis* **2014**, *234*, 162–168. [CrossRef]
57. Morganroth, J.; Levy, R.I.; McMahon, A.E.; Gotto, A.M., Jr. Pseudohomozygous type II hyperlipoproteinemia. *J. Pediatr.* **1974**, *85*, 639–643. [CrossRef]
58. Park, J.H.; Chung, I.H.; Kim, D.H.; Choi, M.H.; Garg, A.; Yoo, E.G. Sitosterolemia presenting with severe hypercholesterolemia and intertriginous xanthomas in a breastfed infant: Case report and brief review. *J. Clin. Endocrinol. Metab.* **2014**, *99*, 1512–1518. [CrossRef]
59. Niu, D.M.; Chong, K.W.; Hsu, J.H.; Wu, T.J.; Yu, H.C.; Huang, C.H.; Lo, M.Y.; Kwok, C.F.; Kratz, L.E.; Ho, L.T. Clinical observations, molecular genetic analysis, and treatment of sitosterolemia in infants and children. *J. Inherit. Metab. Dis.* **2010**, *33*, 437–443. [CrossRef] [PubMed]
60. Patel, M.D.; Thompson, P.D. Phytosterols and vascular disease. *Atherosclerosis* **2006**, *186*, 12–19. [CrossRef] [PubMed]
61. Kwiterovich, P.O.; Chen, S.C., Jr.; Virgil, D.G.; Schweitzer, A.; Arnold, D.R.; Kratz, L.E. Response of obligate heterozygotes for phytosterolemia to a low-fat diet and to a plant sterol ester dietary challenge. *J. Lipid Res.* **2003**, *44*, 1143–1155. [CrossRef]
62. Stalenhoef, A.F.; Hectors, M.; Demacker, P.N. Effect of plant sterol-enriched margarine on plasma lipids and sterols in subjects heterozygous for phytosterolaemia. *J. Intern. Med.* **2001**, *249*, 163–166. [CrossRef]
63. Zhao, H.L.; Houweling, A.H.; Vanstone, C.A.; Jew, S.; Trautwein, E.A.; Duchateau, G.S.; Jones, P.J. Genetic variation in ABC G5/G8 and NPC1L1 impact cholesterol response to plant sterols in hypercholesterolemic men. *Lipids* **2008**, *43*, 1155–1164. [CrossRef]
64. Fuentes, F.; Lopez-Miranda, J.; Garcia, A.; Perez-Martinez, P.; Moreno, J.; Cofan, M.; Caballero, J.; Paniagua, J.A.; Ros, E.; Perez-Jimenez, F. Basal plasma concentrations of plant sterols can predict LDL-C response to sitosterol in patients with familial hypercholesterolemia. *Eur. J. Clin. Nutr.* **2008**, *62*, 495–501. [CrossRef]
65. Dumolt, J.H.; Rideout, T.C. The Lipid-Lowering Effects and Associated Mechanisms of Dietary Phytosterol Supplementation. *Curr. Pharm. Des.* **2017**, *23*, 5077–5085. [CrossRef]
66. Nomura, A.; Emdin, C.A.; Won, H.H.; Peloso, G.M.; Natarajan, P.; Ardissino, D.; Danesh, J.; Schunkert, H.; Correa, A.; Bown, M.J.; et al. Heterozygous ABCG5 Gene Deficiency and Risk of Coronary Artery Disease. *Circ. Genom. Precis. Med.* **2020**, *13*, 417–423. [CrossRef]

67. Helske, S.; Miettinen, T.; Gylling, H.; Mäyränpää, M.; Lommi, J.; Turto, H.; Werkkala, K.; Kupari, M.; Kovanen, P.T. Accumulation of cholesterol precursors and plant sterols in human stenotic aortic valves. *J. Lipid Res.* **2008**, *49*, 1511–1518. [CrossRef] [PubMed]
68. Vanmierlo, T.; Weingärtner, O.; van der Pol, S.; Husche, C.; Kerksiek, A.; Friedrichs, S.; Sijbrands, E.; Steinbusch, H.; Grimm, M.; Hartmann, T.; et al. Dietary intake of plants sterols stably increases plant sterols in the murine brain. *J. Lipid Res.* **2012**, *53*, 726–735. [CrossRef]
69. Miettinen, T.A.; Nissinen, M.; Lepäntalo, M.; Albäck, A.; Railo, M.; Vikatmaa, P.; Kaste, M.; Mustanoja, S.; Gylling, H. Non-cholesterol sterols in serum and endarterectomized carotid arteries after a short-term plant stanol and sterol ester challenge. *Nutr. Metab. Cardiovasc. Dis.* **2011**, *21*, 182–188. [CrossRef] [PubMed]
70. Horenstein, R.B.; Braxton, D.M.; Post, W.S.; Lütjohann, D.; von Bergmann, K.; Ryan, K.A.; Terrin, M.; Shuldiner, A.R.; Steinle, N.I. The ABCG8 G574R Variant, Serum Plant Sterol Levels, and Cardiovascular Disease Risk in the Old Order Amish. *Arterioscler. Thromb. Vasc. Biol.* **2013**, *33*, 413–419. [CrossRef]
71. Miettinen, T.A.; Klett, E.L.; Gylling, H.; Isoniemi, H.; Patel, S.B. Liver transplantation in a patient with sitosterolemia and cirrhosis. *Gastroenterology* **2006**, *130*, 542–547. [CrossRef] [PubMed]
72. Kidambi, S.; Patel, S.B. Sitosterolaemia: Pathophysiology, clinical presentation and laboratory diagnosis. *J. Clin. Pathol.* **2008**, *61*, 588–594. [CrossRef] [PubMed]
73. Hidaka, H.; Nakamura, T.; Aoki, T.; Kojima, H.; Nakajima, Y.; Kosugi, K.; Hatanaka, I.; Harada, M.; Kobayashi, M.; Tamura, A.; et al. Increased plasma plant sterol levels in heterozygotes with sitosterolemia and xanthomatosis. *J. Lipid Res.* **1990**, *31*, 881–888. [CrossRef]
74. Hatanaka, I.; Yasuda, H.; Hidaka, H.; Harada, N.; Kobayashi, M.; Okabe, H.; Matsumoto, K.; Hukuda, S.; Shigeta, Y. Spinal cord compression with paraplegia in xanthomatosis due to normocholesterolemic sitosterolemia. *Ann. Neurol.* **1990**, *28*, 390–393. [CrossRef]
75. Lütjohann, D.; von Bergmann, K. Phytosterolaemia: Diagnosis, characterization and therapeutical approaches. *Ann. Med.* **1997**, *29*, 181–184. [CrossRef]
76. Ratnayake, W.M.; L'Abbe, M.R.; Mueller, R.; Hayward, S.; Plouffe, L.; Hollywood, R.; Trick, K. Vegetable oils high in phytosterols make erythrocytes less deformable and shorten the life span of stroke-prone spontaneously hypertensive rats. *J. Nutr.* **2000**, *130*, 1166–1178. [CrossRef]
77. Yu, L.; Hammer, R.E.; Li-Hawkins, J.; Von Bergmann, K.; Lütjohann, D.; Cohen, J.C.; Hobbs, H.H. Disruption of Abcg5 and Abcg8 in mice reveals their crucial role in biliary cholesterol secretion. *Proc. Natl. Acad. Sci. USA* **2002**, *99*, 16237–16242. [CrossRef]
78. Klett, E.L.; Lu, K.; Kosters, A.; Vink, E.; Lee, M.H.; Altenburg, M.; Shefer, S.; Batta, A.K.; Yu, H.; Chen, J.; et al. A mouse model of sitosterolemia: Absence of Abcg8/sterolin-2 results in failure to secrete biliary cholesterol. *BMC Med.* **2004**, *2*, 5. [CrossRef]
79. Plösch, T.; Bloks, V.W.; Terasawa, Y.; Berdy, S.; Siegler, K.; Van Der Sluijs, F.; Kema, I.P.; Groen, A.K.; Shan, B.; Kuipers, F.; et al. Sitosterolemia in ABC-transporter G5-deficient mice is aggravated on activation of the liver-X receptor. *Gastroenterology* **2004**, *126*, 290–300. [CrossRef]
80. Su, Y.; Wang, Z.; Yang, H.; Cao, L.; Liu, F.; Bai, X.; Ruan, C. Clinical and molecular genetic analysis of a family with sitosterolemia and co-existing erythrocyte and platelet abnormalities. *Haematologica* **2006**, *91*, 1392–1395. [PubMed]
81. Solca, C.; Tint, G.S.; Patel, S.B. Dietary xenosterols lead to infertility and loss of abdominal adipose tissue in sterolin-deficient mice. *J. Lipid Res.* **2013**, *54*, 397–409. [CrossRef] [PubMed]
82. Patel, S.B. Recent advances in understanding the STSL locus and ABCG5/ABCG8 biology. *Curr. Opin. Lipidol.* **2014**, *25*, 169–175. [CrossRef]
83. McDaniel, A.L.; Alger, H.M.; Sawyer, J.K.; Kelley, K.L.; Kock, N.D.; Brown, J.M.; Temel, R.E.; Rudel, L.L. Phytosterol feeding causes toxicity in ABCG5/G8 knockout mice. *Am. J. Pathol.* **2013**, *182*, 1131–1138. [CrossRef]
84. Wang, J.; Mitsche, M.A.; Lütjohann, D.; Cohen, J.C.; Xie, X.S.; Hobbs, H.H. Relative roles of ABCG5/ABCG8 in liver and intestine. *J. Lipid Res.* **2015**, *56*, 319–330. [CrossRef] [PubMed]
85. Othman, R.A.; Myrie, S.B.; Mymin, D.; Merkens, L.S.; Roullet, J.B.; Steiner, R.D.; Jones, P.J. Ezetimibe reduces plant sterol accumulation and favorably increases platelet count in sitosterolemia. *J. Pediatr.* **2015**, *166*, 125–131. [CrossRef] [PubMed]
86. Salen, G.; Starc, T.; Sisk, C.M.; Patel, S.B. Intestinal cholesterol absorption inhibitor ezetimibe added to cholestyramine for sitosterolemia and xanthomatosis. *Gastroenterology* **2006**, *130*, 1853–1857. [CrossRef]
87. Hung, C.N.; Lee, C.Y. A case of sitosterolaemia with stomatocyticanaemia and thrombocytopenia treated with Ezetimibe with good response. *Int. J. Pediatr. Endocrinol.* **2013**, *2013* (Suppl. S1), 174. [CrossRef]
88. de Jong, A.; Plat, J.; Mensink, R.P. Plant sterol or stanol consumption does not affect erythrocyte osmotic fragility in patients on statin treatment. *Eur. J. Clin. Nutr.* **2006**, *60*, 985–990. [CrossRef] [PubMed]
89. Kritchevsky, D.; Chen, S.C. Phytosterols-health benefits and potential concerns: A review. *J. Nutr. Res.* **2005**, *25*, 413–428. [CrossRef]
90. Wilund, K.R.; Yu, L.; Xu, F.; Vega, G.L.; Grundy, S.M.; Cohen, J.C.; Hobbs, H.H. No association between plasma levels of plant sterols and atherosclerosis in mice and men. *Arterioscler. Thromb. Vasc. Biol.* **2004**, *24*, 2326–2332. [CrossRef] [PubMed]
91. Plat, J.; Beugels, I.; Gijbels, M.J.; de Winther, M.P.; Mensink, R.P. Plant sterol or stanol esters retard lesion formation in LDL receptor-deficient mice independent of changes in serum plant sterols. *J. Lipid Res.* **2006**, *47*, 2762–2771. [CrossRef]
92. Weingärtner, O.; Lütjohann, D.; Ji, S.; Weisshoff, N.; List, F.; Sudhop, T.; von Bergmann, K.; Gertz, K.; König, J.; Schäfers, H.J.; et al. Vascular effects of diet supplementation with plant sterols. *J. Am. Coll. Cardiol.* **2008**, *51*, 1553–1561. [CrossRef]

93. Bombo, R.P.; Afonso, M.S.; Machado, R.M.; Lavrador, M.S.; Nunes, V.S.; Quintão, E.R.; Koike, M.; Catanozi, S.; Lin, C.J.; Nakandakare, E.R.; et al. Dietary phytosterol does not accumulate in the arterial wall and prevents atherosclerosis of LDLr-KO mice. *Atherosclerosis* **2013**, *231*, 442–447. [CrossRef]
94. Cannon, C.P.; Blazing, M.A.; Giugliano, R.P.; McCagg, A.; White, J.A.; Theroux, P.; Darius, H.; Lewis, B.S.; Ophuis, T.O.; Jukema, J.W.; et al. Ezetimibe added to statin therapy after acute coronary syndromes. *N. Engl. J. Med.* **2015**, *372*, 2387–2397. [CrossRef] [PubMed]
95. Hibi, K.; Sonoda, S.; Kawasaki, M.; Otsuji, Y.; Murohara, T.; Ishii, H.; Sato, K.; Koshida, R.; Ozaki, Y.; Sata, M.; et al. Effects of Ezetimibe-Statin Combination Therapy on Coronary Atherosclerosis in Acute Coronary Syndrome. *Circ. J.* **2017**, *82*, 757–766. [CrossRef] [PubMed]
96. Getz, G.S.; Reardon, C.A. Animal models of atherosclerosis. *Arterioscler. Thromb. Vasc. Biol.* **2012**, *32*, 1104–1115. [CrossRef] [PubMed]
97. Wouters, K.; Shiri-Sverdlov, R.; van Gorp, P.J.; van Bilsen, M.; Hofker, M.H. Understanding hyperlipidemia and atherosclerosis: Lessons from genetically modified apoe and ldlr mice. *Clin. Chem. Lab. Med.* **2005**, *43*, 470–479. [CrossRef] [PubMed]
98. Assmann, G.; Cullen, P.; Erbey, J.; Ramey, D.R.; Kannenberg, F.; Schulte, H. Plasma sitosterol elevations are associated with an increased incidence of coronary events in men: Results of a nested case-control analysis of the Prospective Cardiovascular Münster (PROCAM) study. *Nutr. Metab. Cardiovasc. Dis.* **2006**, *16*, 13–21. [CrossRef]
99. Fassbender, K.; Lütjohann, D.; Dik, M.G.; Bremmer, M.; König, J.; Walter, S.; Liu, Y.; Letièmbre, M.; von Bergmann, K.; Jonker, C. Moderately elevated plant sterol levels are associated with reduced cardiovascular risk–the LASA study. *Atherosclerosis* **2008**, *196*, 283–288. [CrossRef]
100. Windler, E.; Zyriax, B.C.; Kuipers, F.; Linseisen, J.; Boeing, H. Association of plasma phytosterol concentrations with incident coronary heart disease. Data from the CORA study, a case-control study of coronary artery disease in women. *Atherosclerosis* **2009**, *203*, 284–290. [CrossRef]
101. Strandberg, T.E.; Gylling, H.; Tilvis, R.S.; Miettinen, T.A. Serum plant and other noncholesterol sterols, cholesterol metabolism and 22-year mortality among middle-aged men. *Atherosclerosis* **2010**, *210*, 282–287. [CrossRef]
102. Escurriol, V.; Cofán, M.; Moreno-Iribas, C.; Larrañaga, N.; Martínez, C.; Navarro, C.; Rodríguez, L.; González, C.A.; Corella, D.; Ros, E. Phytosterol plasma concentrations and coronary heart disease in the prospective Spanish EPIC cohort. *J. Lipid Res.* **2010**, *51*, 618–624. [CrossRef] [PubMed]
103. Weingärtner, O.; Pinsdorf, T.; Rogacev, K.S.; Blömer, L.; Grenner, Y.; Gräber, S.; Ulrich, C.; Girndt, M.; Böhm, M.; Fliser, D.; et al. The relationships of markers of cholesterol homeostasis with carotid intima-media thickness. *PLoS ONE* **2010**, *5*, e13467. [CrossRef]
104. Silbernagel, G.; Fauler, G.; Renner, W.; Landl, E.M.; Hoffmann, M.M.; Winkelmann, B.R.; Boehm, B.O.; März, W. The relationships of cholesterol metabolism and plasma plant sterols with the severity of coronary artery disease. *J. Lipid Res.* **2009**, *50*, 334–341. [CrossRef] [PubMed]
105. Silbernagel, G.; Fauler, G.; Hoffmann, M.M.; Lütjohann, D.; Winkelmann, B.R.; Boehm, B.O.; März, W. The associations of cholesterol metabolism and plasma plant sterols with all-cause and cardiovascular mortality. *J. Lipid Res.* **2010**, *51*, 2384–2393. [CrossRef]
106. Matthan, N.R.; Pencina, M.; LaRocque, J.M.; Jacques, P.F.; D'Agostino, R.B.; Schaefer, E.J.; Lichtenstein, A.H. Alterations in cholesterol absorption/synthesis markers characterize Framingham offspring study participants with CHD. *J. Lipid Res.* **2009**, *50*, 1927–1935. [CrossRef]
107. Genser, B.; Silbernagel, G.; De Backer, G.; Bruckert, E.; Carmena, R.; Chapman, M.J.; Deanfield, J.; Descamps, O.S.; Rietzschel, E.R.; Dias, K.C.; et al. Plant sterols and cardiovascular disease: A systematic review and meta-analysis. *Eur. Heart J.* **2012**, *33*, 444–451. [CrossRef]
108. Jaceldo-Siegl, K.; Lütjohann, D.; Sirirat, R.; Mashchak, A.; Fraser, G.E.; Haddad, E. Variations in dietary intake and plasma concentrations of plant sterols across plant-based diets among North American adults. *Mol. Nutr. Food Res.* **2017**, *61*. [CrossRef]
109. Miettinen, T.A.; Gylling, H.; Lindbohm, N.; Miettinen, T.E.; Rajaratnam, R.A.; Relas, H. Finnish Treat-to-Target Study Investigators. *Serum noncholesterol sterols during inhibition of cholesterol synthesis by statins. J. Lab. Clin. Med.* **2003**, *141*, 131–137. [PubMed]
110. Nunes, V.S.; de Campos, E.V.S.; Baracat, J.; França, V.; Gomes, É.I.L.; Coelho, R.P.; Nakandakare, E.R.; Zago, V.H.S.; de Faria, E.C.; Quintão, E.C.R. Plasma Campesterol Is Positively Associated with Carotid Plaques in Asymptomatic Subjects. *Int. J. Mol. Sci.* **2022**, *23*, 11997. [CrossRef] [PubMed]
111. Teupser, D.; Baber, R.; Ceglarek, U.; Scholz, M.; Illig, T.; Gieger, C.; Holdt, L.M.; Leichtle, A.; Greiser, K.H.; Huster, D.; et al. Genetic regulation of serum phytosterol levels and risk of coronary artery disease. *Circ. Cardiovasc. Genet.* **2010**, *3*, 331–339. [CrossRef]
112. Silbernagel, G.; Chapman, M.J.; Genser, B.; Kleber, M.E.; Fauler, G.; Scharnagl, H.; Grammer, T.B.; Boehm, B.O.; Mäkelä, K.M.; Kähönen, M.; et al. High intestinal cholesterol absorption is associated with cardiovascular disease and risk alleles in ABCG8 and ABO: Evidence from the LURIC and YFS cohorts and from a meta-analysis. *J. Am. Coll. Cardiol.* **2013**, *62*, 291–299. [CrossRef]
113. CARDIoGRAMplusC4D Consortium; Deloukas, P.; Kanoni, S.; Willenborg, C.; Farrall, M.; Assimes, T.L.; Thompson, J.R.; Ingelsson, E.; Saleheen, D.; Erdmann, J.; et al. Large-scale association analysis identifies new risk loci for coronary artery disease. *Nat. Genet.* **2013**, *45*, 25–33. [CrossRef] [PubMed]
114. Weingärtner, O.; Teupser, D.; Patel, S.B. The atherogenicity of plant sterols: The evidence from genetics to clinical trials. *J. AOAC Int.* **2015**, *98*, 742–749. [CrossRef] [PubMed]

115. Helgadottir, A.; Thorleifsson, G.; Alexandersson, K.F.; Tragante, V.; Thorsteinsdottir, M.; Eiriksson, F.F.; Gretarsdottir, S.; Björnsson, E.; Magnusson, O.; Sveinbjornsson, G.; et al. Genetic variability in the absorption of dietary sterols affects the risk of coronary artery disease. *Eur. Heart J.* **2020**, *41*, 2618–2628. [CrossRef]
116. Helgadottir, A.; Thorleifsson, G.; Stefansson, K. Increased absorption of phytosterols is the simplest and most plausible explanation for coronary artery disease risk not accounted for by non-HDL cholesterol in high cholesterol absorbers. *Eur. Heart J.* **2021**, *42*, 283–284. [CrossRef] [PubMed]
117. Plat, J.; Strandberg, T.E.; Gylling, H. Intestinal cholesterol and phytosterol absorption and the risk of coronary artery disease. *Eur. Heart J.* **2021**, *42*, 281–282. [CrossRef]
118. Scholz, M.; Horn, K.; Pott, J.; Gross, A.; Kleber, M.E.; Delgado, G.E.; Mishra, P.P.; Kirsten, H.; Gieger, C.; Müller-Nurasyid, M.; et al. Genome-wide meta-analysis of phytosterols reveals five novel loci and a detrimental effect on coronary atherosclerosis. *Nat. Commun.* **2022**, *13*, 1122. [CrossRef]
119. Stender, S.; Frikke-Schmidt, R.; Nordestgaard, B.G.; Tybjaerg-Hansen, A. The ABCG5/8 cholesterol transporter and myocardial infarction versus gallstone disease. *J. Am. Coll. Cardiol.* **2014**, *63*, 2121–2128. [CrossRef]
120. Silbernagel, G.; Duarte, K.; Sadiku, S.; Fauler, G.; März, W.; Schmieder, R.E.; Jardine, A.G.; Massy, Z.A.; Girerd, N.; Fellström, B.; et al. High cholesterol absorption is associated with increased cardiovascular risk in haemodialysis patients: Insights from the AURORA study. *Eur. J. Prev. Cardiol.* **2022**, *29*, 1731–1739. [CrossRef]
121. Otto, S.; Lütjohann, D.; Kerksiek, A.; Friedrichs, S.; Christian Schulze, P.; Möbius-Winkler, S.; Pörner, T.C.; Weingärtner, O. Increased cholesterol absorption is associated with In-stent-restenosis after stent implantation for stable coronary artery disease. *Steroids* **2022**, *187*, 109079. [CrossRef]
122. Plat, J.; Baumgartner, S.; Vanmierlo, T.; Lütjohann, D.; Calkins, K.L.; Burrin, D.G.; Guthrie, G.; Thijs, C.; Te Velde, A.A.; Vreugdenhil, A.C.E.; et al. Plant-based sterols and stanols in health & disease: Consequences of human development in a plant-based environment? *Prog. Lipid Res.* **2019**, *74*, 87–102. [CrossRef] [PubMed]
123. Eli Lilly Package Insert. *Package Insert M100 Suspension Cytellin (Beta and Di-hydro beta Sitosterols)—A Drug Product Indicated for the Reduction of Hypercholesterolemia*; Eli Lilly Package Insert: Indianapolis, IN, USA, 1954.
124. Lees, R.S.; Lees, A.M. Effects of Sitosterol on Plasma Lipid and Lipoprotein Concentration. In *Lipoprotein Metabolism*; Springer: Berlin/Heidelberg, Germany, 1976; pp. 119–130.
125. Oster, P.; Schlierf, G.; Heuck, C.C.; Greten, H.; Gundert-Remy, U.; Haase, W.; Klose, G.; Nothelfer, A.; Raetzer, H.; Schellenberg, B.; et al. Sitosterin bei familiärer Hyperlipoproteinämie Typ II. Eine randomisierte gekreuzte Doppelblindstudie. *Dtsch. Med. Wochenschr.* **1976**, *101*, 1308–1311. [CrossRef] [PubMed]
126. Musa-Veloso, K.; Poon, T.H.; Elliot, J.A.; Chung, C. A comparison of the LDL-cholesterol lowering efficacy of plant stanols and plant sterols over a continuous dose range: Results of a meta-analysis of randomized, placebo-controlled trials. *Prostaglandins Leukot. Essent. Fat. Acids* **2011**, *85*, 9–28. [CrossRef] [PubMed]
127. Ras, R.T.; Geleijnse, J.M.; Trautwein, E.A. LDL-cholesterol-lowering effect of plant sterols and stanols across different dose ranges: A meta-analysis of randomised controlled studies. *Br. J. Nutr.* **2014**, *112*, 214–219. [CrossRef]
128. Ras, R.T.; Hiemstra, H.; Lin, Y.; Vermeer, M.A.; Duchateau, G.S.; Trautwein, E.A. Consumption of plant sterol-enriched foods and effects on plasma plant sterol concentrations–a meta-analysis of randomized controlled studies. *Atherosclerosis* **2013**, *230*, 336–346. [CrossRef]
129. Kratz, M.; Kannenberg, F.; Gramenz, E.; Berning, B.; Trautwein, E.; Assmann, G.; Rust, S. Similar serum plant sterol responses of human subjects heterozygous for a mutation causing sitosterolemia and controls to diets enriched in plant sterols or stanols. *Eur. J. Clin. Nutr.* **2007**, *61*, 896–905. [CrossRef]
130. Racette, S.B.; Lin, X.; Ma LOstlund, R.E.J.R. Natural dietary phytosterols. *J. AOAC Int.* **2015**, *98*, 679–684. [CrossRef]
131. Hagiwara, N.; Kawada-Watanabe, E.; Koyanagi, R.; Arashi, H.; Yamaguchi, J.; Nakao, K.; Tobaru, T.; Tanaka, H.; Oka, T.; Endoh, Y.; et al. Low-density lipoprotein cholesterol targeting with pitavastatin + ezetimibe for patients with acute coronary syndrome and dyslipidaemia: The HIJ-PROPER study, a prospective, open-label, randomized trial. *Eur. Heart J.* **2017**, *38*, 2264–2276. [CrossRef]
132. Päivä, H.; Thelen, K.; Van Coster, R.; Smet, J.; De Paepe, B.; Mattila, K.M.; Laakso, J.; Lehtimäki, T.; von Bergmann, K.; Lütjohann, D.; et al. High-dose statins and skeletal muscle metabolism in humans: A randomized, controlled trial. *Clin. Pharm. Ther.* **2005**, *78*, 60–68. [CrossRef]
133. Van Himbergen, T.M.; Matthan, N.R.; Resteghini, N.A.; Otokozawa, S.; Ai, M.; Stein, E.A.; Jones, P.H.; Schaefer, E.J. Comparison of the effects of maximal dose atorvastatin and rosuvastatin therapy on cholesterol synthesis and absorption markers. *J. Lipid Res.* **2009**, *50*, 730–739. [CrossRef] [PubMed]

nutrients

MDPI

Review

Severe Dyslipidemia Mimicking Familial Hypercholesterolemia Induced by High-Fat, Low-Carbohydrate Diets: A Critical Review

Veera Houttu [1,2], Aldo Grefhorst [2], Danny M. Cohn [1], Johannes H. M. Levels [2], Jeanine Roeters van Lennep [3], Erik S. G. Stroes [1], Albert K. Groen [2] and Tycho R. Tromp [1,*]

1 Department of Vascular Medicine, Amsterdam UMC, Location AMC, Meibergdreef 9, 1105 AZ Amsterdam, The Netherlands
2 Department of Experimental Vascular Medicine, Amsterdam UMC, Location AMC, Meibergdreef 9, 1105 AZ Amsterdam, The Netherlands
3 Department of Internal Medicine, Erasmus MC, University Medical Center, Doctor Molewaterplein 40, 3015 GD Rotterdam, The Netherlands
* Correspondence: t.r.tromp@amsterdamumc.nl; Tel.: +31-20-5668791

Abstract: Emerging studies in the literature describe an association between high-fat, low-carbohydrate diets and severe hypercholesterolemia consistent with the levels observed in patients with (homozygous) familial hypercholesterolemia (FH). High levels of low-density lipoprotein cholesterol (LDL-C) may result from the reduced clearance of LDL particles from the circulation, the increased production of their precursor, or a combination of both. The increased intake of (saturated) fat and cholesterol, combined with limited to no intake of carbohydrates and fiber, are the main features of diets linked to hypercholesterolemia. However, several observations in previous studies, together with our observations from our lipid clinic, do not provide a definitive pathophysiological explanation for severe hypercholesterolemia. Therefore, we review these findings and possible pathophysiological explanations as well as opportunities for future research. Altogether, clinicians should rule out high-fat, low-carbohydrate diets as a possible cause for hypercholesterolemia in patients presenting with clinical FH in whom no mutation is found and discuss dietary modifications to durably reduce LDL-C levels and cardiovascular disease risk.

Keywords: carnivore diet; ketogenic diet; low-carbohydrate diet; high-fat diet; LDL cholesterol; familial hypercholesterolemia; cardiovascular disease

Citation: Houttu, V.; Grefhorst, A.; Cohn, D.M.; Levels, J.H.M.; Roeters van Lennep, J.; Stroes, E.S.G.; Groen, A.K.; Tromp, T.R. Severe Dyslipidemia Mimicking Familial Hypercholesterolemia Induced by High-Fat, Low-Carbohydrate Diets: A Critical Review. *Nutrients* **2023**, *15*, 962. https://doi.org/10.3390/nu15040962

Academic Editor: Frans Stellaard

Received: 4 January 2023
Revised: 4 February 2023
Accepted: 7 February 2023
Published: 15 February 2023

1. Introduction

The notion that circulating low-density lipoprotein cholesterol (LDL-C) causes atherosclerotic cardiovascular disease (ASCVD) is firmly rooted in evidence from genetic, epidemiological, and clinical studies [1]. Interestingly, some of the earliest lines of evidence were derived from experimental studies involving the modification of plasma cholesterol levels through diet. The pioneering groundwork for cholesterol's involvement in atherosclerosis can be traced back to St. Petersburg in 1913, when a young experimental pathologist named Anitschkow induced atherosclerosis in rabbits by feeding them purified cholesterol dissolved in sunflower oil. Control animals fed only the sunflower oil showed no lesions [2]. In the following decades, evidence accumulated that unraveled the relationship between (dietary) cholesterol, atherosclerosis, and potential options for treatment [3]. Over a century after Anitchkow's seminal observations, the interplay between diet and (severe) dyslipidemia is still of special scientific interest, specifically concerning high-fat, low-carbohydrate diets, which is the topic of the narrative review presented here.

Epidemiological observations in large, cross-cultural prospective studies from the general population have been instrumental in linking the intake of fat and cholesterol with plasma cholesterol levels and the incidence of coronary events [4,5]. The first recommendations on dietary intervention to prevent cardiovascular disease were proposed by large professional medical associations in the 1960s [6], and it has now been widely acknowledged that diet has a modest yet meaningful influence on LDL-C levels on a population level [7,8]. The first data to show that cholesterol uptake in the intestine was amenable to a pharmacotherapeutic intervention were derived from the Coronary Primary Prevention Trial, published in 1984. This landmark placebo-controlled randomized clinical trial showed that the bile acid sequestrant cholestyramine, which helps remove cholesterol from the enterohepatic circulation, lowered LDL-C levels and reduced cardiovascular events in 3806 asymptomatic middle-aged men with primary hypercholesterolemia [9].

Bile acid sequestrants have since then been surpassed by statins and other more effective pharmacological lipid-lowering therapies, but dietary modifications continue to be highlighted in the most recent clinical guidelines for the prevention of cardiovascular disease [10–12], partly due to their LDL-lowering effect [7,13]. More specifically, guidelines recommend plant-based and Mediterranean-type diets containing vegetables, nuts, whole grains, and fish rich in unsaturated fats and dietary fibers, while the intake of processed meat, deep-fried foods, ice cream, high-fat dairy products, refined carbohydrates, and sweetened beverages high in saturated fatty acids (SFAs), cholesterol, trans-fatty acids (TFAs), sodium, and glucose should be avoided or kept to a minimum.

Besides observational and interventional data on cholesterol intake or reuptake in large study populations, understanding the link between plasma LDL-C, atherosclerosis, and treatment thereof has hugely progressed through the study of patients with extremely elevated plasma LDL-C levels due to genetic causes [3,14]. Patients with familial hypercholesterolemia (FH), an autosomal dominant inherited disorder caused by variants in the genes involved in lipoprotein metabolism [15], have elevated LDL-C levels from birth, which are more than double the LDL-C levels observed in patients from the general population [16,17]. In general, such high LDL-C levels are not observed in healthy individuals who adhere to a dietary pattern rich in fat and cholesterol.

Sparked by striking observations from our own clinical practice, we review the available evidence of high-fat, low-carbohydrate diets linked to severe dyslipidemia that may be mistaken for (clinical) FH. We discuss this phenomenon in light of possible pathophysiological explanations, point out knowledge gaps for future research, and provide practical recommendations to clinicians who encounter such patients.

2. High-Fat, Low-Carbohydrate Diets: Examples from the Lipid Clinic

2.1. Carnivorous Diet

Two brothers aged 33 years old (patient 1) and 28 years old (patient 2) were referred by their general practitioner to a university lipid clinic because of severe hypercholesterolemia. Upon referral, the LDL-C levels reported by the general practitioner were 15 and 12 mmol/L, respectively.

Both patients had an unremarkable medical history, used no medication, and had no family history of dyslipidemia or (premature) cardiovascular disease. Patients reported exercising regularly (approximately 4×/week, including resistance training) and had a muscular physique with a body mass index (BMI) of 26.2 and 27.3, respectively.

Physical examination was normal and did not specifically show visible cholesterol depositions in the form of xanthomas, corneal arcus, or xanthelasmata.

A fasting blood sample was obtained, and laboratory evaluations showed elevated LDL-C of 15.02 mmol/L and 8.59 mmol/L (patients 1 and 2, respectively, Table 1). We ruled out hypothyroidism and proteinuria as secondary causes of hypercholesterolemia. Liver enzymes were within the upper limits of normal. We suspected familial hypercholesterolemia (FH), possibly even in a homozygous form, as the underlying genetic cause and performed the next-generation sequencing of 27 lipid-related genes [16]. However, no pathogenic variants in any of these genes were found.

Dietary anamnesis revealed that both patients had started a carnivorous diet approximately one year prior to presentation to the lipid clinic. This diet consists solely of meat (mostly red) and high-fat dairy products. The calculated energy intake from carbohydrates on a typical day was <3 E-% (6 g/d), meaning that this diet was strongly ketogenic. The intake from protein and fat was 37 E-% and 61 E-%, respectively (see Appendix A).

Although no lipid profiles were available prior to the initiation of the carnivore diet, we suspected this diet to cause an abnormal lipid profile. We strongly advised the patients to adopt a regular, balanced dietary pattern (including carbohydrates and vegetables) and explained the risks of prolonged exposure to such elevated LDL-C levels. However, both decided to continue the carnivore diet and decided against starting lipid-lowering therapy as an alternative way to reduce the elevated LDL-C levels.

We performed fast protein liquid chromatography (FPLC) using stored plasma. This showed disproportionately elevated very-low-density lipoprotein (VLDL) and intermediate-density lipoprotein (IDL) fractions, which suggested that the overproduction of VLDL contributed to hypercholesterolemia (Figure 1).

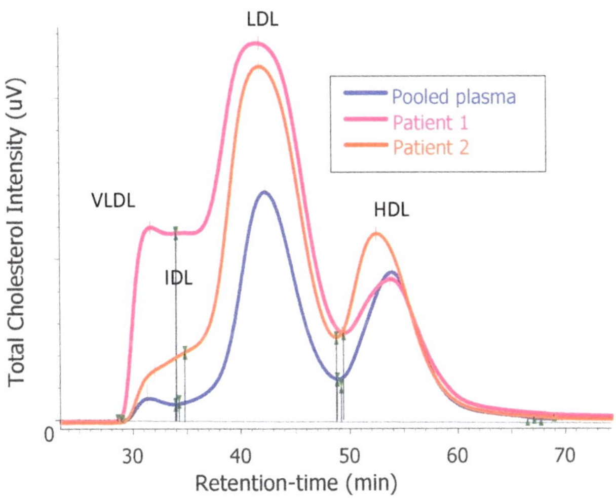

Figure 1. Results of the total cholesterol distribution in plasma after fast protein liquid chromatography (FPLC) profiling from patient 1 and patient 2 compared with that of pooled plasma from healthy subjects. The elution of the main lipoprotein classes is indicated for very-low-density lipoprotein (VLDL), intermediate-density lipoprotein (IDL), low-density lipoprotein (LDL), and high-density lipoprotein (HDL), respectively.

To better understand the effect of this diet on glucose metabolism, we measured fasting levels of glucose, insulin, and c-peptide and calculated the HOMA-IR to gauge the level of insulin sensitivity. Insulin and c-peptide were close to the lower limit of normal, and HOMA-IR was found to approximate >95th percentile, compared with a relevant reference population [18], which suggested good to very good insulin sensitivity.

We investigated whether a copious intake of dietary fatty acids and cholesterol would overwhelm the hepatic triglyceride and cholesterol pools and manifest as hepatic steatosis. We performed vibration-controlled transient elastography (VCTE) using FibroScan® 530 Compact (Echosens, France) estimated liver stiffness measurement (LSM) and controlled attenuation parameter (CAP) in the right liver lobe of both patients but found no signs of liver fibrosis or steatosis. Liver MR spectroscopy subsequently confirmed the near absence of liver fat in both patients.

We performed a carotid ultrasound, which revealed intima–media thickening suggestive of the early development of atherosclerosis.

Table 1. Summary of laboratory and imaging results derived from patients 1 and 2 adhering to a carnivore diet.

	Patient 1	Patient 2	Reference Range
BMI (kg/m^2)	26.2	27.3	18.5–25
Total cholesterol (mmol/L)	16.81	10.88	
HDL cholesterol (mmol/L)	1.53	2.00	
LDL cholesterol (mmol/L)	15.02	8.59	
Triglycerides (mmol/L)	0.57	0.65	<1.80
ApoB (mg/dL)	321	186	<120
ApoA1 (mg/dL)	156	188	
Lp(a) (nmol/L)	18	7	<105
Glucose (mmol/L)	4.3	5.0	<6.0
Insulin (pmol/L)	16.4	17.5	12–96
C-peptide (nmol/L)	0.19	0.19	0.2–0.8
HOMA-IR	0.5	0.6	<2.0
FibroScan CAP (dB/m)	185	185	156–288
FibroScan LSM (kPa)	6.2	6.2	<7
Liver fat content (%)	<2	2.7	<5
Carotid IMT thickness (percentile)	>97th	90th	Based on age and sex, according to [19]

2.2. *Journey from Zero-Carb to Raw-Food Diet*

Patient 3 was a 23-year-old male referred to the academic hospital lipid clinic for the analysis of severe hypercholesterolemia, with an LDL-C level of 12.2 mmol/L reported by his general practitioner.

The patient had no relevant medical history and used no medication. Besides his maternal grandmother, who had started statin therapy at an elderly age, it was unknown if family members had hypercholesterolemia. The family history of cardiovascular disease was negative. The patient told us he had specially requested a lipid panel measurement from his general practitioner because he was interested to learn his blood cholesterol levels after having recently adopted a new diet. The patient reported that, since his youth, he had struggled to maintain a healthy and balanced diet and that he had experienced that his dietary pattern was closely linked to his psychological well-being. The patient had, therefore, experimented with progressively eliminating food products from his diet. This led to the point that, at the time of referral, the patient was on a zero-carb diet rich in animal proteins and fats in the form of eggs and bacon, supplemented with raw minced meat, raw liver, and a limited amount of vegetables.

Physical examination showed a lean young male (BMI 21 kg/m^2) without signs of xanthoma, xanthelasmata, or corneal arcus, and was otherwise unremarkable.

A repeat lipid profile was obtained, which showed somewhat lower, albeit still severely elevated, levels of LDL-C at 7.9 mmol/L, which could well be consistent with heterozygous FH. However, genetic analysis using next-generation sequencing showed no FH-causing variant. It was suspected that the patient's diet contributed to the elevated LDL-C levels. Dietary changes as well as the lowering of LDL-C by starting statin therapy were discussed, and the patient opted to modify his diet first.

Three months later, the patient had significantly reduced his intake of animal fats (mostly cut his intake of eggs and bacon), and a repeat lipid panel showed a near normalization of LDL-C levels (Table 2), after which he was referred back to the care of the general practitioner. The most recent lipid panel was completely normal for age and gender. At this time, the patient reported being on a balanced raw food diet with approximately 400 g of raw meat or fish daily, supplemented with leafy vegetables, nuts, and fruit (usually 2 or 3 mangos daily). The patient reported no intake of starchy carbohydrates.

Table 2. Lipid panels of patient 3.

	Referral	Lipid Clinic, First Visit	Lipid Clinic, Second Visit	Study Visit
Timepoint	**T = 0**	**+3 Months**	**+6 Months**	**+4 Years**
Total cholesterol (mmol/L)	15.0	10.07	5.11	3.79
HDL cholesterol (mmol/L)	2.22	1.76	1.55	1.17
LDL cholesterol (mmol/L)	12.15	7.93	3.40	2.29
Triglycerides (mmol/L)	1.40	0.85	0.35	0.74
ApoB (mg/dL)	N.A.	N.A.	N.A.	64

2.3. Heterozygous Familial Hypercholesterolemia Derailed

Patient 4 was a male diagnosed with FH (variant p.R3500Q in *APOB*) at the age of 31 years old, with no relevant medical history. Following the diagnosis of FH, a statin combined with ezetimibe had been prescribed as lipid-lowering therapy, and the patient was regularly followed up by his internist at the lipid clinic.

At the age of 41 years old, the patient started a low-carb high-fat diet along with his spouse, who intended to lose weight. At the latest lipid panel obtained before the patient had started this diet, LDL-C levels were relatively well controlled at 2.9 mmol/L, with the patient taking rosuvastatin 40 mg and ezetimibe 10 mg daily. However, when the patient presented for his next yearly follow-up visit after having adopted his new diet, LDL-C levels were severely elevated at 8.39 mmol/L (Table 3). Adherence to lipid-lowering medication was unchanged, and secondary causes of hypercholesterolemia (hypothyroidism and proteinuria) were ruled out. The estimated untreated LDL-C levels would well be within the range observed in patients with homozygous FH.

Table 3. Lipid panels of patient 4.

	Pre-Diet	On Diet	Modified Diet
Timepoint	**T = 0**	**+6 Months**	**+9 Months**
Total cholesterol (mmol/L)	4.7	10.8	5.1
HDL cholesterol (mmol/L)	1.4	2.15	1.4
LDL cholesterol (mmol/L)	2.9	8.39	2.8
Triglycerides (mmol/L)	0.9	1.78	0.8
ApoB (mg/dL)	79	194	79

The patient was referred to a dietician for dietary assessment and counseling. His diet was shown to be low in carbohydrates and high in dairy fats (Appendix B), and recommendations were made for a more balanced diet. His BMI dropped from 23.4 to 21.4 kg/m^2 with diet. With the help of his dietician, the patient modified his diet by, amongst others, reducing his intake of dairy fat. At the next follow-up visit, LDL-C had completely normalized to pre-diet levels (Table 3), with unchanged adherence to lipid-lowering therapy.

3. Mechanism of Dietary Induced Hypercholesterolemia

Recent observations from our clinical practice add to the emerging literature describing the association between high-fat, low-carbohydrate diets and severe hypercholesterolemia. We reviewed these findings and possible pathophysiological explanations for this phenomenon as well as opportunities for future research. We argue that clinicians should rule out adherence to a high-fat, low-carbohydrate diet in patients presenting with clinical FH.

Several other studies have reported patients consuming a high-fat ketogenic diet who present with high LDL-C levels consistent with FH, which was shown to be reversible with the normalization of the diet. Goldberg and colleagues reported five cases of LDL-C levels ranging from 6.3 to as high as 17.7 mmol/L. Genetic analysis revealed no pathogenic FH variants in any of the patients, although the patient with the highest cholesterol levels was found to have dysbetalipoproteinemia (confirmed by identification of *APOE* E2/E2 genotype) [20]. Schaffer et al. reported three patients with severe hypercholesterolemia (LDL-C levels of 8.9, 11.5, and 15.5 mmol/L), whose LDL-C levels decreased after making dietary modifications [21]. Norwitz and colleagues reported a series of five patients on a ketogenic diet whose LDL-C levels ranged from 6.2 to up to 17.2 mmol/L. All tested negative for an FH-causing variant and after the reported moderate reintroduction of carbohydrates (50–100 g/day), hypercholesterolemia attenuated and even normalized in one patient [22]. The same authors also recently described a 26-year-old male with ulcerative colitis who, in an attempt to relieve its symptoms, initiated a ketogenic diet on which LDL-C levels peaked at 14.1 mmol/L [23]. Interestingly, a coronary CT angiography did not show (non)calcified plaque after two years of exposure to LDL-C levels in a similar range, as seen in patients with homozygous FH [23,24].

The fact that this phenomenon is more widespread than the subject of academic curiosity, only described in rare case descriptions, is exemplified by a recent online survey that collected patient-reported data, including lipid panels, from several hundred adults following a carbohydrate-restricted diet [22]. Of all the 903 participants who participated in the survey, 42% and 22% reported having LDL-C levels higher than 6.5 mmol/L and 8.5 mmol/L, respectively. These values are thresholds in the Dutch Lipid Clinic Network (DLCN) criteria used in the (clinical) diagnosis of FH. Notably, 5% of the respondents entered LDLC levels > 13 mmol/L, which is considered to be consistent with homozygous FH, the most severe form of inherited dyslipidemia, which can cause cardiovascular mortality as early as in childhood if left untreated [24,25].

These observations, combined with the cases we described, are examples that both individuals adhering to these diets as well as their clinicians, should be cautious of their potential to cause or exacerbate severe hypercholesterolemia. High-fat, low-carbohydrate ('keto') diets may be considered in a line of fad diets known for exaggerated health claims and are frequently propagated through social media, where health misinformation is widespread, and the quality is difficult to assess [26]. Support for these diets is often provided in the form of opinionated, absolute statements that lack the backing of good-quality evidence highlighting harms and benefits [27,28]. Nevertheless, the number of individuals who follow a high-fat, low-carbohydrate diet is considerable and increasing. This diet is reported among the most frequently followed dietary patterns in the United States, with comparable prevalence to the commonly recommended Mediterranean and dietary approaches to stopping hypertension (DASH) diet [29,30]. There is currently no evidence to support that prolonged exposure to high levels of LDL-C in the context of high-fat, low-carbohydrate diets is not atherogenic (and therefore 'safe'), which should be balanced against the totality of evidence on atherogenic lipoproteins and their causal role in the development of ASCVD [1].

The observations of extreme hypercholesterolemia described here and by others beg a discussion of possible pathophysiological explanations for this phenomenon. High levels of LDL-C may result from the reduced clearance of LDL particles from the circulation, the increased production of their VLDL precursor, or a combination of both. The increased intake of dietary fat and cholesterol, combined with limited to no intake of carbohydrates, are salient features of the diets consumed by the patients we described. In the following sections, we discuss these factors and their potential roles in the pathophysiology of hypercholesterolemia individually, but it is likely that these factors act in concert.

3.1. Increased Intake of Dietary Cholesterol

Cholesterol stores in the human body are in a constant state of flux. Cholesterol in LDL particles either originates from intestinal absorption or de novo synthesis in the liver. After an LDL particle is taken up by the liver, its cholesterol content enters the hepatic cholesterol pool and may either be secreted back into the bloodstream packed in lipoproteins or be excreted in bile directly or after conversion to bile acid. Combined with the intake from diet, this cholesterol is partly taken up by the intestine and transported back to the liver and partly exits the body through feces [31]. It could be hypothesized that the hypercholesterolemia observed in our patients is partly due to their cholesterol-rich diet.

However, cholesterol absorption, synthesis, and biliary excretion appear to be balanced in a way that the dietary intake of cholesterol, under normal circumstances, only modestly translates into the cholesterol levels found in the circulation. For example, vegans consuming 90% less dietary cholesterol than omnivores were found to have plasma LDL-C levels that were only 13% lower [32]. A meta-analysis of intervention studies that supplemented cholesterol through diet showed only moderate increases in LDL-C levels (up to 0.22 mmol/L) [33]. The apparent 'resilience' of the plasma cholesterol to high dietary cholesterol intake is exemplified by a case report of an 88-year-old male who ate 25 eggs per day but had a normal plasma LDL-C level of 3.68 mmol/L and no clinically important atherosclerosis. Isotope studies revealed that, compared with 10 healthy controls, the subject had a markedly reduced rate of cholesterol absorption and hepatic cholesterol synthesis but greatly increased synthesis of bile acids [34].

In healthy adults, the average fractional absorption rate of cholesterol in the intestine is approximately 50% but with great inter-individual variation ranging from 20% to 80% [35,36]. Although stable isotope studies were beyond the scope of our study, it is possible that the increased dietary intake of cholesterol exceeded the relative capacity to downregulate its intestinal absorption in our patients. The overload of dietary cholesterol delivered to hepatocytes may contribute to the downregulation of the LDL receptor (LDLR) and thus increase the circulating LDL-C levels [37–39].

The influence of gut microbiota on bile acid metabolism is another factor influencing cholesterol absorption [40]. Gut microbiota influence the circulating LDL-C levels via the conjugation of primary bile acids to secondary bile acids [41] and by facilitating bile acid excretion into feces [40]. Specific bacterial phyla, such as *Lactobacillus* and *Clostridium*, can prevent the reabsorption of bile acids into the enterohepatic circulation through the deconjugation of bile acids. Since most bile acids are taken up in a conjugated form, deconjugation by gut microbes may be an important factor in mediating cholesterol levels. It is possible that a high-fat, low-carbohydrate diet alters gut microbiota composition in an unfavorable way so that the intestinal reabsorption of bile acids and cholesterol is enhanced, and hence, cholesterol accumulates more quickly in the circulation. Unfortunately, fecal samples were unavailable from the cases we described.

3.2. Increased Intake of Dietary Fatty Acids

In the second half of the 20th century, large epidemiological studies linked the composition of dietary fat with plasma LDL-C levels and cardiovascular outcomes [42]. It is now clearly established that the intake of saturated fatty acids (SFAs) increases the circulating LDL-C levels, although the overall effect of limiting SFA intake on cardiovascular health remains controversial [43,44]. Meta-regression analysis of 84 feeding studies showed that replacing the daily energy intake from SFAs with carbohydrates or unsaturated fatty acids considerably lowers LDL-C levels [45]. The LDL-increasing mechanism of SFAs was investigated in various in vitro and in vivo models [46] and likely involves decreased mRNA and protein expression of LDLR, as well as decreased LDLR activity [47]. It has to be noted that this effect depends on SFA type: 12–16 carbon fatty acids, found in dairy and red meat, have the largest effect to raise LDL-C [43,44]. This notion is supported by feeding trials in which increased SFA intake in the form of red meat resulted in higher LDL-C levels than the same SFA intake through nonmeat protein sources [48]. High SFA intake by our patients likely contributed to their hypercholesterolemia, and the fact that LDL-C levels dropped when patient 4 exchanged his intake of butter, meat, and dairy products for more vegetables and vegetable oils supports this notion.

Furthermore, an increased flux of fatty acids to the liver stimulates the assembly and secretion of VLDL particles [49], which would be in line with the abundance of VLDL-C observed with FPLC analysis in patients 1 and 2. However, the absence of triglycerides suggests other factors are also at play, such as extremely efficient lipolytic activity or the active exchange of cholesteryl esters.

3.3. Decreased Intake of Carbohydrates

Diets are generally considered very low in carbohydrates when the energy intake from carbohydrates is <10% of the total energy intake or <50 g per day [50]. These diets have been recommended as a treatment for patients with specific medical conditions, such as rare metabolic diseases or epilepsy, where hyperlipidemia is a well-known side effect of the ketogenic diet [51,52]. The ketogenic diet is gaining increasing scientific attention in the field of sports medicine, where multiple feeding trials have been conducted in healthy athletes that consistently show increases in LDL-C levels [53–57]. However, individual responses are variable, and LDL-C levels do not reach the levels consistent with homozygous FH, as observed in the extreme cases we presented. Carbohydrate-restricted diets have further gained attention through popular scientific news outlets and social media channels to reduce weight, as was the case for patient 4.

A decreased intake of carbohydrates inadvertently means that the relative intake of protein and/or fat is increased, but the relative importance of 'low-carbohydrate' vs. 'high-fat' diet intake in causing hypercholesterolemia remains to be established. It has been hypothesized that carbohydrate restriction leads to increased dependence on fat as a metabolic substrate, which drives the increased hepatic secretion of triglyceride rich VLDL. It is thought that triglycerides are taken up very rapidly by peripheral tissues, which have come to rely heavily on fats as their metabolic substrate. The resulting lipid profile is characterized by markedly elevated levels of LDL-C and HDL-C, yet low triglycerides [58]. Our observation of elevated VLDL-C, LDL-C, and HDL-C, combined with relatively low levels of triglycerides found in patients 1 and 2, are in support of this 'lipid energy model'. The authors further hypothesize that low levels of insulin, as we observed in patients 1 and 2, in combination with low hepatic glycogen stores (not measured by us), contribute to increased VLDL secretion rates. There is anecdotal evidence that the reintroduction of carbohydrates reverses the hypercholesterolemia seen in these subjects [22,23], but this could not be accurately assessed in our patients. Future studies, such as stable isotope studies under controlled feeding conditions, are required to unravel the relative importance of carbohydrate restriction on the observed increases in LDL-C levels. Such studies can also be used to determine both the excretion and absorption rates of cholesterol in the gut, as well as the production and uptake of lipoproteins by the liver in the context of high-fat, low-carbohydrate diets.

4. Conclusions and Future Research

In conclusion, mounting evidence describes the relationship between high-fat, low-carbohydrate diets and severely elevated plasma LDL-C levels generally considered to be consistent with (homozygous) FH. This phenomenon provides unique opportunities to study the fundamental (patho)physiological mechanisms involving cholesterol and lipoprotein homeostasis. Clinicians should rule out high-fat, low-carbohydrate diets as a possible cause for hypercholesterolemia in patients presenting with clinical FH in whom no mutation is found and discuss dietary modifications to durably reduce LDL-C levels and ASCVD risk.

Author Contributions: Conceptualization, T.R.T., V.H. and A.K.G.; resources, E.S.G.S.; data curation, T.R.T., J.H.M.L. and J.R.v.L.; writing—original draft preparation, T.R.T. and V.H.; writing—review and editing, A.G., D.M.C., J.R.v.L., A.K.G. and E.S.G.S.; visualization, T.R.T. and V.H.; supervision, A.K.G. All authors have read and agreed to the published version of the manuscript.

Funding: This research was funded by research grants from the Netherlands Organization for Scientific Research (vidi 016.156.445) and Klinkerpad funds. VH is appointed by the European Union's Horizon 2020 research and innovation program under the Marie Skłodowska-Curie Grant Agreement No. 813781 ITN BestTreat.

Institutional Review Board Statement: The study was conducted in accordance with the Declaration of Helsinki and approved by the Institutional Review Board (or Ethics Committee) of Amsterdam UMC, location AMC (protocol code NL62407.018.17) for studies involving humans.

Informed Consent Statement: Written informed consent has been obtained from the patient(s) to publish this paper.

Data Availability Statement: Not applicable.

Acknowledgments: The authors thank the participants for their contribution to this research.

Conflicts of Interest: V.H., A.G., D.M.C., J.H.M.L., A.K.G. and T.R.T. declare no conflict of interest. J.R.v.L. reports research grants from Amryt and Novartis (all paid to the institution). E.S.G.S. reports consulting fees from Amgen, Sanofi, Regeneron, Esperion, Novartis, and Ionis/Akcea (all paid to the institution).

Appendix A

Dietary Pattern of Patient 1 on an Example Day

- Breakfast
  - Black coffee (filtered).
- Lunch
  - Eggs (6 to 8) and bacon fried in butter.
- Dinner
  - Approximately 750 g of meat, such as red meat steak pork chops (fat still on), chicken thighs, fatty fish, meatballs (three parts 18% minced meat mixed with one part minced liver), fried in butter.

Beverages

- Water with meals, black coffee, approximately two beers weekly;
- No protein shakes/beverages or sugar-sweetened beverages.

Cooking/baking

- 500 g of butter weekly used for baking;
- Explicitly no vegetables oils or margarines; no vegetables or fruits; no rice, pasta, or potatoes.

Table A1. Calculated energy (kcal) and macronutrient (g and E-%), fiber (g), salt (g), and sodium (mg) intake on an example day of case 1 on a carnivore diet.

	Patient 1
Energy (kcal/d)	3275
Fat (g/d)	220.5
Fat (E-%)	60.6
Saturated fat (g/d)	99.0
Saturated fat (E-%)	27.2
Protein (g/d)	304.1
Protein E-%	37.1
Carbohydrate (g/d)	5.5
Carbohydrate (E-%)	0.7
Alcohol (g/d)	6
Fiber (g/d)	0.4
Salt (g/d)	12.8
Sodium (mg/d)	5097

E-%, percentage of energy intake; g/d, grams per day; kcal/d, kilocalories per day; mg/d, milligrams per day.

Appendix B

Dietary Pattern of Patient 4 on Example Days

- Breakfast
  - 10% Greek yogurt with home-baked seeds in a little coconut fat, honey, and cinnamon (no other sugar or sweeteners);
  - Three eggs and bacon fried in butter;
  - Oven-baked ricotta cheese with raspberries, oranges, pecan nuts, and an egg.
- Lunch
  - Alternatives for lunch were beet salad with herring in olive oil or fatty mayonnaise; chicory salad with salmon and a scoop of 10% Greek yogurt; Cesar salad with fatty mayonnaise; zucchini soup with fatty cooking cream; OR low-carbohydrate sandwich with fried egg, cheese, and raw vegetables.
- Dinner
  - Alternatives for dinner were leek disk codfish in high-fat cooking cream; chicory dish in cheese and ham, in butter; spinach with egg and cream; Nasi (typical Indonesian dish) of cauliflower rice with rice and chicken fried in coconut fat; lightly cooked beans with chicken, tomato and some extra herbs, stir-fried in butter; OR pizza made with chicken breast with lots of vegetables covered in tomato sauce and mozzarella.
- Cooking/baking
  - Nearly all dishes are cooked based on recipes in the books by "The New Food": https://thenewfood.nl/ (accessed on 26 November 2021)

Table A2. Calculated energy (kcal) and macronutrient (g and E-%), fiber (g), salt (g), and sodium (mg) intake on an example day of case 4 on a low-carbohydrate, high-fat diet.

	Patient 4
Energy (kcal/d)	1912
Fat (g/d)	147.2
Fat (E-%)	69.3
Saturated fat (g/d)	74.1
Saturated fat (E-%)	34.9
Protein (g/d)	85.2
Protein (E-%)	17.8
Carbohydrate (g/d)	55.3
Carbohydrate (E-%)	11.6
Alcohol (g/d)	0.0
Fiber (g/d)	17.2
Salt (g/d)	3.9
Sodium (mg/d)	1558

E-%, percentage of energy intake; g/d, grams per day; kcal/d, kilocalories per day; mg/d, milligrams per day.

References

1. Ference, B.A.; Ginsberg, H.N.; Graham, I.; Ray, K.K.; Packard, C.J.; Bruckert, E.; Hegele, R.A.; Krauss, R.M.; Raal, F.J.; Schunkert, H.; et al. Low-density lipoproteins cause atherosclerotic cardiovascular disease. 1. Evidence from genetic, epidemiologic, and clinical studies. A consensus statement from the European Atherosclerosis Society Consensus Panel. *Eur. Heart J.* **2017**, *38*, 2459–2472. [CrossRef]
2. Anitschkow, N.N.; Chalatov, S. Ueber experimentelle Choleserinsteatose und ihre Bedeutung fur die Entstehung einiger pathologischer Prozesse. *Zentralbl. Allg. Pathol.* **1913**, *24*, 1–9.
3. Goldstein, J.L.; Brown, M.S. A century of cholesterol and coronaries: From plaques to genes to statins. *Cell* **2015**, *161*, 161–172. [CrossRef] [PubMed]
4. Kromhout, D.; Menotti, A.; Bloemberg, B.; Aravanis, C.; Blackburn, H.; Buzina, R.; Dontas, A.S.; Fidanza, F.; Giaipaoli, S.; Jansen, A.; et al. Dietary saturated and trans fatty acids and cholesterol and 25-year mortality from coronary heart disease: The Seven Countries Study. *Prev. Med.* **1995**, *24*, 308–315. [CrossRef] [PubMed]

5. Menotti, A.; Puddu, P.E.; Adachi, H.; Tolonen, H.; Kafatos, A. Association of serum cholesterol with coronary heart disease mortality during 50-year follow-up in ten cohorts of the seven countries study. *Nutr. Metab. Cardiovasc. Dis.* **2020**, *30*, 1337–1346. [CrossRef]
6. Steinberg, D. An interpretive history of the cholesterol controversy: Part I. *J. Lipid. Res.* **2004**, *45*, 1583–1593. [CrossRef] [PubMed]
7. Poli, A.; Barbagallo, C.M.; Cicero, A.F.G.; Corsini, A.; Manzato, E.; Trimarco, B.; Bernini, F.; Visioli, F.; Bianchi, A.; Canzone, G.; et al. Nutraceuticals and functional foods for the control of plasma cholesterol levels. An intersociety position paper. *Pharmacol. Res.* **2018**, *134*, 51–60. [CrossRef]
8. Schoeneck, M.; Iggman, D. The effects of foods on LDL cholesterol levels: A systematic review of the accumulated evidence from systematic reviews and meta-analyses of randomized controlled trials. *Nutr. Metab. Cardiovasc. Dis.* **2021**, *31*, 1325–1338. [CrossRef]
9. The Lipid Research Clinics Coronary Primary Prevention Trial results. I. Reduction in incidence of coronary heart disease. *JAMA* **1984**, *251*, 351–364. [CrossRef]
10. Arnett, D.K.; Blumenthal, R.S.; Albert, M.A.; Buroker, A.B.; Goldberger, Z.D.; Hahn, E.J.; Himmelfarb, C.D.; Khera, A.; Lloyd-Jones, D.; McEvoy, J.W.; et al. 2019 ACC/AHA Guideline on the Primary Prevention of Cardiovascular Disease: A Report of the American College of Cardiology/American Heart Association Task Force on Clinical Practice Guidelines. *J. Am. Coll. Cardiol.* **2019**, *74*, e177–e232. [CrossRef]
11. Mach, F.; Baigent, C.; Catapano, A.L.; Koskinas, K.C.; Casula, M.; Badimon, L.; Chapman, M.J.; De Backer, G.G.; Delgado, V.; Ference, B.A.; et al. 2019 ESC/EAS Guidelines for the management of dyslipidaemias: Lipid modification to reduce cardiovascular risk. *Eur. Heart J.* **2020**, *41*, 111–188. [CrossRef] [PubMed]
12. Visseren, F.L.J.; Mach, F.; Smulders, Y.M.; Carballo, D.; Koskinas, K.C.; Bäck, M.; Benetos, A.; Biffi, A.; Boavida, J.-M.; Capodanno, D.; et al. 2021 ESC Guidelines on cardiovascular disease prevention in clinical practice. *Eur. Heart J.* **2021**, *42*, 3227–3337. [CrossRef] [PubMed]
13. Hollænder, P.L.B.; Ross, A.B.; Kristensen, M. Whole-grain and blood lipid changes in apparently healthy adults: A systematic review and meta-analysis of randomized controlled studies1-3. *Am. J. Clin. Nutr.* **2015**, *102*, 556–572. [CrossRef] [PubMed]
14. Brown, M.S.; Goldstein, J.L. Familial hypercholesterolemia: Defective binding of lipoproteins to cultured fibroblasts associated with impaired regulation of 3 hydroxy 3 methylglutaryl coenzyme A reductase activity. *Proc. Natl. Acad. Sci. USA* **1974**, *71*, 788–792. [CrossRef] [PubMed]
15. Nordestgaard, B.G.; Chapman, M.J.; Humphries, S.E.; Ginsberg, H.N.; Masana, L.; Descamps, O.S.; Wiklund, O.; Hegele, R.A.; Raal, F.J.; Defesche, J.C.; et al. Familial hypercholesterolaemia is underdiagnosed and undertreated in the general population: Guidance for clinicians to prevent coronary heart disease: Consensus Statement of the European Atherosclerosis Society. *Eur. Heart J.* **2013**, *34*, 3478–3490. [CrossRef]
16. Reeskamp, L.F.; Tromp, T.R.; Defesche, J.C.; Grefhorst, A.; Stroes, E.S.; Hovingh, G.K.; Zuurbier, L. Next-generation sequencing to confirm clinical familial hypercholesterolemia. *Eur. J. Prev. Cardiol.* **2020**, *28*, 875–883. [CrossRef]
17. Balder, J.W.; de Vries, J.K.; Nolte, I.M.; Lansberg, P.J.; Kuivenhoven, J.A.; Kamphuisen, P.W. Lipid and lipoprotein reference values from 133,450 Dutch Lifelines participants: Age- and gender-specific baseline lipid values and percentiles. *J. Clin. Lipidol.* **2017**, *11*, 1055–1064.e6. [CrossRef]
18. Matli, B.; Schulz, A.; Koeck, T.; Falter, T.; Lotz, J.; Rossmann, H.; Pfeiffer, N.; Beutel, M.; Münzel, T.; Strauch, K.; et al. Distribution of HOMA-IR in a population-based cohort and proposal for reference intervals. *Clin. Chem. Lab. Med.* **2021**, *59*, 1844–1851. [CrossRef]
19. Engelen, L.; Ferreira, I.; Stehouwer, C.D.; Boutouyrie, P.; Laurent, S. Reference intervals for common carotid intima-medi thickness measured with echotracking: Relation with risk factors. *Eur. Heart J.* **2013**, *34*, 2368–2380. [CrossRef]
20. Goldberg, I.J.; Ibrahim, N.; Bredefeld, C.; Foo, S.; Lim, V.; Gutman, D.; Huggins, L.-A.; Hegele, R.A. Ketogenic diets, not for everyone. *J. Clin. Lipidol.* **2021**, *15*, 61–67. [CrossRef]
21. Schaffer, A.E.; D'Alessio, D.A.; Guyton, J.R. Extreme elevations of low-density lipoprotein cholesterol with very low carbohydrate, high fat diets. *J. Clin. Lipidol.* **2021**, *15*, 525–526. [CrossRef]
22. Norwitz, N.G.; Feldman, D.; Soto-Mota, A.; Kalayjian, T.; Ludwig, D.S. Elevated LDL Cholesterol with a Carbohydrate-Restricted Diet: Evidence for a "lean Mass Hyper-Responder" Phenotype. *Curr. Dev. Nutr.* **2022**, *6*, nzab144. [CrossRef] [PubMed]
23. Norwitz, N.G.; Soto-Mota, A.; Feldman, D.; Parpos, S.; Budoff, M. Case Report: Hypercholesterolemia "Lean Mass Hyper-Responder" Phenotype Presents in the Context of a Low Saturated Fat Carbohydrate-Restricted Diet. *Front. Endocrinol.* **2022**, *13*, 830325. [CrossRef] [PubMed]
24. Tromp, T.R.; Hartgers, M.L.; Hovingh, G.K.; Vallejo-Vaz, A.J.; Ray, K.K.; Soran, H.; Freiberger, T.; Bertolini, S.; Harada-Shiba, M.; Blom, D.J.; et al. Worldwide experience of homozygous familial hypercholesterolaemia: Retrospective cohort study. *Lancet* **2022**, *399*, 719–728. [CrossRef]
25. Cuchel, M.; Bruckert, E.; Ginsberg, H.N.; Raal, F.J.; Santos, R.D.; Hegele, R.A.; Kuivenhoven, J.A.; Nordestgaard, B.G.; Descamps, O.S.; Steinhagen-Thiessen, E.; et al. Homozygous familial hypercholesterolaemia: New insights and guidance for clinicians to improve detection and clinical management. A position paper from the Consensus Panel on Familial Hypercholesterolaemia of the European Atherosclerosis Society. *Eur. Heart J.* **2014**, *35*, 2146–2157. [CrossRef]
26. Suarez-Lledo, V.; Alvarez-Galvez, J. Prevalence of health misinformation on social media: Systematic review. *J. Med. Internet Res.* **2021**, *23*, e17187. [CrossRef] [PubMed]

27. Tahreem, A.; Rakha, A.; Rabail, R.; Nazir, A.; Socol, C.T.; Maerescu, C.M.; Aadil, R.M. Fad Diets: Facts and Fiction. *Front. Nutr.* **2022**, *9*, 960922. [CrossRef]
28. O'Neill, B.; Raggi, P. The ketogenic diet: Pros and cons. *Atherosclerosis* **2020**, *292*, 119–126. [CrossRef]
29. International Food Information Council. 2021 Food & Health Survey. Available online: https://foodinsight.org/wp-content/uploads/2021/05/IFIC-2021-Food-and-Health-Survey.May-2021-1.pdf (accessed on 26 November 2022).
30. International Food Information Council. 2022 Food and Health Survey. 18 May 2022. Available online: https://foodinsight.org/wp-content/uploads/2022/05/IFIC-2022-Food-and-Health-Survey-Report.pdf (accessed on 26 November 2022).
31. Stellaard, F. From Dietary Cholesterol to Blood Cholesterol, Physiological Lipid Fluxes, and Cholesterol Homeostasis. *Nutrients* **2022**, *14*, 1643. [CrossRef]
32. Lütjohann, D.; Meyer, S.; von Bergmann, K.; Stellaard, F. Cholesterol Absorption and Synthesis in Vegetarians and Omnivores. *Mol. Nutr. Food Res.* **2018**, *62*, 1700689. [CrossRef]
33. Berger, S.; Raman, G.; Vishwanathan, R.; Jacques, P.F.; Johnson, E.J. Dietary cholesterol and cardiovascular disease: A systematic review and meta-analysis. *Am. J. Clin. Nutr.* **2015**, *102*, 276–294. [CrossRef] [PubMed]
34. Kern, F. Normal Plasma Cholesterol in an 88-Year-Old Man Who Eats 25 Eggs a Day. *N. Engl. J. Med.* **1991**, *324*, 896–899. [CrossRef] [PubMed]
35. Bosner, M.S.; Ostlund, R.E.; Osofisan, O.; Grosklos, J.; Fritschle, C.; Lange, L.G. Assessment of percent cholesterol absorption in humans with stable isotopes. *J. Lipid. Res.* **1993**, *34*, 1047–1053. [CrossRef] [PubMed]
36. Lutjohann, D.; Meese, C.O.; Crouse, J.R.; Von Bergmann, K. Evaluation of deuterated cholesterol and deuterated sitostanol for measurement of cholesterol absorption in humans. *J. Lipid. Res.* **1993**, *34*, 1039–1046. [CrossRef]
37. Grundy, S.M. Does Dietary Cholesterol Matter? *Curr. Atheroscler. Rep.* **2016**, *18*, 68. [CrossRef]
38. Goldstein, J.L.; Brown, M.S. The LDL receptor. *Arterioscler. Thromb. Vasc. Biol.* **2009**, *29*, 431–438. [CrossRef]
39. Dandan, M.; Han, J.; Mann, S.; Kim, R.; Mohammed, H.; Nyangau, E.; Hellerstein, M. Turnover Rates of the Low-Density Lipoprotein Receptor and PCSK9: Added Dimension to the Cholesterol Homeostasis Model. *Arterioscler. Thromb. Vasc. Biol.* **2021**, *41*, 2866–2876. [CrossRef]
40. Vourakis, M.; Mayer, G.; Rousseau, G. The role of gut microbiota on cholesterol metabolism in atherosclerosis. *Int. J. Mol. Sci.* **2021**, *22*, 8074. [CrossRef]
41. Russell, D.W. The enzymes, regulation, and genetics of bile acid synthesis. *Annu. Rev. Biochem.* **2003**, *72*, 137–174. [CrossRef]
42. Menotti, A.; Kromhout, D.; Puddu, P.E.; Alberti-Fidanza, A.; Hollman, P.; Kafatos, A.; Tolonen, H.; Adachi, H.; Jacobs, D.R., Jr. Baseline fatty acids, food groups, a diet score and 50-year all-cause mortality rates. An ecological analysis of the Seven Countries Study. *Ann. Med.* **2017**, *49*, 718–727. [CrossRef]
43. Maki, K.C.; Dicklin, M.R.; Kirkpatrick, C.F. Saturated fats and cardiovascular health: Current evidence and controversies. *J. Clin. Lipidol.* **2021**, *15*, 765–772. [CrossRef] [PubMed]
44. Astrup, A.; Magkos, F.; Bier, D.M.; Brenna, J.T.; De Oliveira Otto, M.C.; Hill, J.O.; King, J.C.; Mente, A.; Ordovas, J.M.; Volek, J.S.; et al. Saturated Fats and Health: A Reassessment and Proposal for Food-Based Recommendations: JACC State-of-the-Art Review. *J. Am. Coll. Cardiol.* **2020**, *76*, 844–857. [CrossRef] [PubMed]
45. Mensink, R.P.; World Health Organization. *Effects of Saturated Fatty Acids on Serum Lipids and Lipoproteins: A Systematic Review and Regression Analysis*; World Health Organization: Geneva, Switzerland, 2016; pp. 1–63.
46. Fernandez, M.L.; West, K.L. Mechanisms by which dietary fatty acids modulate plasma lipids. *J. Nutr.* **2005**, *135*, 2075–2078. [CrossRef]
47. Woollett, L.A.; Spady, D.K.; Dietschy, J.M. Saturated and unsaturated fatty acids independently regulate low density lipoprotein receptor activity and production rate. *J. Lipid. Res.* **1992**, *33*, 77–88. [CrossRef]
48. Bergeron, N.; Chiu, S.; Williams, P.T.; MKing, S.; Krauss, R.M. Effects of red meat, white meat, and nonmeat protein sources on atherogenic lipoprotein measures in the context of low compared with high saturated fat intake: A randomized controlled trial. *Am. J. Clin. Nutr.* **2019**, *110*, 24–33. [CrossRef]
49. Ginsberg, H.N. Insulin resistance and cardiovascular disease. *J. Clin. Invest.* **2000**, *106*, 453–458. [CrossRef] [PubMed]
50. Kirkpatrick, C.F.; Bolick, J.P.; Kris-Etherton, P.M.; Sikand, G.; Aspry, K.E.; Soffer, D.E.; Willard, K.-E.; Maki, K.C. Review of current evidence and clinical recommendations on the effects of low-carbohydrate and very-low-carbohydrate (including ketogenic) diets for the management of body weight and other cardiometabolic risk factors: A scientific statement from the Nati. *J. Clin. Lipidol.* **2019**, *13*, 689–711.e1. [CrossRef]
51. Kossoff, E.H.; Zupec-Kania, B.A.; Auvin, S.; Ballaban-Gil, K.R.; Christina Bergqvist, A.G.; Blackford, R.; Buchhalter, J.R.; Caraballo, R.H.; Cross, J.H.; Dahlin, M.G.; et al. Optimal clinical management of children receiving dietary therapies for epilepsy: Updated recommendations of the International Ketogenic Diet Study Group. *Epilepsia Open.* **2018**, *3*, 175–192. [CrossRef]
52. Kwiterovich, P.O.; Vining, E.P.G.; Pyzik, P.; Skolasky, R.; Freeman, J.M. Effect of a High-Fat Ketogenic Diet on Plasma Levels of Lipids, Lipoproteins, and Apolipoproteins in Children. *J. Am. Med. Assoc.* **2003**, *290*, 912–920. [CrossRef] [PubMed]
53. Lee, H.S.; Lee, J. Influences of ketogenic diet on body fat percentage, respiratory exchange rate, and total cholesterol in athletes: A systematic review and meta-analysis. *Int. J. Environ. Res. Public Health* **2021**, *18*, 2912. [CrossRef]
54. Buga, A.; Welton, G.L.; Scott, K.E.; Atwell, A.D.; Haley, S.J.; Esbenshade, N.J.; Abraham, J.; Buxton, J.D.; Ault, D.L.; Raabe, A.S.; et al. The Effects of Carbohydrate versus Fat Restriction on Lipid Profiles in Highly Trained, Recreational Distance Runners: A Randomized, Cross-Over Trial. *Nutrients* **2022**, *14*, 1135. [CrossRef]

55. Creighton, B.C.; Hyde, P.N.; Maresh, C.M.; Kraemer, W.J.; Phinney, S.D.; Volek, J.S. Paradox of hypercholesterolaemia in highly trained, keto-adapted athletes. *BMJ Open Sport Exerc. Med.* **2018**, *4*, e000429. [CrossRef] [PubMed]
56. Burén, J.; Ericsson, M.; Damasceno, N.; Sjödin, A. A Ketogenic Low-Carbohydrate High-Fat Diet Increases LDL Cholesterol in Healthy, Young, Normal-Weight Women: A Randomized Controlled Feeding Trial. *Nutrients* **2021**, *13*, 814. [CrossRef] [PubMed]
57. Retterstøl, K.; Svendsen, M.; Narverud, I.; Holven, K.B. Effect of low carbohydrate high fat diet on LDL cholesterol and gene expression in normal-weight, young adults: A randomized controlled study. *Atherosclerosis* **2018**, *279*, 52–61. [CrossRef]
58. Norwitz, N.G.; Soto-Mota, A.; Kaplan, B.; Ludwig, D.S.; Budoff, M.; Kontush, A.; Feldman, D. The Lipid Energy Model: Reimagining Lipoprotein Function in the Context of Carbohydrate-Restricted Diets. *Metabolites* **2022**, *12*, 460. [CrossRef] [PubMed]

nutrients

MDPI

Review

Emerging Roles of Gut Microbial Modulation of Bile Acid Composition in the Etiology of Cardiovascular Diseases

Tess Yntema [1], Debby P. Y. Koonen [1] and Folkert Kuipers [1,2,*]

[1] Department of Pediatrics, University of Groningen, University Medical Center Groningen, 9713 GZ Groningen, The Netherlands; t.yntema@umcg.nl (T.Y.); d.p.y.koonen@umcg.nl (D.P.Y.K.)
[2] European Research Institute for the Biology of Ageing (ERIBA), University of Groningen, University Medical Center Groningen, 9713 GZ Groningen, The Netherlands
* Correspondence: f.kuipers@umcg.nl

Abstract: Despite advances in preventive measures and treatment options, cardiovascular disease (CVD) remains the number one cause of death globally. Recent research has challenged the traditional risk factor profile and highlights the potential contribution of non-traditional factors in CVD, such as the gut microbiota and its metabolites. Disturbances in the gut microbiota have been repeatedly associated with CVD, including atherosclerosis and hypertension. Mechanistic studies support a causal role of microbiota-derived metabolites in disease development, such as short-chain fatty acids, trimethylamine-N-oxide, and bile acids, with the latter being elaborately discussed in this review. Bile acids represent a class of cholesterol derivatives that is essential for intestinal absorption of lipids and fat-soluble vitamins, plays an important role in cholesterol turnover and, as more recently discovered, acts as a group of signaling molecules that exerts hormonal functions throughout the body. Studies have shown mediating roles of bile acids in the control of lipid metabolism, immunity, and heart function. Consequently, a picture has emerged of bile acids acting as integrators and modulators of cardiometabolic pathways, highlighting their potential as therapeutic targets in CVD. In this review, we provide an overview of alterations in the gut microbiota and bile acid metabolism found in CVD patients, describe the molecular mechanisms through which bile acids may modulate CVD risk, and discuss potential bile-acid-based treatment strategies in relation to CVD.

Keywords: gut microbiota; bile acids; cardiovascular disease; atherosclerosis

Citation: Yntema, T.; Koonen, D.P.Y.; Kuipers, F. Emerging Roles of Gut Microbial Modulation of Bile Acid Composition in the Etiology of Cardiovascular Diseases. *Nutrients* **2023**, *15*, 1850. https://doi.org/10.3390/nu15081850

Academic Editor: Frans Stellaard

Received: 20 March 2023
Revised: 4 April 2023
Accepted: 6 April 2023
Published: 12 April 2023

1. Introduction

Cardiovascular diseases (CVD) represent the number one cause of death globally, taking approximately 17.9 million lives each year, i.e., 32% of all deaths in 2019 [1]. Although CVD mainly affects the elderly, its incidence in younger people is increasing [2]. In accordance with the continuous ageing of the global population, CVD prevalence is steadily rising [3,4]. Atherosclerosis, a chronic low-grade inflammatory disorder of the vascular wall, is the main pathophysiological condition to result in CVD [5]. Atherosclerosis is characterized by the build-up of cholesterol-engulfed macrophages and other inflammatory immune cells in the vascular wall [2], causing coronary artery disease (CAD), ischemic stroke, or peripheral artery disease, through luminal narrowing of the blood vessels, plaque rupture, and thrombus formation [6].

Traditional risk factors of atherosclerosis include hyperlipidemia, hypertension, and chronic inflammation [7]. Although the scientific community has made tremendous progress in understanding, preventing, and modulating these risk factors, a significant residual cardiovascular risk remains, highlighting the urgent need for additional treatment strategies [8]. In recent years, non-traditional drivers of atherosclerosis, such as the gut microbiota, have gained attention [2,9]. An imbalanced gut microbiota, or dysbiosis, has been associated with cardiometabolic diseases, i.e., atherosclerosis, obesity, type 2 diabetes

(T2D), and non-alcoholic steatohepatitis (NASH) [10,11]. Dysbiosis can lead to increased intestinal permeability, which results in the translocation of the bacteria and endotoxins that promote chronic inflammation [12]. In addition, the gut microbiota has been implicated in CVD through the actions of microbiota-derived metabolites, such as short-chain fatty acids (SCFAs) [13], trimethylamine (TMA) [14], and bile acids (BAs) [15]; the latter will be elaborated on in this review.

BAs comprise a class of cholesterol metabolites, whose hepatic synthesis and eventual fecal excretion constitute the major pathway of cholesterol removal from the human body. Over 40 years ago, BA sequestrants (BAS) were among the first FDA approved drugs to lower cholesterol levels in CVD [16]. BAS bind to BAs and prevent their reabsorption from the intestine to promote their fecal loss. To compensate those that are lost, the liver produces more BAs from cholesterol, leading to upregulation of the hepatic low-density lipoprotein (LDL) receptor expression and lowering of the plasma LDL levels. Other than the lipid-lowering effects, clinical studies have shown an increase in coronary lumen diameter and lower cardiovascular risk upon BAS administration in CAD patients [17–19]. BAs function physiologically as detergents of lipids and fat-soluble vitamins to facilitate their intestinal absorption. In the last decennia, BAs were also shown to exert endocrine functions within and outside the enterohepatic circulation through the activation of a variety of BA receptors (BAR). Upon BAR activation, BAs can mediate lipids, as well as glucose metabolism and inflammation [20]. Interestingly, the expression of several BAR in cardiovascular tissues, such as the vasculature and immune system, highlights the potential of BA modulation in CVD [21]. In this review, we aim to provide an overview of (microbiota-modulated) BA metabolism and describe their potential biological impact thereof, especially in relation to CVD risk. Finally, we will discuss novel BA-based therapeutic targets that could be useful in the management of CVD.

2. Bile Acids Are Synthesized by the Liver and Extensively Metabolized by the Microbiota

BAs are amphipathic cholesterol derivates, characterized by one or more hydroxyl groups at their steroid nucleus (in mammals, commonly at positions 3, 7, and/or 12) and a shortened side chain bearing a carboxylic acid group [22] (Figure 1A). BAs are exclusively synthesized in hepatocytes in a process that involves a cascade of enzymatic steps in different compartments of the cell [23–25]. The products of this multi-step process are the so-called primary BAs, i.e., cholic acid (CA) and chenodeoxycholic acid (CDCA) in humans, which are products of the well-established "classical" or "alternative" pathways [26].

The first step of the classical pathway is catalyzed by the rate-controlling enzyme cytochrome P450 cholesterol 7α-hydroxylase (CYP7A1), which yields 7α-hydroxycholesterol [23,27] (Figure 1B). The alternative pathway is initiated by sterol 27α-hydroxylase (CYP27A1), and further processed by oxysterol 7α-hydrolase (CYP7B1) [27]. The 7α-hydroxylated intermediates from both the classical and alternative pathway undergo several sterol ring modifications, as well as side chain oxidation and shortening, leading to the production of CDCA [26,27]. In addition, the classical pathway also generates CA, which involves the activity of sterol 12α-hydroxylase (CYP8B1) [28]. Of note, rodents produce additional forms of primary BAs in the liver—α- and β muricholic acids (MCAs) and ursodeoxycholic acid (UDCA)—which are derived from CDCA by mouse/rat-specific cytochrome P450 2C70 (Cyp2c70) [26,29].

Primary BAs are conjugated with either taurine or glycine at the C24 position, which allows for their active secretion into the bile and decreases their pKa, and hence prevents passive reabsorption in the upper intestine [25]. BAs are secreted into the bile canaliculi by the bile salt-export pump (BSEP) and stored in the gallbladder in high mM range [26,30] (Figure 1B).

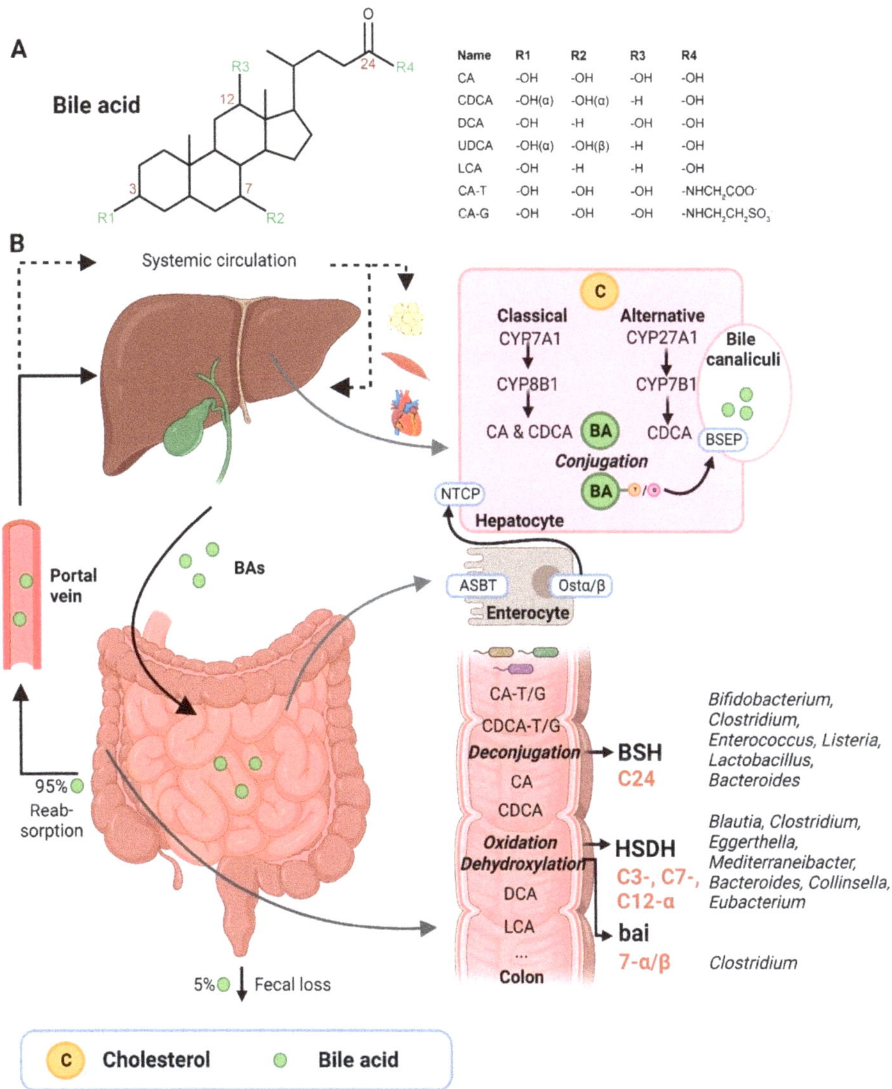

Figure 1. (**A**) Structural formula of bile acid. (**B**) Hepatic BA synthesis, enterohepatic circulation, and microbial BA modifications in the human body. BA = bile acid; CA = cholic acid; CDCA = chenodeoxycholic acid; DCA = deoxycholic acid; UDCA = ursodeoxycholic acid; LCA = lithocholic acid; TCA = taurocholic acid; GCA = glycocholic acid; C = cholesterol; BSEP = bile salt-export pump; NTCP = Na^+-taurocholic acid co-transporting polypeptide; ASBT = apical sodium-dependent BA transporter; Ostα/β = organic solute transporter α/β; BSH = bile salt hydrolases; HSDH = hydroxysteroid dehydrogenase; bai = BA inducible genes. This illustration was created with Biorender.com (accessed on 1 February 2023).

Upon meal intake, cephalic-phase-induced cholecystokinin secretion from the duodenum stimulates gallbladder contraction, leading to the discharge of stored BAs into the intestine [25]. In the intestinal lumen, BAs solubilize lipids, including cholesterol, and fat-soluble vitamins. Most primary BAs are actively reabsorbed in the ileum by the apical sodium-dependent BA transporter (ASBT) [25]. A small proportion of BAs escape uptake and reach the colon, where the gut microbiota can convert primary into secondary BAs (lithocholic acid (LCA) and deoxycholic acid (DCA)), a process that involves three major groups of bacterial enzymes (discussed in the subsequent section) [24]. Unconjugated secondary BAs are relatively hydrophobic and can thus be passively reabsorbed from the colon [20]. Most BAs (95%) are reabsorbed in the ileum or colon, and via the portal vein they are delivered back to the liver, whereafter they are taken up by Na^+-taurocholic acid co-transporting polypeptide (NTCP), in order to finalize a process called the enterohepatic circulation. In the liver, unconjugated primary and secondary BAs are re-conjugated and secreted into bile. Only a minor amount (~5%) of BAs, which are not taken up in the ileum or colon, are excreted in feces per cycle [20]. This loss of BAs is compensated by the de novo synthesis of BAs in the liver to maintain the size of the circulating BA pool [20].

A relatively small fraction of BAs escape first pass clearance by the liver and flow into the systemic circulation, with concentrations in the μM range [30]. From the peripheral circulation, BAs can reach multiple organs, including adipose tissue, muscle, and the heart [31,32]. BAs exert hormone-like functions by acting as signaling molecules and activate several BARs, i.e., nuclear receptors—farnesoid X receptor (FXR), vitamin D receptor (VDR), pregnane R receptor (PXR)—and membrane-bound receptors—Takeda G protein-coupled receptor (TGR5) and muscarinic receptors (MRs). Through these actions, BAs regulate their own homeostasis, as well as lipid and glucose metabolism, intestinal barrier function, cardiovascular functions, and inflammation [21,31,33]. Importantly, different BA species have dissimilar affinities for the activation of BARs (FXR: CDCA > LCA = DCA > CA; TGR5: LCA > DCA > CDCA > CA; VDR: LCA; PXR: LCA = CDCA = DCA = CA) [30,34]. Differences in BA pool composition, which are prominent across individuals [35], will thus affect BA signaling in a personalized manner.

3. Bacteria Involved in Bile Acid Metabolism

The three major groups of bacterial enzymes—bile salt hydrolases (BSHs), hydroxysteroid dehydrogenases (HSDHs), and bile-acid-inducible (bai) genes—are responsible for the generation of secondary BAs in the colon, leading to a major increase of the diversity of the BA pool. The major structural modifications include deconjugation, which is an obligatory first step; oxidation of hydroxy groups at the C3, C7, and C12 position; and 7α/β-dehydroxylation [24,25]. In addition, (7α/β, 3α/β) isomerization- and (5-H α/β) epimerization modifications give rise to UDCA and iso- and allo-BAs, respectively [26].

BSHs catalyze the deconjugation of the *N*-acylamide bond between primary BAs and taurine or glycine at the C24 position [25]. BSHs have been identified in several microbial genera, including *Bifidobacterium* [30,36], *Clostridium* [30,37], *Enterococcus* [30], *Listeria* [30,38,39], *Lactobacillus* [30,40,41], and *Bacteroides* [42]. Recently, computational analyses have shown that the human gut microbiota contains 591 intestinal bacterial strains within 117 genera with BSHs sequences [43]. BSHs, encompass seven [44] or eight sub-groups [43], showing differences in deconjugation ability. BSH-T3 shows the highest enzyme and deconjugation activity and is only found in *Lactobacillus* [43]. Recently, research has shown that, after deconjugation, the gut bacteria can also mediate the conjugation of the CA backbone with the amino acids phenylalanine, tyrosine, or leucine [45]. The microbial enzyme responsible for these BA modifications remains unknown. Interestingly, these amino acid BA conjugates are found in humans and are enriched in patients with inflammatory bowel disease or cystic fibrosis [45].

The second major group of bacterial enzymes are HSDHs, which oxidize and epimerize C3, C7, and C12 hydroxy groups of BAs. HSDH enzymes have been identified in the microbial genera *Blautia* (3α), *Clostridium* (3-, 7-, 12α), *Eggerthella* (3-, 12α), *Mediterraneibacter* (3α), *Bacteroides* (7α), *Collinsella* (7α), and *Eubacterium* (7α) [30]. Epimerization of hydroxy groups leads to a reversible change from the α to the β configuration, generating stable oxo-BAs as intermediates [25]. The reaction depends, in part, on the redox potential of the environment. For example, oxo-BAs are more present at the mucosal surface, where the redox potential is high, whereas less oxo-BAs are present in the lumen of the large intestine, where the redox potential is low [24]. Interestingly, the production of 12-oxoCDCA may reduce the formation of secondary BA DCA [46], which has been implicated in liver [47] and colon cancer [48], cholesterol gallstone formation [49], and CVD [50].

Bacteria that carry the bai operon produce enzymes that carry out 7-α/β dihydroxylation, resulting in the major secondary BAs DCA and LCA. Surprisingly, this metabolic pathway is only found in 0.0001% of colonic gut microbiota, belonging to the genera *Clostridium* [24,25,51]. Moreover, 7-α/β dehydroxylation only takes place after deconjugation, implicating a functional interplay between deconjugation and dehydroxylation [24].

Thus, the gut microbiota is responsible for diversifying the BA pool. This strongly affects BA signaling, as BAs have different affinities towards BARs [20]. Importantly, the regulation between the gut microbiota and BAs is reciprocal, meaning that BAs can also modulate the gut microbiota either by direct or indirect effects. For example, BAs can disrupt bacterial membranes or bind to intestinal FXR, promoting the expression of antimicrobial agents [52]. Moreover, conjugated BAs play an important role in the prevention of bacterial overgrowth in the proximal small intestine, which is relatively devoid of microbes under normal conditions [53]. Studies have shown that replenishing intestinal BA concentrations in BA-deficient rats abolished bacterial overgrowth in the small intestine [53]. On the other hand, bacteria that inhabit the intestinal tract must have specific resistance mechanisms to protect themselves against bile [54]. For example, *Lactobacillus* and *Bifidobacterium* produce proteins that are devoted to the efflux of BAs [54].

Of note, gut bacteria can also directly metabolize cholesterol in the intestine via dehydrogenase activity encoded by intestinal sterol metabolism A (*ismA*) genes, producing cholestanone and coprostanol [55]. These genes are found in the human gut microbiota in geographically diverse human cohorts and show a negative correlation with circulating cholesterol levels [55]. In addition, recent studies highlight the bacterial potential to sulfonate cholesterol and related steroids in the gut. The sulfotransferase enzyme is identified in the microbial species *Bacteroides thetaiotaomicron* [56,57]. The new (direct) lipid-metabolizing functions of the gut microbiota, which was originally thought to be performed by host enzymes, represent a great breakthrough in the understanding of cholesterol homeostasis.

4. Gut Microbiota Signatures in Cardiovascular Disease

Since the late 1980s, researchers have implicated a role of bacteria in atherogenesis [58]. In these early studies, the bacterium *Chlamydia pneumoniae* was associated with CAD and myocardial infarction. A few years later, bacterial DNA from many other bacteria genera/species were found in human atherosclerotic plaques [59–61], whereas healthy tissue (e.g., non-transplanted hearts) does not contain bacterial DNA [62]. Bacterial DNA has also been linked to inflammation, as the amount of bacterial DNA was found to correlate with the number of leukocytes in the plaque [59]. This suggests that the underlying pathophysiology of atherosclerosis may involve bacterial activation of the immune system. In addition, bacteria originating from the gut and the oral cavity matched with bacterial DNA present in atherosclerotic plaques and correlated with disease biomarkers [59]. Thus, the re-allocation of bacteria from the intestinal tract to the heart may contribute to disease development, which has sparked interest to evaluate the role of the gut microbiota in CVD.

Multiple cross-sectional studies have assessed the association of the gut microbiota with atherosclerosis (Table 1) [63–71] and hypertension (extensively reviewed in [72]). Despite differences in sequencing methods (16S versus shotgun) and downstream analyses, five out of nine studies showed a reduction in the genera *Roseburia* and *Faecalibacterium* in atherosclerosis patients [64–67,69]. These genera produce SCFAs, especially butyrate, which affect colonic motility and immunity, among other functions [73,74]. In addition, higher abundances of Gram-negative bacteria, such as *Enterobacter* [66,67], *Klebsiella*, and *Veillonella* [69], as well as *Streptococcus*, were found in atherosclerosis patients [67,70]. Gram-negative bacteria have lipopolysaccharides (LPS) on their outer membrane, which are pro-inflammatory. *Veillonella* and *Streptococcus* genera have been found in most human atherosclerotic plaques [59], and have both been implicated as pathogens [75–78].

Specifically looking at the major groups of bacterial enzymes, studies classified bacteria containing 3α-HSDH, 7α-HSDH, and/or BSH genes (indicated by [30,43]) in the gut microbiota of atherosclerosis patients, regardless of the direction of change (increased or decreased) (Table 1). For example, Toya et al. observed a higher abundance of *Ruminococcus gnavus* in atherosclerosis patients [68]. This species expresses 3α-HSDH enzymes and is an important producer of iso-DCA [79]. Moreover, the genera *Collinsella, Eubacterium,* and *Bacteroides* carry both 7α-HSDH and BSH genes, whereas *Roseburia, Streptococcus, Enterococcus,* and *Clostridium* only carry BSH genes (Table 1). Interestingly, BSH phylotypes are related to CVD [43,44]. For example, Karlsson et al. observed an increased abundance of *Collinsella* in the gut microbiota of atherosclerosis patients [64]. This genus carries BSH-T4 genes, whose relative abundance is significantly higher in atherosclerosis populations [43]. Moreover, multiple studies observed a reduced abundance of *Bacteroides* in the gut microbiota of atherosclerosis patients [66,71]. This genus carries BSH-T5/BSH-T6 genes, whose relative abundance are significantly lower in atherosclerosis populations [43]. This suggests that changes in BA-modifying bacteria in the gut microbiota may play a role in the pathophysiology of atherosclerosis.

Regarding the microbial alpha diversity, reflecting the number of bacterial species, the results are inconsistent across the studies enlisted in Table 1. In general, gut microbiota diversity has been found to negatively correlate with risk factors of atherosclerosis, such as obesity, hyperinsulinemia, hypertension, and dyslipidemia [72,80,81]. However, others have demonstrated that alpha diversity was either not different or increased in atherosclerosis patients compared with the controls [63,66,67,69,71]. These discrepancies could be due to limitations in the respective studies, such as omitting important confounders in their analyses (e.g., age, sex, and BMI) or assessing diversity with an estimation-based method (Chao1) [63]. Moreover, the temporal dynamics and intra- and interindividual heterogeneity of the gut microbiota underscores the difficulty in studying and comparing cross-sectional studies. Rodent models (especially germ-free models), which are more experimentally controlled, represent an invaluable tool for studying microbe–host interactions in the context of CVD.

In recent years, several studies have shown causal effects of the gut microbiota in the pathology of CVD [82–85]. For example, we recently showed that atherosclerosis was increased by 30% in (atherosclerosis-prone) low-density lipoprotein receptor knockout ($Ldlr^{-/-}$) mice following transplantation with a pro-inflammatory gut microbiota and feeding a high-fat, cholesterol-enriched diet [85]. In addition, an in vitro screening protocol identified specific peptides that selectively modified bacterial growth [83]. These peptides can selectively partition into bacterial membranes, interrupt the transmembrane potential, and impair cell growth. Interestingly, Western-type diet-associated gut microbiota dysbiosis was remodeled by cyclic d,L-α-peptides towards the chow-diet microbial state, which was accompanied with rebalanced BA and SCFA homeostasis, suppressed production of pro-inflammatory cytokines and improved gut barrier integrity in $Ldlr^{-/-}$ mice [83]. Liu et al. transplanted germ-free mice with feces from CAD patients or healthy controls [84]. CAD patients were characterized by an altered gut microbiota composition and elevated serum levels of secondary BAs. Mice colonized with CAD microbiota showed an imbalanced BA composition with increased secondary BAs, worsened gut barrier permeability, and vascular dysfunction. In another study, supplementation with glycoursodeoxycholic acid (GUDCA) was shown to partly normalize Western diet-associated gut microbiota dysbiosis, which improved cholesterol homeostasis and local chronic inflammation and protected against atherosclerosis progression in (atherosclerosis-prone) apolipoprotein E-deficient ($ApoE^{-/-}$) mice [82]. Changes in bacteria genera *Alloprevotella* and *Parabacteroides* positively, and *Turicibacter* and *Alistipes* negatively, were modulated by GUDCA and correlated with the plaque area in mice aortas [82]. Combined, these studies provide causal evidence that there is a functional interplay between gut bacteria and BAs, which affects CVD in mice.

Table 1. Cross-sectional studies on the gut microbiota composition in atherosclerotic patients.

Author	Population	Atherosclerosis Definition	Sequencing Method	Higher Abundance in Atherosclerosis	Lower Abundance in Atherosclerosis	Microbial Diversity in Atherosclerosis	Covariates in Analyses
Zheng et al., 2020 [63]	152 patients 105 controls	Atherosclerosis (≥50% stenosis in one or more vessels)	16S	*Bulleidia*, *Comamonas*, *Enhydrobacter*	*Agrobacterium*, *Delftia*, *Enterobacter*, *Morganella*	Increased	Unadjusted
Karlsson et al., 2012 [64]	12 patients 13 controls	Symptomatic atherosclerosis (who had undergone CAE)	Shotgun	*Collinsella* (7α-HSDH BSH-T4)	*Roseburia* (BSH-T1), *Eubacterium* (7α-HSDH/BSH-T1)	NR	Smoking, diabetes, age and BMI
Zhu et al., 2018 [65]	70 patients 98 controls	Atherosclerosis (confirmed by coronary angiography)	16S	*Escherichia-Shigella*, Enterococcus (BSH)	*Roseburia* (BSH-T1), *Eubacterium* rectale (7α-HSDH/BSH-T1), *Faecalibacterium*, *Enterococcus* (BSH-T0)	Decreased	Unadjusted
Yin et al., 2015 [66]	141 patients 94 controls	Symptomatic atherosclerosis (with TIA)	16S	*Enterobacter*, *Megaspaera*, *Desulfovibrio*	*Bacteroides* (7α-HSDH/BSH-T5/6), *Prevotella*, *Faecalibacterium*	Increased	Unadjusted
Jie et al., 2017 [67]	218 patients 187 controsl	Atherosclerosis (≥50% stenosis in one or more vessels)	Shotgun	*Enterobacteriaceae*, *Streptococcus* spp. (BSH-T2)	*Roseburia* (BSH-T1), *Faecalibacterium Prausnitzii*	No difference	Unadjusted
Toya et al., 2020 [68]	53 patients 53 controls	Atherosclerosis (≥50% stenosis in one or more vessels)	16S	*Ruminococcus gnavus* (3α-HSDH BSH-T1)	*Lachnospiraceae NK4B4*, *Ruminococcus Gauvreauii* (BSH-T1)	Decreased	Age, sex, race, BMI, DM, dyslipidemia
Liu et al., 2019 [69]	161 patients 40 controls	Atherosclerosis (≥50% stenosis in one or more vessels)	16S	*Veillonella*, *Haemophilus*, *Klebsiella*	*Roseburia* (BSH-T1), *Faecalibacterium*	No difference	Unadjusted
Feng et al., 2016 [70]	59 patients 43 controls	Atherosclerosis (confirmed by coronary angiography)	Shotgun	*Clostridium* sp. *HGF2* (BSH-T0), *Streptococcus* sp. *M334/M143* (BSH-T2)	NR	NR	Unadjusted
Yoshida et al., 2018 [71]	30 patients 30 controls with risk factors	Atherosclerosis (≥75% stenosis in one or more vessels) AND stable angina pectoris, MI	16S	*Faecalibacterium prausnitzii*, *Prevotella copri*	*Bacteroides vulgatus*, *Bacteroides dorei* (BSH-T5/6)	No difference	Age, sex

5. Altered Bile Acid Metabolism in Cardiovascular Disease

Examining the composition of the gut microbiota at one single point in time ignores the complex nature of the gut microbiota as an ecosystem. Investigating functional shifts of the gut microbiota, including changes in gut-derived metabolites, helps to identify measurable read-outs of bacterial functions in health and disease. BAs might be a relevant gut metabolite in relation to CVD, despite controversy across studies [86–89]. Lower plasma BA levels have been reported in CVD [86,87]. Low serum BAs appeared to be independently associated with the presence and severity of CAD, especially for the presence of myocardial infarction (MI) [87]. These results were largely confirmed by Nguyen et al., although serum BA concentrations in both the control and CAD patients were lower compared with the previous study [86]. The latter study also observed doubled serum BAs in patients receiving statin therapy, suggesting that serum BAs levels are amendable by statin administration in CAD patients. Statins are a commonly described drug to lower cholesterol via the inhibition of β-Hydroxy β-methylglutaryl-CoA (HMG-CoA) reductase, the rate limiting enzyme in the cholesterol biosynthesis pathway [90]. Moreover, glyco-CDCA was two-fold higher in CAD patients [86] and, together with the total serum BAs, were predictors of CAD [86,87]. In contrast, a previous study comparing CAD and non-CAD patients did not demonstrate a significant association between serum BAs and CAD [88]. In addition, the total serum BAs are known to increase in patients with liver cirrhosis, which is associated with cardiac dysfunction [89]; total serum BAs are elevated up to 100 times the normal values in patients with cirrhotic cardiomyopathy [89], i.e., much more compared with CAD patients. Although inconsistency is found between studies, it appears that either low or extreme high serum BAs can be associated with CVD.

For zooming in on small-molecule metabolites in relation to CVD, untargeted metabolomics is a powerful tool to discover novel metabolites. Zhang et al. discovered that six metabolites were significantly altered in CAD patients [91]. LCA, together with 4-pyridocix acid and phosphatidylglycerol (20:3/2:0), showed the strongest positive correlation with CAD, defined as >80% stenosis in at least one artery. Of note, Chen et al. observed large inter-individual variability in plasma BA profiles in human obesity [35]. This variability suggests a more personalized approach to finding biomarkers and future therapeutic applications of BAs in CVD, although participants with recent cardiovascular events were excluded from this study [35]. Nevertheless, secondary BAs, i.e., DCA and LCA, were associated with diabetes and liver fat content, which are two risk factors of CVD, in these obese subjects.

In addition to plasma BAs, fecal bile acid excretion (BAE), which equals hepatic BA synthesis under steady state conditions, has also been associated with CAD [92–94]. CAD-patients were found to excrete less BAs, particularly DCA and LCA, compared with non-CAD controls [92–94]. A historical follow-up of 20 years showed that BAE was a significant independent parameter that predicted CAD in humans, in which BAE < 415 mg/day was associated with a higher long-term mortality due to CAD [94]. More specifically, 75% of the patients with BAE < 262.4/day developed a stroke relative to none of the patients with BAE > 622 mg/day [93]. BAE can thus serve as an interesting biomarker of CAD.

Thus, these studies provide evidence that measuring plasma and fecal BAs may aid in the assessment of the gut microbiota contributions to CVD. Understanding determinants of BA pool/metabolism and its reflection in CVD is important to rationalize their use as potential biomarkers and therapeutic targets.

6. Bile Acids as Mediators of Cardiovascular Disease Risk

In this section, the focus switches from association to causality regarding the potential roles of BAs in CVD. The multifaceted roles of BAs in lipid homeostasis, immunity, and heart function indicate the ability to mediate CVD, as discussed in the following sections (Figure 2).

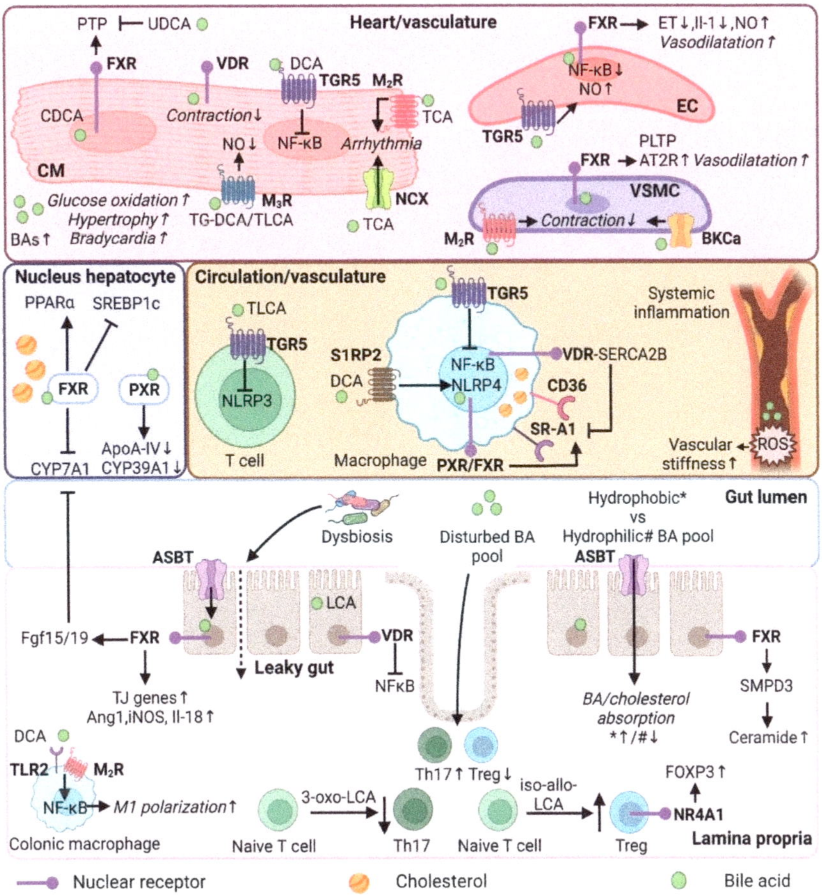

Figure 2. Bile acid regulation in lipid and immune metabolism, and heart function. Mechanistic effects of BAs in different organs (heart/vasculature/circulation/colon/lamina propria) or cell types (cardiomyocyte, endothelial cell, vascular smooth muscle cell, immune cells, enterocyte) in relation to lipid and immune metabolism, and heart function. CM = cardiomyocyte; EC = endothelial cell; VSMC = vascular smooth muscle cell; Th17—T helper 17 cells; Treg = Regulatory T cells; FXR = farnesoid X receptor; VDR = vitamin D receptor; PXR = pregnane R receptor; NR4A1 = nuclear receptor 4A1; TGR5 = Takeda G protein-coupled receptor; MR = muscarinic receptor; S1RP2 = sphingosine 1-phosphate receptor 2; SR-A1 = Class A1 scavenger receptors; AT2R = angiotensin II receptor type 2; TLR2—toll-like receptor 2; NCX = Na^+/Ca^{2+} exchange protein; BKCa = large conductance Ca^{2+}-activated K^+; ASBT = apical sodium-dependent bile acid transporter; BA = bile acid; UDCA = ursodeoxycholic acid; (T/G)DCA = (tauro/glycol)deoxycholic acid; (T)LCA = Taurolithocholic acid; TCA = Taurocholic acid; CDCA = chenodeoxycholic acid; NF-κB = nuclear factor kappa B; PTP = Mitochondrial permeability transition pore; PLTP = phospholipid transfer protein; ET = endothelin-1; NO = nitric oxide; ROS = reactive oxygen species; PPARα = peroxisome proliferator-activated receptor alpha; SREBP1c = sterol regulatory element binding protein 1c; CYP7A1 = cytochrome P450 7A1; CYP39A1 = cytochrome P450 39A1; ApoA-IV = Apolipoprotein A-IV; NLRP3/4 = NLR family pyrin; FOXP3 = forkhead box P3; SMPD3 = Sphingomyelin Phosphodiesterase 3; SERCA2B = Sarcoendoplasmic reticulum calcium ATPase 2b; FGF15/19 = fibroblast growth factor 15/19; Ang1 = angiopoietin 1; iNOS = nitric oxide synthase; IL = interleukin; TJ = tight junction; * = hydrophobic bile acid pool; # = hydrophilic bile acid pool. This illustration was created with Biorender.com.

6.1. Regulation of Lipid Metabolism

Recent studies have shown that the elevation of plasma triglycerides (TG) comprises the major lipid abnormality in patients with atherosclerosis, changing the focus from LDL cholesterol to TG as being causal in atherosclerosis [2]. Interestingly, BAs appear to display beneficial effects in hypertriglyceridemia [95]. BA activation of the FXR-SHP pathway interferes with the regulation of fatty acid biosynthesis genes, mediated by the liver X receptor (LXR) and sterol regulatory element binding protein 1c (SREBP-1c) [95]. Feeding mice for 8 weeks with 0.5% CA was associated with a reduction in hepatic SREBP-1c expression and its lipogenic target genes in mouse models [95]. It has been proposed that the rate of hepatic lipogenesis is a major determinant of very-low-density lipoprotein (VLDL)-TG production, although it has been reported that massive lipogenesis in obese ob/ob mice does not lead to increased VLDL production [96]. While SREBP controls fatty acid synthesis, peroxisome proliferator-activated receptor α (PPARα) promotes fatty acid β-oxidation [97–99]. Th activation of PPARα by fenofibrate reduced the plasma TG's, adiposity, and atherosclerosis development in high-fat diet-fed Ldlr$^{-/-}$ mice [100,101]. Lipid accumulation in the aorta was prevented upon PPARα activation, probably by enhanced fatty oxidation. BAs can also induce PPARα via the activation of FXR [102].

In contrast with these beneficial effects in hypertriglyceridemia, BA activation of PXR, a well-known transcription regulator in the control of lipid metabolism [50,103], increased levels of total cholesterol, VLDL, and LDL; decreased high-density lipoprotein (HDL) levels; and increased atherosclerosis in ApoE$^{-/-}$ mice [104], while PXR deletion reduced atherosclerosis in ApoE$^{-/-}$ mice [105]. The genes involved in lipoprotein transport and cholesterol metabolism, including apolipoprotein A-IV (ApoA-IV), cytochrome P450 family 39 subfamily A member 1 (Cyp39a1), and cluster of differentiation 36 (Cd36), were affected upon PXR activation [104]. ApoA-IV and Cyp39a1 were down-regulated, whereas Cd36 was upregulated. Studies have indicated that CD36 plays an important role in atherosclerosis by mediating oxidized LDL (oxLDL) uptake by macrophages, leading to the formation of foam cells [106]. PXR activation in peritoneal macrophages led to increased Cd36 expression, which was consistent with increased lipid accumulation in these cells [104]. In addition, GUDCA was shown to inhibit macrophage foam cell formation by down-regulating scavenger receptor A1 gene expression, whereas Cd36 expression was not affected, implicating a different mode of action [82].

Of note, compensatory mechanisms preventing lipid accumulation in the circulation have been observed while directly targeting BA synthesis in CYP27A1/ApoE double-knockout (DKO) mice [50]. Zurkinden et al. showed that loss of the BA synthesis gene Cyp27a1 in ApoE$^{-/-}$ mice fed a Western-type diet (WD) was associated with an upregulation of Cyp7a1 and cytochrome P450 3A1 (Cyp3a1), resulting in increased BA synthesis and excretion (i.e., accelerated cholesterol turnover), and protection against atherosclerosis [107]. In this study, they also observed differential cardiovascular outcomes of CA and CDCA feeding, in addition to the WD. Despite the observation that both feeding regimes led to reduced Cyp7a1 and Cyp3a1 expression, only CA-WD resulted in a strong increase in atherosclerosis, together with increased LDL and reduced HDL in the serum, increased intestinal absorption of cholesterol, and decreased faecal cholesterol output [107]. Although in humans CDCA is a more potent FXR agonist than CA, in rodents, FXR is mostly activated by CA as CDCA is rapidly converted in MCAs (FXR antagonists) [108]. Macrophage FXR activation by CA takes part in reverse cholesterol transport and reduces the HDL efflux [107]. In addition, the hydrophilic MCAs in the BA pool in CDCA-WD fed mice are likely responsible for the reduced cholesterol absorption [107], as the 12α-OH group of C7 appears to be essential for efficient cholesterol uptake by enterocytes [109].

Similar to BA sequestration, directly blocking intestinal BA absorption also affects BA synthesis. Targeting ASBT leads to the increased expression of BA synthesis genes and HMG-CoA reductase in ApoE$^{-/-}$ mice [110]. To fuel the liver with additional cholesterol, hepatocytes increase both the de novo synthesis and expression of the cell surface LDL receptors, which results in reduced plasma cholesterol and less prominent aortic lesions.

FXR and TGR5 are both expressed in the vasculature. Whereas FXR is found in vascular smooth muscle cells (VSMCs) and endothelial cells of the vascular wall [111] (and possibly in macrophages [112]), TGR5 is expressed on the surface of macrophages [113], which are present in atherosclerotic plaque. Studies have shown that the activation of FXR in rat and human VSMCs reduces migration and inflammation [114], and affects lipid metabolism [111]; the latter particularly via phospholipid transfer protein (PLTP), an important regulator of reverse cholesterol transport. Moreover, FXR activation has been shown to mediate vasodilation in endothelial cells [115] and TGR5 activation to reduce inflammatory responses in macrophages [116].

Studies have evaluated the effect of TGR5 and/or FXR activation on atherogenesis [116,117]. The activation of TGR5 by INT-777 attenuated atherosclerosis development in $Ldlr^{-/-}$ mice. INT-777 treatment led to reduced macrophage lipid loading and pro-inflammatory cytokine production, an effect mediated by altered cAMP-signalling and nuclear factor-κB (NF-kB) inhibition [116]. In the absence of TGR5, these anti-atherosclerotic effects were abolished [116]. Similarly, simultaneous activation of FXR and TGR5 by the dual agonist INT-767 in $Ldlr^{-/-}$ mice reduced atherosclerosis via anti-inflammatory and lipid-lowering effects [117]. FXR deficiency alone completely blocked the lipid-lowering effects of INT-767: this was not found in TGR5-deficient mice. Interestingly, both FXR- and TGR5-deficient mice show reduced atherosclerosis and aortic inflammation upon INT-767 administration [117].

Thus, the anti-atherosclerotic effect of FXR and TGR5 activation is driven by inflammation, and the loss of one receptor is compensated by the other. Moreover, these anti-atherosclerotic effects of INT-767 were abolished by the dual deficiency of FXR and TGR5 in $Ldlr^{-/-}$ mice [117]. In addition, Jadhav et al. found that INT-767 induced thermogenesis genes and reduced hepatic de novo lipogenesis genes, suggesting an additional role in energy homeostasis on top of the anti-atherosclerotic effects of this compound [118].

Intriguingly, both the deficiency and activation of FXR leads to similar results in atherosclerosis mouse models. FXR loss of function results in a decreased atherosclerotic plaque surface in $Ldlr^{-/-}$ and $ApoE^{-/-}$ mice [119,120]. Although the serum cholesterol levels were reduced, the TGs were elevated in these models. Peritoneal macrophages showed a reduced Cd36 gene expression and decreased lipid accumulation in FXR-LDLR DKO mice [120]. This suggests an indirect effect of FXR on macrophages. FXR activation also showed an atheroprotective effect in atherosclerosis models, which was in part the result of an improved lipid profile [117,118,121,122]. TG-lowering effects can be controlled by FXR-induced lipoprotein lipase and SREPB-1c activity [121]. Other anti-atherosclerotic effects could be the result of increased fecal cholesterol excretion and macrophage reverse cholesterol transport due to reduced BA pool size and composition [122]. Thus, both loss of function and activation of FXR attenuates atherogenesis in mice, which complicates the interpretation of these results.

Recent research has demonstrated that patients with hypercholesterolemia have elevated fibroblast growth factor 19 (FGF19), which is positively correlated with pro-atherogenic ceramide levels [123]. Intestinal FXR activation releases FGF19 (or fibroblast growth factor 15 (FGF15) in rodents) in the ileum, which is a negative regulator of hepatic BA synthesis. Intestinal FXR-deficient $ApoE^{-/-}$ mice showed decreased atherosclerotic lesions in the aorta and heart, and reduced serum levels of ceramides after high-cholesterol diet feeding [123]. The enzyme sphingomyelin phosphodiesterase 3 (SMPD3) was identified as an intestinal FXR target, which is involved in ceramide synthesis. The overexpression of SMPD3 eliminated the anti-atherosclerotic effects in intestinal FXR-deficient $ApoE^{-/-}$ mice [123].

Thus, BAs play an important role in lipid metabolism, as BAs affect BA and lipid synthesis, absorption, and excretion, but are also involved in foam cell formation and ceramide synthesis, which are both associated with CVD. Although it is not always clear whether the lipid-lowering (anti-atherosclerotic) effects of BAs are caused by direct or indirect effects of BA signalling, BAs represent an interesting therapeutic target for CVD.

6.2. Regulation of Immune Functions by Secondary Bile Acids

Inflammation plays a central role in CVD, involving both the innate and adaptive immune systems [12,124–130]. Alterations in the gut microbiota are associated with impaired gut integrity, leading to increased leakage of microbiota-derived LPS, which promotes systemic inflammation [12]. Targeting innate immunity with antibodies against interleukin-1 beta (IL-1β) led to reduced recurrent cardiovascular events in patients with previous myocardial infarction, independent of lipid lowering [131], suggesting that CVD is driven by inflammation. Interestingly, among the continuously growing list of metabolites regulating the immune system, secondary BAs are recognized to possess immuno-regulatory properties that affect CVD development [132].

FXR signalling is evidently involved in the maintenance of intestinal integrity and the regulation of inflammatory processes. For example, intestinal FXR activation has been shown to protect mice from intestinal inflammation through the downregulation of key pro-inflammatory cytokines [133], which are known to increase the permeability of the intestinal epithelial monolayer [133,134]. A different study showed that the genes encoding tight junction proteins, zonulin-1 and occludin, were increased upon FXR activation in a cirrhotic rat model [135]. In addition, FXR has been shown to induce the expression of genes such as angiopoietin 1 (Ang1), inducible NO synthase (iNOS), and interleukin-18 (Il-18), which show anti-microbial effects and protect the mucosal intestinal barrier in mice [136]. VDR is also involved in maintaining gut integrity and can be activated by LCA and iso-LCA [34,137,138]. The activation of VDR by LCA showed a protective effect on tumor necrosis factor alpha (TNFα)-induced injury of intestinal barrier function in Caco-2 cells, possibly through the NFκB signalling pathway [137]. Moreover, cell experiments demonstrated that LCA can induce adhesion molecule expression in endothelial cells through the activation of reactive oxygen species formation, NF-κB, and p38 mitogen-activated protein kinase (MAPK) signalling [139]. Increased surface adhesion molecules can attract immune cells to the vascular wall. Controversially, studies have shown anti-inflammatory effects upon VDR activation in CVD [50,140–142].

On the contrary, in vivo and in vitro studies have shown that the prolonged presence of excess DCA can reduce gut barrier function and promote intestinal inflammation [143]. DCA may serve as an endogenous danger-associated molecular pattern (DAMP) and activate inflammasome NOD-, LRR-, and pyrin domain-containing protein 3 (NLRP3), a cytosolic multiprotein of the innate immune system, which promotes the secretion of pro-inflammatory cytokines [144]. DCA binds to the sphingosine-1-phosphate receptor 2 (S1PR2) in macrophages that mediates NLRP3 activation [144]. Moreover, Wang et al. found that HFD-induced dysbiosis promotes DCA production, leading to increased pro-inflammatory colonic macrophage infiltration in mice [145]. DCA dose-dependently promoted M1 macrophage polarization and cytokine production via the NF-kB signalling pathway, partially through toll-like receptor 2 (TLR2) and muscarinic 2 receptor (M_2R) activation.

TGR5 activation in CD4 T cells inhibits inflammation by regulating the recruitment of CD4 and CD8 T cells after myocardial infarction [146]. In line with this anti-inflammatory effect, the administration of TLCA led to NLRP3 inflammasome inhibition via the TGR5-cAMP-PKA axis in T cells and macrophages. PKA can induce the phosphorylation of NLRP3, thereby preventing the activation of the inflammasome [147]. Interestingly, SHP is a negative regulator of NLPR3 inflammasome activation and its deficiency showed increased pro-inflammatory cytokines in vivo, including Il-1β and Il-18 [148].

Similar to FXR and TGR5, VDR activation exhibits strong anti-inflammatory effects in macrophages [149–151]. VDR deletion increased foam cell formation from a lack of the VDR-sarcoendoplasmic reticulum calcium ATPase 2b (SERCA2b) interaction, causing activation of ER stress and the induction of CD36 and class A1 scavenger receptor (SR-A1) in the macrophages [149]. SERCA2b, is a critical enzyme that maintains ER calcium levels to optimize protein production and folding [149].

Accumulating evidence indicates that an altered microbiota can activate both the innate and adaptive immune system [83–85]. Western-type diet-induced dysbiosis [83] and microbes from CAD patients [84] was shown to modulate BA pool composition, worsen gut barrier permeability, and increase systemic and intestinal inflammation, likely by imbalanced T helper and T regulatory cells (Tregs) in the intestinal lamina propria. A higher ratio of retinoid-acid-related orphan receptor-γt (RORγt) T helper 17 cells to (Helios+) regulatory T cells was observed [83,84]. While transcription factor RORγt is required for the differentiation of pro-inflammatory Th17 cells, orchestrating intestinal inflammatory responses, Helios+ cells play a critical role in tissue repair, and in the maintenance of gut barrier function [152]. Chen et al. speculated that BAs and other oxysterols can serve as endogenous RORγt agonists, modulating different T cell subsets [83]. Indeed, fecal LCA is positively correlated with the Th17−Treg ratio in the intestinal lamina propria [84], suggesting that signals from the gut microbiota could drive the differentiation and maintenance of T cell subsets in the intestine. In addition, microbes from CAD patients led to increased reactive oxygen species generation and vascular stiffness in the aorta, probably caused by increased intestinal inflammation and worsened gut barrier permeability [78].

Follow-up studies have shown that derivates from LCA can modulate T cell subsets. While 3-oxoLCA inhibited the differentiation of Th17 through direct binding to RORγt [138,153], iso-alloLCA increased the differentiation of Tregs [138] in a nuclear hormone receptor Nur77-dependent manner [154], enhancing Forkhead Box P3 (FOXP3), a master regulator in the development and functioning of Tregs. Changing the BA pool by manipulating dietary and microbial factors confirmed the role of BAs in modulating FOXP3 Tregs in the colon [155]. The genetic ablation of BA enzymes in the bacteria reduced Tregs, which was resolved after restoring the BA content in the intestine. Moreover, a nutrient minimal diet lowered fecal deconjugated BAs and reduced Tregs in the colon, which was only restored by the administration of LCA/3-oxo-LCA [155]. Investigating different BA receptors in modulating colonic Treg cell populations demonstrated a major role of the BA-VDR pathway [155]. VDR is expressed in FOXP3 Tregs, but also in endothelial and dendritic cells in the colon [155]. The secondary BA isoDCA can increase FOXP3 Treg induction by acting on dendritic cells [156]. Dendritic cell FXR ablation led to increased Treg production, showing the same transcriptional profile compared to isoDCA induction [156]. The role of VDR in this process was not investigated. Interestingly, human VDR genetic variants could affect intestinal inflammation through controlling the Treg pool [157]. Forward genetics studies have observed polymorphisms in the VDR promotor, which controlled VDR expression and T cell activation [157]. The different activation status of VDR could affect intestinal inflammatory susceptibility through the improper control of Tregs in the colon, driving low-grade systemic inflammation [157].

Thus, both the innate and adaptive immune systems can be modulated by BA signalling, either by boosting or inhibiting inflammatory responses. Therefore, (indirectly) targeting BA metabolism to reduce systemic inflammation could be a new therapeutic opportunity for CVD.

6.3. Regulation of Heart Function

Studies have shown a direct role of BAs in features of heart and arteries (cardiovascular tissue). BAs appear to regulate cardiovascular function through the activation of BARs (FXR, TGR5, and VDR) and MRs, as well as through the interaction with ion channels in cardiovascular tissue [31]. This section summarizes BA signalling in relation to cardiovascular function.

FXR is highly expressed in the liver, kidney, and gastrointestinal tract, but is also expressed in heart tissue (i.e., cardiomyocytes and fibroblasts) and in the vasculature (i.e., endothelial and vascular smooth muscle cells) [111]. FXR activation results in cell-type specific responses [111,115,158–161]. FXR activation of isolated neonatal rat cardiomyocytes, either by natural (CDCA) or synthetic (GW4064) FXR activators, induced apoptosis [158]. Interestingly, increased Fxr expression was observed after myocardial ischaemia/reperfusion in mouse hearts, whereas both pharmacological inhibition or the genetic deletion of FXR reduced myocardial apoptosis, decreased infarct size, and improved the cardiac function of ischaemic hearts [158]. Hydrophilic UDCA was shown to protect the myocardium against reperfusion injury in rat hearts by blocking the opening of the cardiac mitochondrial permeability transition pore (PTP) during reperfusion of the heart [162].

Studies have also indicated a role of BAs in regulating vascular tension [115,161,163]. For example, activation of endothelial FXR led to downregulation of Endothelin-1 (Et-1) and Il-1 mRNA expression, both potent vasoconstrictive agents [115]. Additionally, FXR activation resulted in upregulation of endothelial nitric oxide synthase (eNOS) [161]. eNOS-derived nitric oxide (NO) has vasodilatory effects and plays an important role in vasomotor tone, VSMC proliferation, platelet aggregation, and the inhibition of lipid oxidation [164]. Mechanistic studies reported an FXR-responsive element in the eNOS promotor [161]. In vitro and in vivo studies demonstrated that, next to FXR, the activation of TGR5 also increased NO production in aortic endothelial cells, and reduced monocyte adhesion and the activation of NF-kB [163]. In addition, DCA treatment improved cardiac function by inhibiting Il-1β expression in the infarcted hearts of mice in cardiomyocytes via the TGR5-NFκB pathway [165]. In line with these results, serum DCA was reduced in acute myocardial patients [165].

FXR activation in VSMCs stimulates the angiotensin system [159,160]. FXR activation in rat aortic smooth muscle cells led to increased expression of angiotensin (Ang) II type 2 receptor (AT2R) and the inhibition of Ang II-mediated extracellular signal-regulated kinase (ERK) activation and cell proliferation [160]. Whereas angiotensin II Type I Receptor (AT1R) activation has vasopressor effects, AT2R activation has vasodilatory roles in the regulation of blood pressure [166]. These vasodilatory effects were impaired in cultured rabbit mesenteric arteries after chronic FXR activation. The FXR agonist GW4064 dose-dependently impaired endothelium relaxation, caused by the decreased sensibility of VSMCs to NO [159]. Thus, these studies suggest different roles in the regulation of blood pressure upon short- and long-term FXR activation.

Whole body FXR/SHP double knockout mice, a model of BA overload, displayed cardiac hypertrophy, bradycardia, and exercise intolerance [167]. Cardiac fatty acid oxidation was reduced in favour of glucose oxidation. Cholestatic mice also have elevated plasma BAs and show increased hypertrophic signalling in the heart, along with suppressed fatty acid oxidation and increased myocardial glycogen content [168]. Interestingly, reducing plasma BAs by intestinal BA sequestration reversed the observed heart dysfunction in FXR/SHP double knockout mice [167]. These results imply a role of serum BAs on heart function. Of note, mechanistic studies have demonstrated that only the deletion of SHP led to hypertrophy and bradycardia [167], suggesting that SHP is an important antihypertrophic regulator.

VDR signalling is also involved in regulating cardiac function [169,170]. Functional VDR is found in t-tubules of cardiac myocytes [169]. T-tubules regulate intracellular calcium flow and allow the heart to contract more forcefully [171]. Cardiac myocytes isolated from VDR knockout mice show increased rates of contraction, cardiac hypertrophy, and systolic and diastolic dysfunction compared with wild-type mice [169]. In addition, epidemiological studies have observed a link between vitamin D deficiency and CVD [172–174]. Vitamin D deficiency in mouse models leads to increased systolic and diastolic blood pressure, high plasma renin and decreased urinary sodium excretion, and increased atherosclerosis in the aortic arch accompanied by increased macrophage/foam cell infiltration with ER stress activation [150]. Interestingly, vitamin D supplementation improved left ventricular (LV) function and reversed LV remodelling in heart failure (HF) patients [170]. As LCA is a potent endogenous ligand of VDR in T cells, it may be speculated whether these specific BAs can also activate VDR in other cell-types, such as myocytes. If this is the case, LCA-VDR signalling in myocytes could be involved in the modulation of heart function.

BAs also bind to "non-classical BAR", such as muscarinic (acetylcholine) receptors (MR) [31]. MRs are G-protein coupled receptors and are classified into different subtypes, namely M_1R–M_5R [31]. Molecular modelling shows strong similarities in the molecular surface of acetylcholine and BAs [175]. In contrast with acetylcholine, BA-specific MR activation depends on hydrophobic interactions. Hence, conjugates of (hydrophobic) secondary BAs showed favourable binding to M_3R, acting as antagonists [175–178]. Tauro-LCA (TLCA) and conjugates of DCA are bound to M_3R and inhibit acetylcholine-induced increases in inositol phosphate formation and MAPK phosphorylation [176,177]. Moreover, tauro-DCA (TDCA) stimulated vasodilatory actions in rat thoracic aortae rings, in part by an NO-, M_3R-dependent mechanism [179]. Additionally, conjugated BAs were found to be partial agonists of M_2R, slowing the contraction rate in neonatal mesenteric vascular muscle cells [180]. Tauro-CA (TCA) was found to interact with M_2R on neonatal rat cardiomyocytes, lowering intracellular cAMP and inducing arrhythmia in cardiac tissue. Arrhythmia was caused by reduced myocardial cell contraction [181]. Importantly, reduced contraction could also be the consequence of cytotoxicity at higher BA concentrations [180]. The effects of BAs on the remaining MRs in the heart remain to be explored.

In addition to binding to BAR and MRs, BAs can interact with ion channels, such as the large conductance Ca^{2+}-activated K^+ (BKCa) channel [182,183] and Na^+/Ca^{2+} exchange protein (NCX) [184]. Natural BAs and synthetic analogues show direct binding to BKCa channels, increasing their activity and leading to the relaxation of rabbit mesenteric artery smooth muscle cells [182]. For example, LCA was found to mediate BKCa channel activation causing relaxation in the small arteries [183]. Moreover, TCA dose-dependently induced arrhythmias in adult human atrial tissue, probably by depolarization of the resting membrane potential, enhancing the NCX current density, and inducing after polarizations [184]. These effects were prevented after NCX inhibition [184].

To summarize, BAs are recognized signalling molecules in modulating cardiovascular function. Their action is mediated by BARs, MRs, or ion channel interaction and mostly leads to vasodilatory effects. The hydrophobicity, polarity, and/or conjugation state of BAs seems to play an important role in the magnitude of receptor/channel activation. Future studies should uncover BA effects in cardiac tissue to exploit BA-mediated targets in order to control heart disease.

7. Bile-Acid-Based Therapies in Cardiovascular Disease

As summarized above, BAs act as important hormonal signalling molecules that modulate cardiovascular function. Given the new insights of BAs in lipid metabolism, immunity, and heart function, strategies for BA-based treatment of CVD can be considered. Targeting BA metabolism could either be indirectly (via gut microbiota) or directly (BAR modulators). In this section, we briefly discuss potential BA-based therapies for the treatment of CVD.

7.1. Indirect Bile-Acid-Based Therapies

Targeting the gut microbiota (and thereby indirectly BA metabolism) can be accomplished via restoration (fecal microbiota transplantation (FMT)) of the gut flora. However, this strategy is unlikely to be suitable for future applications, given that the efficiency of FMT is still limited and inconclusive (due to non-specificity and engraftments problems) and there is risk of pathogenic infection [60]. Other approaches include prebiotics, probiotics, and synbiotics (mixture of pre and probiotics).

Prebiotics do not contain bacteria, but are substrates that are selectively utilized by host microorganisms, and thereby stimulate their growth [60]. Prebiotics (e.g., fibers and oligosaccharides) can be obtained from various sources, including raw oats, soybeans, and several plants [185]. Interestingly, dietary fibers can also regulate BA levels in the gut lumen by binding to conjugated BAs, and then serve as a platform for gut bacteria that possess BA-metabolizing enzymes [186]. A recent clinical trial demonstrated the health benefit of pea fiber in weight control and blood glucose levels [187], two important factors in managing cardiovascular function. Pea fiber can modulate the gut microbiota and alter fecal SCFAs and BAs [188]. For example, the genus *Oscillospira* was reduced after pea fiber treatment, which was negatively correlated with reduced abundances of BAs, such as DCA and isoLCA [188]. This study also reported an overall decrease in fecal BAs, including CA,

CDCA, and DCA. This study indicates the potential role of prebiotics in modulating the microbiota and its metabolites, which contributes to metabolic health benefits in overweight patients (who are at risk for CVD) [187,188].

Probiotics are live bacteria with the entire molecular machinery of living cells, which have been shown various health benefits when administered in adequate amounts [60]. Preclinical and clinical studies have shown anti-atherogenic effects of probiotics (reviewed extensively in [189]). Probiotics (e.g., species of *Lactobaccillus* and *Bifidobacterium*) often contain a BSH activity, which is related to cholesterol-lowering effects in plasma [190]. Some probiotics can also increase BA synthesis via direct inhibitory actions on FXR and SHP [189,191]. In addition, probiotics seem to be associated with anti-inflammatory and anti-oxidative effects and enhanced endothelial function in arteries [192]. Besides the genera *Lactobacillus* and *Bifidobacterium*, lactic acid bacteria has been shown to have both fermentation productivity and antioxidant properties [193]. Moreover, as previously described, derivates from LCA can modulate T cell subsets (Figure 2)—3-oxoLCA and isoLCA inhibit the differentiation of Th17 cells [138,153], while iso-alloLCA increases the differentiation of Tregs [138]. Higher intestinal levels of these secondary BAs may be beneficial in controlling gut immune homeostasis and reducing systemic inflammation [155]. Screening bacteria in human stool samples for their ability to convert LCA into its derivates (3-oxo-, iso-, and iso-allo-LCA) using UPLC-MC traces, is a useful tool to select bacterial species as potential probiotics [153]. Paik et al. revealed that species *Eggerthella lenta* and *Ruminococcus gnavus* are able to convert LCA to 3-oxoLCA and isoLCA, while species *Bacteroides fragilis* convert 3-oxoLCA to isoLCA. The colonization of these species reduced Th17 cell levels in the colonic lamina propria in mice [153]. This study showed promising results of using probiotics to alter BA metabolism and control immune homeostasis in the gut [153]. Whether these effects could also influence heart function or atherosclerotic progression needs to be further investigated. Of note, probiotics do not always show beneficial effects, partly due to heterogeneity in the human population or engraftment problems [60].

To improve colonization, probiotics can be given as a mixture with prebiotics to promote survival and activity of these specific bacterial strains [60]. For example, administration of *Bifidobacterium animalis substrate Lactis 420* together with prebiotic Litesse Ultra polydextrose (consisting of fibers) show synergistic clinical effects on body fat regulation [194]. Synbiotic treatment also led to alterations in gut microbiota composition, BA composition (reduced conjugated BAs), improved gut barrier function, glucose tolerance, and mitigation of inflammation [194]. In addition, evidence from systematic reviews and meta-analysis of clinical trials lend support for synbiotic treatments to reduce plasma LDL and triglyceride levels and increase HDL, probably through reduced intestinal cholesterol absorption and higher BA excretion [195,196]. The effects of synbiotics were more effective when consumed for longer than 8 weeks [196].

7.2. Direct Bile-Acid-Based Therapies

As previously discussed, preclinical studies have shown lipid-lowering and anti-inflammatory effects for FXR and TGR5 agonists. These benefits make the receptors potential candidates for the treatment of CVD. The development of FXR and TGR5 modulators are in its early phase and clinical studies in patients with CVD are lacking. However, obeticholic acid (OCA), which is the first developed small molecule targeting FXR, was evaluated in clinical trials and showed promising results in patients with metabolic diseases [197]. OCA treatment (6 weeks) was well-tolerated and increased insulin sensitivity and reduced markers of liver inflammation and fibrosis in diabetic and non-alcoholic fatty liver disease (NAFLD) patients. As expected, a side effect of OCA treatment is increased serum cholesterol levels due to FXR activation and reduced hepatic CYP7A1 expression [197]. Evaluation of the lipid profile modulation by OCA treatment is currently ongoing. In addition, TGR5 agonists are in development as therapeutics for cardiometabolic diseases. For example, derivatives of CDCA have been tested to treat obesity, insulin sensitivity, and inflammation [198,199]. However, most TGR5 agonists are still in preclinical phase or showed unsatisfactory results in phase I trials [199]. Whether FXR and/or TGR5 agonists are interesting targets as treatment for CVD still needs to be determined.

Of note, mouse models have been widely used in preclinical studies to investigate the mechanistic roles of BAs in CVD. However, the marked species difference in BA metabolism between humans and mice have hampered the interpretation of the results [200–202]. As mentioned before, mice produce rodent-specific MCAs, which account for 35% of the total BA pool in mice [203]. Because of the hydrophilicity of MCAs, the murine BA pool is more hydrophilic than the human BA pool [203]. As BA species have dissimilar affinities for the activation of BARs, differences in BA composition could differentially affect lipid metabolism, immunity, and heart function. The depletion of the rodent-specific enzyme Cyp2c70, which is responsible for the production of MCAs, has been developed to humanize the BA pool composition in mice and is of great value in this respect [200–202]. In addition, the mouse specific enzyme Cyp2a12 was recently discovered [201]. This enzyme rehydroxylates DCA and LCA upon their arrival in the liver, giving rise to CA and MCAs. The proportion of secondary BAs are thus reduced in the pool in mice. Generation of Cyp2a12 knockout mice show the accumulation of DCA in the pool, resembling a more human-like BA pool [201]. These novel humanized BA mouse models will become of critical importance to bridge the gap between laboratory and clinical application, while preserving the benefits of the mouse as a preclinical model.

In addition, ex vivo models are being developed and validated to study atherosclerosis based on the actual human plaque (obtained from endarterectomy surgery) [204–206]. Although the complexity of the disease underscores the necessity to use in vivo (rodent) models to investigate the mechanisms of atherosclerosis, this ex vivo human atherosclerosis model displays major advantages and opportunities. Previous studies have shown that segments of the human plaque can be maintained for 2 weeks in culture [206], providing an attractive ex vivo model to acquire fundamental knowledge and study the impact of novel treatment strategies, such as FXR and TGR5 modulators, to provide a basis for future innovative therapeutics.

8. Conclusions

Evidence is accumulating that BAs exert a much broader range of biological functions than initially recognized, playing a part in mediating lipid metabolism, immunity, and heart function. Given the fact that human studies have shown disturbances in the gut microbiota and BA metabolism in relation to CVD, there are speculations that BA metabolism could be a potential therapeutic target in the future. However, our understanding of the mechanisms is mainly based on preclinical studies, and translational human studies are much needed. Nevertheless, BA-based therapeutics, either indirect or direct, show the potential to reduce the risk factors of CVD. Of note, due to the heterogeneity in the human population, a "one size fits all" approach is expected to not be successful (especially when targeting the gut microbiota). In conclusion, BAs are important signalling molecules in the human body, acting as integrators and modulators of important cardiometabolic pathways. A better and more comprehensive understanding of BAs in cardiovascular responses will be of great importance in the establishment of novel therapeutic approaches to combat CVD.

Funding: This work was funded by the Netherlands Heart Foundation (IN-CONTROL II, CVON grant 2018-27).

Institutional Review Board Statement: Not applicable for studies not involving humans or animals.

Informed Consent Statement: Not applicable for studies not involving humans or animals.

Data Availability Statement: Not applicable.

Acknowledgments: The illustrations in the figures have been created with Biorender.com.

Conflicts of Interest: The authors declare no conflict of interest.

References

1. Cardiovascular Diseases (CVDs). 2021. Available online: https://www.who.int/news-room/fact-sheets/detail/cardiovascular-diseases-(cvds) (accessed on 9 November 2022).
2. Libby, P. The changing landscape of atherosclerosis. *Nature* **2021**, *592*, 524–533. [CrossRef]
3. Virani, S.S.; Alonso, A.; Aparicio, H.J.; Benjamin, E.J.; Bittencourt, M.S.; Callaway, C.W.; Carson, A.P.; Chamberlain, A.M.; Cheng, S.; Delling, F.S.; et al. Heart Disease and Stroke Statistics—2021 Update: A Report From the American Heart Association. *Circulation* **2021**, *143*, E254–E743. [CrossRef]
4. Arora, S.; Stouffer, G.A.; Kucharska-Newton, A.M.; Qamar, A.; Vaduganathan, M.; Pandey, A.; Porterfield, D.S.; Blankstein, R.; Rosamond, W.D.; Bhatt, D.L.; et al. Twenty Year Trends and Sex Differences in Young Adults Hospitalized with Acute Myocardial Infarction: The ARIC Community Surveillance Study. *Circulation* **2019**, *139*, 1047–1056. [CrossRef]
5. Herrington, W.; Lacey, B.; Sherliker, P.; Armitage, J.; Lewington, S. Epidemiology of Atherosclerosis and the Potential to Reduce the Global Burden of Atherothrombotic Disease. *Circ. Res.* **2016**, *118*, 535–546. [CrossRef]
6. Bentzon, J.F.; Otsuka, F.; Virmani, R.; Falk, E. Mechanisms of plaque formation and rupture. *Circ. Res.* **2014**, *114*, 1852–1866. [CrossRef]
7. Mehta, J.L.; PSaldeen, T.G.; Rand, K. Interactive Role of Infection, Inflammation and Traditional Risk Factors in Atherosclerosis and Coronary Artery Disease. *J. Am. Coll. Cardiol.* **1998**, *31*, 1217–1225. [CrossRef]
8. Dhindsa, D.S.; Sandesara, P.B.; Shapiro, M.D.; Wong, N.D. The Evolving Understanding and Approach to Residual Cardiovascular Risk Management. *Front. Cardiovasc. Med.* **2020**, *7*, 88. [CrossRef]
9. Jonsson, A.L.; Bäckhed, F. Role of gut microbiota in atherosclerosis. *Nat. Rev. Cardiol.* **2017**, *14*, 79–87. [CrossRef]
10. Witkowski, M.; Weeks, T.L.; Hazen, S.L. Gut Microbiota and Cardiovascular Disease. *Circ. Res.* **2020**, *127*, 553–570. [CrossRef]
11. Canfora, E.E.; Meex, R.C.R.; Venema, K.; Blaak, E.E. Gut microbial metabolites in obesity, NAFLD and T2DM. *Nat. Rev. Endocrinol.* **2019**, *15*, 261–273. [CrossRef]
12. Tilg, H.; Zmora, N.; Adolph, T.E.; Elinav, E. The intestinal microbiota fuelling metabolic inflammation. *Nat. Rev. Immunol.* **2020**, *20*, 40–54. [CrossRef]
13. Chambers, E.S.; Preston, T.; Frost, G.; Morrison, D.J. Role of Gut Microbiota-Generated Short-Chain Fatty Acids in Metabolic and Cardiovascular Health. *Curr. Nutr. Rep.* **2018**, *7*, 198–206. [CrossRef]
14. Schiattarella, G.G.; Sannino, A.; Toscano, E.; Giugliano, G.; Gargiulo, G.; Franzone, A.; Trimarco, B.; Esposito, G.; Perrino, C. Gut microbe-generated metabolite trimethylamine-N-oxide as cardiovascular risk biomarker: A systematic review and dose-response meta-analysis. *Eur. Heart J.* **2017**, *38*, 2948–2956. [CrossRef]
15. Rodríguez-Morató, J.; Matthan, N.R. Nutrition and Gastrointestinal Microbiota, Microbial-Derived Secondary Bile Acids, and Cardiovascular Disease. *Curr. Atheroscler. Rep.* **2020**, *22*, 47. [CrossRef]
16. Bays, H.E.; Goldberg, R.B. The Forgotten Bile Acid Sequestrants: Is Now a Good Time to Remember? *Am. J. Ther.* **2007**, *14*, 567–580. [CrossRef]
17. Rifkind, B.M. Lipid research clinics coronary primary prevention trial: Results and implications. *Am. J. Cardiol.* **1984**, *54*, 30–34. [CrossRef]
18. Watts, G.F.; Lewis, B.; Brunt, J.N.H.; Lewis, E.S.; Coltart, D.J.; Smith, L.D.R.; Mann, A.V. Effects on coronary artery disease of lipid-lowering diet, or diet plus cholestyramine, in the St Thomas Atherosclerosis regression study. *Lancet* **1992**, *339*, 563–569. [CrossRef]
19. Watts, G.F.; Pang, J.; Chan, D.C.; Brunt, J.N.; Lewis, B. Angiographic progression of coronary atherosclerosis in patients with familial hypercholesterolaemia treated with non-statin therapy: Impact of a fat-modified diet and a resin. *Atherosclerosis* **2016**, *252*, 82–87. [CrossRef]
20. Kuipers, F.; Bloks, V.W.; Groen, A.K. Beyond intestinal soap—Bile acids in metabolic control. *Nat. Rev. Endocrinol.* **2014**, *10*, 488–498. [CrossRef]
21. Porez, G.; Prawitt, J.; Gross, B.; Staels, B. Bile acid receptors as targets for the treatment of dyslipidemia and cardiovascular disease. *J. Lipid Res.* **2012**, *53*, 1723–1737. [CrossRef]
22. Monte, M.J.; Marin, J.J.G.; Antelo, A.; Vazquez-Tato, J. Bile acids: Chemistry, physiology, and pathophysiology. *World J. Gastroenterol.* **2009**, *15*, 804–816. [CrossRef] [PubMed]
23. Russell, D.W. The enzymes, regulation, and genetics of bile acid synthesis. *Annu. Rev. Biochem.* **2003**, *72*, 137–174. [CrossRef] [PubMed]
24. Ridlon, J.M.; Kang, D.J.; Hylemon, P.B. Bile salt biotransformations by human intestinal bacteria. *J. Lipid Res.* **2006**, *47*, 241–259. [CrossRef]
25. Lefebvre, P.; Cariou, B.; Lien, F.; Kuipers, F.; Staels, B. Role of Bile Acids and Bile Acid Receptors in Metabolic Regulation. *Physiol. Rev.* **2009**, *89*, 147–191. [CrossRef] [PubMed]
26. Wahlström, A.; Sayin, S.I.; Marschall, H.U.; Bäckhed, F. Intestinal Crosstalk between Bile Acids and Microbiota and Its Impact on Host Metabolism. *Cell Metab.* **2016**, *24*, 41–50. [CrossRef]
27. Thomas, C.; Pellicciari, R.; Pruzanski, M.; Auwerx, J.; Schoonjans, K. Targeting bile-acid signalling for metabolic diseases. *Nat. Rev. Drug Discov.* **2008**, *24*, 678–693. [CrossRef]
28. Li-Hawkins, J.; Gåfvels, M.; Olin, M.; Lund, E.G.; Andersson, U.; Schuster, G.; Björkhem, I.; Russell, D.W. Cholic acid mediates negative feedback regulation of bile acid synthesis in mice. *J. Clin. Investig.* **2002**, *110*, 1191–1200. [CrossRef]

29. Sayin, S.I.; Wahlström, A.; Felin, J.; Jäntti, S.; Marschall, H.U.; Bamberg, K.; Angelin, B.; Hyötyläinen, T.; Orešič, M.; Bäckhed, F. Gut microbiota regulates bile acid metabolism by reducing the levels of tauro-beta-muricholic acid, a naturally occurring FXR antagonist. *Cell Metab.* **2013**, *17*, 225–235. [CrossRef]
30. Li, R.; Andreu-Sánchez, S.; Kuipers, F.; Fu, J. Gut microbiome and bile acids in obesity-related diseases. *Best Pract. Res. Clin. Endocrinol. Metab.* **2021**, *35*, 101493. [CrossRef]
31. Khurana, S.; Raufman, J.P.; Pallone, T.L. Bile acids regulate cardiovascular function. *Clin. Transl. Sci.* **2011**, *4*, 210–218. [CrossRef]
32. Chiang, J.Y.L. Bile acid metabolism and signaling. *Compr. Physiol.* **2013**, *3*, 1191–1212.
33. Ahmad, T.R.; Haeusler, R.A. Bile acids in glucose metabolism and insulin signalling—Mechanisms and research needs. *Nat. Rev. Endocrinol.* **2019**, *15*, 701–712. [CrossRef] [PubMed]
34. Sun, R.; Xu, C.; Feng, B.; Gao, X.; Liu, Z. Critical roles of bile acids in regulating intestinal mucosal immune responses. *Ther. Adv. Gastroenterol.* **2021**, *14*, 17562848211018098. [CrossRef]
35. Chen, L.; van den Munckhof, I.C.L.; Schraa, K.; ter Horst, R.; Koehorst, M.; van Faassen, M.; van der Ley, C.; Doestzada, M.; Zhernakova, D.V.; Kurilshikov, A.; et al. Genetic and Microbial Associations to Plasma and Fecal Bile Acids in Obesity Relate to Plasma Lipids and Liver Fat Content. *Cell Rep.* **2020**, *33*, 108212. [CrossRef] [PubMed]
36. Kim, G.B.; Miyamoto, C.M.; Meighen, E.A.; Lee, B.H. Cloning and characterization of the bile salt hydrolase genes (bsh) from Bifidobacterium bifidum strains. *Appl. Environ. Microbiol.* **2004**, *70*, 5603–5612. [CrossRef] [PubMed]
37. Coleman, J.P.; Hudson, L.L. Cloning and Characterization of a Conjugated Bile Acid Hydrolase Gene from Clostridium perfringens. *Appl. Environ. Microbiol.* **1995**, *61*, 2514–2520. [CrossRef] [PubMed]
38. Dussurget, O.; Cabanes, D.; Dehoux, P.; Lecuit, M.; Buchrieser, C.; Glaser, P.; Cossart, P.; European Listeria Genome Consortium. Listeria monocytogenes bile salt hydrolase is a PrfA-regulated virulence factor involved in the intestinal and hepatic phases of listeriosis. *Mol. Microbiol.* **2002**, *45*, 1095–1106. [CrossRef] [PubMed]
39. Begley, M.; Sleator, R.D.; Gahan, C.G.M.; Hill, C. Contribution of three bile-associated loci, bsh, pva, and btlB, to gastrointestinal persistence and bile tolerance of Listeria monocytogenes. *Infect. Immun.* **2005**, *73*, 894–904. [CrossRef]
40. Wang, Z.; Zeng, X.; Mo, Y.; Smith, K.; Guo, Y.; Lin, J. Identification and characterization of a bile salt hydrolase from Lactobacillus salivarius for development of novel alternatives to antibiotic growth promoters. *Appl. Environ. Microbiol.* **2012**, *78*, 8795–8802. [CrossRef]
41. Corzo, G.; Gilliland, S.E. Bile salt hydrolase activity of three strains of Lactobacillus acidophilus. *J. Dairy Sci.* **1999**, *82*, 472–480. [CrossRef]
42. Stellwag, E.J.; Hylemon, P.B. Purification and characterization of bile salt hydrolase from *Bacteroides fragilis* subsp. *fragilis*. *Biochim. Biophys. Acta* **1976**, *452*, 165–176. [CrossRef] [PubMed]
43. Song, Z.; Cai, Y.; Lao, X.; Wang, X.; Lin, X.; Cui, Y.; Kalavagunta, P.K.; Liao, J.; Jin, L.; Shang, J.; et al. Taxonomic profiling and populational patterns of bacterial bile salt hydrolase (BSH) genes based on worldwide human gut microbiome. *Microbiome* **2019**, *7*, 9. [CrossRef] [PubMed]
44. Jia, B.; Park, D.; Hahn, Y.; Jeon, C.O. Metagenomic analysis of the human microbiome reveals the association between the abundance of gut bile salt hydrolases and host health. *Gut Microbes.* **2020**, *11*, 1300–1313. [CrossRef] [PubMed]
45. Quinn, R.A.; Melnik A v Vrbanac, A.; Fu, T.; Patras, K.A.; Christy, M.P.; Bodai, Z.; Belda-Ferre, P.; Tripathi, A.; Chung, L.K.; Dorrestein, P.C.; et al. Global chemical effects of the microbiome include new bile-acid conjugations. *Nature* **2020**, *579*, 123–129. [CrossRef] [PubMed]
46. Doden, H.; Sallam, L.A.; Devendran, S.; Ly, L.; Doden, G.; Daniel, S.L.; Alves, J.M.P.; Ridlon, J.M. Metabolism of oxo-bile acids and characterization of recombinant 12α- hydroxysteroid dehydrogenases from bile acid 7α-dehydroxylating human gut bacteria. *Appl. Environ. Microbiol.* **2018**, *84*, e00235-18. [CrossRef] [PubMed]
47. Yoshimoto, S.; Loo, T.M.; Atarashi, K.; Kanda, H.; Sato, S.; Oyadomari, S.; Iwakura, Y.; Oshima, K.; Morita, H.; Hattori, M.; et al. Obesity-induced gut microbial metabolite promotes liver cancer through senescence secretome. *Nature* **2013**, *499*, 97–101. [CrossRef]
48. Bernstein, C.; Holubec, H.; Bhattacharyya, A.K.; Nguyen, H.; Payne, C.M.; Zaitlin, B.; Bernstein, H. Carcinogenicity of deoxycholate, a secondary bile acid. *Arch. Toxicol.* **2011**, *85*, 863–871. [CrossRef]
49. Berr, F.; Kullak-Ublick, G.A.; Paumgartner, G.; Mu¨nzing, W.; Mu¨nzing, M.; Hylemon, P.B. 7a-Dehydroxylating Bacteria Enhance Deoxycholic Acid Input and Cholesterol Saturation of Bile in Patients With Gallstones. *Gastroenterology* **1996**, *111*, 1611–1620. [CrossRef]
50. Brown, J.M.; Hazen, S.L. Microbial modulation of cardiovascular disease. *Nat. Rev. Microbiol.* **2018**, *16*, 171–181. [CrossRef]
51. Kang, D.J.; Ridlon, J.M.; Moore, D.R.; Barnes, S.; Hylemon, P.B. Clostridium scindens baiCD and baiH genes encode stereo-specific 7α/7β-hydroxy-3-oxo-Δ4-cholenoic acid oxidoreductases. *Biochim. Biophys. Acta Mol. Cell. Biol. Lipids* **2008**, *1781*, 16–25. [CrossRef]
52. Urdaneta, V.; Casadesús, J. Interactions between bacteria and bile salts in the gastrointestinal and hepatobiliary tracts. *Front. Med.* **2017**, *4*, 163. [CrossRef] [PubMed]
53. Hofmann, A.F.; Eckmann, L. How bile acids confer gut mucosal protection against bacteria. *Proc. Natl. Acad. Sci. USA* **2006**, *103*, 4333–4334. [CrossRef] [PubMed]
54. Ruiz, L.; Margolles, A.; Sánchez, B. Bile resistance mechanisms in Lactobacillus and Bifidobacterium. *Front. Microbiol.* **2013**, *4*, 396. [CrossRef] [PubMed]

55. Kenny, D.J.; Plichta, D.R.; Shungin, D.; Koppel, N.; Hall, A.B.; Fu, B.; Vasan, R.S.; Shaw, S.Y.; Vlamakis, H.; Balskus, E.P.; et al. Cholesterol Metabolism by Uncultured Human Gut Bacteria Influences Host Cholesterol Level. *Cell. Host Microbe.* **2020**, *28*, 245–257.e6. [CrossRef]
56. Le, H.H.; Lee, M.T.; Besler, K.R.; Comrie, J.M.C.; Johnson, E.L. Characterization of interactions of dietary cholesterol with the murine and human gut microbiome. *Nat. Microbiol.* **2022**, *7*, 1390–1403. [CrossRef] [PubMed]
57. Yao, L.; D'Agostino, G.D.; Park, J.; Hang, S.; Adhikari, A.A.; Zhang, Y.; Li, W.; Avila-Pacheco, J.; Bae, S.; Clish, C.B.; et al. A biosynthetic pathway for the selective sulfonation of steroidal metabolites by human gut bacteria. *Nat. Microbiol.* **2022**, *7*, 1404–1418. [CrossRef]
58. Davidson, M.; Kuo, C.C.; Middaugh, J.P.; Campbell, L.A.; Wang, S.P.; Newman, W.P.; Finley, J.C.; Grayston, J.T. Confirmed Previous Infection With Chlamydia pneumoniae (TWAR) and Its Presence in Early Coronary Atherosclerosis. *Circulation* **1998**, *98*, 628–633. [CrossRef]
59. Koren, O.; Spor, A.; Felin, J.; Fåk, F.; Stombaugh, J.; Tremaroli, V.; Behre, C.J.; Knight, R.; Fagerberg, B.; Ley, R.E.; et al. Human oral, gut, and plaque microbiota in patients with atherosclerosis. *Proc. Natl. Acad. Sci. USA* **2011**, *108* (Suppl. S1), 4592–4598. [CrossRef]
60. Chakaroun, R.M.; Olsson, L.M.; Bäckhed, F. The potential of tailoring the gut microbiome to prevent and treat cardiometabolic disease. *Nat. Rev. Cardiol.* **2022**, *20*, 217–235. [CrossRef]
61. Lanter, B.B.; Sauer, K.; Davies, D.G. Bacteria present in carotid arterial plaques are found as biofilm deposits which may contribute to enhanced risk of plaque rupture. *MBio* **2014**, *5*, e01206-14. [CrossRef]
62. Ott, S.J.; el Mokhtari, N.E.; Musfeldt, M.; Hellmig, S.; Freitag, S.; Rehman, A.; Kühbacher, T.; Nikolaus, S.; Namsolleck, P.; Blaut, M. Detection of diverse bacterial signatures in atherosclerotic lesions of patients with coronary heart disease. *Circulation* **2006**, *113*, 929–937. [CrossRef] [PubMed]
63. Zheng, Y.Y.; Wu, T.T.; Liu, Z.Q.; Li, A.; Guo, Q.Q.; Ma, Y.Y.; Zhang, Z.L.; Xun, Y.L.; Zhang, J.C.; Wang, W.-R.; et al. Gut Microbiome-Based Diagnostic Model to Predict Coronary Artery Disease. *J. Agric. Food Chem.* **2020**, *68*, 3548–3557. [CrossRef] [PubMed]
64. Karlsson, F.H.; Fåk, F.; Nookaew, I.; Tremaroli, V.; Fagerberg, B.; Petranovic, D.; Bäckhed, F.; Nielsen, J. Symptomatic atherosclerosis is associated with an altered gut metagenome. *Nat. Commun.* **2012**, *3*, 1245. [CrossRef] [PubMed]
65. Zhu, Q.; Gao, R.; Zhang, Y.; Pan, D.; Zhu, Y.; Zhang, X.; Yang, R.; Jiang, R.; Xu, Y.; Qin, H.; et al. Dysbiosis signatures of gut microbiota in coronary artery disease. *Physiol. Genom.* **2018**, *50*, 893–903. [CrossRef]
66. Yin, J.; Liao, S.X.; He, Y.; Wang, S.; Xia, G.H.; Liu, F.T.; Zhu, J.J.; You, C.; Chen, Q.; Zhou, L.; et al. Dysbiosis of gut microbiota with reduced trimethylamine-n-oxide level in patients with large-artery atherosclerotic stroke or transient ischemic attack. *J. Am. Heart Assoc.* **2015**, *4*, e002699. [CrossRef]
67. Jie, Z.; Xia, H.; Zhong, S.L.; Feng, Q.; Li, S.; Liang, S.; Zhong, H.; Liu, Z.; Gao, Y.; Zhao, H.; et al. The gut microbiome in atherosclerotic cardiovascular disease. *Nat. Commun.* **2017**, *8*, 845. [CrossRef]
68. Toya, T.; Corban, M.T.; Marrietta, E.; Horwath, I.E.; Lerman, L.O.; Murray, J.A.; Lerman, A. Coronary artery disease is associated with an altered gut microbiome composition. *PLoS ONE* **2020**, *15*, e0227147. [CrossRef]
69. Liu, H.; Chen, X.; Hu, X.; Niu, H.; Tian, R.; Wang, H.; Pang, H.; Jiang, L.; Qiu, B.; Chen, X.; et al. Alterations in the gut microbiome and metabolism with coronary artery disease severity. *Microbiome* **2019**, *7*, 68. [CrossRef]
70. Feng, Q.; Liu, Z.; Zhong, S.; Li, R.; Xia, H.; Jie, Z.; Wen, B.; Chen, X.; Yan, W.; Fan, Y.; et al. Integrated metabolomics and metagenomics analysis of plasma and urine identified microbial metabolites associated with coronary heart disease. *Sci. Rep.* **2016**, *6*, 22525. [CrossRef]
71. Yoshida, N.; Emoto, T.; Yamashita, T.; Watanabe, H.; Hayashi, T.; Tabata, T.; Hoshi, N.; Hatano, N.; Ozawa, G.; Sasaki, N.; et al. Bacteroides vulgatus and Bacteroides dorei Reduce Gut Microbial Lipopolysaccharide Production and Inhibit Atherosclerosis. *Circulation* **2018**, *138*, 2486–2498. [CrossRef]
72. Verhaar, B.J.H.; Prodan, A.; Nieuwdorp, M.; Muller, M. Gut microbiota in hypertension and atherosclerosis: A review. *Nutrients* **2020**, *12*, 2982. [CrossRef]
73. Tamanai-Shacoori, Z.; Smida, I.; Bousarghin, L.; Loreal, O.; Meuric, V.; Fong, S.B.; Bonnaure-Mallet, M.; Jolivet-Gougeon, A. *Roseburia* spp.: A marker of health? *Future Microbiol.* **2017**, *12*, 157–170. [CrossRef] [PubMed]
74. Ferreira-Halder, C.V.; de Sousa Faria, A.V.; Andrade, S.S. Action and function of Faecalibacterium prausnitzii in health and disease. *Best Pract. Res. Clin. Gastroenterol.* **2017**, *31*, 643–648. [CrossRef] [PubMed]
75. Bhatti, M.A.; Frank, M.O. Veillonella parvula Meningitis: Case Report and Review of Veillonella Infections. *Clin. Infect. Dis.* **2000**, *31*, 839–840. [CrossRef] [PubMed]
76. Kesavalu, L.; Lucas, A.R.; Verma, R.K.; Liu, L.; Dai, E.; Sampson, E.; Progulske-Fox, A. Increased atherogenesis during Streptococcus mutans infection in ApoE-null mice. *J. Dent. Res.* **2012**, *91*, 255–260. [CrossRef]
77. Chalmers, N.I.; Palmer, R.J.; Cisar, J.O.; Kolenbrander, P.E. Characterization of a *Streptococcus* sp.-*Veillonella* sp. community micromanipulated from dental plaque. *J. Bacteriol.* **2008**, *190*, 8145–8154. [CrossRef]
78. Binder, C.J.; Hörkkö, S.; Dewan, A.; Chang, M.K.; Kieu, E.P.; Goodyear, C.S.; Silverman, G.J. Pneumococcal vaccination decreases atherosclerotic lesion formation molecular mimicry between Streptococcus pneumoniae and oxidized LDL. *Nat. Med.* **2003**, *9*, 736–743. [CrossRef]

79. Devlin, A.S.; Fischbach, M.A. A biosynthetic pathway for a prominent class of microbiota-derived bile acids. *Nat. Chem. Biol.* **2015**, *11*, 685–690. [CrossRef]
80. Barlow, G.M.; Yu, A.; Mathur, R. Role of the gut microbiome in obesity and diabetes mellitus. *Nutr. Clin. Pract.* **2015**, *30*, 787–797. [CrossRef]
81. Tilg, H.; Moschen, A.R.; Kaser, A. Obesity and the Microbiota. *Gastroenterology* **2009**, *136*, 1476–1483. [CrossRef]
82. Huang, K.; Liu, C.; Peng, M.; Su, Q.; Liu, R.; Guo, Z.; Chen, S.; Li, Z.; Chang, G. Glycoursodeoxycholic acid ameliorates atherosclerosis and alters gut microbiota in apolipoprotein e–deficient mice. *J. Am. Heart Assoc.* **2021**, *10*, e019820. [CrossRef] [PubMed]
83. Chen, P.B.; Black, A.S.; Sobel, A.L.; Zhao, Y.; Mukherjee, P.; Molparia, B.; Moore, N.E.; Aleman Muench, G.R.; Wu, J.; Chen, W.; et al. Directed remodeling of the mouse gut microbiome inhibits the development of atherosclerosis. *Nat. Biotechnol.* **2020**, *38*, 1288–1297. [CrossRef] [PubMed]
84. Liu, H.; Tian, R.; Wang, H.; Feng, S.; Li, H.; Xiao, Y.; Luan, X.; Zhang, Z.; Shi, N.; Niu, H.; et al. Gut microbiota from coronary artery disease patients contributes to vascular dysfunction in mice by regulating bile acid metabolism and immune activation. *J. Transl. Med.* **2020**, *18*, 382. [CrossRef] [PubMed]
85. Brandsma, E.; Kloosterhuis, N.J.; Koster, M.; Dekker, D.C.; Gijbels, M.J.; van der Velden, S.; Ríos-Morales, M.; van Faassen, M.J.; Loreti, M.G.; de Bruin, A.; et al. A Proinflammatory Gut Microbiota Increases Systemic Inflammation and Accelerates Atherosclerosis. *Circ. Res.* **2019**, *124*, 94–100. [CrossRef]
86. Chong Nguyen, C.; Duboc, D.; Rainteau, D.; Sokol, H.; Humbert, L.; Seksik, P.; Bellino, A.; Abdoul, H.; Bouazza, N.; Treluyer, J.M.; et al. Circulating bile acids concentration is predictive of coronary artery disease in human. *Sci. Rep.* **2021**, *11*, 22661. [CrossRef]
87. Li, W.; Shu, S.; Cheng, L.; Hao, X.; Wang, L.; Wu, Y.; Yuan, Z.; Zhou, J. Fasting serum total bile acid level is associated with coronary artery disease, myocardial infarction and severity of coronary lesions. *Atherosclerosis* **2020**, *292*, 193–200. [CrossRef]
88. Steiner, C.; Othman, A.; Saely, C.H.; Rein, P.; Drexel, H.; von Eckardstein, A.; Rentsch, K.M. Bile acid metabolites in serum: Intraindividual variation and associations with coronary heart disease, metabolic syndrome and diabetes mellitus. *PLoS ONE* **2011**, *6*, e25006. [CrossRef]
89. Neale, G.; Lewis, B.; Weaver, V.; Panveliwalla, D. Serum bile acids in liver disease. *Gut* **1971**, *12*, 145–152. [CrossRef]
90. Sirtori, C.R. The pharmacology of statins. *Pharmacol. Res.* **2014**, *88*, 3–11. [CrossRef]
91. Li, Y.; Zhang, D.; He, Y.; Chen, C.; Song, C.; Zhao, Y.; Bai, Y.; Wang, Y.; Pu, J.; Chen, J.; et al. Investigation of novel metabolites potentially involved in the pathogenesis of coronary heart disease using a UHPLC-QTOF/MS-based metabolomics approach. *Sci. Rep.* **2017**, *7*, 15357. [CrossRef]
92. Charach, G.; Grosskopf, I.; Rabinovich, A.; Shochat, M.; Weintraub, M.; Rabinovich, P. The association of bile acid excretion and atherosclerotic coronary artery disease. *Ther. Adv. Gastroenterol.* **2011**, *4*, 95–101. [CrossRef] [PubMed]
93. Charach, G.; Karniel, E.; Novikov, I.; Galin, L.; Vons, S.; Grosskopf, I.; Charach, L. Reduced bile acid excretion is an independent risk factor for stroke and mortality: A prospective follow-up study. *Atherosclerosis* **2020**, *293*, 79–85. [CrossRef]
94. Charach, G.; Argov, O.; Geiger, K.; Charach, L.; Rogowski, O.; Grosskopf, I. Diminished bile acids excretion is a risk factor for coronary artery disease: 20-year follow up and long-term outcome. *Ther. Adv. Gastroenterol.* **2018**, *11*, 1756283X17743420. [CrossRef] [PubMed]
95. Watanabe, M.; Houten, S.M.; Wang, L.; Moschetta, A.; Mangelsdorf, D.J.; Heyman, R.A.; Moore, D.D.; Auwerx, J. Bile acids lower triglyceride levels via a pathway involving FXR, SHP, and SREBP-1c. *J. Clin. Investig.* **2004**, *113*, 1408–1418. [CrossRef] [PubMed]
96. Wiegman, C.H.; Bandsma, R.H.J.; Ouwens, M.; Van Der Sluijs, F.H.; Havinga, R.; Boer, T.; Reijngoud, D.J.; Romijn, J.A.; Kuipers, F. Hepatic VLDL Production in ob/ob Mice is Not Stimulated by Massive De Novo Lipogenesis but is Less Sensitive to the Suppressive Effects of Insulin. *Diabetes* **2003**, *52*, 1081–1089. [CrossRef] [PubMed]
97. Schoonjans, K.; Staels, B.; Auwerx', J. I review Role of the peroxisome proliferator-activated receptor (PPAR) in mediating the effects of fibrates and fatty acids on gene expression. *J. Lipid Res.* **1996**, *37*, 907–925. [CrossRef]
98. Horton, J.D.; Goldstein, J.L.; Brown, M.S. SREBPs: Activators of the complete program of cholesterol and fatty acid synthesis in the liver. *J. Clin. Investig.* **2002**, *109*, 1125–1131. [CrossRef]
99. Preidis, G.A.; Kim, K.H.; Moore, D.D. Nutrient-sensing nuclear receptors PPAR and FXR control liver energy balance. *J. Clin. Investig.* **2017**, *127*, 1193–1201. [CrossRef]
100. Srivastava, R.A.K.; Jahagirdar, R.; Azhar, S.; Sharma, S.; Bisgaier, C.L. Peroxisome proliferator-activated receptor-α selective ligand reduces adiposity, improves insulin sensitivity and inhibits atherosclerosis in LDL receptor-deficient mice. *Mol. Cell. Biochem.* **2006**, *285*, 35–50. [CrossRef]
101. Lee, Y.; Kim, B.R.; Kang, G.H.; Lee, G.J.; Park, Y.J.; Kim, H.; Jang, H.C.; Choi, S.H. The Effects of PPAR Agonists on Atherosclerosis and Nonalcoholic Fatty Liver Disease in ApoE−/−FXR−/− Mice. *Endocrinol. Metab.* **2021**, *36*, 1243–1253. [CrossRef]
102. Torra, I.P.; Claudel, T.; Duval, C.; Kosykh, V.; Fruchart, J.C.; Staels, B. Bile acids induce the expression of the human peroxisome proliferator-activated receptor α gene via activation of the farnesoid X receptor. *Mol. Endocrinol.* **2003**, *17*, 259–272. [CrossRef] [PubMed]
103. Staudinger, J.L.; Goodwin, B.; Jones, S.A.; Hawkins-Brown, D.; Mackenzie, K.I.; Latour, A.; Liu, Y.; Klaassen, C.D.; Brown, K.K.; Reinhard, J.; et al. The nuclear receptor PXR is a lithocholic acid sensor that protects against liver toxicity. *Proc. Natl. Acad. Sci. USA* **2001**, *98*, 3369–3374. [CrossRef] [PubMed]

104. Zhou, C.; King, N.; Chen, K.Y.; Breslow, J.L. Activation of PXR induces hypercholesterolemia in wild-type and accelerates atherosclerosis in apoE deficient mice. *J. Lipid Res.* **2009**, *50*, 2004–2013. [CrossRef]
105. Sui, Y.; Xu, J.; Rios-Pilier, J.; Zhou, C. Deficiency of PXR decreases atherosclerosis in apoE-deficient mice. *J. Lipid Res.* **2011**, *52*, 1652–1659. [CrossRef] [PubMed]
106. Park, Y.M. CD36, a scavenger receptor implicated in atherosclerosis. *Exp. Mol. Med.* **2014**, *46*, e99. [CrossRef]
107. Zurkinden, L.; Sviridov, D.; Vogt, B.; Escher, G. Downregulation of Cyp7a1 by Cholic Acid and Chenodeoxycholic Acid in Cyp27a1/ApoE Double Knockout Mice: Differential Cardiovascular Outcome. *Front. Endocrinol.* **2020**, *11*, 586980. [CrossRef] [PubMed]
108. Song, P.; Rockwell, C.E.; Cui, J.Y.; Klaassen, C.D. Individual bile acids have differential effects on bile acid signaling in mice. *Toxicol. Appl. Pharmacol.* **2015**, *283*, 57–64. [CrossRef] [PubMed]
109. Li, R.; Palmiotti, A.; de Vries, H.D.; Hovingh, M.V.; Koehorst, M.; Mulder, N.L.; Zhang, Y.; Kats, K.; Bloks, V.W.; Fu, J.; et al. Low production of 12α-hydroxylated bile acids prevents hepatic steatosis in Cyp2c70-/- mice by reducing fat absorption. *J. Lipid Res.* **2021**, *62*, 100134. [CrossRef]
110. Bhat, B.G.; Rapp, S.R.; Beaudry, J.A.; Napawan, N.; Butteiger, D.N.; Hall, K.A.; Null, C.L.; Luo, Y.; Keller, B.T. Inhibition of ileal bile acid transport and reduced atherosclerosis in apoE-/- mice by SC-435. *J. Lipid Res.* **2003**, *44*, 1614–1621. [CrossRef]
111. Bishop-Bailey, D.; Walsh, D.T.; Warner, T.D. Expression and activation of the farnesoid X receptor in the vasculature. *Proc. Natl. Acad. Sci. USA* **2004**, *101*, 3668–3673. [CrossRef]
112. Mencarelli, A.; Renga, B.; Distrutti, E.; Fiorucci, S. Antiatherosclerotic effect of farnesoid X receptor. *Am. J. Physiol. Heart Circ. Physiol.* **2009**, *296*, 272–281. [CrossRef] [PubMed]
113. Kawamata, Y.; Fujii, R.; Hosoya, M.; Harada, M.; Yoshida, H.; Miwa, M.; Fukusumi, S.; Habata, Y.; Itoh, T.; Shintani, Y.; et al. A G protein-coupled receptor responsive to bile acids. *J. Biol. Chem.* **2003**, *278*, 9435–9440. [CrossRef]
114. Li, Y.T.Y.; Swales, K.E.; Thomas, G.J.; Warner, T.D.; Bishop-Bailey, D. Farnesoid X receptor ligands inhibit vascular smooth muscle cell inflammation and migration. *Arterioscler. Thromb. Vasc. Biol.* **2007**, *27*, 2606–2611. [CrossRef] [PubMed]
115. He, F.; Li, J.; Mu, Y.; Kuruba, R.; Ma, Z.; Wilson, A.; Alber, S.; Jiang, Y.; Stevens, T.; Watkins, S.; et al. Downregulation of endothelin-1 by farnesoid X receptor in vascular endothelial cells. *Circ. Res.* **2006**, *98*, 192–199. [CrossRef]
116. Pols, T.W.H.; Nomura, M.; Harach, T.; lo Sasso, G.; Oosterveer, M.H.; Thomas, C.; Rizzo, G.; Gioiello, A.; Adorini, L.; Pellicciari, R.; et al. TGR5 activation inhibits atherosclerosis by reducing macrophage inflammation and lipid loading. *Cell. Metab.* **2011**, *14*, 747–757. [CrossRef]
117. Miyazaki-Anzai, S.; Masuda, M.; Kohno, S.; Levi, M.; Shiozaki, Y.; Keenan, A.L.; Miyazaki, M. Simultaneous inhibition of FXR and TGR5 exacerbates atherosclerotic formation. *J. Lipid Res.* **2018**, *59*, 1709–1713. [CrossRef] [PubMed]
118. Jadhav, K.; Xu, Y.; Xu, Y.; Li, Y.; Xu, J.; Zhu, Y.; Adorini, L.; Lee, Y.K.; Kasumov, T.; Yin, L.; et al. Reversal of metabolic disorders by pharmacological activation of bile acid receptors TGR5 and FXR. *Mol. Metab.* **2018**, *9*, 131–140. [CrossRef]
119. Guo, G.L.; Santamarina-Fojo, S.; Akiyama, T.E.; Amar, M.J.A.; Paigen, B.J.; Brewer, B.; Gonzalez, F.J. Effects of FXR in foam-cell formation and atherosclerosis development. *Biochim. Biophys. Acta Mol. Cell. Biol. Lipids* **2006**, *1761*, 1401–1409. [CrossRef]
120. Zhang, Y.; Wang, X.; Vales, C.; Lee, F.Y.; Lee, H.; Lusis, A.J.; Edwards, P.A. FXR deficiency causes reduced atherosclerosis in Ldlr-/- mice. *Arterioscler. Thromb. Vasc. Biol.* **2006**, *26*, 2316–2321. [CrossRef]
121. Hartman, H.B.; Gardell, S.J.; Petucci, C.J.; Wang, S.; Krueger, J.A.; Evans, M.J. Activation of farnesoid X receptor prevents atherosclerotic lesion formation in LDLR−/− and apoE−/− mice. *J. Lipid Res.* **2009**, *50*, 1090–1100. [CrossRef] [PubMed]
122. Xu, Y.; Li, F.; Zalzala, M.; Xu, J.; Gonzalez, F.J.; Adorini, L.; Lee, Y.L.; Yin, L.; Zhang, Y. Farnesoid X Receptor Activation Increases Reverse Cholesterol Transport by Modulating Bile Acid Composition and Cholesterol Absorption in Mice. *Hepatology* **2016**, *64*, 1072–1085. [CrossRef]
123. Wu, Q.; Sun, L.; Hu, X.; Wang, X.; Xu, F.; Chen, B.; Liang, X.; Xia, J.; Wang, P.; Aibara, D.; et al. Suppressing the intestinal farnesoid X receptor/sphingomyelin phosphodiesterase 3 axis decreases atherosclerosis. *J. Clin. Investig.* **2021**, *131*, e143865. [CrossRef]
124. Frostegård, J. Immunity, atherosclerosis and cardiovascular disease. *BMC Med.* **2013**, *11*, 117. [CrossRef]
125. Zhong, Z.; Zhang, H.; Xu, T.; Hao, J.; Chen, X.; Sun, S.; Yang, J.; Sun, J.; Lin, H.; Guo, H. Identification and verification of immune-related biomarkers and immune infiltration in diabetic heart failure. *Front. Cardiovasc. Med.* **2022**, *9*, 931066. [CrossRef]
126. Hayward, S.L.; Bautista-Lopez, N.; Suzuki, K.; Atrazhev, A.; Dickie, P.; Elliott, J.F. CD4 T Cells Play Major Effector Role and CD8 T Cells Initiating Role in Spontaneous Autoimmune Myocarditis of HLA-DQ8 Transgenic IAb Knockout Nonobese Diabetic Mice 1. *J. Immunol.* **2006**, *176*, 7715–7725. [CrossRef] [PubMed]
127. Kumar, V.; Prabhu, S.D.; Bansal, S.S. CD4+ T-lymphocytes exhibit biphasic kinetics post-myocardial infarction. *Front. Cardiovasc. Med.* **2022**, *9*, 992653. [CrossRef] [PubMed]
128. Lu, Y.; Xia, N.; Cheng, X. Regulatory T Cells in Chronic Heart Failure. *Front. Immunol.* **2021**, *12*, 732794. [CrossRef]
129. Rurik, J.G.; Aghajanian, H.; Epstein, J.A. Immune Cells and Immunotherapy for Cardiac Injury and Repair. *Circ. Res.* **2021**, *128*, 1766–1779. [CrossRef]
130. Rosenzweig, R.; Kumar, V.; Gupta, S.; Bermeo-Blanco, O.; Stratton, M.S.; Gumina, R.J.; Bansal, S.S. Estrogen Receptor-β Agonists Modulate T-Lymphocyte Activation and Ameliorate Left Ventricular Remodeling during Chronic Heart Failure. *Circ. Heart Fail.* **2022**, *15*, E008997. [CrossRef] [PubMed]

131. Ridker, P.M.; Everett, B.M.; Thuren, T.; MacFadyen, J.G.; Chang, W.H.; Ballantyne, C.; Fonseca, F.; Nicolau, J.; Koenig, W.; Anker, S.D.; et al. Antiinflammatory Therapy with Canakinumab for Atherosclerotic Disease. *N. Engl. J. Med.* **2017**, *377*, 1119–1131. [CrossRef]
132. Guan, B.; Tong, J.; Hao, H.; Yang, Z.; Chen, K.; Xu, H.; Wang, A. Bile acid coordinates microbiota homeostasis and systemic immunometabolism in cardiometabolic diseases. *Acta Pharm. Sin. B* **2022**, *12*, 2129–2149. [CrossRef]
133. Gadaleta, R.M.; van Erpecum, K.J.; Oldenburg, B.; Willemsen, E.C.L.; Renooij, W.; Murzilli, S.; Klomp, L.W.J.; Siersema, P.D.; Schipper, M.E.I.; Danese, S.; et al. Farnesoid X receptor activation inhibits inflammation and preserves the intestinal barrier in inflammatory bowel disease. *Gut* **2011**, *60*, 463–472. [CrossRef]
134. Sorribas, M.; Jakob, M.O.; Yilmaz, B.; Li, H.; Stutz, D.; Noser, Y.; de Gottardi, A.; Moghadamrad, S.; Hassan, M.; Albillos, A.; et al. FXR modulates the gut-vascular barrier by regulating the entry sites for bacterial translocation in experimental cirrhosis. *J. Hepatol.* **2019**, *71*, 1126–1140. [CrossRef]
135. Úbeda, M.; Lario, M.; Muñoz, L.; Borrero, M.J.; Rodríguez-Serrano, M.; Sánchez-Díaz, A.M.; Del Campo, R.; Lledó, L.; Pastor, Ó.; García-Bermejo, L.; et al. Obeticholic acid reduces bacterial translocation and inhibits intestinal inflammation in cirrhotic rats. *J. Hepatol.* **2016**, *64*, 1049–1057. [CrossRef]
136. Inagaki, T.; Moschetta, A.; Lee, Y.K.; Peng, L.; Zhao, G.; Downes, M.; Yu, R.T.; Shelton, J.M.; Richardson, J.A.; Repa, J.J.; et al. Regulation of antibacterial defense in the small intestine by the nuclear bile acid receptor. *Proc. Natl. Acad. Sci. USA* **2006**, *103*, 3920–3925. [CrossRef]
137. Yao, B.; He, J.; Yin, X.; Shi, Y.; Wan, J.; Tian, Z. The protective effect of lithocholic acid on the intestinal epithelial barrier is mediated by the vitamin D receptor via a SIRT1/Nrf2 and NF-κB dependent mechanism in Caco-2 cells. *Toxicol. Lett.* **2019**, *316*, 109–118. [CrossRef]
138. Hang, S.; Paik, D.; Yao, L.; Kim, E.; Jamma, T.; Lu, J.; Ha, S.; Nelson, B.N.; Kelly, S.P.; Wu, L.; et al. Bile acid metabolites control TH17 and Treg cell differentiation. *Nature* **2019**, *576*, 143–148. [CrossRef] [PubMed]
139. Qin, P.; Tang, X.; Elloso, M.M.; Harnish, D.C. Bile acids induce adhesion molecule expression in endothelial cells through activation of reactive oxygen species, NF-B, and p38. *Am. J. Physiol. Heart Circ. Physiol.* **2006**, *291*, 741–747. [CrossRef] [PubMed]
140. Lu, S.; Guo, S.; Hu, F.; Guo, Y.; Yan, L.; Ma, W.; Wang, Y.; Wei, Y.; Zhang, Z.; Wang, Z.; et al. The associations between the polymorphisms of Vitamin D receptor and coronary artery disease. *Medicine* **2016**, *95*, e3467. [CrossRef] [PubMed]
141. Szeto, F.L.; Reardon, C.A.; Yoon, D.; Wang, Y.; Wong, K.E.; Chen, Y.; Kong, J.; Liu, S.Q.; Thadhani, R.; Getz, G.S.; et al. Vitamin D receptor signaling inhibits atherosclerosis in Mice. *Mol. Endocrinol.* **2012**, *26*, 1091–1101. [CrossRef]
142. Latic, N.; Erben, R.G. Vitamin D and cardiovascular disease, with emphasis on hypertension, atherosclerosis, and heart failure. *Int. J. Mol. Sci.* **2020**, *21*, 6483. [CrossRef]
143. Stenman, L.K.; Holma, R.; Eggert, A.; Korpela, R. A novel mechanism for gut barrier dysfunction by dietary fat: Epithelial disruption by hydrophobic bile acids. *Am. J. Physiol. Gastrointest. Liver Physiol.* **2013**, *304*, 227–234. [CrossRef]
144. Zhao, S.; Gong, Z.; Du, X.; Tian, C.; Wang, L.; Zhou, J.; Xu, C.; Chen, Y.; Cai, W.; Wu, J. Deoxycholic acid-mediated sphingosine-1-phosphate receptor 2 signaling exacerbates DSS-induced colitis through promoting cathepsin b release. *J. Immunol. Res.* **2018**, *2018*, 2481418. [CrossRef]
145. Wang, L.; Gong, Z.; Zhang, X.; Zhu, F.; Liu, Y.; Jin, C.; Du, X.; Xu, C.; Chen, Y.; Cai, W.; et al. Gut microbial bile acid metabolite skews macrophage polarization and contributes to high-fat diet-induced colonic inflammation. *Gut Microbes* **2020**, *12*, 1819155. [CrossRef]
146. Wang, J.; Xu, T.; Xu, M. Roles and Mechanisms of TGR5 in the Modulation of CD4+ T Cell Functions in Myocardial Infarction. *J. Cardiovasc. Transl. Res.* **2022**, *15*, 350–359. [CrossRef] [PubMed]
147. Guo, C.; Xie, S.; Chi, Z.; Zhang, J.; Liu, Y.; Zhang, L.; Zheng, M.; Zhang, X.; Xia, D.; Ke, Y.; et al. Bile Acids Control Inflammation and Metabolic Disorder through Inhibition of NLRP3 Inflammasome. *Immunity* **2016**, *45*, 802–816. [CrossRef]
148. Yang, C.S.; Kim, J.J.; Kim, T.S.; Lee, P.Y.; Kim, S.Y.; Lee, H.M.; Shin, D.M.; Nguyen, L.T.; Lee, M.S.; Jin, H.S.; et al. Small heterodimer partner interacts with NLRP3 and negatively regulates activation of the NLRP3 inflammasome. *Nat. Commun.* **2015**, *6*, 6115. [CrossRef] [PubMed]
149. Oh, J.; Riek, A.E.; Darwech, I.; Funai, K.; Shao, J.S.; Chin, K.; Sierra, O.L.; Carmeliet, G.; Ostlund, R.E.; Bernal-Mizrachi, C. Deletion of macrophage vitamin D receptor promotes insulin resistance and monocyte cholesterol transport to accelerate atherosclerosis in mice. *Cell. Rep.* **2015**, *10*, 1872–1886. [CrossRef]
150. Weng, S.; Sprague, J.E.; Oh, J.; Riek, A.E.; Chin, K.; Garcia, M.; Bernal-Mizrachi, C. Vitamin D Deficiency Induces High Blood Pressure and Accelerates Atherosclerosis in Mice. *PLoS ONE* **2013**, *8*, e54625. [CrossRef]
151. Dong, B.; Zhou, Y.; Wang, W.; Scott, J.; Kim, K.; Sun, Z.; Guo, Q.; Lu, Y.; Gonzales, N.M.; Wu, H.; et al. Vitamin D Receptor Activation in Liver Macrophages Ameliorates Hepatic Inflammation, Steatosis, and Insulin Resistance in Mice. *Hepatology* **2020**, *71*, 1559–1574. [CrossRef] [PubMed]
152. Schiering, C.; Krausgruber, T.; Chomka, A.; Fröhlich, A.; Adelmann, K.; Wohlfert, E.A.; Pott, J.; Griseri, T.; Bollrath, J.; Hegazy, A.N.; et al. The alarmin IL-33 promotes regulatory T-cell function in the intestine. *Nature* **2014**, *513*, 564–568. [CrossRef] [PubMed]
153. Paik, D.; Yao, L.; Zhang, Y.; Bae, S.; D'Agostino, G.D.; Zhang, M.; Kim, E.; Franzosa, E.A.; Avila-Pacheco, J.; Bisanz, J.E.; et al. Human gut bacteria produce TH17-modulating bile acid metabolites. *Nature* **2022**, *603*, 907–912. [CrossRef] [PubMed]

154. Li, W.; Hang, S.; Fang, Y.; Bae, S.; Zhang, Y.; Zhang, M.; Wang, G.; McCurry, M.D.; Bae, M.; Paik, D.; et al. A bacterial bile acid metabolite modulates Treg activity through the nuclear hormone receptor NR4A1. *Cell. Host Microbe.* **2021**, *29*, 1366–1377.e9. [CrossRef] [PubMed]
155. Song, X.; Sun, X.; Oh, S.F.; Wu, M.; Zhang, Y.; Zheng, W.; Geva-Zatorsky, N.; Jupp, R.; Mathis, D.; Benoist, C.; et al. Microbial bile acid metabolites modulate gut RORγ+ regulatory T cell homeostasis. *Nature* **2020**, *577*, 410–415. [CrossRef] [PubMed]
156. Campbell, C.; McKenney, P.T.; Konstantinovsky, D.; Isaeva, O.I.; Schizas, M.; Verter, J.; Mai, C.; Jin, W.B.; Guo, C.J.; Violante, S.; et al. Bacterial metabolism of bile acids promotes generation of peripheral regulatory T cells. *Nature* **2020**, *581*, 475–479. [CrossRef] [PubMed]
157. Lahore, G.F.; Raposo, B.; Lagerguist, M.; Ohlsson, C.; Sabatier, P.; Xu, B.; Aoun, M.; James, J.; Cai, X.; Zubarev, R.A.; et al. VDR polymorphisms regulated t cells and t cell dependent inflammatory disease. *Proc. Natl. Acad. Sci. USA* **2020**, *117*, 24986–24997. [CrossRef]
158. Pu, J.; Yuan, A.; Shan, P.; Gao, E.; Wang, X.; Wang, Y.; Lau, W.B.; Koch, W.; Ma, X.L.; He, B. Cardiomyocyte-expressed farnesoid-X-receptor is a novel apoptosis mediator and contributes to myocardial ischaemia/reperfusion injury. *Eur. Heart J.* **2013**, *34*, 1834–1845. [CrossRef]
159. Kida, T.; Murata, T.; Hori, M.; Ozaki, H. Chronic stimulation of farnesoid X receptor impairs nitric oxide sensitivity of vascular smooth muscle. *Am. J. Physiol. Heart Circ. Physiol.* **2009**, *296*, H195–H201. [CrossRef]
160. Zhang, Q.; He, F.; Kuruba, R.; Gao, X.; Wilson, A.; Li, J.; Billiar, T.J.; Pitt, B.R.; Xie, W.; Li, S. FXR-mediated regulation of angiotensin type 2 receptor expression in vascular smooth muscle cells. *Cardiovasc. Res.* **2008**, *77*, 560–569. [CrossRef]
161. Li, J.; Wilson, A.; Kuruba, R.; Zhang, Q.; Gao, X.; He, F.; Zhang, L.M.; Pitt, B.R.; Xie, W.; Li, S. FXR-mediated regulation of eNOS expression in vascular endothelial cells. *Cardiovasc. Res.* **2008**, *77*, 169–177. [CrossRef]
162. Rajesh, K.G.; Suzuki, R.; Maeda, H.; Yamamoto, M.; Yutong, X.; Sasaguri, S. Hydrophilic bile salt ursodeoxycholic acid protects myocardium against reperfusion injury in a PI3K/Akt dependent pathway. *J. Mol. Cell. Cardiol.* **2005**, *39*, 766–776. [CrossRef]
163. Thromb, A.; Biol, V. Bile Acid Receptor TGR5 Agonism Induces NO Production and Reduces Monocyte Adhesion in Vascular Endothelial Cells. *Arterioscler. Thromb. Vasc. Biol.* **2013**, *33*, 1663–1669.
164. Sandoo, A.; Veldhuijzen Van Zanten, J.J.C.S.; Metsios, G.S.; Carroll, D.; Kitas, G.D. The Endothelium and Its Role in Regulating Vascular Tone. *Open Cardiovasc. Med. J.* **2010**, *4*, 302–312. [CrossRef]
165. Wang, J.; Zhang, J.; Lin, X.; Wang, Y.; Wu, X.; Yang, F.; Gao, W.; Zhang, Y.; Sun, J.; Jiang, C.; et al. DCA-TGR5 signaling activation alleviates inflammatory response and improves cardiac function in myocardial infarction. *J. Mol. Cell. Cardiol.* **2021**, *151*, 3–14. [CrossRef]
166. Mehta, P.K.; Griendling, K.K.; Griendling, K.K. Angiotensin II cell signaling: Physiological and pathological effects in the cardiovascular system. *Am. J. Physiol. Cell. Physiol.* **2007**, *292*, 82–97. [CrossRef]
167. Desai, M.S.; Mathur, B.; Eblimit, Z.; Vasquez, H.; Taegtmeyer, H.; Karpen, S.J.; Penny, D.J.; Moore, D.D.; Anakk, S. Bile Acid Excess Induces Cardiomyopathy and Metabolic Dysfunctions in the Heart. *Hepatology* **2017**, *65*, 189–201. [CrossRef]
168. Desai, M.S.; Shabier, Z.; Taylor, M.; Lam, F.; Thevananther, S.; Kosters, A.; Karpen, S.J. Hypertrophic cardiomyopathy and dysregulation of cardiac energetics in a mouse model of biliary fibrosis. *Hepatology* **2010**, *51*, 2097–2107. [CrossRef]
169. Tishkoff, D.X.; Nibbelink, K.A.; Holmberg, K.H.; Dandu, L.; Simpson, R.U. Functional vitamin D receptor (VDR) in the T-tubules of cardiac myocytes: VDR knockout cardiomyocyte contractility. *Endocrinology* **2008**, *149*, 558–564. [CrossRef]
170. Rodriguez, A.J.; Mousa, A.; Ebeling, P.R.; Scott, D.; de Courten, B. Effects of Vitamin D supplementation on inflammatory markers in heart failure: A systematic review and meta-analysis of randomized controlled trials. *Sci. Rep.* **2018**, *8*, 1169. [CrossRef]
171. Ibrahim, M.; Gorelik, J.; Yacoub, M.H.; Terracciano, C.M. The structure and function of cardiac t-tubules in health and disease. *Proc. R. Soc. B Biol. Sci.* **2011**, *278*, 2714–2723. [CrossRef]
172. Giovannucci, E.; Liu, Y.; Hollis, B.W.; Rimm, E.B. 25-Hydroxyvitamin D and Risk of Myocardial Infarction in Men a Prospective Study. *Arch. Intern. Med.* **2008**, *168*, 1174–1180. [CrossRef]
173. Dobnig, H.; Pilz, S.; Scharnagl, H.; Renner, W.; Seelhorst, U.; Wellnitz, B.; Kinkeldei, J.; Boehm, B.O.; Weihrauch, G.; Maerz, W. Independent Association of Low Serum 25-Hydroxyvitamin D and 1,25-Dihydroxyvitamin D Levels With All-Cause and Cardiovascular Mortality. *Arch. Intern. Med.* **2008**, *168*, 1340–1349. [CrossRef]
174. Wang, T.J.; Pencina, M.J.; Booth, S.L.; Jacques, P.F.; Ingelsson, E.; Lanier, K.; Benjamin, E.J.; D'Agostino, R.B.; Wolf, M.; Vasan, R.S. Vitamin D deficiency and risk of cardiovascular disease. *Circulation* **2008**, *117*, 503–511. [CrossRef]
175. Raufman, J.P.; Chen, Y.; Cheng, K.; Compadre, C.; Compadre, L.; Zimniak, P. Selective interaction of bile acids with muscarinic receptors: A case of molecular mimicry. *Eur. J. Pharmacol.* **2002**, *457*, 77–84. [CrossRef]
176. Raufman, J.P.; Chen, Y.; Zimniak, P.; Cheng, K. Deoxycholic Acid Conjugates are Muscarinic Cholinergic Receptor Antagonists. *Pharmacology* **2002**, *65*, 215–221. [CrossRef]
177. Cheng, K.; Khurana, S.; Chen, Y.; Kennedy, R.H.; Zimniak, P.; Raufman, J.P. Lithocholylcholine, a bile acid/acetylcholine hybrid, is a muscarinic receptor antagonist. *J. Pharmacol. Exp. Ther.* **2002**, *303*, 29–35. [CrossRef]
178. Shah, N.; Khurana, S.; Cheng, K.; Raufman, J.P. Muscarinic receptors and ligands in cancer. *Am. J. Physiol. Cell. Physiol.* **2009**, *296*, 221–232. [CrossRef]
179. Khurana, S.; Yamada, M.; Wess, J.; Kennedy, R.H.; Raufman, J.P. Deoxycholyltaurine-induced vasodilation of rodent aorta is nitric oxide- and muscarinic M3 receptor-dependent. *Eur. J. Pharmacol.* **2005**, *517*, 103–110. [CrossRef]

180. Ibrahim, E.; Diakonov, I.; Arunthavarajah, D.; Swift, T.; Goodwin, M.; McIlvride, S.; Nikolova, V.; Williamson, C.; Gorelik, J. Bile acids and their respective conjugates elicit different responses in neonatal cardiomyocytes: Role of Gi protein, muscarinic receptors and TGR5. *Sci. Rep.* **2018**, *8*, 7110. [CrossRef]
181. Sheikh Abdul Kadir, S.H.; Miragoli, M.; Abu-Hayyeh, S.; Moshkov A v Xie, Q.; Keitel, V.; Nikolaev, V.O.; Williamson, C.; Gorelik, J. Bile acid-induced arrhythmia is mediated by muscarinic M2 receptors in neonatal rat cardiomyocytes. *PLoS ONE* **2010**, *5*, e9689. [CrossRef]
182. Dopico, A.M.; Walsh J v Singer, J.J. Natural Bile Acids and Synthetic Analogues Modulate Large Conductance Ca 2-activated K (BK Ca) Channel Activity in Smooth Muscle Cells. *J. Gen. Physiol.* **2002**, *119*, 251–273. [CrossRef]
183. Bukiya, A.N.; Liu, J.; Toro, L.; Dopico, A.M. β1 (KCNMB1) subunits mediate lithocholate activation of large-conductance Ca2+-activated K+ channels and dilation in small, resistance-size arteries. *Mol. Pharmacol.* **2007**, *72*, 359–369. [CrossRef]
184. Rainer, P.P.; Primessnig, U.; Harenkamp, S.; Doleschal, B.; Wallner, M.; Fauler, G.; Stojakovic, T.; Wachter, R.; Yates, A.; Groschner, K.; et al. Bile acids induce arrhythmias in human atrial myocardium-implications for altered serum bile acid composition in patients with atrial fibrillation. *Heart* **2013**, *99*, 1685–1692. [CrossRef]
185. Oniszczuk, A.; Oniszczuk, T.; Gancarz, M.; Szymá Nska, J.; Hamaguchi, M. Role of Gut Microbiota, Probiotics and Prebiotics in the Cardiovascular Diseases. *Molecules* **2021**, *26*, 1172. [CrossRef]
186. Bui, T.P.N.; de Vos, W.M. Next-generation therapeutic bacteria for treatment of obesity, diabetes, and other endocrine diseases. *Best Pract. Res. Clin. Endocrinol. Metab.* **2021**, *35*, 101504. [CrossRef]
187. Lambert, J.E.; Parnell, J.A.; Tunnicliffe, J.M.; Han, J.; Sturzenegger, T.; Reimer, R.A. Consuming yellow pea fiber reduces voluntary energy intake and body fat in overweight/obese adults in a 12-week randomized controlled trial. *Clin. Nutr.* **2017**, *36*, 126–133. [CrossRef]
188. Mayengbam, S.; Lambert, J.E.; Parnell, J.A.; Tunnicliffe, J.M.; Nicolucci, A.C.; Han, J.; Sturzenegger, T.; Shearer, J.; Mickiewicz, B.; Vogel, H.J.; et al. Impact of dietary fiber supplementation on modulating microbiota–host–metabolic axes in obesity. *J. Nutr. Biochem.* **2019**, *64*, 228–236. [CrossRef]
189. O'Morain, V.L.; Ramji, D.P. The Potential of Probiotics in the Prevention and Treatment of Atherosclerosis. *Mol. Nutr. Food Res.* **2020**, *64*, e1900797. [CrossRef]
190. Begley, M.; Hill, C.; Gahan, C.G.M. Bile salt hydrolase activity in probiotics. *Appl. Environ. Microbiol.* **2006**, *72*, 1729–1738. [CrossRef]
191. Hassan, A.; Din, A.U.; Zhu, Y.; Zhang, K.; Li, T.; Wang, Y.; Luo, Y.; Wang, G. Updates in understanding the hypocholesterolemia effect of probiotics on atherosclerosis. *Appl. Microbiol. Biotechnol.* **2019**, *103*, 5993–6006. [CrossRef]
192. Mahdavi-Roshan, M.; Salari, A.; Kheirkhah, J.; Ghorbani, Z. The Effects of Probiotics on Inflammation, Endothelial Dysfunction, and Atherosclerosis Progression: A Mechanistic Overview. *Heart Lung Circ.* **2022**, *31*, e45–e71. [CrossRef]
193. Kim, K.T.; Kim, J.W.; Kim S il Kim, S.; Nguyen, T.H.; Kang, C.H. Antioxidant and anti-inflammatory effect and probiotic properties of lactic acid bacteria isolated from canine and feline feces. *Microorganisms* **2021**, *9*, 1971. [CrossRef]
194. Hibberd, A.A.; Yde, C.C.; Ziegler, M.L.; Honoré, A.H.; Saarinen, M.T.; Lahtinen, S.; Stahl, B.; Jensen, H.M.; Stenman, L.K. Probiotic or synbiotic alters the gut microbiota and metabolism in a randomised controlled trial of weight management in overweight adults. *Benef. Microbes* **2019**, *10*, 121–135. [CrossRef]
195. Ghorbani, Z.; Kazemi, A.; UPBartolomaeus, T.; Martami, F.; Noormohammadi, M.; Salari, A.; Löber, U.; Balou, H.A.; Forslund, S.K.; Mahdavi-Roshan, M. The effect of probiotic and synbiotic supplementation on lipid parameters among patients with cardiometabolic risk factors: A systematic review and meta-analysis of clinical trials. *Cardiovasc. Res.* **2022**. [CrossRef]
196. Hadi, A.; Ghaedi, E.; Khalesi, S.; Pourmasoumi, M.; Arab, A. Effects of synbiotic consumption on lipid profile: A systematic review and meta-analysis of randomized controlled clinical trials. *Eur. J. Nutr.* **2020**, *59*, 2857–2874. [CrossRef]
197. Mudaliar, S.; Henry, R.R.; Sanyal, A.J.; Morrow, L.; Marschall, H.U.; Kipnes, M.; Adorini, L.; Sciacca, C.I.; Clopton, P.; Castelloe, E.; et al. Efficacy and safety of the farnesoid x receptor agonist Obeticholic acid in patients with type 2 diabetes and nonalcoholic fatty liver disease. *Gastroenterology* **2013**, *145*, 574–582.e1. [CrossRef]
198. Hodge, R.J.; Lin, J.; Vasist Johnson, L.S.; Gould, E.P.; Bowers, G.D.; Nunez, D.J. Safety, pharmacokinetics, and pharmacodynamic effects of a selective TGR5 Agonist, SB-756050, in Type 2 Diabetes. *Clin. Pharmacol. Drug Dev.* **2013**, *2*, 213–222. [CrossRef]
199. Bhimanwar, R.S.; Mittal, A. TGR5 agonists for diabetes treatment: A patent review and clinical advancements (2012-present). *Expert Opin. Ther. Pat.* **2022**, *32*, 191–209. [CrossRef]
200. de Boer, J.F.; de Vries, H.D.; Palmiotti, A.; Li, R.; Doestzada, M.; Hoogerland, J.A.; Fu, J.; La Rose, A.M.; Westerterp, M.; Mulder, N.L.; et al. Cholangiopathy and Biliary Fibrosis in Cyp2c70-Deficient Mice Are Fully Reversed by Ursodeoxycholic Acid. *Cell. Mol. Gastroenterol. Hepatol.* **2021**, *11*, 1045–1069. [CrossRef]
201. Honda, A.; Miyazaki, T.; Iwamoto, J.; Hirayama, T.; Morishita, Y.; Monma, T.; Ueda, H.; Mizuno, S.; Sugiyama, F.; Takahashi, S.; et al. Regulation of bile acid metabolism in mouse models with hydrophobic bile acid composition. *J. Lipid Res.* **2020**, *61*, 54–69. [CrossRef]
202. Straniero, S.; Laskar, A.; Savva, C.; Härdfeldt, J.; Angelin, B.; Rudling, M. Of mice and men: Murine bile acids explain species differences in the regulation of bile acid and cholesterol metabolism. *J. Lipid Res.* **2020**, *61*, 480–491. [CrossRef]
203. Yang, J.; Palmiotti, A.; Kuipers, F. Emerging roles of bile acids in control of intestinal functions. *Curr. Opin. Clin. Nutr. Metab. Care* **2021**, *24*, 127–133. [CrossRef]

204. Kritikou, E.; Depuydt, M.A.C.; de Vries, M.R.; Mulder, K.E.; Govaert, A.M.; Smit, M.D.; van Duijn, J.; Foks, A.C.; Wezel, A.; Smeets, H.J.; et al. Flow Cytometry-Based Characterization of mast Cells in Human Atherosclerosis. *Cells* **2019**, *8*, 334. [CrossRef]
205. Erbel, C.; Okuyucu, D.; Akhavanpoor, M.; Zhao, L.; Wangler, S.; Hakimi, M.; Doesch, A.; Dengler, T.J.; Katus, H.A.; Gleissner, C.A. A human ex vivo atherosclerotic plaque model to study lesion biology. *J. Vis. Exp.* **2014**, *87*, e50542.
206. Lebedeva, A.; Vorobyeva, D.; Vagida, M.; Ivanova, O.; Felker, E.; Fitzgerald, W.; Danilova, N.; Gontarenko, V.; Shpektor, A.; Vasilieva, E.; et al. Ex vivo culture of human atherosclerotic plaques: A model to study immune cells in atherogenesis. *Atherosclerosis* **2017**, *267*, 90–98. [CrossRef]

nutrients

Review

Measurement of Serum Low Density Lipoprotein Cholesterol and Triglyceride-Rich Remnant Cholesterol as Independent Predictors of Atherosclerotic Cardiovascular Disease: Possibilities and Limitations

Dieter Lütjohann [1,*], Hans-Ulrich Klör [2] and Frans Stellaard [1]

[1] Institute of Clinical Chemistry and Clinical Pharmacology, University Hospital Bonn, 53127 Bonn, Germany
[2] Department of Internal Medicine III, University of Gießen, 35392 Gießen, Germany
* Correspondence: to: dieter.luetjohann@ukbonn.de

Abstract: The serum low density lipoprotein cholesterol (LDL-C) concentration is the dominant clinical parameter to judge a patient's risk of developing cardiovascular disease (CVD). Recent evidence supports the theory that cholesterol in serum triglyceride-rich lipoproteins (TRLs) contributes significantly to the atherogenic risk, independent of LDL-C. Therefore, combined analysis of both targets and adequate treatment may improve prevention of CVD. The validity of TRL-C calculation is solely dependent on the accuracy of the LDL-C measurement. Direct measurement of serum LDL- C is more accurate than established estimation procedures based upon Friedewald, Martin–Hopkins, or Sampson equations. TRL-C can be easily calculated as total C minus high density lipoprotein C (HDL-C) minus LDL-C. Enhanced serum LDL-C or TRL-C concentrations require different therapeutic approaches to lower the atherogenic lipoprotein C. This review describes the different atherogenic lipoproteins and their possible analytical properties and limitations.

Keywords: chylomicrons; remnants; cholesterol; Friedewald equation; Martin–Hopkins equation; Sampson equation

Citation: Lütjohann, D.; Klör, H.-U.; Stellaard, F. Measurement of Serum Low Density Lipoprotein Cholesterol and Triglyceride-Rich Remnant Cholesterol as Independent Predictors of Atherosclerotic Cardiovascular Disease: Possibilities and Limitations. *Nutrients* **2023**, *15*, 2202. https://doi.org/10.3390/nu15092202

Academic Editor: Arrigo Cicero

Received: 12 April 2023
Revised: 27 April 2023
Accepted: 3 May 2023
Published: 5 May 2023

1. Introduction

Increased serum total cholesterol (TC) is associated with an increased risk of developing atherosclerotic cardiovascular disease (ASCVD) [1–3]. High serum low density lipoprotein C (LDL-C) is generally considered as the predominant cause of ASCVD progression [4–6]. For decades, serum LDL-C has been the main target to be lowered with statins, either alone or in combination with ezetimibe [7]. Recently, bempedoic acid has been introduced as a possible replacement for statins when these cannot be tolerated by the patient [8]. In a considerable number of patients, LDL-C lowering targets are not reached [9–11]. Additional LDL-C lowering therapies have been developed, such as inhibition of proprotein convertase subtilisin/kexin type 9 (PCSK9), which intercellularly degrades the LDL-receptor (LDLR) [12,13]. LDLR is the carrier protein that enables LDL to enter the cell. LDL is formed by very low density lipoprotein (VLDL) through intermediate density lipoprotein (IDL) after progressive removal of triglyceride (TG) by lipoprotein lipase (LPL) and hepatic lipase (HL) [14] (Figure 1). VLDL, IDL, and LDL differ in particle size and density [15].

It is important to realize that VLDL, IDL, and LDL are density classes only composed of differently sized particles. Since VLDL secretion and LPL activity may vary over time, a large variety of lipoprotein particles with a range of TG and C contents is simultaneously present in serum [13,16–18]. These particles, which are larger and more TG-rich than LDL, are called TG-rich lipoproteins (TRLs). Additionally, highly TG-rich chylomicrons (CM) are continuously released from the intestine two and four hours after a fat-rich meal [19–22]. CMs are gradually converted to chylomicron remnants (CMR) via loss of TG by LPL. The sum of C in VLDL derived TRLs and CMs plus CMRs may be called total TRL-C. Recently, the measurement of remnant cholesterol (remnant C) has attracted considerable attention and its abundance has been proven to be associated with the development of various types of atherosclerotic events independent of LDL-C [23]. However, the definition of remnant-C and the determination of serum remnant-C concentration are subjects of discussion. According to the generally applied calculation, remnant C equals TRL-C. TRLs contain lipoprotein particles in between VLDL and LDL including IDL in combination with CMR [24]. All of these remnants and LDL are supposedly atherogenic. Graduation in atherogenicity cannot be clearly defined, despite the fact that small, highly dense LDL particles ("small-dense LDL") are more atherogenic than larger ones [25]. In addition, while no clear general description of atherogenic lipoproteins can be provided, the presence of apolipoprotein B (ApoB) as carrier protein is at least characteristic. ApoB is represented as ApoB100 in VLDL-derived particles and ApoB48 for chylomicron-derived particles. This separates atherogenic particles from high density lipoprotein (HDL) which carries apolipoprotein A1 (ApoA1) as the unifying apolipoprotein. Additionally, the cholesterol ester (CE) content adds further to the atherogenicity [26–28]. Different risk indicators have been introduced to predict atherosclerotic risk. Apart from clinical indicators, such as obesity, smoking and/or diabetes, these include the serum concentrations of LDL-C, VLDL-remnant C (TRL-C), non-HDL-C and ApoB. ApoB has been introduced as a risk indicator based on the knowledge that CMs, CMRs, TRLs, and LDLs contain one ApoB molecule per particle and that the number of particles may be more conclusive than the concentration of lipoprotein C [29,30]. To date, clinicians tend to rely on LDL-C as the best marker for pro-atherogenic lipoproteins and HDL-C as the marker for anti-atherogenic lipoproteins [31–33]. It should be realized that HDL particles exchange CE with ApoB-containing particles in exchange for TG [34] mediated by cholesterol ester transfer protein (CETP). Reduction of CETP activity is considered a potential target for increased reversed cholesterol transport [35]. Thus, depending on CETP activity the TG and CE proportions in HDL and ApoB lipoproteins may vary. HDL mainly delivers phospholipids and CE to the liver, whereas ApoB remnants and LDL are taken up to some extent by extrahepatic tissues, but predominantly by the liver via the LDL-receptor (LDLR) and the LDL-receptor related protein (LRP) [36]. One particular lipoprotein is the lipoprotein (a) (Lp(a)). It represents the densest ApoB-100-containing particle with a density higher than LDL. The measured LDL-C contains C originating from Lp(a). The lipoprotein lipid metabolism is presented in Figure 2.

VLDL is formed in the liver and transports TG and CE into the blood. It is gradually converted into LDL via intermediate formation of IDL. VLDL remnants and IDL may partly return to the liver before being converted to LDL. LDL is extracted into extrahepatic cells, but predominantly into the liver. CMs are produced in the enterocyte, transporting TG and CE from absorbed fatty acids from the diet and FC from the diet and from bile. They are converted to CMRs by the action of LPL and then delivered to the liver. The TG content of VLDL is provided by TG derived from CMR, fatty acids (FA) synthesized in the liver and FA taken up from blood (Figures 2 and 3). The hepatic C pool is composed of C derived from extracted HDL, LDL, VLDL remnants, IDL and CMR as well as from synthesized C. Hepatic TG is secreted into the blood in VLDL particles. Hepatic C is secreted in VLDL as CE, secreted into bile as FC, and as bile acids. The distribution of C divided over these three fluxes is largely unknown. HDL-C appears to be dominantly secreted into bile [37,38] and converted to bile acids [39]. The flux distribution may be dependent on the hepatic C concentration. In this review we critically evaluate the proposed predictive markers according to the characteristics of various lipoproteins, their metabolism, and their analysis or calculation procedures.

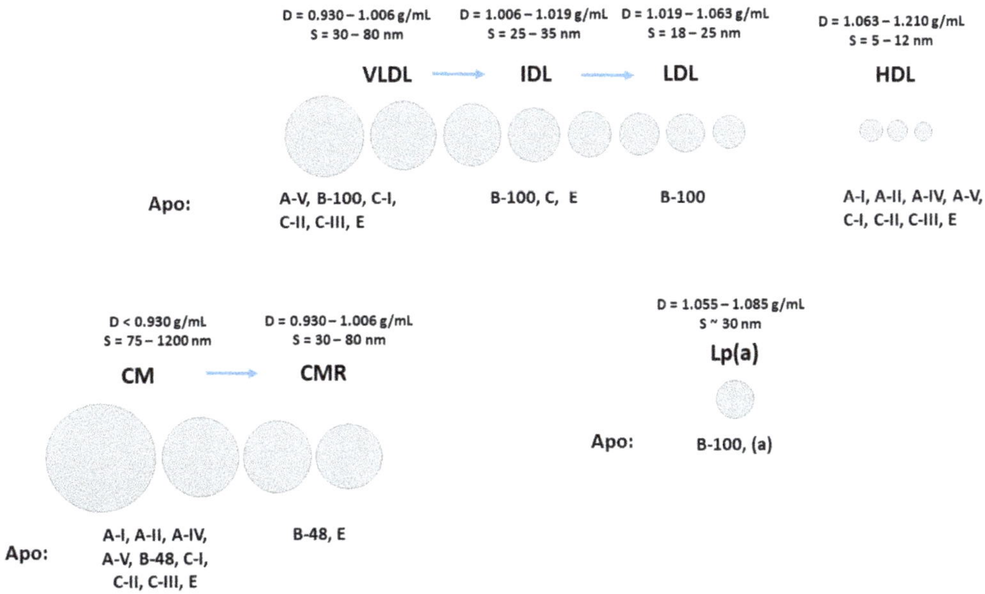

Figure 1. Lipoprotein distribution originating from the liver, i.e., very low density lipoprotein (VLDL), intermediate density lipoprotein (IDL), low density lipoprotein (LDL), lipoprotein (a) [Lp(a)], high density lipoprotein (HDL), and the intestine, i.e., chylomicron(CM) and CM remnants (R). The characteristics related to density (D; g/mL), size (S; nm) and lipoprotein composition are presented. Apolipoprotein (Apo) [15].

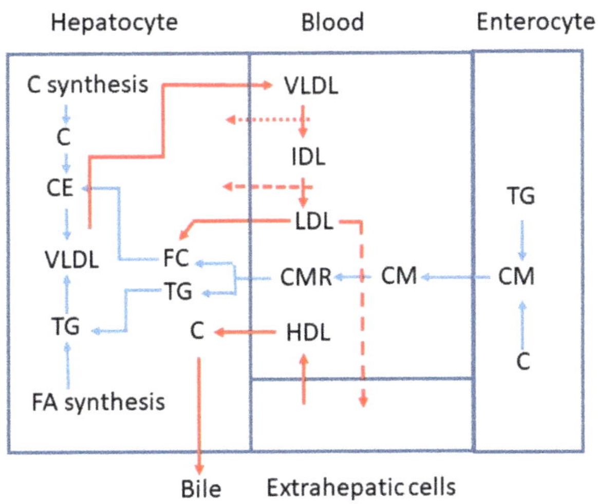

Figure 2. Formation of very low density lipoproteins (VLDL) and chylomicron (CM) lipid cores, secretion of VLDL and CM particles into blood and the blood metabolism of lipoproteins. Cholesterol, C; cholesterol ester, CE; triglyceride (TG); fatty acid (FA); intermediate density lipoproteins (IDL); chylomicron remnants (CMR); high density lipoproteins (HDL).

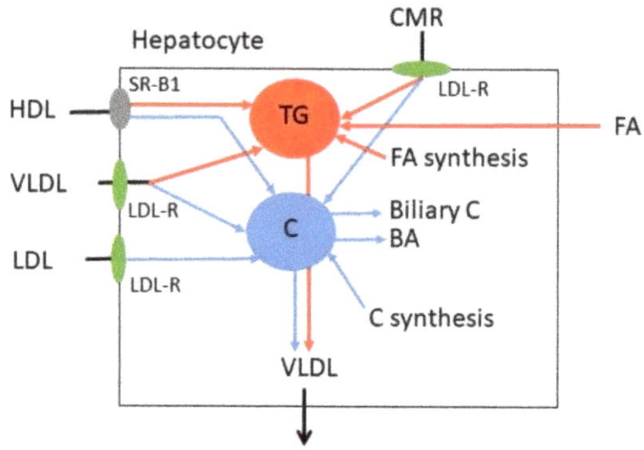

Figure 3. Hepatic TG and C metabolism. High density lipoprotein, HDL; very low density lipoprotein, VLDL; low density lipoprotein, LDL; scavenger receptor, class B type 1, SR-B1; LDL receptor, LDL-R; cholesterol, C; bile acids, BA; chylomicron remnants, CMR; fatty acid, FA.

2. Atherogenic Lipoprotein C Concentrations as Indicators of Enhanced Risk for Atherosclerosis Development

The pro-atherogenic character of lipoproteins is not yet fully understood. In principle, HDL particles are considered anti-atherogenic. Accepted atherogenic characteristics are the presence of ApoBs (ApoB100 and ApoB48), reduced size and increased density beyond undefined limits, and a high load of CE. To actually cause atherosclerosis, the turnover of lipoproteins must be delayed. This increases the exposure of the arterial wall to the toxic lipoproteins, thus enhancing the atherosclerotic process. A delay of turnover may be caused by reduced activity of LPL, HL, and/or LDLR. During the day, the presence of intestinal-derived CMs and CMRs is maximized with fat-rich meals. Therefore, lipoprotein analysis in fasting blood will not represent the daily exposure to atherogenic lipoproteins. The time to transport CMs, convert CM to CMR, and transport CMR to the liver affect the time-dependent contribution of CM-derived lipoproteins in the postprandial phase. Thus, the residence time of TRLs in plasma determines an individual's atherosclerotic risk. Patients with obesity, diabetes mellitus, kidney disease, and/or familial history of cardiovascular disease have an increased risk of developing ASCVD. This risk must be conveyed to the patient and followed by treatments for risk reduction. After cardiovascular events have been treated, the residual risk for repeated events must be considered. Continuous treatment is required to reduce any residual risk. The degree of atherogenicity is associated with the level of circulating C-rich lipoproteins (LDLs, remnants, Lp(a)). Initially, serum TC was used as a predictive marker and serum C lowering therapies were developed as the treatment of choice. Thereafter, the target became the lowering of serum LDL-C concentration. In the last decade, initiatives were undertaken to extend the focus to the other potentially atherogenic lipoprotein C, by applying non-HDL-C as the predictive marker [40]. Non-HDL-C is calculated as TC minus HDL-C and contains C present in all potentially atherogenic lipoproteins, including LDLs, remnants and Lp(a). Additional evidence identified the C content of the TRLs, i.e., all ApoB containing lipoproteins, excluding LDL-C, were an additional risk marker in patients with normal LDL-C levels or those with a sufficiently reduced LDL-C level following C-lowering treatment [41]. A third suggestion has been to measure total ApoB, i.e., ApoB-100 plus ApoB-48 as a risk marker [42,43]. Lp (a) is an independent risk marker in specific patients and should always be measured during the original diagnosis. Next, the validity of various markers is discussed.

3. LDL-C

For many years, a high serum LDL-C concentration has been considered the central factor associated with an increased risk for development of atherosclerosis and cardiovascular events [1,3,5]. The definition of LDL-C is the lipoprotein consisting of Apo B-100 as the associated apolipoprotein (20% of total content), with a density of 1.019 to 1.063 g/mL and a diameter of 20 to 25 nm. The lipid core contains 12% TG and 59% cholesteryl esters. Small LDL particles are more atherogenic than larger ones [44]. The LDL-C concentration is determined in every hospital all over the world. For the most accurate analysis of LDL-C, serum must be treated with ultracentrifugation [45,46] in order to isolate the LDL density fraction for C analysis. This is the official reference method, which is considered optimally selective. A second approach to separate lipoproteins is via electrophoresis [47] and a third approach is via nuclear magnetic resonance (NMR) [48]. In addition, liquid mass spectrometry methods are now being developed [49]. However, for routine clinical laboratories, these techniques are time consuming, laborious, and expensive. In 1972, Friedewald established a simple estimation procedure [50]. In fasting serum, total C is predominantly composed of HDL-C, LDL-C, and VLDL-C (plus IDL-C). From measurement of VLDL-C and VLDL-TG after isolation with ultracentrifugation, Friedewald found that the mean TG/C ratio in fasting serum was 5.0 and considered this number to reflect the ratio in the healthy population. Thus, he expressed the LDL-C calculation as: LDL-C = TC minus HDL-C minus TG/5. Obviously, the Friedewald formula means that the calculated LDL-C strongly depends on the TG concentration. TG is bound to the above-mentioned lipoproteins and is not limited to VLDL. VLDL contains about 55% TG. CMs carry 88% TG, which decreases during conversion to CMR. The TG content of lipoproteins is dependent on VLDL and chylomicron production and release as well as LPL activity. Furthermore, CE is transferred from HDL into LDL in exchange for TG. From VLDL, CM, and CMR, low LPL activity leads to high TRLs. It has been observed that the Friedewald equation starts to lose accuracy when the LDL-C concentration is low (<1.8 mmol/L, <70 mg/dL) and the TG concentration is high (>1.69 mmol/L, >150 mg/dL). Generally, it is acceptable to use the Friedewald equation when LDL-C > 1.4 mmol/L (>54 mg/dL) and the TG concentration < 4.5 mmol/L (<400 mg/dL) and Lp(a) is within the reference range [51]. This makes the equation unsuitable in patients with hypertriglyceridemia and mixed hyperlipidemia and increased Lp(a) levels. Furthermore, recent literature advocates lipoprotein profiling to be performed in the postprandial phase, when TG is at higher levels and CMs and CMRs are present at variable higher concentrations [52–55]. Recent evidence has been provided indicating that production and release of chylomicron particles are slow processes. Fat is temporarily stored as liquid droplets in the intestinal cells [56]. The supply of chylomicrons and as a consequence chylomicron remnants are spread out over time. This way, the body is protected against an excessive load of fat after meal consumption. Depending on the dietary fat intake, the time point of the last meal, and the delay of CM secretion, the presence of CMs and CMRs in fasting serum may become relevant. Approaches have been made to improve the weakness of the Friedewald equation. The most accepted improvements are the approaches of Martin [57] and Sampson [58]. They correct the LDL-C value according to the combination of TG and HDL-C in the sample. Both approaches extend the range of TG concentrations at least up to 9 mmol/L (800 mg/dL). The Martin–Hopkins approach also provides more accuracy at low LDL-C concentrations. However, direct measurement of LDL-C is highly recommended. Many commercially available direct homogeneous LDL-C assays are on the market [59], enabling a rapid and selective analysis. These assays are based on the fact that they exclude HDL, VLDL, and CMs from the C measurement. However, CM- and VLDL remnants with reduced TG content—if present in fasting serum—may potentially interfere with the measurement. Another potentially interfering factor is LP(a). LP(a) equals LDL in size and density and may be included in the measurement of LDL-C if present.

4. TG Rich Lipoprotein C (TRL-C) or Remnant C

Almost three decades ago it was indicated that TG enriched lipoproteins (TRL-C) in serum correlate with the severity of coronary artery disease [60]. Serum TRL-C, also called remnant C, received much attention as an atherogenic component associated with cardiovascular events and independent of serum LDL-C. In 2022, over 2000 hits were obtained when searching the Pubmed data base for "remnant cholesterol". A small extract is shown here [61–63]. It has frequently been proposed that "remnant C" should be determined in the individual patient at risk and that remnant-C lowering therapies need to be established. Interestingly, according to the calculation procedure, remnant-C equals TRL-C in fasting serum. Therefore, we will continue using the term TRL. Elevated serum TRL concentrations may be caused by various factors, such as excess dietary TG intake, high secretion rates of CMs, high hepatic VLDL secretion, and most importantly, by reduced efficiency of LPL [14,64,65]. A high serum TRL concentration is most likely the result of the combination of enhanced secretion and reduced lipolysis. This will initially result in relatively large TG rich particles that may be less atherogenic. In the extreme situation of genetically caused inhibition of LPL, hyperlipidemic pancreatitis is more common than ASVD [66,67]. As indicated before, TRL-C is calculated as TC minus HDL-C minus LDL- C. When the Friedewald formula is used for the LDL-C calculation, the inaccuracy in the determination of LDL-C affects the TRL-C calculation. As a matter of fact, the equation can then be rewritten as TRL-C = TG/5. In a healthy situation, the TRL-C concentration calculated via the Friedewald equation is on average about 10% of the LDL-C concentration. The LDL-C concentration calculated by the Friedewald equation tends to underestimate LDL-C by about 10% when compared to LDL-C measurement after LDL isolation using ultracentrifugation [68]. Correcting LDL-C for a potential 10% underestimation leads to about 50% reduction in TRL-C. This suggests using direct measurement of LDL- C to calculate a reliable TRL-C concentration. Recently Varbo et al. [69] described an alternative technique to measure TRL-C independent of LDL-C and HDL-C. Using a commercial assay (Denka, TRL-C, Denka Company Limited, Tokyo, Japan), LDL and HDL are degraded and removed. Thereafter C is measured. It was found that directly measured TRL-C identified 5% more patients with increased risk of cardiovascular disease than calculated TRL-C applying the Martin–Hopkins equation. The question arises as to how the patients involved should be treated. Probably, their clinical and nutritional status must be closely studied. An obese patient with a high fat intake may be successfully treated by reduction of dietary fat intake. This may be achieved with a low fat, fiber rich diet, eventually combined with orlistat, which binds to pancreatic lipase reducing fat digestion and promotes weight loss [70–72]. A patient with high sugar intake may limit sugar intake and thereby potential endogenous fat synthesis. New therapies are under development such as pemafibrate [73] and the omega-3 fatty acid icosapent ethyl [74]. Decreased serum TG was observed under statin treatment and more pronounced under combination of statin with ezetimibe [75,76]. The mechanism for this serum TG reduction during LDL-C reduction therapy is unclear.

5. Non HDL-C and ApoB

Interestingly, discussion has focused serum LDL-C and TRL-C as determinants of increased cardiovascular risk. Apparently, a subgroup of patients develops cardiovascular disease despite a normal LDL-C concentration [1]. Other research groups promote non-HDL-C as the ultimate marker of atherogenic cardiovascular disease risk. Non-HDL-C is a calculated parameter obtained as non-HDL-C = TC − HDL-C. TC as well as HDL-C are measured directly with generally accepted methods. The difference is well defined and without discussion. It reflects LDL-C plus TRL-C. Thus, non-HDL-C contains all atherogenic components. However, large and potentially less atherogenic TRL components may be included, particularly when LPL activity is low. This may decrease the prognostic efficiency. Furthermore, at TG > 400 mg/dL, HDL-C measurement is inaccurate since TRLs are not sufficiently precipitated and thus TRL-C is partly included in the HDL-C value. Normally, LDL-C comprises the majority of non-HDL-C. However, non-HDL-C is considered a better predictor for a residual risk for cardiovascular disease than LDL-C [40]. It is also known that an undefined subgroup of ApoB-containing lipoproteins expresses the highest atherogenic action and it has been established that small dense

LDL particles are more atherogenic than larger ones. Thus, the number of lipoprotein particles reflects the atherogenicity better than the lipoprotein concentration. Since each ApoB-containing lipoprotein carries only one ApoB molecule, it has been proposed to determine the total ApoB concentration as a measure of atherogenicity [43,77], also under statin treatment [78]. This uncouples atherogenicity from C. Apo-B and Apo-AI can be assayed using commercial test kits based on automated immunoturbidimetric methods (Randox, Crumlin, United Kingdom). First, Apo-B-containing particles are precipitated from serum by phosphotungstic acid–$MgCl_2$. ApoA1 is measured in this fraction while ApoB in the residual fraction. For optimal differentiation, the separation of ApoB100 from ApoB48 may be considered in distinguishing between liver-derived and gut-derived ApoB containing TRL particles.

6. Personalized Diagnostics and Therapy

This review highlights a discrepancy between available research findings and daily clinical routine. It may take some time before the measurement of TRL-C and ApoB concentrations in serum become routine analysis in the clinical laboratory. Routine daily measurements include TC, TG, HDL-C, and LDL-C. Measurement of LDL-C via direct methodology is slowly being introduced and must be further standardized in all clinical laboratories. While LDL-C may remain a leading predictive parameter, the additionally acquired data for TRL-C should also be considered. Table 1 outlines a diagnostic and personalized treatment strategy.

Table 1. A proposed scheme of diagnosing the cause of development of atherosclerosis via elevated serum LDL-C or serum TRL-C. Personalized treatment can be applied.

Total C	LDL-C	TRL-C	Therapy
Normal	Elevated	Low	LDL-C lowering
Normal	Low	Elevated	TG-lowering
Elevated	Elevated	Normal	LDL-C lowering
Elevated	Normal	Elevated	TG-lowering
Elevated	Elevated	Elevated	LDL-C and TG-lowering

C, cholesterol; LDL, low density lipoprotein; TRL, triglyceride-rich lipoproteins; TG, triglyceride.

It must be realized that Lp(a) is included in LDL. Therefore, this has to be measured in each patient at least once in a lifetime. LDL-C lowering should consist of combined statin or bempedoic acid and ezetimibe treatment in order to obtain the maximal response. When LDL-C lowering is insufficient, PCSK9 inhibition should be added to the combination treatment.

7. Limitations

The proposed extended diagnosis procedure of the atherogenic lipoprotein components depends to a large extend on the quality of the LDL-C measurement. Isolation of LDL using ultracentrifugation followed by C measurement will ensure ultimate quality. However, this technique is too time consuming and complex to be incorporated into daily clinical routine. Homogeneous direct assays are now available to isolate LDL by chemical means [59,79]. However, these various commercial assays may produce different results. In addition, the validity of the assay appears high in healthy subjects and lower in patients with cardiovascular disease [80]. Furthermore, the analytical result may differ when measured in fresh serum or frozen serum. Therefore, the most reliable assay must be chosen and applied under controlled conditions. Any reasoning for applying the measurement must be well defined. When the risk of development of cardiovascular disease needs to be determined by measuring atherogenic lipoprotein C in documented patients, LDL-C measurement must be performed exactly under cardiovascular conditions. A concentration above the cut off level of the normal range is the criterion for treatment. It will suffice to apply the same assay and quality control criteria continuously. The patient should be followed over time during treatment.

8. Summary of Results

VLDL and CM-derived remnants including CMR, IDL, and LDL in serum are considered atherogenic. Their C concentrations in serum are documented as predictive atherogenic indicators of cardiovascular disease risk. VLDL, IDL, and LDL are the dominant lipoproteins in fasting serum. CMRs are added to postprandial serum in amounts depending on the dietary fat intake. Serum LDL-C is used as the gold standard for risk prediction and treatment is focused on lowering serum LDL-C. VLDL-C, IDL-C, and CMR-C are called TRLs. Atherosclerosis may develop in patients with low LDL-C and high TRL-C concentrations. The accuracy of LDL-C and TRL-C determinations is procedure-dependent, i.e., on direct measurement or estimation procedures. It is unknown whether all TRLs are equally atherogenic. Non-HDL-C combines TRL-C and LDL-C and thus all potentially atherogenic lipoprotein species. Non-HDL-C may be considered the best and simplest marker of lipoprotein atherogenicity. At higher serum TG concentrations (>300 mg/dL) HDL-C also includes TRL-C because those particles are not completely precipitated by the HDL-determination method. LDL-C and TRL-C concentrations in fasting serum do not usually reflect the daily exposure to atherogenic lipoproteins, which is highest in the postprandial phases. Potentially atherogenic lipoproteins all contain ApoB. The serum ApoB concentration reflects the number of atherogenic particles and thus the cardiovascular risk. Patients with elevated serum TRL-C concentrations may be detected when serum LDL-C is measured directly with sufficient accuracy. Homogeneous, direct assays are available allowing rapid analysis in a clinical routine setting. However, selection of the preferred assay must be performed carefully.

9. Conclusions

Epidemiologic and genetic studies have established TRL and their remnants as important contributors to ASCVD. Combinations of LDL-C, non-HDL-C, TRL-C, and ApoB concentrations must be evaluated as the utmost predictive risk marker for development of cardiovascular disease and are recommended in the current guidelines. For clinical routine, direct measurements of TGs, TC, HDL-C, and LDL-C allow semi accurate calculation of TRL-C and non HDL-C. Patients with elevated LDL-C may be treated with conventional C lowering therapies. Patients with elevated TRL-C should be detected and treated specifically. The first step of treatment is the implementation of lifestyle interventions. Second, LDL-C lowering with statins or bempedoic acid—in case of statin intolerance—with or without ezetimibe are recommended to reduce vascular risk, independent of statin-associated lowering of TRL itself. Novel and emerging data, e.g., on omega-3 fatty acids (high-dose icosapent ethyl) and new generations of selective peroxisome proliferator-activated receptor (PPAR) modulator pemafibrate may identify patients who will benefit from TRL lowering.

Author Contributions: Literature research, F.S. and D.L.; Conceptualization, F.S. and D.L.; writing—original draft preparation, F.S.; writing—review and editing, F.S., H.-U.K. and D.L.; visualization F.S. and D.L. All authors have read and agreed to the published version of the manuscript.

Funding: This research received no external funding.

Conflicts of Interest: The authors declare no conflict of interest.

References

1. Gitt, A.K.; Parhofer, K.G.; Laufs, U.; Marz, W.; Paar, W.D.; Bramlage, P.; Marx, N. Hypercholesterolemia diagnosis, treatment patterns and target achievement in patients with acute coronary syndromes in Germany. *Clin. Res. Cardiol.* **2023**, *112*, 299–311. [CrossRef] [PubMed]
2. Holme, I.; Solberg, L.A.; Weissfeld, L.; Helgeland, A.; Hjermann, I.; Leren, P.; Strong, J.P.; Williams, O.D. Coronary risk factors and their pathway of action through coronary raised lesions, coronary stenoses and coronary death. Multivariate statistical analysis of an autopsy series: The Oslo Study. *Am. J. Cardiol.* **1985**, *55*, 40–47. [CrossRef] [PubMed]

3. Kannel, W.B.; Castelli, W.P.; Gordon, T. Cholesterol in the prediction of atherosclerotic disease. New perspectives based on the Framingham study. *Ann. Intern. Med.* **1979**, *90*, 85–91. [CrossRef] [PubMed]
4. Newman, W.P., 3rd; Freedman, D.S.; Voors, A.W.; Gard, P.D.; Srinivasan, S.R.; Cresanta, J.L.; Williamson, G.D.; Webber, L.S.; Berenson, G.S. Relation of serum lipoprotein levels and systolic blood pressure to early atherosclerosis. The Bogalusa Heart Study. *N. Engl. J. Med.* **1986**, *314*, 138–144. [CrossRef] [PubMed]
5. Tertov, V.V.; Orekhov, A.N.; Martsenyuk, O.N.; Perova, N.V.; Smirnov, V.N. Low-density lipoproteins isolated from the blood of patients with coronary heart disease induce the accumulation of lipids in human aortic cells. *Exp. Mol. Pathol.* **1989**, *50*, 337–347. [CrossRef] [PubMed]
6. Werner, C.; Laufs, U. Moving beyond the "LDL hypothesis". *Vasa* **2015**, *44*, 333–340. [CrossRef]
7. Bruckert, E. New advances in lipid-modifying therapies for reducing cardiovascular risk. *Cardiology* **2002**, *97*, 59–66. [CrossRef]
8. Gaine, S.P.; Quispe, R.; Patel, J.; Michos, E.D. New Strategies for Lowering Low Density Lipoprotein Cholesterol for Cardiovascular Disease Prevention. *Curr. Cardiovasc. Risk Rep.* **2022**, *16*, 69–78. [CrossRef]
9. Makhmudova, U.; Samadifar, B.; Maloku, A.; Haxhikadrija, P.; Geiling, J.A.; Romer, R.; Lauer, B.; Mobius-Winkler, S.; Otto, S.; Schulze, P.C.; et al. Intensive lipid-lowering therapy for early achievement of guideline-recommended LDL-cholesterol levels in patients with ST-elevation myocardial infarction ("Jena auf Ziel"). *Clin. Res. Cardiol.* 2023, *Online ahead of print*. [CrossRef]
10. Patnaik, S.; Pollevick, M.E.; Lara-Breitinger, K.M.; Stone, N.J. Inter-Individual Variability in Lipid Response: A Narrative Review. *Am. J. Med.* **2022**, *135*, 1427–1433.e7. [CrossRef]
11. Sun, L.; Wolska, A.; Amar, M.; Zubiran, R.; Remaley, A.T. Approach to the Patient with a Suboptimal Statin Response: Causes and Algorithm for Clinical Management. *J. Clin. Endocrinol. Metab.* 2023, *Online ahead of print*. [CrossRef]
12. O'Donoghue, M.L.; Giugliano, R.P.; Wiviott, S.D.; Atar, D.; Keech, A.; Kuder, J.F.; Im, K.; Murphy, S.A.; Flores-Arredondo, J.H.; Lopez, J.A.G.; et al. Long-Term Evolocumab in Patients With Established Atherosclerotic Cardiovascular Disease. *Circulation* **2022**, *146*, 1109–1119. [CrossRef]
13. Laufs, U.; Birkenfeld, A.L.; Fraass, U.; Hohenstein, B.; Siegert, C.; Klotsche, J.; Steinhagen-Thiessen, E.; Pittrow, D.; Dexl, S.; Salmen, S.; et al. Novel Insights into the Management of Patients with Very High Cardiovascular Risk Eligible for PCSK9 Inhibitor Treatment: Baseline Findings from the PERI-DYS Study. *Cardiovasc. Drugs Ther.* 2022, *Online ahead of print*. [CrossRef]
14. Kumari, A.; Kristensen, K.K.; Ploug, M.; Winther, A.L. The Importance of Lipoprotein Lipase Regulation in Atherosclerosis. *Biomedicines* **2021**, *9*, 782. [CrossRef] [PubMed]
15. Feingold, K.R. Lipid and Lipoprotein Metabolism. *Endocrinol. Metab. Clin. N. Am.* **2022**, *51*, 437–458. [CrossRef] [PubMed]
16. Jansen, M.; Contini, C. LDL retention time in plasma can be -based on causation- estimated by the lipid composition of LDL and other lipoproteins. *PLoS ONE* **2022**, *17*, e0272050. [CrossRef] [PubMed]
17. Kuklenyik, Z.; Jones, J.I.; Gardner, M.S.; Schieltz, D.M.; Parks, B.A.; Toth, C.A.; Rees, J.C.; Andrews, M.L.; Carter, K.; Lehtikoski, A.K.; et al. Core lipid, surface lipid and apolipoprotein composition analysis of lipoprotein particles as a function of particle size in one workflow integrating asymmetric flow field-flow fractionation and liquid chromatography-tandem mass spectrometry. *PLoS ONE* **2018**, *13*, e0194797. [CrossRef]
18. Morton, R.E.; Liu, Y. The lipid transfer properties of CETP define the concentration and composition of plasma lipoproteins. *J. Lipid Res.* **2020**, *61*, 1168–1179. [CrossRef]
19. Bjornson, E.; Packard, C.J.; Adiels, M.; Andersson, L.; Matikainen, N.; Soderlund, S.; Kahri, J.; Sihlbom, C.; Thorsell, A.; Zhou, H.; et al. Investigation of human apoB48 metabolism using a new, integrated non-steady-state model of apoB48 and apoB100 kinetics. *J. Intern. Med.* **2019**, *285*, 562–577. [CrossRef]
20. Higgins, V.; Adeli, K. Postprandial Dyslipidemia: Pathophysiology and Cardiovascular Disease Risk Assessment. *EJIFCC* **2017**, *28*, 168–184.
21. Levy, E.; Poinsot, P.; Spahis, S. Chylomicron retention disease: Genetics, biochemistry, and clinical spectrum. *Curr. Opin. Lipidol.* **2019**, *30*, 134–139. [CrossRef]
22. Lo, C.C.; Coschigano, K.T. ApoB48 as an Efficient Regulator of Intestinal Lipid Transport. *Front. Physiol.* **2020**, *11*, 796. [CrossRef]
23. Varbo, A.; Nordestgaard, B.G. Remnant lipoproteins. *Curr. Opin. Lipidol.* **2017**, *28*, 300–307. [CrossRef]
24. Laufs, U.; Parhofer, K.G.; Ginsberg, H.N.; Hegele, R.A. Clinical review on triglycerides. *Eur. Heart J.* **2020**, *41*, 99–109c. [CrossRef] [PubMed]
25. Tomo, S.; Sankanagoudar, S.; Shukla, R.; Sharma, P. Calculated small dense LDL-cholesterol and its correlation with the atherogenic index of plasma. *Ann. Clin. Biochem.* **2022**, *59*, 454–456. [CrossRef]
26. Havel, R.J. Remnant lipoproteins as therapeutic targets. *Curr. Opin. Lipidol.* **2000**, *11*, 615–620. [CrossRef] [PubMed]
27. Miller, M.; Ginsberg, H.N.; Schaefer, E.J. Relative atherogenicity and predictive value of non-high-density lipoprotein cholesterol for coronary heart disease. *Am. J. Cardiol.* **2008**, *101*, 1003–1008. [CrossRef]
28. Sniderman, A.D.; Thanassoulis, G.; Glavinovic, T.; Navar, A.M.; Pencina, M.; Catapano, A.; Ference, B.A. Apolipoprotein B Particles and Cardiovascular Disease: A Narrative Review. *JAMA Cardiol.* **2019**, *4*, 1287–1295. [CrossRef] [PubMed]
29. Cantey, E.P.; Wilkins, J.T. Discordance between lipoprotein particle number and cholesterol content: An update. *Curr. Opin. Endocrinol. Diabetes Obes.* **2018**, *25*, 130–136. [CrossRef] [PubMed]

30. Jacobson, T.A. Opening a new lipid "apo-thecary": Incorporating apolipoproteins as potential risk factors and treatment targets to reduce cardiovascular risk. *Mayo Clin. Proc.* **2011**, *86*, 762–780. [CrossRef] [PubMed]
31. Feingold, K.R.; Chait, A. Approach to patients with elevated low-density lipoprotein cholesterol levels. *Best Pract. Res. Clin. Endocrinol. Metab.* **2023**, *37*, 101658. [CrossRef]
32. Liu, H.; Li, J.; Liu, F.; Huang, K.; Cao, J.; Chen, S.; Li, H.; Shen, C.; Hu, D.; Huang, J.; et al. Efficacy and safety of low levels of low-density lipoprotein cholesterol: Trans-ancestry linear and non-linear Mendelian randomization analyses. *Eur. J. Prev. Cardiol.* 2023; *Online ahead of print*. [CrossRef]
33. White-Al Habeeb, N.M.A.; Higgins, V.; Wolska, A.; Delaney, S.R.; Remaley, A.T.; Beriault, D.R. The Present and Future of Lipid Testing in Cardiovascular Risk Assessment. *Clin. Chem.* **2023**, *69*, 456–469. [CrossRef]
34. Groener, J.E.; Van Rozen, A.J.; Erkelens, D.W. Cholesteryl ester transfer activity. Localization and role in distribution of cholesteryl ester among lipoproteins in man. *Atherosclerosis* **1984**, *50*, 261–271. [CrossRef]
35. Nelson, A.J.; Sniderman, A.D.; Ditmarsch, M.; Dicklin, M.R.; Nicholls, S.J.; Davidson, M.H.; Kastelein, J.J.P. Cholesteryl Ester Transfer Protein Inhibition Reduces Major Adverse Cardiovascular Events by Lowering Apolipoprotein B Levels. *Int. J. Mol. Sci.* **2022**, *23*, 9417. [CrossRef] [PubMed]
36. Calvier, L.; Herz, J.; Hansmann, G. Interplay of Low-Density Lipoprotein Receptors, LRPs, and Lipoproteins in Pulmonary Hypertension. *JACC Basic Transl. Sci.* **2022**, *7*, 164–180. [CrossRef] [PubMed]
37. Bakkeren, H.F.; Kuipers, F.; Vonk, R.J.; Van Berkel, T.J. Evidence for reverse cholesterol transport in vivo from liver endothelial cells to parenchymal cells and bile by high-density lipoprotein. *Biochem. J.* **1990**, *268*, 685–691. [CrossRef] [PubMed]
38. Schwartz, C.C.; Halloran, L.G.; Vlahcevic, Z.R.; Gregory, D.H.; Swell, L. Preferential utilization of free cholesterol from high-density lipoproteins for biliary cholesterol secretion in man. *Science* **1978**, *200*, 62–64. [CrossRef] [PubMed]
39. Pieters, M.N.; Schouten, D.; Bakkeren, H.F.; Esbach, B.; Brouwer, A.; Knook, D.L.; van Berkel, T.J. Selective uptake of cholesteryl esters from apolipoprotein-E-free high-density lipoproteins by rat parenchymal cells in vivo is efficiently coupled to bile acid synthesis. *Biochem. J.* **1991**, *280*, 359–365. [CrossRef]
40. Sniderman, A.; McQueen, M.; Contois, J.; Williams, K.; Furberg, C.D. Why is non-high-density lipoprotein cholesterol a better marker of the risk of vascular disease than low-density lipoprotein cholesterol? *J. Clin. Lipidol.* **2010**, *4*, 152–155. [CrossRef]
41. Balling, M.; Langsted, A.; Afzal, S.; Varbo, A.; Davey Smith, G.; Nordestgaard, B.G. A third of nonfasting plasma cholesterol is in remnant lipoproteins: Lipoprotein subclass profiling in 9293 individuals. *Atherosclerosis* **2019**, *286*, 97–104. [CrossRef]
42. Brown, W.V.; Myers, G.L.; Sniderman, A.D.; Stein, E. Should we use apoB for risk assessment and as a target for treatment? *J. Clin. Lipidol.* **2010**, *4*, 144–151. [CrossRef]
43. Langlois, M.R.; Sniderman, A.D. Non-HDL Cholesterol or apoB: Which to Prefer as a Target for the Prevention of Atherosclerotic Cardiovascular Disease? *Curr. Cardiol. Rep.* **2020**, *22*, 67. [CrossRef]
44. Babiak, J.; Rudel, L.L. Lipoproteins and atherosclerosis. *Baillieres Clin. Endocrinol. Metab.* **1987**, *1*, 515–550. [CrossRef]
45. Potts, J.L.; Fisher, R.M.; Humphreys, S.M.; Gibbons, G.F.; Frayn, K.N. Separation of lipoprotein fractions by ultracentrifugation: Investigation of analytical recovery with sequential flotation and density gradient procedures. *Clin. Chim. Acta* **1994**, *230*, 215–220. [CrossRef]
46. Redgrave, T.G.; Roberts, D.C.; West, C.E. Separation of plasma lipoproteins by density-gradient ultracentrifugation. *Anal. Biochem.* **1975**, *65*, 42–49. [CrossRef] [PubMed]
47. Cheng, M.L.; Kammerer, C.M.; Lowe, W.F.; Dyke, B.; VandeBerg, J.L. Method for quantitating cholesterol in subfractions of serum lipoproteins separated by gradient gel electrophoresis. *Biochem. Genet.* **1988**, *26*, 657–681. [CrossRef] [PubMed]
48. Otvos, J. Measurement of triglyceride-rich lipoproteins by nuclear magnetic resonance spectroscopy. *Clin. Cardiol.* **1999**, *22*, II21–II27. [CrossRef]
49. Hirowatari, Y.; Yoshida, H. Innovatively Established Analysis Method for Lipoprotein Profiles Based on High-Performance Anion-Exchange Liquid Chromatography. *J. Atheroscler. Thromb.* **2019**, *26*, 1027–1040. [CrossRef] [PubMed]
50. Friedewald, W.T.; Levy, R.I.; Fredrickson, D.S. Estimation of the concentration of low-density lipoprotein cholesterol in plasma, without use of the preparative ultracentrifuge. *Clin. Chem.* **1972**, *18*, 499–502. [CrossRef]
51. Martins, J.; Rossouw, H.M.; Pillay, T.S. How should low-density lipoprotein cholesterol be calculated in 2022? *Curr. Opin. Lipidol.* **2022**, *33*, 237–256. [CrossRef] [PubMed]
52. Kolovou, G.D.; Watts, G.F.; Mikhailidis, D.P.; Perez-Martinez, P.; Mora, S.; Bilianou, H.; Panotopoulos, G.; Katsiki, N.; Ooi, T.C.; Lopez-Miranda, J.; et al. Postprandial Hypertriglyceridaemia Revisited in the Era of Non-Fasting Lipid Profile Testing: A 2019 Expert Panel Statement, Narrative Review. *Curr. Vasc. Pharmacol.* **2019**, *17*, 515–537. [CrossRef] [PubMed]
53. Kolovou, G.D.; Watts, G.F.; Mikhailidis, D.P.; Perez-Martinez, P.; Mora, S.; Bilianou, H.; Panotopoulos, G.; Katsiki, N.; Ooi, T.C.; Lopez-Miranda, J.; et al. Postprandial Hypertriglyceridaemia Revisited in the Era of Non-Fasting Lipid Profile Testing: A 2019 Expert Panel Statement, Main Text. *Curr. Vasc. Pharmacol.* **2019**, *17*, 498–514. [CrossRef]
54. Liu, M.M.; Peng, J.; Cao, Y.X.; Guo, Y.L.; Wu, N.Q.; Zhu, C.G.; Gao, Y.; Li, J.J. The difference between fasting and non-fasting lipid measurements is not related to statin treatment. *Ann. Transl. Med.* **2021**, *9*, 386. [CrossRef]

55. Nakajima, K.; Tokita, Y.; Sakamaki, K.; Shimomura, Y.; Kobayashi, J.; Kamachi, K.; Tanaka, A.; Stanhope, K.L.; Havel, P.J.; Wang, T.; et al. Triglyceride content in remnant lipoproteins is significantly increased after food intake and is associated with plasma lipoprotein lipase. *Clin. Chim. Acta* **2017**, *465*, 45–52. [CrossRef] [PubMed]
56. D'Aquila, T.; Hung, Y.H.; Carreiro, A.; Buhman, K.K. Recent discoveries on absorption of dietary fat: Presence, synthesis, and metabolism of cytoplasmic lipid droplets within enterocytes. *Biochim. Biophys. Acta* **2016**, *1861*, 730–747. [CrossRef] [PubMed]
57. Martin, S.S.; Blaha, M.J.; Elshazly, M.B.; Toth, P.P.; Kwiterovich, P.O.; Blumenthal, R.S.; Jones, S.R. Comparison of a novel method vs the Friedewald equation for estimating low-density lipoprotein cholesterol levels from the standard lipid profile. *JAMA* **2013**, *310*, 2061–2068. [CrossRef] [PubMed]
58. Sampson, M.; Ling, C.; Sun, Q.; Harb, R.; Ashmaig, M.; Warnick, R.; Sethi, A.; Fleming, J.K.; Otvos, J.D.; Meeusen, J.W.; et al. A New Equation for Calculation of Low-Density Lipoprotein Cholesterol in Patients With Normolipidemia and/or Hypertriglyceridemia. *JAMA Cardiol.* **2020**, *5*, 540–548. [CrossRef] [PubMed]
59. Miller, W.G.; Myers, G.L.; Sakurabayashi, I.; Bachmann, L.M.; Caudill, S.P.; Dziekonski, A.; Edwards, S.; Kimberly, M.M.; Korzun, W.J.; Leary, E.T.; et al. Seven direct methods for measuring HDL and LDL cholesterol compared with ultracentrifugation reference measurement procedures. *Clin. Chem.* **2010**, *56*, 977–986. [CrossRef] [PubMed]
60. Koren, E.; Corder, C.; Mueller, G.; Centurion, H.; Hallum, G.; Fesmire, J.; McConathy, W.D.; Alaupovic, P. Triglyceride enriched lipoprotein particles correlate with the severity of coronary artery disease. *Atherosclerosis* **1996**, *122*, 105–115. [CrossRef] [PubMed]
61. Jae, S.Y.; Kim, H.J.; Kunutsor, S.K.; Bunsawat, K.; Kurl, S.; Laukkanen, J.A.; Choi, Y.H. Associations of Cardiorespiratory Fitness With Estimated Remnant Cholesterol and Non-High-Density Lipoprotein Cholesterol in Healthy Men. *Am. J. Cardiol.* **2023**, *186*, 170–175. [CrossRef] [PubMed]
62. Tian, Y.; Wu, W.; Qin, L.; Yu, X.; Cai, L.; Wang, H.; Zhang, Z. Prognostic value of remnant cholesterol in patients with coronary heart disease: A systematic review and meta-analysis of cohort studies. *Front. Cardiovasc. Med.* **2022**, *9*, 951523. [CrossRef]
63. Yang, J.; Wang, Y.; Xi, Z.; Ma, Y.; Shao, C.; Wang, W.; Tang, Y.D. Remnant-Like Particle Cholesterol and the Risk of Major Adverse Cardiovascular Events: A Systematic Review and Meta-Analysis. *J. Cardiovasc. Dev. Dis.* **2022**, *9*, 452. [CrossRef]
64. Moon, J.H.; Kim, K.; Choi, S.H. Lipoprotein Lipase: Is It a Magic Target for the Treatment of Hypertriglyceridemia. *Endocrinol. Metab.* **2022**, *37*, 575–586. [CrossRef]
65. Su, X. ANGPLT3 in cardio-metabolic disorders. *Mol. Biol. Rep.* **2021**, *48*, 2729–2739. [CrossRef] [PubMed]
66. Okazaki, H.; Gotoda, T.; Ogura, M.; Ishibashi, S.; Inagaki, K.; Daida, H.; Hayashi, T.; Hori, M.; Masuda, D.; Matsuki, K.; et al. Current Diagnosis and Management of Primary Chylomicronemia. *J. Atheroscler. Thromb.* **2021**, *28*, 883–904. [CrossRef] [PubMed]
67. Paquette, M.; Amyot, J.; Fantino, M.; Baass, A.; Bernard, S. Rare Variants in Triglycerides-Related Genes Increase Pancreatitis Risk in Multifactorial Chylomicronemia Syndrome. *J. Clin. Endocrinol. Metab.* **2021**, *106*, e3473–e3482. [CrossRef] [PubMed]
68. Steyn, N.; Muller Rossouw, H.; Pillay, T.S.; Martins, J. Comparability of calculated LDL-C with directly measured LDL-C in selected paediatric and adult cohorts. *Clin. Chim. Acta* **2022**, *537*, 158–166. [CrossRef]
69. Varbo, A.; Nordestgaard, B.G. Directly measured vs. calculated remnant cholesterol identifies additional overlooked individuals in the general population at higher risk of myocardial infarction. *Eur. Heart J.* **2021**, *42*, 4833–4843. [CrossRef] [PubMed]
70. Avenell, A.; Robertson, C.; Skea, Z.; Jacobsen, E.; Boyers, D.; Cooper, D.; Aceves-Martins, M.; Retat, L.; Fraser, C.; Aveyard, P.; et al. Bariatric surgery, lifestyle interventions and orlistat for severe obesity: The REBALANCE mixed-methods systematic review and economic evaluation. *Health Technol. Assess.* **2018**, *22*, 1–246. [CrossRef] [PubMed]
71. Kumar, A.; Chauhan, S. Pancreatic lipase inhibitors: The road voyaged and successes. *Life Sci.* **2021**, *271*, 119115. [CrossRef]
72. Paccosi, S.; Cresci, B.; Pala, L.; Rotella, C.M.; Parenti, A. Obesity Therapy: How and Why? *Curr. Med. Chem.* **2020**, *27*, 174–186. [CrossRef]
73. Arai, H.; Yamashita, S.; Yokote, K.; Araki, E.; Suganami, H.; Ishibashi, S.; Group, K.S. Efficacy and safety of K-877, a novel selective peroxisome proliferator-activated receptor alpha modulator (SPPARMalpha), in combination with statin treatment: Two randomised, double-blind, placebo-controlled clinical trials in patients with dyslipidaemia. *Atherosclerosis* **2017**, *261*, 144–152. [CrossRef]
74. Orringer, C.E. Icosapent ethyl: Where will it fit into guideline-based medical therapy for high risk atherosclerotic cardiovascular disease? *Trends Cardiovasc. Med.* **2020**, *30*, 151–157. [CrossRef]
75. Rodriguez-Saldana, J.; Padilla-Padilla, F.; Cardona-Munoz, E.G.; Romero-Antonio, Y.; Arguedas-Nunez, M.M.; Sander-Padilla, J.G.; Martinez-Munoz, A.; Lugo-Sanchez, L.A.; Rodriguez-Vazquez, I.C.; Gonzalez-Canudas, J. Real-World Evidence Evaluation on the Lipid Profile, Therapeutic Goals, and Safety of the Fixed-Dose Combination of Rosuvastatin/Ezetimibe (Trezete(R)) in Dyslipidemia Patients. *Cardiol. Res. Pract.* **2022**, *2022*, 9464733. [CrossRef] [PubMed]
76. Sudhop, T.; Reber, M.; Tribble, D.; Sapre, A.; Taggart, W.; Gibbons, P.; Musliner, T.; von Bergmann, K.; Lutjohann, D. Changes in cholesterol absorption and cholesterol synthesis caused by ezetimibe and/or simvastatin in men. *J. Lipid Res.* **2009**, *50*, 2117–2123. [CrossRef] [PubMed]
77. Carr, S.S.; Hooper, A.J.; Sullivan, D.R.; Burnett, J.R. Non-HDL-cholesterol and apolipoprotein B compared with LDL-cholesterol in atherosclerotic cardiovascular disease risk assessment. *Pathology* **2019**, *51*, 148–154. [CrossRef] [PubMed]
78. Johannesen, C.D.L.; Mortensen, M.B.; Langsted, A.; Nordestgaard, B.G. Apolipoprotein B and Non-HDL Cholesterol Better Reflect Residual Risk Than LDL Cholesterol in Statin-Treated Patients. *J. Am. Coll. Cardiol.* **2021**, *77*, 1439–1450. [CrossRef]

79. Miida, T.; Nishimura, K.; Okamura, T.; Hirayama, S.; Ohmura, H.; Yoshida, H.; Miyashita, Y.; Ai, M.; Tanaka, A.; Sumino, H.; et al. A multicenter study on the precision and accuracy of homogeneous assays for LDL-cholesterol: Comparison with a beta-quantification method using fresh serum obtained from non-diseased and diseased subjects. *Atherosclerosis* **2012**, *225*, 208–215. [CrossRef]
80. Wolska, A.; Remaley, A.T. LDL Cholesterol: What Is the Best Way to Measure It? *Clin. Chem.* **2019**, *65*, 1067–1069. [CrossRef]

nutrients

MDPI

Article

Activation of Liver X Receptors and Peroxisome Proliferator-Activated Receptors by Lipid Extracts of Brown Seaweeds: A Potential Application in Alzheimer's Disease?

Nikita Martens [1,2], Na Zhan [1,3], Gardi Voortman [1], Frank P. J. Leijten [1], Connor van Rheenen [1], Suzanne van Leerdam [1], Xicheng Geng [3], Michiel Huybrechts [4], Hongbing Liu [3], Johan W. Jonker [5], Folkert Kuipers [5,6], Dieter Lütjohann [7], Tim Vanmierlo [1,2,8] and Monique T. Mulder [1,*]

1 Department of Internal Medicine, Section Pharmacology and Vascular Medicine, Erasmus University Medical Center, 3015 CN Rotterdam, The Netherlands; n.martens@erasmusmc.nl (N.M.); n.zhan@erasmusmc.nl (N.Z.); g.voortman@erasmusmc.nl (G.V.); f.leijten@erasmusmc.nl (F.P.J.L.); connorvanrheenen@gmail.com (C.v.R.); suzannevanleerdam@hotmail.com (S.v.L.); tim.vanmierlo@uhasselt.be (T.V.)

2 Department of Neuroscience, Biomedical Research Institute, European Graduate School of Neuroscience, Hasselt University, B-3590 Hasselt, Belgium

3 Key Laboratory of Marine Drugs, Chinese Ministry of Education, School of Medicine and Pharmacy, Ocean University of China, Qingdao 266003, China; gengxicheng@stu.ouc.edu.cn (X.G.); liuhongb@ouc.edu.cn (H.L.)

4 Department of Environmental Biology, Center for Environmental Sciences, Hasselt University, B-3590 Diepenbeek, Belgium; michiel.huybrechts@uhasselt.be

5 Department of Pediatrics, Section of Molecular Metabolism and Nutrition, University of Groningen, University Medical Center Groningen, 9713 GZ Groningen, The Netherlands; j.w.jonker@umcg.nl (J.W.J.); f.kuipers@umcg.nl (F.K.)

6 European Research Institute for the Biology of Ageing (ERIBA), University of Groningen, University Medical Center Groningen, 9713 AV Groningen, The Netherlands

7 Institute of Clinical Chemistry and Clinical Pharmacology, University Hospital Bonn, D-53127 Bonn, Germany; dieter.luetjohann@ukbonn.de

8 Department of Psychiatry and Neuropsychology, School for Mental Health and Neurosciences, Division Translational Neuroscience, Maastricht University, 6200 MD Maastricht, The Netherlands

* Correspondence: m.t.mulder@erasmusmc.nl; Tel.: +31-107-032-707

Citation: Martens, N.; Zhan, N.; Voortman, G.; Leijten, F.P.J.; van Rheenen, C.; van Leerdam, S.; Geng, X.; Huybrechts, M.; Liu, H.; Jonker, J.W.; et al. Activation of Liver X Receptors and Peroxisome Proliferator-Activated Receptors by Lipid Extracts of Brown Seaweeds: A Potential Application in Alzheimer's Disease?. *Nutrients* **2023**, *15*, 3004. https://doi.org/10.3390/nu15133004

Academic Editor: Tatsuhiro Hisatsune

Received: 17 May 2023
Revised: 25 June 2023
Accepted: 26 June 2023
Published: 30 June 2023

Abstract: The nuclear liver X receptors (LXRα/β) and peroxisome proliferator-activated receptors (PPARα/γ) are involved in the regulation of multiple biological processes, including lipid metabolism and inflammation. The activation of these receptors has been found to have neuroprotective effects, making them interesting therapeutic targets for neurodegenerative disorders such as Alzheimer's Disease (AD). The Asian brown seaweed *Sargassum fusiforme* contains both LXR-activating (oxy)phytosterols and PPAR-activating fatty acids. We have previously shown that dietary supplementation with lipid extracts of *Sargassum fusiforme* prevents disease progression in a mouse model of AD, without inducing adverse effects associated with synthetic pan-LXR agonists. We now determined the LXRα/β- and PPARα/γ-activating capacity of lipid extracts of six European brown seaweed species (*Alaria esculenta*, *Ascophyllum nodosum*, *Fucus vesiculosus*, *Himanthalia elongata*, *Saccharina latissima*, and *Sargassum muticum*) and the Asian seaweed Sargassum fusiforme using a dual luciferase reporter assay. We analyzed the sterol and fatty acid profiles of the extracts by GC-MS and UPLC MS/MS, respectively, and determined their effects on the expression of LXR and PPAR target genes in several cell lines using quantitative PCR. All extracts were found to activate LXRs, with the *Himanthalia elongata* extract showing the most pronounced efficacy, comparable to *Sargassum fusiforme*, for LXR activation and transcriptional regulation of LXR-target genes. Extracts of Alaria esculenta, Fucus vesiculosus, and Saccharina latissima showed the highest capacity to activate PPARα, while extracts of *Alaria esculenta*, *Ascophyllum nodosum*, *Fucus vesiculosus*, and *Sargassum muticum* showed the highest capacity to activate PPARγ, comparable to *Sargassum fusiforme* extract. In CCF-STTG1 astrocytoma cells, all extracts induced expression of cholesterol efflux genes (*ABCG1*, *ABCA1*, and *APOE*) and suppressed expression of cholesterol and fatty acid synthesis genes (*DHCR7*, *DHCR24*, *HMGCR* and *SREBF2*, and *SREBF1*, *ACACA*, *SCD1* and *FASN*, respectively). Our data show that lipophilic fractions of European brown seaweeds activate LXRs and PPARs and thereby modulate lipid metabolism. These results support the potential of brown seaweeds in the prevention and/or treatment of neurodegenerative diseases and possibly cardiometabolic and inflammatory diseases via concurrent activation of LXRs and PPARs.

Keywords: nuclear receptor superfamily; liver X receptors; peroxisome proliferator-activated receptors; lipid metabolism; phytosterols; seaweed; Alzheimer's Disease

1. Introduction

Liver X receptors (LXRs) α (NR1H3) and β (NR1H2) and peroxisome proliferator-activated receptors (PPARs) α (NR1C1) and γ (NR1C3) are members of the nuclear receptor superfamily of ligand-activated transcription factors and are implicated in transcriptional control of a wide range of biological processes [1]. In response to binding their specific ligands, LXRs and PPARs form heterodimers with retinoid X receptors (RXRs) and bind to the LXR or PPAR response elements (LXREs/PPREs) in the promoter region of target genes. LXRs and PPARs are key players in the control of lipid and glucose metabolism and inflammation [2,3]. Disturbances in these biological processes contribute to metabolic, inflammatory, and neurodegenerative disorders, including Alzheimer's Disease (AD). AD is the most common form of dementia. It is a progressive neurodegenerative disorder manifested by cognitive loss and defined by the presence of amyloid-β (Aβ)-containing plaques, tau-containing neurofibrillary tangles, synaptic degeneration, and neuroinflammation [4–6]. LXRs and PPARs are abundantly expressed in metabolically active tissues, including the brain, and their activation has been found to exert neuroprotective effects, making these receptors interesting targets in the treatment of neurodegenerative disorders, such as AD [1,7–9].

LXRs (α and β) are oxysterol sensors implicated in the regulation of cholesterol and lipid homeostasis, including lipogenesis and reverse cholesterol transport, and in the regulation of glucose homeostasis and inflammatory processes. Activation of LXRs is thought to protect against AD pathologies by promoting cholesterol turnover in the brain, reducing inflammation, and possibly through its anti-amyloidogenic effects [8,10–16]. We and others have shown that synthetic LXRα/β agonists, such as T0901317, protect against cognitive decline in animal models of AD, without affecting the amyloid pathology [8,13–19]. However, the induction of lipogenesis by T0901317 and most other synthetic LXR-agonists also results in adverse effects such as hepatic steatosis and hypertriglyceridemia which has limited their clinical application [20–22]. Such adverse effects are not induced by endogenous (oxy)sterols or (oxy)phytosterols [23]. Lipid extracts of the seaweed *Sargassum fusiforme* (*S. fusiforme*), which contain LXR-activating (oxy)phytosterols such as saringosterol, have been found to activate LXRs and prevent cognitive decline in AD mice [24]. Recently, we also identified (3β,22E)-3-hydroxycholesta-5,22-dien-24-one and fucosterol-24,28 epoxide as LXR agonists in *S. fusiforme* [25]. Unlike synthetic LXR agonists, these *S. fusiforme* extracts did not have adverse effects on hepatic or serum lipid levels, making them a promising alternative for clinical use [24,26]. Recently, we found that purified 24(S)-saringosterol can also prevent cognitive decline in AD mice [26], providing further evidence of the potential health benefits of saringosterol-rich macroalgae.

PPARs (α, δ/β, and γ) are also lipid sensors involved in multiple biological processes, including lipid and glucose metabolism, inflammatory processes, and cell differentiation and migration. We here focus on PPARα and PPARγ. Both PPARα and PPARγ are activated by endogenous and dietary fatty acids (FAs) and their derivatives, and by numerous other synthetic or natural ligands [27–30]. PPARα is the target of the fibrate class of lipid-lowering drugs used as a therapeutic strategy against dyslipidemia [3,31,32], PPARγ is a target of antidiabetic thiazolidinediones [3] and anti-inflammatory compounds [33]. Activation of PPARs may protect against AD by repression of pro-inflammatory pathways or by its anti-amyloidogenic, anti-oxidative, and insulin-sensitizing effects [7,34–44]. PPARα activators were demonstrated to diminish memory decline, reduce Aβ aggregation (WY-14643, 4-phenylbutyrate, gemfibrozil, and cinnamic acid), and reduce tau phosphorylation, astrogliosis, microgliosis, and postsynaptic protein loss (WY-14643 and 4-phenylbutyrate) in AD mouse models [39,45–47]. The PPARγ agonist rosiglitazone was also found to diminish memory deficits, reduce Aβ levels and p-Tau aggregates, and ameliorate the cytotoxic amoeboid morphology of microglia in AD mice [34,48,49]. Beneficial effects of PPARγ agonists pioglitazone and rosiglitazone on the cognitive functioning of patients with AD, especially in those with co-morbid diabetes, have also been reported [50–54]. Altogether, these data underline the potential of LXR and PPAR agonists as well as their concurrent activation for therapeutic applications in AD and other metabolic and inflammatory disorders.

While a *S. fusiforme* lipid extract and 24(S)-saringosterol both prevented cognitive decline in AD mice, a reduction in amyloid deposition was observed exclusively after *S. fusiforme* extract administration [24,26]. Based on the literature reporting a reduction in Aβ levels via activation of PPARα or PPARγ, components in *S. fusiforme* other than saringosterol may activate PPARα or PPARγ. The application of *S. fusiforme* originating from the East Asian coast in Europe is complicated by legislative reasons and by the required specific growth conditions. Therefore, we aim to identify European brown seaweeds with LXR and PPAR-activating capacities comparable to that of *S. fusiforme*. In this study, we assessed the efficacy of lipid extracts of six European brown seaweed species for activation of LXR and PPAR and their effect on target gene expression in cultured cells.

2. Materials and Methods

2.1. Seaweed Species

The brown seaweed species *Himanthalia elongata* (*H. elongata*), *Sargassum muticum* (*S. muticum*), *Alaria esculenta* (*A. esculenta*), *Ascophyllum nodosum* (*A. nodosum*), *Fucus vesiculosus* (*F. vesiculosus*), and *Saccharina latissima* (*S. latissima*) were selected based on their European origin and their saringosterol content. *H. elongata* and *A. esculenta* were harvested in Ireland and provided by The Seaweed Company (Schiedam, The Netherlands). *S. muticum*, *A. nodosum*, *F. vesiculosus*, and *S. latissima* were harvested in The Netherlands and provided by Stichting Zeeschelp (Kamperland, The Netherlands). *S. fusiforme* was harvested in Japan and purchased from Terrasana BV (Leimuiden, The Netherlands). After harvest, the seaweeds were washed in seawater and dried by air.

2.2. Preparation of Seaweed Extracts

The dried seaweed samples were finely powdered in a mixer and soaked overnight in a 2:1 (v/v) chloroform/methanol mixture upon exposure to Ultraviolet-C (UVC) light (wavelengths between 200–280 nm) at room temperature. After 10 min of sonification, these mixtures were filtered using Whatman filter paper. The filtrates were evaporated in a vacuum rotary evaporator at 40 °C. The remaining lipid fractions were washed with 100% ethanol, again evaporated in the rotary evaporator, and dissolved in 100% ethanol to obtain the final lipid extracts.

2.3. Sterol Analysis

In the crude seaweeds and seaweed extracts, the concentrations of saringosterol and its precursor fucosterol were determined using gas chromatography/mass spectrometry as previously described [55]. In summary, the seaweed samples were first dried using a speed vacuum dryer to relate sterol concentrations to dry weight (DW). Then, the sterols were extracted from the dried tissues by adding a mixture of chloroform-methanol. A volume of 1 mL of the extracts was evaporated to dryness and mixed with 1 mL distilled water. To extract the neutral sterols, 3 mL of cyclohexane was added twice. The combined cyclohexane phases were evaporated under a stream of nitrogen, and the sterols were dissolved in n-decane. The sterols were then converted to trimethylsilyl ethers (TMSis) and incubated at 60 °C for 1 h [56]. Saringosterol and fucosterol levels were then determined using gas chromatography-mass spectrometry (GC-MS).

2.4. Lipomics Analysis

A UPLC MS/MS method was used for the lipidomics analysis of the seaweed extracts. Chromatographic experiments were performed on a column of InfinityLab Poroshell 120 EC-C18 (2.1 × 150 mm, 2.7 μm) using a Thermo UltiMate 3000 UPLCTM system (Thermo Fisher Scientific, Waltham, USA). The column temperature was set at 35 °C with an injection volume of 1 μL, and the mobile phase consisted of solvent A (0.1% formic acid and 10 mM ammonium formate in acetonitrile/water = 60:40) and solvent B (0.1% formic acid and 10 mM ammonium formate in isopropanol/acetonitrile = 90:10) at a flow rate of 0.3 mL/min. The gradient elution procedure was as follows: 0–1.5 min, 40% B; 1.5–10.5 min, 40–85% B; 10.5–14 min, 85% B; 14–14.1 min, 85–100% B; 14.1–15 min, 100% B; 15–15.2 min, 100–40% B; and 15.2–18 min, 40% B [57]. Mass spectrometry was performed on a Q Exactive™ Focus Orbitrap™ (Thermo Fisher Scientific). The instrument was operated using a full MS/dd-MS2 mode detection, in accordance with our previous study [58]. The UPLC-MS/MS raw data file were imported into Progenesis QI (Waters, Milford, CT, USA) for matching, alignment, and normalization. The relative content in this manuscript refers to the relative content of a molecular species in its sub class.

2.5. Arsenic and Cadmium Analysis

Seaweed samples were pulverized with zirconium oxide balls in a jar using a Retsch Vibration mill MM 2000 and divided into three replicates for each species. Next, approximately 100 mg for each sample was placed in open heat-resistant glass tubes (SCHOTT DURAN®, Rye Brook, NY, USA) and digested at 110 °C in a heating block using 69% HNO_3 (ARISTAR®, Leicestershire, England for trace analysis) three times and the last time using 37% HCl (ARISTAR® for trace analysis). Lastly, the samples were dissolved in a 2% HCl solution (diluted with Milli-Q H_2O). The concentrations of arsenic (As) and cadmium (Cd) were quantified via inductively coupled plasma-optical emission spectrometry (ICP-OES 710, Agilent Technologies, Amstelveen, The Netherlands).

2.6. Cell Culture

Immortalized human endothelial kidney cells (HEK293; Merck, Amsterdam, The Netherlands), human astrocytoma cells (CCF-STTG1; Merck), human neuroblastoma cells (SH-SY5Y; American Type Culture Collection (ATCC)), and human microglia cells (CHME3; a kind gift from prof. Dr. M. Tardieu, Université Paris-Sud, France) were used for the reporter assays and gene expression studies. All cell lines were cultured in DMEM/F-12 medium (ThermoFisher Scientific, Waltham, MA, USA) supplemented with 10% heat-inactivated fetal calf serum (FCS) (Thermo Fisher Scientific) and 1% 10,000 U penicillin/10,000 μg streptomycin/mL (Thermo Fisher Scientific) at 37 °C and 5% CO_2.

2.7. Cell Transfection

The LXR and PPAR-activating capacity of the seaweed extracts was determined in a cell-based reporter assay previously described by Zwarts et al. [59]. For this purpose, 1.0×10^6 cells were plated in T-25 culture flasks and after 24 h transfected by exposing the cells for 24 h to 1000 ng of pcDNA3.1/V5H6 vector containing clones of the full-length cDNAs for the murine nuclear receptors LXRα, LXRβ, PPARα or PPARγ, 1000 ng of vector encoding RXRα and 4000 ng of vectors encoding LXRE or PPRE using FuGENE® 6 reagent (Promega, Leiden, The Netherlands) according to the manufacturer's instructions. Control conditions included cells transfected with 2000 ng of the RXRα-containing vector and 4000 ng of the LXRE- or PPRE-containing vector, and cells transfected with 2000 ng of an empty pcDNA3.1/V5-HisA vector (Invitrogen, Carlsbad, CA, USA) and 4000 ng of the LXRE- or PPRE-containing vector. All cells were co-transfected with 1000 ng of Renilla to normalize for variation in transfection efficiency.

2.8. LXR and PPAR Reporter Assays

Transfected cells were seeded in a 96-well luminescence plate. After 24 h, the cells were incubated for 24 h in phenol red-free DMEM/F-12 medium (ThermoFisher Scientific) with the seaweed extracts (dosages based on saringosterol content), the LXRα/β agonist T0901317 (1 μM; #293754-55-9; Cayman, Ann Arbor, MI, USA), PPARα agonist WY-14643 (50 μM; #50892-23-4; Merck), PPARγ agonist pioglitazone (10 μM; #111025-46-8; Cayman), or the extract/compound solvents ethanol or DMSO. Cells were lysed in 25 μL lysis buffer and the Firefly and Renilla luminescent signals were measured using the Dual-Luciferase® Reporter assay system (Promega) and a Victor X4 plate reader (PerkinElmer, Groningen, The Netherlands). The relative receptor activity was defined as the ratio of Firefly luminescence to Renilla luminescence. The fold change was defined as the ratio of the relative receptor activity of seaweed- or agonist-exposed cells to the relative receptor activity of ethanol-exposed cells. The experiments, with the stimulation performed in triplicate, were repeated $\geq$ three times.

2.9. Quantitative Real-Time PCR

Cells were incubated for 24 h with the seaweed extracts (dosages based on saringosterol content), RXR agonist bexarotene (1 μM; #153559-49-0, Merck), the LXRα/β agonist T0901317 (1 μM), PPARα agonist WY-14643 (50 μM), PPARγ agonist pioglitazone (10 μM), or the extract or compound solvent ethanol or DMSO. Cells were washed with cold phosphate-buffered saline and RNA was isolated using Trizol (Thermo Fisher Scientific) and reverse transcribed to cDNA using the Maxima H Minus First Strand cDNA Synthesis Kit with dsDNase (Thermo Fisher Scientific), according to the manufacturer's instructions. Quantitative real-time PCR (qPCR) was conducted in duplicate with 10 ng cDNA on a CFX384 Thermal Cycler (Bio-Rad Laboratories) using the PowerTrack™ SYBR Green Master Mix (Applied Biosystems) and the following cycling conditions: 95 °C for 2 min and 40 cycles of [95 °C for 15 s, 60 °C for 60 s]. The intron-spanning primers for qPCR were designed with Primer-BLAST [60]. Primer sequences are listed in Table 1. Relative quantification of the gene expression was accomplished with the comparative Ct method. The data were normalized to five reference genes (*ACTB*, *B2M*, *HPRT1*, *SDHA*, and *YWHAZ*) and expressed as fold change relative to the EtOH or DMSO control. The experiments were performed three times.

Table 1. Primers and their corresponding forward and reverse nucleotide sequences.

Gene	Gene Name	Primer Sequence
ABCA1	ATP Binding Cassette Subfamily A Member 1	F: TCTCTGTTCGGCTGAGCTAC R: TGCAGAGGGCATGGCTTTAT
ABCG1	ATP Binding Cassette Subfamily G Member 1	F: GGTCGCTCCATCATTTGCAC R: GCAGACTTTTCCCCGGTACA
ACACA	Acetyl-CoA Carboxylase Alpha	F: GGGTCAAGTCCTTCCTGCTC R: GGACTGTCGAGTCACCTTAAGTA
ACTB	Actin Beta	F: CTCCCTGGAGAAGAGCTACG R: GAAGGAAGGCTGGAAGAGTG
APOE	Apolipoprotein E	F: ACCCAGGAACTGAGGGC R: CTCCTTGGACAGCCGTG
B2M	Beta-2-Microglobulin	F: CTCCGTGGCCTTAGCTGTG R: TTTGGAGTACGCTGGATAGCCT
DHCR7	7-Dehydrocholesterol Reductase	F: TGGGCCAAGACTCCACCTAT R: ACGTGTACAGAAGCACCTGG
DHCR24	24-Dehydrocholesterol Reductase	F: GTCTCACTACGTGTCGGGAA R: CTCCACACGGACAATCTGTTTC
FASN	Fatty Acid Synthase	F: CACAGACGAGAGCACCTTTGA R: CAGGTCTATGAGGCCTATCTGG

Table 1. *Cont.*

Gene	Gene Name	Primer Sequence
GFAP	Glial Fibrillary Acidic Protein	F: GGCCCGCCACTTGCA R: GGGAATGGTGATCCGGTTCT
HMGCR	3-Hydroxy-3-Methylglutaryl-CoA Reductase	F: GCAGGACCCCTTTGCTTAGA R: GCACCTCCACCAAGACCTAT
HPRT1	Hypoxanthine Phosphoribosyltransferase 1	F: TGACACTGGCAAAACAATGCA R: GGTCCTTTTCACCAGCAAGCT
SCD1	Stearoyl-CoA Desaturase 1	F: GCTGTCAAAGAGAAGGGGAGT R: AGCCAGGTTTGTAGTACCTCCT
SDHA	Succinate Dehydrogenase Complex Flavoprotein Subunit A	F: TGGGAACAAGAGGGCATCTG R: CCACCACTGCATCAAATTCATG
SREBF1	Sterol Regulatory Element Binding Transcription Factor 1	F: ACAGCCATGAAGACAGACGG R: CAAGATGGTTCCGCCACTCA
SREBF2	Sterol Regulatory Element Binding Transcription Factor 2	F: GATCACGCCAACATTCAGCA R: GACTTGAGGCTGAAGGACTTGAA
YWHAZ	Tyrosine 3-Monooxygenase/Tryptophan 5-Monooxygenase Activation Protein Zeta	F: ACTTTTGGTACATTGTGGCTTCAA R: CCGCCAGGACAAACCAGTAT

2.10. Statistical Analysis

The data are presented as mean ± SD. Extreme values were excluded using Dixon's principles of exclusion of extreme values [61,62]. Statistical analyses were performed on the data of the LXR and PPAR reporter assays using GraphPad Prism 8. The D'Agostino-Pearson normality test was used to test normal distribution. The fold change values (treatment vs. ethanol control) were analyzed using a Kruskal–Wallis test. Significance levels are denoted as follows: * $p \leq 0.05$, ** $p \leq 0.01$, *** $p \leq 0.001$.

3. Results

3.1. Characteristics of the Tested Seaweeds and Seaweed Extracts

Because of the observed beneficial effects of *S. fusiforme* in models for AD and atherosclerosis [24,63], we searched for European brown seaweeds with similar effects. Six European brown seaweed species were analyzed for their saringosterol and fucosterol concentrations and compared with *S. fusiforme* (Table 2). Crude *F. vesiculosus*, *S. muticum*, and *H. elongata* contained the highest saringosterol concentrations, comparable to that of *S. fusiforme*. These seaweeds as well as *A. nodosum* also contained the highest fucosterol concentrations. The extract of *H. elongata* was found to contain the highest concentration of saringosterol, comparable to the concentrations found in *S. fusiforme* extract, while extracts of *F. vesiculosus*, *S. muticum*, and *A. nodosum* contained the highest fucosterol concentrations. Next, we conducted a targeted analysis of the composition of glycerolipids. The molecular characteristics of phospholipids, glyceroglycolipids, diglycerides, triglycerides, and fatty acids in each seaweed extract are summarized in Supplementary Tables S1–S3.

Predominant FAs found in the extracts were the saturated FAs myristic acid (C14:0) and palmitic acid (C16:0), monounsaturated variants palmitoleic acid (C16:1) and oleic acid (C18:1), and polyunsaturated FAs (PUFAs) linoleic acid (C18:2/C18:3), parinaric acid (C18:4), arachidonic acid (C20:4) and EPA (C20:5). The odd-chain saturated fatty acid heptadecanoic acid (C17:0) was also found to be one of the predominant molecular species in all the seaweed extracts tested, with the exception of the *A. esculenta* extract. These natural FAs and eicosanoids, and other lipids found in the extracts, can serve as ligands for PPARs (Supplementary Table S4).

Table 2. Saringosterol and fucosterol concentrations in crude seaweed and seaweed lipid extracts.

Seaweed Species	Crude Seaweed		Extract	
	Saringosterol (μg/mg DW)	Fucosterol (μg/mg DW)	Saringosterol (mM)	Fucosterol (mM)
Alaria esculenta	0.008	0.130	0.2	9.9
Ascophyllum nodosum	0.002	0.495	0.2	12.2
Fucus vesiculosus	0.034	0.407	0.7	13.0
Himanthalia elongata	0.018	0.771	1.8	7.3
Saccharina latissima	0.002	0.037	0.7	6.1
Sargassum fusiforme	0.026	0.209	1.1	7.0
Sargassum muticum	0.032	0.325	0.1	12.4

Because of the known high arsenic concentrations in *S. fusiforme* that limits the amount that can be consumed safely, we determined arsenic concentrations in all seaweeds. As expected, *S. fusiforme* contained the highest concentration of arsenic (67.26 ± 3.34 mg/kg DW), followed by *S. muticum* (28.95 ± 1.14 mg/kg DW), *S. latissima* (24.62 ± 1.59 mg/kg DW), *A. esculenta* (16.79 ± 1.7 mg/kg DW), *H. elongata* (13.58 ± 0.37 mg/kg DW), *F. vesiculosus* (10.37 ± 0.49 mg/kg DW), and *A. nodosum* (7.64 ± 0.32 mg/kg DW). Cadmium was detected in *S. fusiforme* (1.03 ± 0.02 mg/kg DW), but not in the other seaweeds.

3.2. LXR Activating Capacity

The dosage of the lipid extracts in the cell experiments was based on the saringosterol concentrations, with the highest dose being the maximal dose tolerated by the cells (dilution factors presented in Supplementary Table S5). The lipid extracts of all the seaweeds activated LXRα and LXRβ, although to a different extent and in a cell type-specific manner (Figures 1–3 and Supplementary Figure S1). LXRs were mostly activated by *H. elongata* comparable to *S. fusiforme*, followed by *S. muticum* and *S. latissima*. The extracts of *F. vesiculosus*, *A. esculenta*, and *A. nodosum* showed LXR activation in HEK and CHME3 cells, exclusively.

3.3. PPARα and PPARγ Activation by the Seaweed Extracts

PPARα was most strongly activated by extracts of *A. esculenta*, *F. vesiculosus*, and *S. latissima* (Figure 4 and Supplementary Figure S2), while activation of PPARγ was strongest by the extracts of *A. esculenta*, *A. nodosum*, *F. vesiculosus*, and *S. muticum*, overall, to a lesser extent than PPARα (Figure 5 and Supplementary Figure S2). The relative increase in activation of both PPARα and PPARγ was found to be most pronounced in SH-SY5Y cells (Figures 4–6).

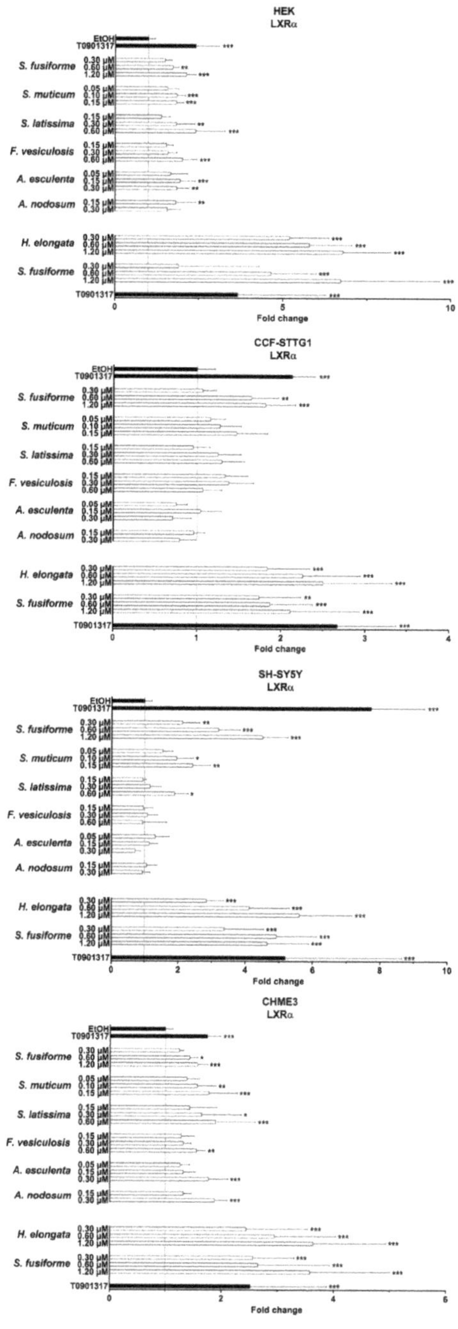

Figure 1. LXRα activation by seaweed extracts. The seaweed extracts were screened for their LXRα activating ability in HEK, CCF-STTG1, SH-SY5Y, and CHME3 cells. Saringosterol concentrations are presented on the Y-axis. The fold change values are presented as mean ± SD of three experiments performed in triplicate (n = 9). Significance relative to the ethanol control (Kruskal–Wallis test): * $p \leq 0.05$, ** $p \leq 0.01$, *** $p \leq 0.001$.

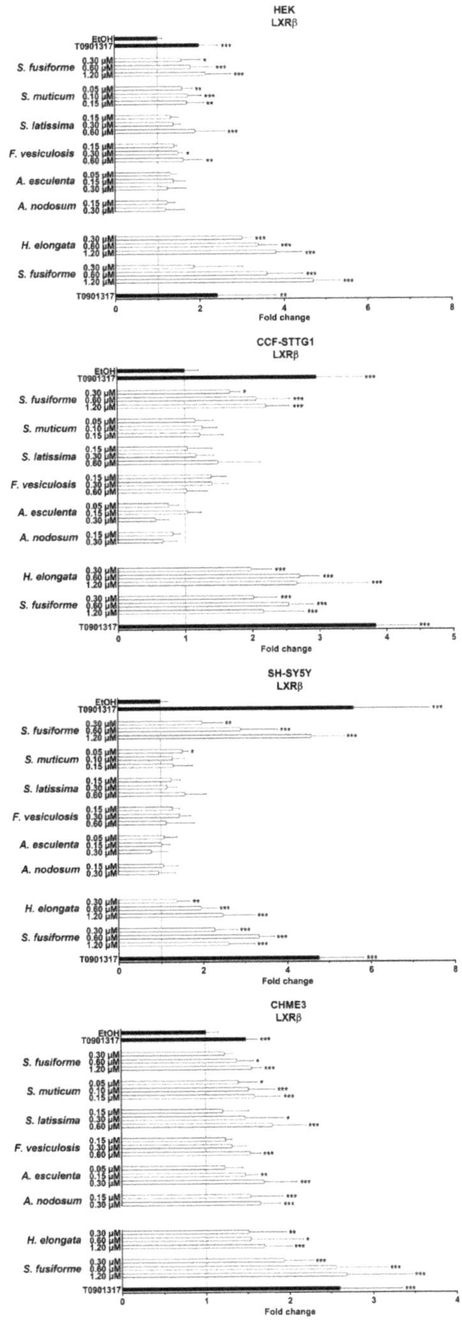

Figure 2. LXRβ activation by seaweed extracts. The seaweed extracts were screened for their LXRβ activating ability in HEK, CCF-STTG1, SH-SY5Y, and CHME3 cells. Saringosterol concentrations are presented on the *Y*-axis. The fold change values are presented as mean ± SD of three experiments performed in triplicate (n = 9). Significance relative to the ethanol control (Kruskal–Wallis test): * $p \leq 0.05$, ** $p \leq 0.01$, *** $p \leq 0.001$.

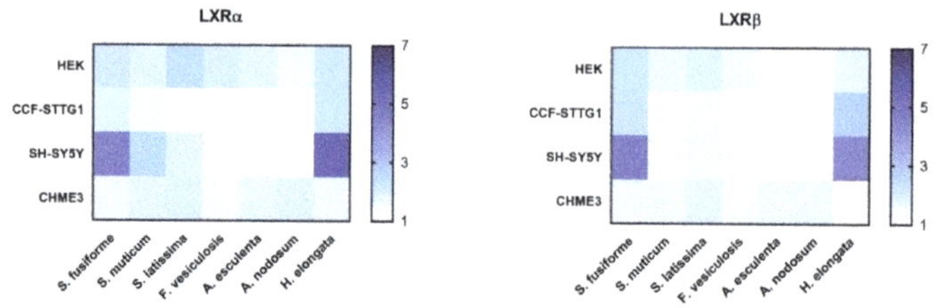

Figure 3. LXRα and LXRβ activation by seaweed extracts. The figures present the fold changes of the highest extract dose of each extract, with the fold change of *H. elongata* corrected for variation in the fold change of *S. fusiforme* in the two experiments.

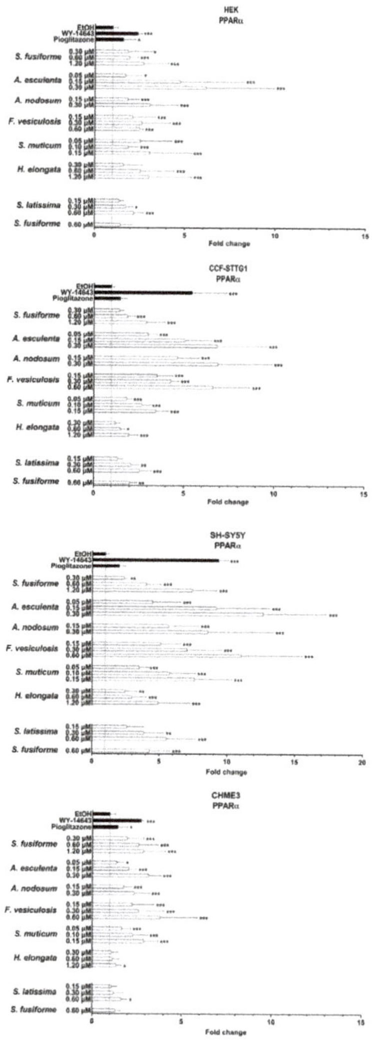

Figure 4. PPARα activation by seaweed extracts. The seaweed extracts were screened for their PPARα activating capacity in HEK, CCF-STTG1, SH-SY5Y, and CHME3 cells. Saringosterol concentrations

are presented on the Y-axis. The fold change values are presented as mean ± SD of ≥three experiments performed in triplicate (n ≥ 9). Significance relative to the ethanol control (Kruskal–Wallis test): * $p \leq 0.05$, ** $p \leq 0.01$, *** $p \leq 0.001$.

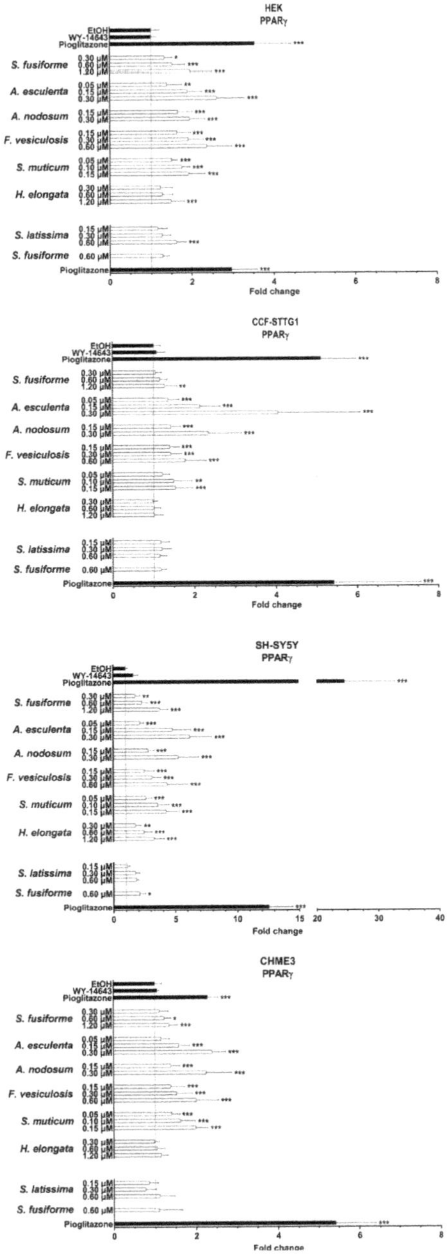

Figure 5. PPARγ activation by seaweed extracts. The seaweed extracts were screened for their PPARγ activating capacity in HEK, CCF-STTG1, SH-SY5Y, and CHME3 cells. Saringosterol concentrations are presented on the Y-axis. The fold change values are presented as mean ± SD of ≥three experiments performed in triplicate (n ≥ 9). Significance relative to the ethanol control (Kruskal–Wallis test): * $p \leq 0.05$, ** $p \leq 0.01$, *** $p \leq 0.001$.

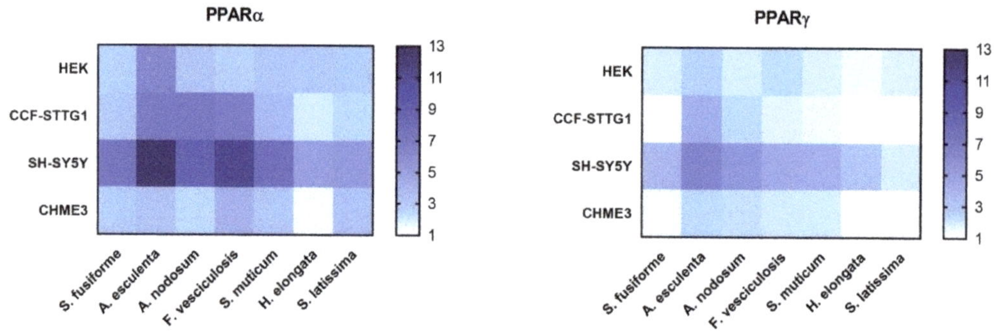

Figure 6. PPARα and PPARγ activation by seaweed extracts. The figures present the fold changes of the highest extract dose of each extract, with the fold change of *S. latissima* corrected for variation in the fold change of *S. fusiforme* in the two experiments.

3.4. Effect of the Seaweed Extracts on the Expression of LXR and PPAR-Target Genes

We examined the effects of the seaweed extracts on gene expression involved in cholesterol efflux and lipid synthesis in astrocytoma cells (CCF-STTG1), given the crucial role of astrocytes in the cerebral cholesterol metabolism and their contribution to the supply of cholesterol to neurons. All extracts, although to a different extent, induced the expression of the LXR target genes *ABCG1*, *ABCA1*, and *APOE* involved in cholesterol efflux (Figure 7). The extracts suppressed the expression of *DHCR7*, *DHCR24*, *HMGCR*, and *SREBF2* involved in cholesterol synthesis (Figure 8) as well as *SREBF1*, *ACACA*, *SCD1*, and *FASN* involved in fatty acid synthesis (Figure 9). The expression of *DHCR24* and *HMGCR* was also reduced by bexarotene and T0901317 (Figure 8); however, as expected, both bexarotene and T0901317 increased the expression of the tested fatty acid synthesis genes (Figure 9). The expression of the glial fibrillary acidic protein (*GFAP*), the gene encoding an astrocytic structural protein indicative of astrogliosis, was decreased by the extracts, while T0901317, WY14643, and pioglitazone increased its expression (Figure 10).

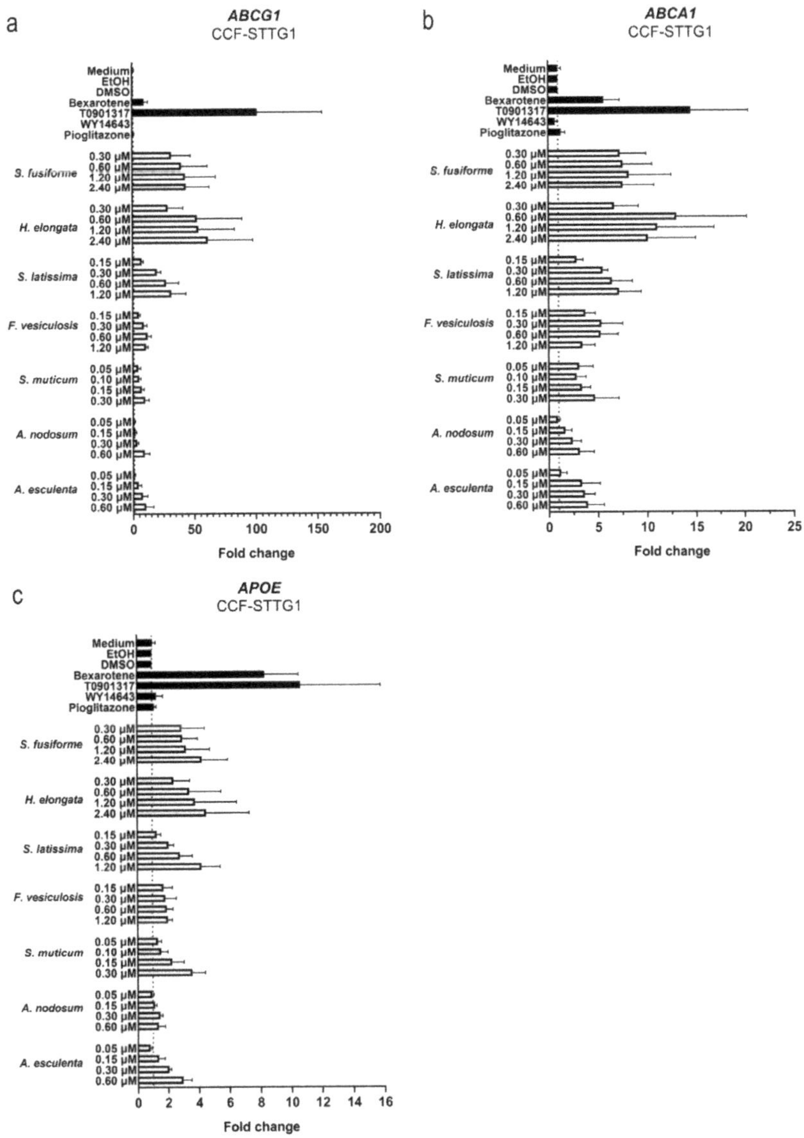

Figure 7. Effect of seaweed extracts on the expression of LXR target genes involved in cholesterol efflux. The expression of cholesterol efflux genes *ABCG1* (**a**), *ABCA1* (**b**), and *APOE* (**c**) in CCF-STTG1 cells was increased by all tested seaweed extracts. Saringosterol concentrations are presented on the *Y*-axis. The experiments were performed three times (n = 3).

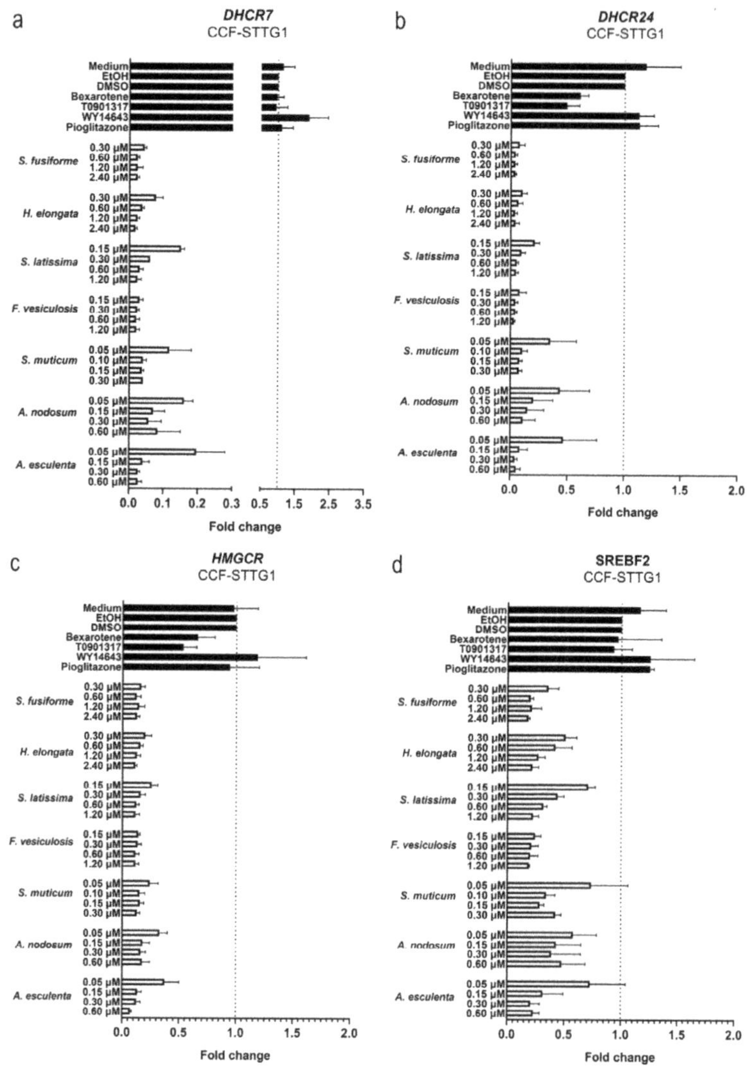

Figure 8. Effect of seaweed extracts on the expression of LXR target genes involved in cholesterol synthesis. The expression of cholesterol synthesis genes *DHCR7* (**a**), *DHCR24* (**b**), *HMGCR* (**c**), and *SREBF2* (**d**) in CCF-STTG1 cells was decreased by all tested seaweed extracts. Saringosterol concentrations are presented on the *Y*-axis. The experiments were performed three times (n = 3).

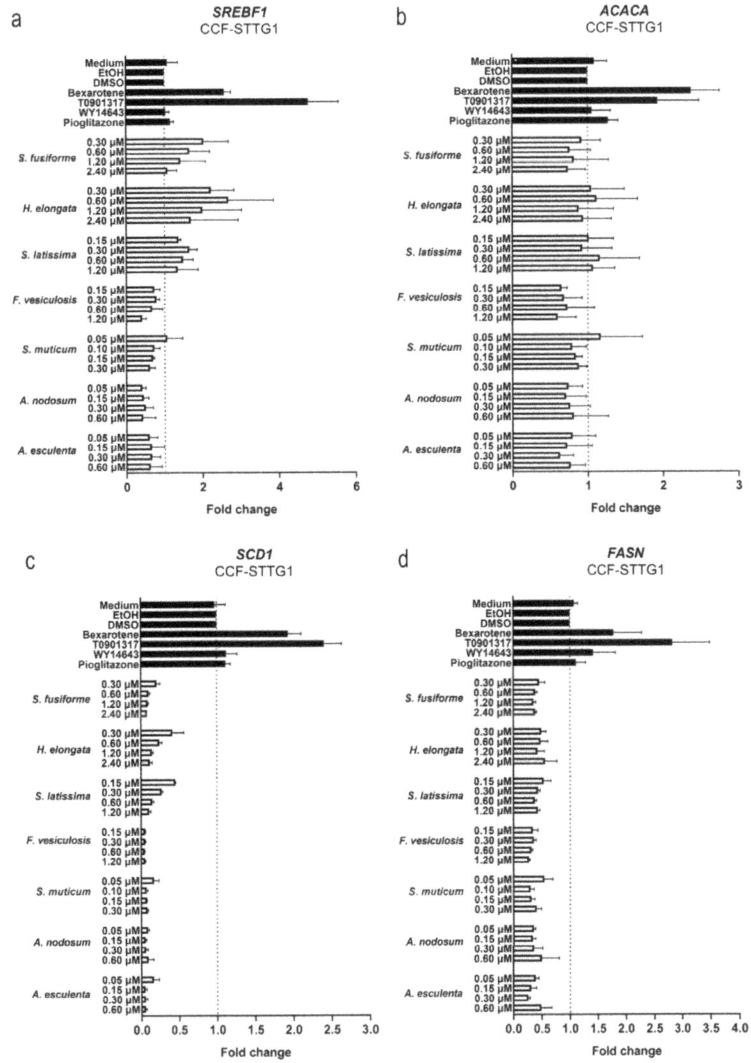

Figure 9. Effect of seaweed extracts on the expression of LXR target genes involved in fatty acid synthesis. The expression of fatty acid synthesis genes *SREBF1* (**a**), *ACACA* (**b**), *SCD1* (**c**), and *FASN* (**d**) in CCF-STTG1 cells was decreased by tested seaweed extracts. Saringosterol concentrations are presented on the *Y*-axis. The experiments were performed three times (n = 3).

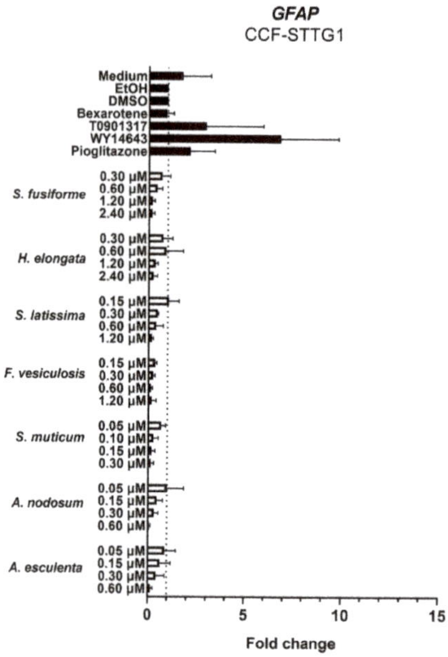

Figure 10. Extracts decreased the expression of *GFAP* in CCF-STTG1 cells. Saringosterol concentrations are presented on the *Y*-axis. The experiments were performed three times (n = 3).

4. Discussion

LXRs and PPARs are recognized as interesting therapeutic targets for cardiometabolic, inflammatory, and neurodegenerative disorders such as AD because their activation can beneficially impact the pathology of these diseases via modulation of lipid metabolism and/or inflammatory processes. Therefore, we have screened lipid extracts of six European seaweeds for their ability to activate LXRs and PPARs. We demonstrate that all six seaweeds tested contained the LXR-activating oxyphytosterol saringosterol as well as multiple PPAR-activating FAs. The *H. elongata* extract with the highest saringosterol content, but not the highest fucosterol concentration, displayed the highest efficacy for activation of LXRs and subsequently for modulation of LXR-target gene expression, comparable to the *S. fusiforme* extract. *S. muticum*, *S. latissima*, *F. vesiculosus*, *A. esculenta*, and *A. nodosum* also activated LXRs and regulated LXR-target gene expression, albeit to a lesser extent than *H. elongata*. PPARα was most strongly activated by *A. esculenta*, *F. vesiculosus* and *S. latissima*, and PPARγ by *A. esculenta*, *A. nodosum*, *F. vesiculosus* and *S. muticum*. The concurrent activation of LXRs and PPARs by the seaweed extracts may combine the beneficial effects of both. These data are supportive of the potential of brown seaweeds for prevention and/or treatment of neurodegenerative, and possibly also cardiometabolic and inflammatory diseases.

Accumulating evidence suggests that disturbances in brain cholesterol homeostasis are linked to AD pathogenesis. An altered cholesterol turnover rate and altered intra- and intercellular distribution of cholesterol in the brain rather than altered cholesterol levels seem to be involved in neuropathologies [64,65]. LXR activation is believed to prevent AD progression by enhancing cholesterol efflux through the secretion of ApoE-containing lipoprotein-like particles, which provide neurons with cholesterol and other lipids that support synaptic plasticity and neuronal regeneration after injury [66]. These processes are promoted by 24-hydroxycholesterol, which is an endogenous LXR agonist that is formed in neurons from cholesterol via the enzyme CYP46A1. Upregulation of the conversion of cholesterol into 24-hydroxycholesterol via activation of CYP46A1 has been shown to protect against pathologies involved in Alzheimer's, Huntington's, and Parkinson's disease, multiple sclerosis, and amyotrophic lateral sclerosis [67], and it is currently being tested in a phase I trial in AD patients (NCT03706885). A 24-oxidized sterol, 24(S)-Saringosterol, similar to 24(S)-hydroxycholesterol, has the ability to cross the blood–brain barrier, and has previously been shown to prevent cognitive decline in AD mice, possibly by exerting effects similar to 24-hydroxycholesterol [24,26]. Fucosterol can also cross the blood–brain barrier, although to a lesser extent than saringosterol, and can also activate LXRs likely indirectly via upregulation of endogenous LXR agonist desmosterol (unpublished observation). FAs are continuously transported in and out of the brain [68]. Our current data show that LXRs were activated by lipid extracts of brown seaweeds, mostly by the extracts of *S. fusiforme*, *H. elongata*, *S. muticum*, and *S. latissima*. In CCF-STTG1 astrocytoma cells, the extracts increased the expression of genes involved in cholesterol efflux, with the most pronounced effects observed with extracts of *S. fusiforme* and *H. elongata*, which is consistent with the relatively high concentrations of saringosterol in these extracts. By increasing the ABCA1- and ABCG1-promoted secretion of ApoE-containing particles by astrocytes, the extracts may thus promote protective functions related to neuronal regeneration, synaptogenesis, Aβ clearance, and neuroinflammation [69–71].

While the extracts induced genes related to cholesterol efflux, they decreased the expression of genes involved in cholesterol synthesis and fatty acid synthesis in CCF-STTG1 cells. The inhibition of SREBP activation is in line with previous findings showing that (oxy)sterols facilitate the binding of SREBP cleavage-activating protein (SCAP) to Insigs preventing SCAP/SREBP binding and the subsequent SREBP activation. In this way, (oxy)sterols reduce lipogenesis [72]. This feedback mechanism can be mediated by endogenous (oxy)sterols such as cholesterol and 25-hydroxycholesterol [73], but possibly also by exogenous (oxy)sterols such as those contained in seaweed. We cannot rule out a contribution of RXR homodimers to the observed effects of the seaweed extracts because the RXR agonist bexarotene also upregulated cholesterol efflux and downregulated cholesterol synthesis genes *DHCR24* and *HMGCR* similar to T0901317 and the extracts.

Although purified 24(S)-saringosterol was able to prevent cognitive decline in AD mice [26], PPARα and PPARγ activation may also have contributed to the previously reported prevention of amyloid pathology and cognitive decline in AD mice [7,34,39,45–49,74]. Natural PPAR agonists may exert fewer side effects than full-blown synthetic PPAR agonists such as thiazolidinediones which have serious adverse effects [30]. PPARα and PPARγ were activated by the extracts of *A. esculenta, A. nodosum* and *F. vesiculosus* and to a lesser extent by *S. fusiforme, H. elongata* and *S. latissima*. PPARα is the most promiscuous isoform of the PPARs and interacts with saturated and unsaturated FAs [75]. The FA binding profile of PPARγ is the most restricted of the three isoforms, interacting most efficiently with PUFAs and only weakly with monounsaturated FAs [75]. We found a variety of saturated and unsaturated FAs in the extracts that are known for their PPAR-activating capacity. However, other unknown natural ligands may also be present in the extracts and activate PPARs. The *A. esculenta* extract was the most effective PPARα and -γ activator and also contained the highest relative amount of total PUFAs, including known PPAR-activating FAs. The cell lines may have responded differently due to variations in the presence of co-factors.

LXR and PPAR activation has been demonstrated to have anti-inflammatory effects via transrepression of inflammatory transcription factors such as NFκB and AP-1 [76–78]. GFAP is a commonly used marker of astrogliosis related to inflammation and is demonstrated to be elevated in plasma of AD patients [79–81]. The reduction in *GFAP* expression observed in CCF-STTG1 cells incubated with seaweed extracts suggests that the extracts may decrease astrogliosis and inflammation, associated with AD and other neurological conditions. However, we have not yet further investigated the potential anti-inflammatory effects of the seaweed extracts induced via LXRα/β and PPARα/γ activation.

One limitation of current experiments is that central nervous system (CNS) cell lines were incubated with the entire seaweed extracts, which contain both known and unknown lipids. However, it is known that only certain lipids, such as (oxy)sterols and FAs, can cross the blood–brain barrier and reach the brain. Therefore, further studies are necessary to verify the effects of tested seaweed extracts on lipid metabolism in the CNS in an in vivo setting, where peripheral effects as well as the crosstalk between the brain and periphery can be explored. It should be noted that extracts were added based on saringosterol concentrations, resulting in comparable but not identical dilutions. While our results demonstrate that the extracts can activate the nuclear receptors LXR and PPAR and affect the expression of several target genes, we cannot draw conclusions on their relative efficiencies. Additionally, we did not consider the harvest period of the seaweeds, which may have resulted in seasonal differences affecting our results, and only total arsenic levels have been analyzed, not the toxic inorganic form separately.

5. Conclusions

Our findings indicate that the lipophilic fractions of six different European brown seaweed species possess the ability to activate LXRs and PPARs in both peripheral and CNS cell lines and can influence lipid metabolism in astrocytoma cells. These findings provide a basis for further investigation into their application in the prevention and/or treatment of neurodegenerative, metabolic, and inflammatory diseases. Among the European seaweed species studied, *Himanthalia elongata* demonstrated the highest efficacy for LXR activation, comparable to that of *Sargassum fusiforme*, and comparably regulated cholesterol efflux and lipid synthesis pathways in astrocytoma CCF-STTG1 cells. Previous studies have demonstrated that *Sargassum fusiforme* has a positive impact on cognitive performance and AD pathology in an AD mouse model [24], and therefore, *Himanthalia elongata* could be a promising alternative to *Sargassum fusiforme* for the prevention of AD pathology.

Supplementary Materials: The following supporting information can be downloaded at: https://www.mdpi.com/article/10.3390/nu15133004/s1, Table S1. The molecular composition and relative content of phospholipid in seaweed extracts. Table S2. The molecular composition and relative content of glyceroglycolipid, diglyceride and triglyceride in seaweed extracts. Table S3. The molecular composition and relative content of fatty acids in seaweed extracts. Table S4. Literature on the PPAR-activating capacities of lipids found in the seaweed extracts. Table S5. Dilution factors for lipid extract and corresponding saringosterol concentrations. Figure S1. LXRα and -β activation by seaweed extracts. Lipid extracts of the selected seaweed species were screened for their LXRα- and LXRβ-activating capacity in HEK (a), CCF-STTG1 (b), SH-SY5Y (c), and CHME3 (d) cells transfected with pcDNA3.1/V5H6 vectors with or without RXR, LXRα, and LXRβ. Extract dosages were based on saringosterol content (concentrations noted on the X-axis). Data are corrected for transfection efficiency (by correcting for differences in Renilla signal) and expressed as fold change compared to the EtOH control. The fold change values are presented as mean ± SD of three experiments performed in triplicate (n = 9). Significance relative to the EtOH was tested with a Kruskal-Wallis test: * $p \leq 0.05$, ** $p \leq 0.01$, *** $p \leq 0.001$. Figure S2: PPARα and -γ activation by seaweed extracts. Lipid extracts of the selected seaweed species were screened for their PPARα- and PPARγ-activating capacity in HEK (a), CCF-STTG1 (b), SH-SY5Y (c), and CHME3 (d) cells transfected with pcDNA3.1/V5H6 vectors with or without RXR, PPARα, and PPARγ. Extract dosages were based on saringosterol content (concentrations noted on the Y-axis). Data are corrected for transfection efficiency (by correcting for differences in Renilla signal) and ex-pressed as fold change compared to the EtOH control. The fold change values are presented as mean ± SD of ≥three experiments performed in triplicate (n ≥ 9). Signifi-cance relative to the EtOH was tested with a Kruskal-Wallis test: * $p \leq 0.05$, ** $p \leq 0.01$, *** $p \leq 0.001$. References [82–97] are cited in the Supplementary Materials.

Author Contributions: M.T.M. and T.V. conceived and supervised the study; N.M., G.V., C.v.R. and S.v.L. performed the reporter assays; F.P.J.L. and N.Z. performed the qPCR experiments; X.G. performed the lipidomics analysis; M.H. performed the Arsenic and Cadmium analysis; H.L. was responsible for the lipid analysis of the extracts; J.W.J. and F.K. provided the reporter assay protocol and the plasmids for cell transfection, and were involved in data analysis and editing of the manuscript; D.L. was responsible for sterol analysis of the extracts; N.M. and M.T.M. were responsible for the experimental design, interpretation of the data, and writing the manuscript. All authors have read and agreed to the published version of the manuscript.

Funding: This research was funded by the Dutch Research Council (NWO-TTW) (#16437) and the Alzheimer Nederland and Alzheimer Forschung Initiative (#AFI-18024CB, #AFI-22034CB, #WE.03-2018-06 AN and # WE.03-2022-06, and #WE. 15-2021-08).

Data Availability Statement: Data are available upon request.

Acknowledgments: We thank Carine Put for her help in the Arsenic and Cadmium analysis. We thank The Seaweed Company and Stichting Zeeschelp for providing the European seaweeds. The authors acknowledge the financial support by the Dutch Research Council (NWO) and Alzheimer Nederland.

Conflicts of Interest: The authors declare that they have no conflicts of interest. The funders had no role in the design of the study; in the collection, analysis, or interpretation of data; or the writing of the manuscript.

Abbreviations

Aβ	Amyloid-β
ABCA1	ATP Binding Cassette Subfamily A Member 1
ABCG1	ATP Binding Cassette Subfamily G Member 1
ACACA	Acetyl-CoA Carboxylase Alpha
ACTB	Actin Beta
AD	Alzheimer's Disease
APOE	Apolipoprotein E
B2M	Beta-2-Microglobulin
cDNA	Complementary Deoxyribonucleic Acid
CNS	Central Nervous System
Ct	Cycle Threshold
DG	Diacylglycerol
DGDG	Digalactosyldiacylglycerol
DHCR7	7-Dehydrocholesterol Reductase
DHCR24	24-Dehydrocholesterol Reductase
DMSO	Dimethyl sulfoxide
DW	Dry Weight
EtOH	Ethyl Alcohol, Ethanol
FA	Fatty Acids
FASN	Fatty Acid Synthase
GFAP	Glial Fibrillary Acidic Protein
HMGCR	3-Hydroxy-3-Methylglutaryl-CoA Reductase
HPRT1	Hypoxanthine Phosphoribosyltransferase 1
IA	Index of Atherogenicity
IT	Index of Thrombogenicity
LPA	Lysophosphatidic Acid
LPC	Lysophosphatidylcholine
LPE	Lysophosphatidylethanolamine
LPG	Lysophosphatidylglycerol
LPI	Lysophosphatidylinositol
LTP	Long-term Potentiation
LXR	Liver X Receptor
LXRE	Liver X Receptor Response Element
MGDG	Monogalactosyldiacylglycerol
MGMG	Monogalactosylmonoacylglycerol
MUFA	Monounsaturated Fatty Acid
NR1C1	Nuclear Receptor Subfamily 1 Group C Member 1
NR1C3	Nuclear Receptor Subfamily 1 Group C Member 3
NR1H2	Nuclear Receptor Subfamily 1 Group H Member 2
NR1H3	Nuclear Receptor Subfamily 1 Group H Member 3
PA	Phosphatidic Acid
PC	Phosphatidylcholine
PCR	Polymerase Chain Reaction

PE	Phosphatidylethanolamine
PG	Phosphatidylglycerol
PI	Phosphatidylinositol
PPAR	Peroxisome proliferator-Activated receptors
PPRE	Peroxisome Proliferator Response Element
PS	Phosphatidylserine
PUFA	Polyunsaturated Fatty Acid
qPCR	Quantitative Polymerase Chain Reaction
RC	Relative Content
RT	Retention Time
RXR	Retinoid X Receptor
SCAP	Sterol Regulatory Element-Binding Protein Cleavage-Activating Protein
SCD1	Stearoyl-CoA Desaturase 1
SD	Standard Deviation
SDHA	Succinate Dehydrogenase Complex Flavoprotein Subunit A
SFA	Saturated Fatty Acid
SQDG	Sulfoquinovosyl Diacylglycerol
SREBF1	Sterol Regulatory Element Binding Transcription Factor 1
SREBF2	Sterol Regulatory Element Binding Transcription Factor 2
SREBP	Sterol Regulatory Element-Binding Protein
TG	Triacylglycerol
UI	Unsaturation Index
UV	Ultraviolet
YWHAZ	Tyrosine 3-Monooxygenase/Tryptophan 5-Monooxygenase Activation Protein Zeta

References

1. Dixon, E.D.; Nardo, A.D.; Claudel, T.; Trauner, M. The Role of Lipid Sensing Nuclear Receptors (PPARs and LXR) and Metabolic Lipases in Obesity, Diabetes and NAFLD. *Genes* **2021**, *12*, 645. [CrossRef]
2. Hong, C.; Tontonoz, P. Coordination of inflammation and metabolism by PPAR and LXR nuclear receptors. *Curr. Opin. Genet. Dev.* **2008**, *18*, 461–467. [CrossRef]
3. Monsalve, F.A.; Pyarasani, R.D.; Delgado-Lopez, F.; Moore-Carrasco, R. Peroxisome proliferator-activated receptor targets for the treatment of metabolic diseases. *Mediat. Inflamm.* **2013**, *2013*, 549627. [CrossRef]
4. Perl, D.P. Neuropathology of Alzheimer's disease. *Mt. Sinai J. Med.* **2010**, *77*, 32–42. [CrossRef]
5. Swerdlow, R.H. Pathogenesis of Alzheimer's disease. *Clin. Interv. Aging* **2007**, *2*, 347–359.
6. Luchsinger, J.A.; Tang, M.X.; Shea, S.; Mayeux, R. Hyperinsulinemia and risk of Alzheimer disease. *Neurology* **2004**, *63*, 1187–1192. [CrossRef]
7. Pierrot, N.; Ris, L.; Stancu, I.C.; Doshina, A.; Ribeiro, F.; Tyteca, D.; Baugé, E.; Lalloyer, F.; Malong, L.; Schakman, O.; et al. Sex-regulated gene dosage effect of PPARα on synaptic plasticity. *Life Sci. Alliance* **2019**, *2*, e201800262. [CrossRef]
8. Vanmierlo, T.; Rutten, K.; Dederen, J.; Bloks, V.W.; van Vark-van der Zee, L.C.; Kuipers, F.; Kiliaan, A.; Blokland, A.; Sijbrands, E.J.G. Liver X receptor activation restores memory in aged AD mice without reducing amyloid. *Neurobiol. Aging* **2011**, *32*, 1262–1272. [CrossRef]
9. Medrano-Jiménez, E.; Jiménez-Ferrer Carrillo, I.; Pedraza-Escalona, M.; Ramírez-Serrano, C.E.; Álvarez-Arellano, L.; Cortés-Mendoza, J.; Herrera-Ruiz, M.; Jiménez-Ferrer, E.; Zamilpa, A.; Tortoriello, J.; et al. Malva parviflora extract ameliorates the deleterious effects of a high fat diet on the cognitive deficit in a mouse model of Alzheimer's disease by restoring microglial function via a PPAR-γ-dependent mechanism. *J. Neuroinflamm.* **2019**, *16*, 143. [CrossRef]
10. Burns, M.P.; Vardanian, L.; Pajoohesh-Ganji, A.; Wang, L.; Cooper, M.; Harris, D.C.; Duff, K.; Rebeck, G.W. The effects of ABCA1 on cholesterol efflux and Abeta levels in vitro and in vivo. *J. Neurochem.* **2006**, *98*, 792–800. [CrossRef]
11. Koldamova, R.P.; Lefterov, I.M.; Staufenbiel, M.; Wolfe, D.; Huang, S.; Glorioso, J.C.; Walter, M.; Roth, M.G.; Lazo, J.S. The liver X receptor ligand T0901317 decreases amyloid beta production in vitro and in a mouse model of Alzheimer's disease. *J. Biol. Chem.* **2005**, *280*, 4079–4088. [CrossRef]
12. Lefterov, I.; Bookout, A.; Wang, Z.; Staufenbiel, M.; Mangelsdorf, D.; Koldamova, R. Expression profiling in APP23 mouse brain: Inhibition of Abeta amyloidosis and inflammation in response to LXR agonist treatment. *Mol. Neurodegener.* **2007**, *2*, 20. [CrossRef]
13. Riddell, D.R.; Zhou, H.; Comery, T.A.; Kouranova, E.; Lo, C.F.; Warwick, H.K.; Ring, R.H.; Kirksey, Y.; Aschmies, S.; Xu, J.; et al. The LXR agonist TO901317 selectively lowers hippocampal Abeta42 and improves memory in the Tg2576 mouse model of Alzheimer's disease. *Mol. Cell. Neurosci.* **2007**, *34*, 621–628. [CrossRef]
14. Carter, A.Y.; Letronne, F.; Fitz, N.F.; Mounier, A.; Wolfe, C.M.; Nam, K.N.; Reeves, V.L.; Kamboh, H.; Lefterov, I.; Koldamova, R. Liver X receptor agonist treatment significantly affects phenotype and transcriptome of $APOE_3$ and $APOE_4$ Abca1 haplo-deficient mice. *PLoS ONE* **2017**, *12*, e0172161. [CrossRef]

15. Donkin, J.J.; Stukas, S.; Hirsch-Reinshagen, V.; Namjoshi, D.; Wilkinson, A.; May, S.; Chan, J.; Fan, J.; Collins, J.; Wellington, C.L. ATP-binding cassette transporter A1 mediates the beneficial effects of the liver X receptor agonist GW3965 on object recognition memory and amyloid burden in amyloid precursor protein/presenilin 1 mice. *J. Biol. Chem.* **2010**, *285*, 34144–34154. [CrossRef]
16. Wesson, D.W.; Borkowski, A.H.; Landreth, G.E.; Nixon, R.A.; Levy, E.; Wilson, D.A. Sensory network dysfunction, behavioral impairments, and their reversibility in an Alzheimer's β-amyloidosis mouse model. *J. Neurosci.* **2011**, *31*, 15962–15971. [CrossRef]
17. Xu, X.; Xiao, X.; Yan, Y.; Zhang, T. Activation of liver X receptors prevents emotional and cognitive dysfunction by suppressing microglial M1-polarization and restoring synaptic plasticity in the hippocampus of mice. *Brain Behav. Immun.* **2021**, *94*, 111–124. [CrossRef]
18. Jiang, Q.; Lee, C.D.; Mandrekar, S.; Wilkinson, B.; Cramer, P.; Zelcer, N.; Mann, K.; Lamb, B.; Willson, T.M. ApoE promotes the proteolytic degradation of Abeta. *Neuron* **2008**, *58*, 681–693. [CrossRef]
19. Navas Guimaraes, M.E.; Lopez-Blanco, R.; Correa, J.; Fernandez-Villamarin, M.; Bistué, M.B.; Martino-Adami, P.; Morelli, L.; Kumar, V.; Wempe, M.F.; Cuello, A.C.; et al. Liver X Receptor Activation with an Intranasal Polymer Therapeutic Prevents Cognitive Decline without Altering Lipid Levels. *ACS Nano* **2021**, *15*, 4678–4687. [CrossRef]
20. Grefhorst, A.; Elzinga, B.M.; Voshol, P.J.; Plo, T.; Kok, T.; Bloks, V.W.; van der Sluijs, F.H.; Havekes, L.M.; Romijn, J.A.; Verkade, H.J.; et al. Stimulation of lipogenesis by pharmacological activation of the liver X receptor leads to production of large, triglyceride-rich very low density lipoprotein particles. *J. Biol. Chem.* **2002**, *277*, 34182–34190. [CrossRef]
21. Repa, J.J.; Liang, G.; Ou, J.; Bashmakov, Y.; Lobaccaro, J.M.; Shimomura, I.; Shan, B.; Brown, M.S.; Goldstein, J.L.; Mangelsdorf, D.J. Regulation of mouse sterol regulatory element-binding protein-1c gene (SREBP-1c) by oxysterol receptors, LXRalpha and LXRbeta. *Genes Dev.* **2000**, *14*, 2819–2830. [CrossRef]
22. Schultz, J.R.; Tu, H.; Luk, A.; Repa, J.J.; Medina, J.C.; Li, L.; Schwendner, S.; Wang, S.; Thoolen, M.; Mangelsdorf, D.J.; et al. Role of LXRs in control of lipogenesis. *Genes Dev.* **2000**, *14*, 2831–2838. [CrossRef]
23. Vanmierlo, T.; Bogie, J.F.; Mailleux, J.; Vanmol, J.; Lütjohann, D.; Mulder, M.; Hendriks, J.J. Plant sterols: Friend or foe in CNS disorders? *Prog. Lipid Res.* **2015**, *58*, 26–39. [CrossRef]
24. Bogie, J.; Hoeks, C.; Schepers, M.; Tiane, A.; Cuypers, A.; Leijten, F.; Chintapakorn, Y.; Suttiyut, T.; Pornpakakul, S.; Struik, D.; et al. Dietary Sargassum fusiforme improves memory and reduces amyloid plaque load in an Alzheimer's disease mouse model. *Sci. Rep.* **2019**, *9*, 4908. [CrossRef]
25. Zhan, N.; Wang, B.; Martens, N.; Liu, Y.; Zhao, S.; Voortman, G.; van Rooij, J.; Leijten, F.; Vanmierlo, T.; Kuipers, F.; et al. Identification of Side Chain Oxidized Sterols as Novel Liver X Receptor Agonists with Therapeutic Potential in the Treatment of Cardiovascular and Neurodegenerative Diseases. *Int. J. Mol. Sci.* **2023**, *24*, 1290. [CrossRef]
26. Martens, N.; Schepers, M.; Zhan, N.; Leijten, F.; Voortman, G.; Tiane, A.; Rombaut, B.; Poisquet, J.; Sande, N.V.; Kerksiek, A.; et al. 24(S)-Saringosterol Prevents Cognitive Decline in a Mouse Model for Alzheimer's Disease. *Mar. Drugs* **2021**, *19*, 190. [CrossRef]
27. Desvergne, B.; Wahli, W. Peroxisome proliferator-activated receptors: Nuclear control of metabolism. *Endocr. Rev.* **1999**, *20*, 649–688.
28. Kliewer, S.A.; Sundseth, S.S.; Jones, S.A.; Brown, P.J.; Wisely, G.B.; Koble, C.S.; Devchand, P.; Wahli, W.; Willson, T.M.; Lenhard, J.M.; et al. Fatty acids and eicosanoids regulate gene expression through direct interactions with peroxisome proliferator-activated receptors alpha and gamma. *Proc. Natl. Acad. Sci. USA* **1997**, *94*, 4318–4323. [CrossRef]
29. Dussault, I.; Forman, B.M. Prostaglandins and fatty acids regulate transcriptional signaling via the peroxisome proliferator activated receptor nuclear receptors. *Prostaglandins Other Lipid Mediat.* **2000**, *62*, 1–13. [CrossRef]
30. Wang, L.; Waltenberger, B.; Pferschy-Wenzig, E.M.; Blunder, M.; Liu, X.; Malainer, C.; Blazevic, T.; Schwaiger, S.; Rollinger, J.M.; Heiss, E.H.; et al. Natural product agonists of peroxisome proliferator-activated receptor gamma (PPARγ): A review. *Biochem. Pharmacol.* **2014**, *92*, 73–89. [CrossRef]
31. Wójtowicz, S.; Strosznajder, A.K.; Jeżyna, M.; Strosznajder, J.B. The Novel Role of PPAR Alpha in the Brain: Promising Target in Therapy of Alzheimer's Disease and Other Neurodegenerative Disorders. *Neurochem. Res.* **2020**, *45*, 972–988. [CrossRef]
32. Staels, B.; Dallongeville, J.; Auwerx, J.; Schoonjans, K.; Leitersdorf, E.; Fruchart, J.C. Mechanism of action of fibrates on lipid and lipoprotein metabolism. *Circulation* **1998**, *98*, 2088–2093. [CrossRef]
33. Heneka, M.T.; Landreth, G.E. PPARs in the brain. *Biochim. Biophys. Acta* **2007**, *1771*, 1031–1045. [CrossRef]
34. Pedersen, W.A.; McMillan, P.J.; Kulstad, J.J.; Leverenz, J.B.; Craft, S.; Haynatzki, G.R. Rosiglitazone attenuates learning and memory deficits in Tg2576 Alzheimer mice. *Exp. Neurol.* **2006**, *199*, 265–273. [CrossRef]
35. De Felice, F.G.; Lourenco, M.V.; Ferreira, S.T. How does brain insulin resistance develop in Alzheimer's disease? *Alzheimer's Dement.* **2014**, *10* (Suppl. 1), S26–S32. [CrossRef]
36. Santos, M.J.; Quintanilla, R.A.; Toro, A.; Grandy, R.; Dinamarca, M.C.; Godoy, J.A.; Inestrosa, N.C. Peroxisomal proliferation protects from beta-amyloid neurodegeneration. *J. Biol. Chem.* **2005**, *280*, 41057–41068. [CrossRef]
37. Esmaeili, M.A.; Yadav, S.; Gupta, R.K.; Waggoner, G.R.; Deloach, A.; Calingasan, N.Y.; Beal, M.F.; Kiaei, M. Preferential PPAR-α activation reduces neuroinflammation, and blocks neurodegeneration in vivo. *Hum. Mol. Genet.* **2016**, *25*, 317–327. [CrossRef]
38. Bougarne, N.; Weyers, B.; Desmet, S.J.; Deckers, J.; Ray, D.W.; Staels, B.; De Bosscher, K. Molecular Actions of PPARα in Lipid Metabolism and Inflammation. *Endocr. Rev.* **2018**, *39*, 760–802. [CrossRef]
39. Inestrosa, N.C.; Carvajal, F.J.; Zolezzi, J.M.; Tapia-Rojas, C.; Serrano, F.; Karmelic, D.; Toledo, E.M.; Toro, A.; Toro, J.; Santos, M.J. Peroxisome proliferators reduce spatial memory impairment, synaptic failure, and neurodegeneration in brains of a double transgenic mice model of Alzheimer's disease. *J. Alzheimer's Dis.* **2013**, *33*, 941–959. [CrossRef]

40. Kummer, M.P.; Heneka, M.T. PPARs in Alzheimer's Disease. *PPAR Res.* **2008**, *2008*, 403896. [CrossRef]
41. Yan, Q.; Zhang, J.; Liu, H.; Babu-Khan, S.; Vassar, R.; Biere, A.L.; Citron, M.; Landreth, G. Anti-inflammatory drug therapy alters beta-amyloid processing and deposition in an animal model of Alzheimer's disease. *J. Neurosci.* **2003**, *23*, 7504–7509. [CrossRef]
42. Sastre, M.; Dewachter, I.; Landreth, G.E.; Willson, T.M.; Klockgether, T.; van Leuven, F.; Heneka, M.T. Nonsteroidal anti-inflammatory drugs and peroxisome proliferator-activated receptor-gamma agonists modulate immunostimulated processing of amyloid precursor protein through regulation of beta-secretase. *J. Neurosci.* **2003**, *23*, 9796–9804. [CrossRef]
43. Luna-Medina, R.; Cortes-Canteli, M.; Alonso, M.; Santos, A.; Martínez, A.; Perez-Castillo, A. Regulation of inflammatory response in neural cells in vitro by thiadiazolidinones derivatives through peroxisome proliferator-activated receptor gamma activation. *J. Biol. Chem.* **2005**, *280*, 21453–21462. [CrossRef]
44. d'Angelo, M.; Castelli, V.; Catanesi, M.; Antonosante, A.; Dominguez-Benot, R.; Ippoliti, R.; Benedetti, E.; Cimini, A. PPARγ and Cognitive Performance. *Int. J. Mol. Sci.* **2019**, *20*, 5068. [CrossRef]
45. Luo, R.; Su, L.Y.; Li, G.; Yang, J.; Liu, Q.; Yang, L.X.; Zhang, D.F.; Zhou, H.; Xu, M.; Fan, Y.; et al. Activation of PPARA-mediated autophagy reduces Alzheimer disease-like pathology and cognitive decline in a murine model. *Autophagy* **2020**, *16*, 52–69. [CrossRef]
46. Chandra, S.; Roy, A.; Jana, M.; Pahan, K. Cinnamic acid activates PPARα to stimulate Lysosomal biogenesis and lower Amyloid plaque pathology in an Alzheimer's disease mouse model. *Neurobiol. Dis.* **2019**, *124*, 379–395. [CrossRef]
47. Chandra, S.; Pahan, K. Gemfibrozil, a Lipid-Lowering Drug, Lowers Amyloid Plaque Pathology and Enhances Memory in a Mouse Model of Alzheimer's Disease via Peroxisome Proliferator-Activated Receptor α. *J. Alzheimer's Dis. Rep.* **2019**, *3*, 149–168. [CrossRef]
48. Escribano, L.; Simón, A.M.; Gimeno, E.; Cuadrado-Tejedor, M.; Lopez de Maturana, R.; García-Osta, A.; Ricobaraza, A.; Pérez-Mediavilla, A.; Del Río, J.; Frechilla, D. Rosiglitazone rescues memory impairment in Alzheimer's transgenic mice: Mechanisms involving a reduced amyloid and tau pathology. *Neuropsychopharmacology* **2010**, *35*, 1593–1604. [CrossRef]
49. Denner, L.A.; Rodriguez-Rivera, J.; Haidacher, S.J.; Jahrling, J.B.; Carmical, J.R.; Hernandez, C.M.; Zhao, Y.; Sadygov, R.G.; Starkey, J.M.; Spratt, H.; et al. Cognitive enhancement with rosiglitazone links the hippocampal PPARγ and ERK MAPK signaling pathways. *J. Neurosci.* **2012**, *32*, 16725–16735a. [CrossRef]
50. Sato, T.; Hanyu, H.; Hirao, K.; Kanetaka, H.; Sakurai, H.; Iwamoto, T. Efficacy of PPAR-γ agonist pioglitazone in mild Alzheimer disease. *Neurobiol. Aging* **2011**, *32*, 1626–1633. [CrossRef]
51. Hanyu, H.; Sato, T.; Kiuchi, A.; Sakurai, H.; Iwamoto, T. Pioglitazone improved cognition in a pilot study on patients with Alzheimer's disease and mild cognitive impairment with diabetes mellitus. *J. Am. Geriatr. Soc.* **2009**, *57*, 177–179. [CrossRef]
52. Risner, M.E.; Saunders, A.M.; Altman, J.F.; Ormandy, G.C.; Craft, S.; Foley, I.M.; Zvartau-Hind, M.E.; Hosford, D.A.; Roses, A.D. Efficacy of rosiglitazone in a genetically defined population with mild-to-moderate Alzheimer's disease. *Pharmacogenom. J.* **2006**, *6*, 246–254. [CrossRef]
53. Watson, G.S.; Cholerton, B.A.; Reger, M.A.; Baker, L.D.; Plymate, S.R.; Asthana, S.; Fishel, M.A.; Kulstad, J.J.; Green, P.S.; Cook, D.G.; et al. Preserved cognition in patients with early Alzheimer disease and amnestic mild cognitive impairment during treatment with rosiglitazone: A preliminary study. *Am. J. Geriatr. Psychiatry* **2005**, *13*, 950–958. [CrossRef]
54. Sagheddu, C.; Melis, M.; Muntoni, A.L.; Pistis, M. Repurposing Peroxisome Proliferator-Activated Receptor Agonists in Neurological and Psychiatric Disorders. *Pharmaceuticals* **2021**, *14*, 1025. [CrossRef]
55. Lütjohann, D.; Brzezinka, A.; Barth, E.; Abramowski, D.; Staufenbiel, M.; von Bergmann, K.; Beyreuther, K.; Multhaup, G.; Bayer, T.A. Profile of cholesterol-related sterols in aged amyloid precursor protein transgenic mouse brain. *J. Lipid Res.* **2002**, *43*, 1078–1085. [CrossRef]
56. Loera-Valencia, R.; Goikolea, J.; Parrado-Fernandez, C.; Merino-Serrais, P.; Maioli, S. Alterations in cholesterol metabolism as a risk factor for developing Alzheimer's disease: Potential novel targets for treatment. *J. Steroid Biochem. Mol. Biol.* **2019**, *190*, 104–114. [CrossRef]
57. Yin, M.; Chen, M.; Matsuoka, R.; Song, X.; Xi, Y.; Zhang, L.; Wang, X. UHPLC-Q-Exactive Orbitrap MS/MS based untargeted lipidomics reveals fatty acids and lipids profiles in different parts of capelin (*Mallotus villosus*). *J. Food Compos. Anal.* **2023**, *116*, 105096. [CrossRef]
58. Ma, W.-P.; Yin, S.-N.; Chen, J.-P.; Geng, X.-C.; Liu, M.-F.; Li, H.-H.; Liu, M.; Liu, H.-B. Stimulating the hematopoietic effect of simulated digestive product of fucoidan from Sargassum fusiforme on cyclophosphamide-induced hematopoietic damage in mice and its protective mechanisms based on serum lipidomics. *Mar. Drugs* **2022**, *20*, 201. [CrossRef]
59. Zwarts, I.; van Zutphen, T.; Kruit, J.K.; Liu, W.; Oosterveer, M.H.; Verkade, H.J.; Uhlenhaut, N.H.; Jonker, J.W. Identification of the fructose transporter GLUT5 (SLC2A5) as a novel target of nuclear receptor LXR. *Sci. Rep.* **2019**, *9*, 9299. [CrossRef]
60. Ye, J.; Coulouris, G.; Zaretskaya, I.; Cutcutache, I.; Rozen, S.; Madden, T.L. Primer-BLAST: A tool to design target-specific primers for polymerase chain reaction. *BMC Bioinform.* **2012**, *13*, 134. [CrossRef]
61. Dixon, W.J. Analysis of extreme values. *Ann. Math. Stat.* **1950**, *21*, 488–506. [CrossRef]
62. Dixon, W.J. Ratios involving extreme values. *Ann. Math. Stat.* **1951**, *22*, 68–78. [CrossRef]
63. Yan, Y.; Niu, Z.; Wang, B.; Zhao, S.; Sun, C.; Wu, Y.; Li, Y.; Ying, H.; Liu, H. Saringosterol from sargassum fusiforme modulates cholesterol metabolism and alleviates atherosclerosis in ApoE-deficient mice. *Mar. Drugs* **2021**, *19*, 485. [CrossRef]

64. Blanchard, J.W.; Akay, L.A.; Davila-Velderrain, J.; von Maydell, D.; Mathys, H.; Davidson, S.M.; Effenberger, A.; Chen, C.Y.; Maner-Smith, K.; Hajjar, I.; et al. APOE4 impairs myelination via cholesterol dysregulation in oligodendrocytes. *Nature* **2022**, *611*, 769–779. [CrossRef]
65. Varma, V.R.; Büşra Lüleci, H.; Oommen, A.M.; Varma, S.; Blackshear, C.T.; Griswold, M.E.; An, Y.; Roberts, J.A.; O'Brien, R.; Pletnikova, O.; et al. Abnormal brain cholesterol homeostasis in Alzheimer's disease-a targeted metabolomic and transcriptomic study. *NPJ Aging Mech. Dis.* **2021**, *7*, 11. [CrossRef]
66. Zhang, J.; Liu, Q. Cholesterol metabolism and homeostasis in the brain. *Protein Cell* **2015**, *6*, 254–264. [CrossRef]
67. Sodero, A.O. 24S-hydroxycholesterol: Cellular effects and variations in brain diseases. *J. Neurochem.* **2021**, *157*, 899–918. [CrossRef]
68. Mitchell, R.W.; Hatch, G.M. Fatty acid transport into the brain: Of fatty acid fables and lipid tails. *Prostaglandins Leukot. Essent. Fat. Acids* **2011**, *85*, 293–302. [CrossRef]
69. Kim, J.; Yoon, H.; Basak, J.; Kim, J. Apolipoprotein E in synaptic plasticity and Alzheimer's disease: Potential cellular and molecular mechanisms. *Mol. Cells* **2014**, *37*, 767–776. [CrossRef]
70. Koldamova, R.; Fitz, N.F.; Lefterov, I. ATP-binding cassette transporter A1: From metabolism to neurodegeneration. *Neurobiol. Dis.* **2014**, *72*, 13–21. [CrossRef]
71. Serrano-Pozo, A.; Das, S.; Hyman, B.T. APOE and Alzheimer's disease: Advances in genetics, pathophysiology, and therapeutic approaches. *Lancet Neurol.* **2021**, *20*, 68–80. [CrossRef]
72. Radhakrishnan, A.; Sun, L.P.; Kwon, H.J.; Brown, M.S.; Goldstein, J.L. Direct binding of cholesterol to the purified membrane region of SCAP: Mechanism for a sterol-sensing domain. *Mol. Cell* **2004**, *15*, 259–268. [CrossRef]
73. Radhakrishnan, A.; Ikeda, Y.; Kwon, H.J.; Brown, M.S.; Goldstein, J.L. Sterol-regulated transport of SREBPs from endoplasmic reticulum to Golgi: Oxysterols block transport by binding to Insig. *Proc. Natl. Acad. Sci. USA* **2007**, *104*, 6511–6518. [CrossRef]
74. Roy, A.; Jana, M.; Corbett, G.T.; Ramaswamy, S.; Kordower, J.H.; Gonzalez, F.J.; Pahan, K. Regulation of cyclic AMP response element binding and hippocampal plasticity-related genes by peroxisome proliferator-activated receptor α. *Cell Rep.* **2013**, *4*, 724–737. [CrossRef]
75. Xu, H.E.; Lambert, M.H.; Montana, V.G.; Parks, D.J.; Blanchard, S.G.; Brown, P.J.; Sternbach, D.D.; Lehmann, J.M.; Wisely, G.B.; Willson, T.M.; et al. Molecular recognition of fatty acids by peroxisome proliferator–activated receptors. *Mol. Cell* **1999**, *3*, 397–403. [CrossRef]
76. Ricote, M.; Li, A.C.; Willson, T.M.; Kelly, C.J.; Glass, C.K. The peroxisome proliferator-activated receptor-γ is a negative regulator of macrophage activation. *Nature* **1998**, *391*, 79–82. [CrossRef]
77. Tugal, D.; Liao, X.; Jain, M.K. Transcriptional control of macrophage polarization. *Arterioscler. Thromb. Vasc. Biol.* **2013**, *33*, 1135–1144. [CrossRef]
78. Thomas, D.G.; Doran, A.C.; Fotakis, P.; Westerterp, M.; Antonson, P.; Jiang, H.; Jiang, X.-C.; Gustafsson, J.-Å.; Tabas, I.; Tall, A.R. LXR suppresses inflammatory gene expression and neutrophil migration through cis-repression and cholesterol efflux. *Cell Rep.* **2018**, *25*, 3774–3785.e4. [CrossRef]
79. Jesse, S.; Steinacker, P.; Cepek, L.; von Arnim, C.A.F.; Tumani, H.; Lehnert, S.; Kretzschmar, H.A.; Baier, M.; Otto, M. Glial fibrillary acidic protein and protein S-100B: Different concentration pattern of glial proteins in cerebrospinal fluid of patients with Alzheimer's disease and Creutzfeldt-Jakob disease. *J. Alzheimer's Dis.* **2009**, *17*, 541–551. [CrossRef]
80. Elahi, F.M.; Casaletto, K.B.; La Joie, R.; Walters, S.M.; Harvey, D.; Wolf, A.; Edwards, L.; Rivera-Contreras, W.; Karydas, A.; Cobigo, Y.; et al. Plasma biomarkers of astrocytic and neuronal dysfunction in early-and late-onset Alzheimer's disease. *Alzheimer's Dement.* **2019**, *16*, 681–695. [CrossRef]
81. Oeckl, P.; Halbgebauer, S.; Anderl-Straub, S.; Steinacker, P.; Huss, A.M.; Neugebauer, H.; von Arnim, C.A.F.; Diehl-Schmid, J.; Grimmer, T.; Kornhuber, J.; et al. Glial fibrillary acidic protein in serum is increased in Alzheimer's disease and correlates with cognitive impairment. *J. Alzheimer's Dis.* **2019**, *67*, 481–488. [CrossRef] [PubMed]
82. Liberato, M.V.; Nascimento, A.S.; Ayers, S.D.; Lin, J.Z.; Cvoro, A.; Silveira, R.L.; Martínez, L.; Souza, P.C.; Saidemberg, D.; Deng, T.; et al. Medium chain fatty acids are selective peroxisome proliferator activated receptor (PPAR) γ activators and pan-PPAR partial agonists. *PLoS ONE* **2012**, *7*, e36297. [CrossRef] [PubMed]
83. Göttlicher, M.; Widmark, E.; Li, Q.; Gustafsson, J.A. Fatty acids activate a chimera of the clofibric acid-activated receptor and the glucocorticoid receptor. *Proc. Natl. Acad. Sci. USA* **1992**, *89*, 4653–4657. [CrossRef] [PubMed]
84. Venn-Watson, S.; Lumpkin, R.; Dennis, E.A. Efficacy of dietary odd-chain saturated fatty acid pentadecanoic acid parallels broad associated health benefits in humans: Could it be essential? *Sci. Rep.* **2020**, *10*, 8161. [CrossRef]
85. Coccia, E.; Varricchio, E.; Vito, P.; Turchini, G.M.; Francis, D.S.; Paolucci, M. Fatty acid-specific alterations in leptin, PPARα, and CPT-1 gene expression in the rainbow trout. *Lipids* **2014**, *49*, 1033–1046. [CrossRef]
86. Ren, B.; Thelen, A.P.; Peters, J.M.; Gonzalez, F.J.; Jump, D.B. Polyunsaturated Fatty Acid Suppression of Hepatic Fatty Acid Synthase and S14 Gene Expression Does Not Require Peroxisome Proliferator-activated Receptor α*. *J. Biol. Chem.* **1997**, *272*, 26827–26832. [CrossRef]
87. Forman, B.M.; Chen, J.; Evans, R.M. Hypolipidemic drugs, polyunsaturated fatty acids, and eicosanoids are ligands for peroxisome proliferator-activated receptors alpha and delta. *Proc. Natl. Acad. Sci. USA* **1997**, *94*, 4312–4317. [CrossRef]
88. Yu, K.; Bayona, W.; Kallen, C.B.; Harding, H.P.; Ravera, C.P.; McMahon, G.; Brown, M.; Lazar, M.A. Differential Activation of Peroxisome Proliferator-activated Receptors by Eicosanoids (∗). *J. Biol. Chem.* **1995**, *270*, 23975–23983. [CrossRef]

89. Hontecillas, R.; Wannemeulher, M.J.; Zimmerman, D.R.; Hutto, D.L.; Wilson, J.H.; Ahn, D.U.; Bassaganya-Riera, J. Nutritional regulation of porcine bacterial-induced colitis by conjugated linoleic acid. *J. Nutr.* **2002**, *132*, 2019–2027. [CrossRef]
90. de Almeida, M.M.; Luquetti, S.C.; Sabarense, C.M.; Corrêa, J.O.; dos Reis, L.G.; Conceição, E.P.; Lisboa, P.C.; de Moura, E.G.; Gameiro, J.; da Gama, M.A.; et al. Butter naturally enriched in cis-9, trans-11 CLA prevents hyperinsulinemia and increases both serum HDL cholesterol and triacylglycerol levels in rats. *Lipids Health Dis.* **2014**, *13*, 200. [CrossRef]
91. Berger, J.; Moller, D.E. The mechanisms of action of PPARs. *Annu. Rev. Med.* **2002**, *53*, 409–435. [CrossRef]
92. Yu, Y.; Correll, P.H.; Vanden Heuvel, J.P. Conjugated linoleic acid decreases production of pro-inflammatory products in macrophages: Evidence for a PPAR gamma-dependent mechanism. *Biochim. Biophys. Acta* **2002**, *1581*, 89–99. [CrossRef] [PubMed]
93. Marion-Letellier, R.; Butler, M.; Déchelotte, P.; Playford, R.J.; Ghosh, S. Comparison of cytokine modulation by natural peroxisome proliferator-activated receptor gamma ligands with synthetic ligands in intestinal-like Caco-2 cells and human dendritic cells--potential for dietary modulation of peroxisome proliferator-activated receptor gamma in intestinal inflammation. *Am. J. Clin. Nutr.* **2008**, *87*, 939–948. [PubMed]
94. Salakhutdinov, N.F.; Laev, S.S. Triglyceride-lowering agents. *Bioorganic Med. Chem.* **2014**, *22*, 3551–3564. [CrossRef] [PubMed]
95. Sheu, S.-H.; Kaya, T.; Waxman, D.J.; Vajda, S. Exploring the Binding Site Structure of the PPARγ Ligand-Binding Domain by Computational Solvent Mapping. *Biochemistry* **2005**, *44*, 1193–1209. [CrossRef]
96. Altinoz, M.A.; Ozpinar, A. PPAR-δ and erucic acid in multiple sclerosis and Alzheimer's Disease. Likely benefits in terms of immunity and metabolism. *Int. Immunopharmacol.* **2019**, *69*, 245–256. [CrossRef]
97. Altinoz, M.A.; Ozpinar, A.; Ozpinar, A.; Hacker, E. Erucic acid, a nutritional PPARδ-ligand may influence Huntington's disease pathogenesis. *Metab. Brain Dis.* **2020**, *35*, 1–9. [CrossRef]

MDPI
St. Alban-Anlage 66
4052 Basel
Switzerland
www.mdpi.com

Nutrients Editorial Office
E-mail: nutrients@mdpi.com
www.mdpi.com/journal/nutrients

www.ingramcontent.com/pod-product-compliance
Lightning Source LLC
Chambersburg PA
CBHW061846210926
51850CB00006B/986
9783036585123